Special Effects on the Screen

Special Effects on the Screen

Faking the View from Méliès to Motion Capture

Edited by
Martin Lefebvre and
Marc Furstenau

Routledge
Taylor & Francis Group
LONDON AND NEW YORK

The publication of this book is made possible by funds from the Laboratoire CinéMédias at Université de Montréal and by Concordia University's Faculty of Fine Arts.

First published in 2022 by Amsterdam University Press Ltd.

Published 2025 by Routledge
4 Park Square, Milton Park, Abingdon, Oxon OX14 4RN
605 Third Avenue, New York, NY 10158

Routledge is an imprint of the Taylor & Francis Group, an informa business

ISBN: 9789462980730 (hbk)
ISBN: 9781041186533 (pbk)
ISBN: 9781003704225 (ebk)
NUR 670

Cover illustration: *The Invisible Man* (James Whale, 1933)

Cover design: Coördesign, Leiden

DOI: 10.5117/9789462980730

Table of Contents

Films

Envoi

Introduction

Martin Lefebvre and Marc Furstenau

Long relegated to the margins of film studies, special effects have recently become the object of a burgeoning field of scholarship, where some new questions are being asked about the status of the cinema as an art and as an industry, and some old questions are being posed anew about the formal and technical history of filmmaking. Indeed, the last few years have seen more historical and critical approaches to special effects — in books and articles — than ever before in the history of film studies.[1] Perhaps the most significant feature of special effects studies is the intersection of history and theory. Prompted specifically by the emergence of a digital cinema and the development of computerized visual effects, film theorists have been reconsidering the traditional accounts of cinematic representation. Film historians have revisited the history of filmic effects, finding many significant historical antecedents to contemporary digital effects and revealing the degree to which the problems posed by the digital have deep historical roots. The very question of what counts exactly as a "special effect" is at once historical and theoretical, and it is the goal of this collection of essays to address the problem from both perspectives.

There are many good reasons to study special effects as an integral aspect of cinematic representation. Indeed, the authors of the essays collected here

1 Recent books on the topic include: Michele Pierson, *Special Effects: Still in Search of Wonder* (New York: Columbia University Press, 2002); Scott Bukatman, *Matters of Gravity: Special Effects and Supermen in the 20th Century* (Durham, NC: Duke University Press, 2003); Shilo T. McClean, *Digital Storytelling: The Narrative Power of Visual Effects* (Cambridge, MA: MIT Press, 2007); Dan North, *Performing Illusions: Cinema, Special Effects, and the Virtual Actor* (London: Wallflower, 2008); Stephen Prince, *Digital Visual Effects in Cinema: The Seduction of Reality* (New Brunswick, NJ: Rutgers University Press, 2012); Kristen Whissel, *Spectacular Digital Effects: CGI and Contemporary Cinema* (Durham, NC: Duke University Press, 2014); Julie Turnock, *Plastic Reality: Special Effects, Technology, and the Emergence of 1970s Blockbuster Aesthetics* (New York: Columbia University Press, 2015); Dan North, Bob Rehak, and Michael S. Duffy, eds., *Special Effects: New Histories/Theories/Contexts* (London: British Film Institute, 2015); and Charlie Keil and Kristen Whissel, eds., *Editing and Special/Visual Effects* (New Brunswick, NJ: Rutgers University Press, 2016).

Lefebvre, M. & M. Furstenau (eds.), *Special Effects on the Screen: Faking the View from Méliès to Motion Capture*. Taylor & Francis Group, 2022
DOI 10.5117/9789462980730_INTRO

argue that the study of special effects can be conceptually and theoretically productive towards understanding what the cinema is and how it works (and more specifically *how it works on its viewers*). Investigating special effects helps us gain knowledge about film as a material and highly technical artefact, all the while offering a perspective from which to study the style and meaning of certain films and the historical and cultural contexts in which they were made. *Concepts, Techniques, Films*: these, then, are the three axes around which this collection is organized, making the argument for the importance of fully integrating special effects in our study of cinema.

Referred to initially as trick shots or more recently as visual effects, what we can describe generally as special effects have been a staple of filmmaking since the earliest years of the commercial exploitation of cinema. Yet given their long history and their importance in the design of countless films, it is surprising how little attention has been paid to them by film scholars until recently. We would like to offer three hypotheses, or better yet perhaps three strands of a single argument, to help account for this situation. This argument rests on three historical pillars of film theory and criticism: an ideological commitment to realism; an ideological commitment to the seamless narrative text; and an ideological commitment to art and authorship.

1. ***Ideological Commitment to Realism* (*or, Now You See Them, Now You Don't*).** The terms "special" or "visual effects" can sometimes be used ambiguously. A film shows a superhero flying through the air, and we say to ourselves: "this is a special effect." The ambiguity resides in what the verbal index "this" refers to. Does it refer to the *means* used to give the impression that a character can fly, how the image is *effected* (e.g., an actor hung on cables and moving before a green screen and then composited into a moving background), or does it refer to the impression itself, the effect of those means on viewers (as in the advertising tagline for *Superman* [Richard Donner, 1978]: "you'll believe a man can fly")? Do special effects reside on the production side of filmmaking or on the reception side of spectatorship?

As several authors in this collection argue, early trick shots — which belonged to the cinema of attractions' mode of filmmaking — were often made so as to be noticed by viewers by creating an *effect* of awe or surprise at seeing heads severed from their bodies or objects appearing or disappearing in the blink of an eye (as in many of Georges Méliès's films). This trend continued well into the cinema of narrative integration (James Whale's 1933 *The Invisible Man* is a case in point), yet as Katharina Loew reminds us in her chapter, not all special effects from the first decade of filmmaking were equally conspicuous, meant simply to attract the viewers' attention as effects and to be visible as such. When Porter used multiple exposures to extend the studio set of *The Great Train Robbery* (1903), it is fair to say he was using such a special effect for a different — indeed opposite — end, not as a visible attraction but rather as an (ideally) imperceptible aspect of cinematic composition. As Loew argues, between 1910 and 1930, special effects started becoming less noticeable, thus, we might add, bolstering a distinction between means (production) and effects (reception) with regards to special effects.

Apart from obvious popular genres like science fiction, films such as *The Invisible Man*, horror films like *Frankenstein* (James Whale, 1931), or fantastic adventure films like *King Kong* (Merian C. Cooper and Ernest B. Schoedsack, 1933) required effects whose visibility for the viewer lay chiefly either in how what is depicted departs from what they know to be the case in the world (there is no such thing as an invisible man, a reanimated cadaver, or a 30-plus-foot gorilla) or in how these effects altered the look or behavior of things/beings (e.g., the jerkiness of stop-motion animation). Advancements, though, in such processes as glass shots, matte shots (including traveling mattes), front and rear projections, optical printing, miniatures, mechanical effects — in conjunction with developments in film stocks, lenses, lighting, etc. — made it possible in some instances, even before the digital age, to manipulate the image more or less without the viewer's awareness (at least when what was shown did not include impossible locations or actions).[2]

2 Writing on the use of optical printing and rear projection in the pages of *American Cinematographer* in the November issue of 1936, Cecil B. DeMille praises the work of effects specialist Farciot Edouart in harmonizing the material shot for background plates for his film *The Plainsman* (1936):

> In running over the background-shots in the projection room, it was noticed that there were nice pictorial clouds in some of the shots and bare, "bald-headed" skies in others. This was natural for the location-unit had enjoyed nice skies the first day of their work and unpleasantly cloudless ones the rest of the time. But it would not be very convincing to see a bunch of Indians charging under white, fluffy clouds in one shot, and in the next, three seconds later, see them coming under a barren, cloudless sky.
>
> So Farciot, by means of his big Optical Printer, proceeded to put clouds in every one of the cloudless backgrounds. And they were natural clouds, too, which could not be distinguished from the real ones. He refuses to tell me just how he did it: but he did a most remarkable job; not a trace of a matte-line shows in any of the shots. (458)

Edouart was the son of a photographer, and the seamless adding of clouds was a longstanding tradition of black and white landscape photography (especially before the advent of isochromatic plates in the late 1880s). He became known for his rear-projection work at Paramount (receiving ten Academy Awards between 1937 and 1955) and headed the Special-Effects Department. Yet, certainly to audiences today, the rear projections of the *Plainsman*, although beautifully timed (events in the back-projected images seemingly reacting to those shot in the studio), are not as

Thus, effects could be created (on the filmmaking side) without obvious *effects* (on the viewing side).

Being unaware that a manipulation has taken place might explain in part why the bulk of film critics and, later, theorists did not concern themselves very much with special effects — mostly interested as they were with what can be seen or consciously experienced (especially with regard to *mise en scène*, montage, acting, plot, meaning, or even the affects they produce). Yet this would pertain only to those manipulations that were completely imperceptible to viewers, a set that, until the arrival of digital effects, did not include the majority of effects shots. For the fact is that throughout the classical era, and for some time after, rear projections, traveling mattes, optical printing, blue and green screens, techniques for color compositing, etc., were often accompanied by visual "defects" of some sort (e.g., trembling, lighting and depth inconsistencies, inadequate range of focus, ghosting, fringing or edging, etc.) when compared to "straightforward" professional cinematography, thus signaling that some special effect had been used in either production or post-production. Even today, when computerized effects have become more "seamless," many digital effects are easily detected by moviegoers (the unnatural and uncanny "bouncy," "elastic," or "weightless" movements of some computer-generated actors in superhero action scenes come to mind). Given that so many effects were indeed noticeable, why is it that they garnered so little attention in critical accounts of cinema?

Notice that in distinguishing between effects that draw attention to themselves — the rings of light around the female robot in *Metropolis* as she acquires human features (Fritz Lang, 1927), the parting of the Red Sea in *The Ten Commandments* (Cecil B. DeMille,1956), the space ships and star-gate of *2001 A Space Odyssey* (Stanley Kubrick, 1969), the liquid-metal cyborg in *Terminator 2: Judgment Day* (James Cameron, 1991) — and those that are not meant to draw attention to themselves, even though many of them were nonetheless noticeable (and in some instances of digital effects still are), we are merely extending the earlier division of *effects* and *means*. These categories can bleed into one another (e.g., the blue screen travelling mattes, such as developed by Larry Butler for the 1940 remake of *The Thief of Bagdad* [Ludwig Berger, Michael Powell, et al.], a film where the fantastic sights of the giant genie and magic carpet rides obviously draw attention to themselves, while the limitations of the technique are also quite noticeable),

imperceptible (quite the opposite) as the optically printed clouds. Cecil B. DeMille, "A Director Looks at 'Process-Shots'," in *American Cinematographer*, November 1936, p. 458.

but answering the question just posed requires we consider the matter in terms of both film aesthetics and film history.

It follows from the above that if special effects were a "blind spot" of film theory and criticism for so long, it cannot be as a result of being unnoticeable, even if in bulk they grew less attention-grabbing during the classical era (which for a long time was certainly the most studied period in filmmaking). Could it be, then, that special effects such as rear projections — think of dialogue scenes in moving cars and their background plates in classical films — were deliberately overlooked by critics and scholars of classical cinema, either as brief but necessary annoyances (because of their "imperfections") or else because they benefitted from a sort of perceptual denial or "willful suspension of disbelief" such as one could also find in classical theater with its blatantly artificial and sparse set design (when compared to what film can offer)? Or perhaps, to put it differently, they were understood and accepted merely as conventions of filmmaking (the way classical audiences accepted that drivers getting out of a car would often slide over and exit from the passenger side; or else, to paraphrase Godard, that Technicolor movie blood typically looks more like "red" — the color — than it does blood)? We might even think of them as possessing an almost *illustrative* role, somewhat like the woodcut prints that accompanied Jules Verne's novels in the original Hetzel editions. These were visual aids that interrupted the flow of words that otherwise made up the novels. This is not to say that noticeable special effects interrupted the narrative in like fashion, but they interrupted the visual texture of the film, and their presence could have equally implied a similar sort of hybridity in the spectator's experience: "I can see that some effect is being used here, but I understand it serves to illustrate or approximate what a given situation would look like if, like the rest of the film, it had been shot without effects."

In any case, it could be argued that if viewers were not physiologically blind to noticeable "defects" (why would they be?), they may well have been "culturally" or "aesthetically" blind to them in the (changing, moving) context of classical film "realism." Indeed, even after filmmakers began leaving behind the studio environment as new conventions of realism emerged in the wake of Neo-Realism, the development of lightweight cameras (and sound-recording equipment), and the competition of television, few viewers or critics seem to have paid much attention to the potentially jarring nature of special effects or to the prospect that they might throw into sharp relief, if only for a brief moment, the materiality of the medium or the conventions of classical realism. The same holds, we would argue, for those "state of the art" digital effects today that can still fall short as perfect renditions of what (we believe) things would look like were they recorded directly, i.e., without the use of any effects. (Viewers today seem mostly fascinated by the achievements of seamless or even imperceptible effects, the latter ones only revealed in "making-of" supplements or effects companies' "show and tell" advertising demos).

It would seem that such momentary breaks in the visual homogeneity of a film's mode of representation that these noticeable effects introduce are not registered so much aesthetically (as would a character addressing the camera, and the audience, say in an otherwise "realist" dramatic film)[3] as they are on the grounds of technology, i.e., as a limit of the medium and its technology that one needs to accept and contend with in order to enjoy the "realism" of a film. And it is on this ground, arguably, that viewers can "overlook" them, as they also overlook, for example, lens flares (which are now often added as visual effects just like film grain!). If such cultural blindness exists, then it must be tied to what Dudley Andrew, in his *envoi* that concludes this book, alludes to as the history of audience expectations regarding special effects: "what counts as 'realistic,'' writes Andrew, "and what provokes credible illusions should change with each subsequent generation, even if the demand for realism and effects remain the same." As a result — and leaving aside for now all matters of content, like impossible worlds or events — until such moment as a given special effect becomes perceptually indistinguishable from non-effects footage (i.e., from what reality is thought to actually look like), it seems plausible that newer generations are always more prone to acknowledge or even to resist a convention that was accepted in the past but is no longer in use; willful blindness or acceptance of the stop-motion effects for the original *King Kong* are likely more difficult to achieve for audiences today than was the case in the early 1930s, even though the limitations in the effects' ability to smoothly reproduce movement were noticed when the film was first released.[4] In fact, the special effects of *King Kong* did not deter a staff writer from *The Hollywood Reporter* noting that "the sets and locations added greatly to the *realism* of this fantastic story" (14 February 1933; emphasis added). Enjoying *King Kong* meant a willingness not only to accept that a larger-than-life gorilla and dinosaurs exist in the fictional world of the film (their screen presence implying attention-grabbing effects) but also

3 We all know that the situation is different with comedies (and some musicals), where the characters' metaleptic engagement with the viewer and address to the camera is allowed, even during the classical period. Typically, however, such moments are brief and only momentarily break down the impression of a self-contained world, an illusion the viewer is quick to return to.

4 See Martin Lefebvre's essay in this collection for an early review of the film.

a willingness to overlook as much as possible the technical limitations of the special effects and accept the film's "visual fiction" or conventions of make-believe so that its drama does not fall flat.

We understand cinematic "realism" here in a rather broad and formal sense. As such, we conceive of it as a representational or aesthetic system that can govern genres as distinct as social dramas, historical films, comedies, as well as science-fiction or fantastic films in the way it uses the resources of the medium to create a world that, while it may differ in content from ours (as happens in most fiction films, though most spectacularly in some science-fiction or fantastic films), does so in accordance to what is generally considered in any given era to be an unobtrusive[5] way (or the least obtrusive way possible) for forming a self-contained world with coherent and identifiable spatial and temporal relations for characters and action. When special effects serve this representational system — say, when rear-projection serves plot development while seemingly seeking to create a continuous, coherent, "realist" space between studio foreground and location background — their noticeability does not appear to provide them with any *aesthetic* visibility of their own. Thus, unless they draw attention to themselves as spectacle, as attractions, all other special effects, even when noticeable through pictorial or behavioral "defects," have long lain on the margins of film spectatorship and scholarship as "non-objects," willfully forgotten, serving (like a tool) rather than rupturing the economy of realist representation; seen with the eyes, yet more or less invisible to the mind. We may think of this invisibility as game playing: in the service of the film, its plot, and its mode of realism, the viewer is willing to "go along," to "play the game," and to accept momentary lapses in visual texture (the technique of "day for night" would be another example of a willingly overlooked visual incongruity related to a technical difficulty: capturing quality images at night). In light of such observations, our account of realism itself may need to be revised, as Marc Furstenau suggests in his essay in this collection.

As for those effects that draw attention to themselves in realist films, they phenomenologically stand out for viewers, no less than in the original *King Kong*, as highly spectacular *individualized moments*, *aspects*, or *fragments* to be distinguished from default "non-special" aspects.[6] This is still

5 Admittedly, "unobtrusive" is a vague term. It refers to a feeling that is not always shared equally among all viewers and that is also subject to change as filmmaking conventions change. What it denotes is no less real, however.

6 Such default, "non-special" aspects pertain to images or even regions in the image that do not seem to have been manipulated, even though they may have been produced by invisible and imperceptible effects. It was Christian Metz who first paved the way for a phenomenology

the case today, although our scholarly attitude toward special effects has started to change as more films make use of them in this fashion (not just popular superhero, sci-fi, and action films but also hybrid popular/art films like those of David Fincher [*The Curious Case of Benjamin Button* 2008], Christopher Nolan [*Inception* 2010], Alfonso Cuarón [*Gravity* 2013], and Wes Anderson [*The Grand Budapest Hotel* 2014]). However, when the technical means in use entail perceived defects, "going along" or "playing the game" may require greater willingness from viewers. Yet, as spectacular as they might be, accepting the rules of the game means that the effects can remain marginal and subservient to the needs of creating a self-contained fictional world. In this regard, it might be tempting to think of special effects — and to explain past critical blindness to them — in terms of the Derridean concept of the *supplément*.[7]

Indeed, it might be argued that the first special effects, the ones used by Méliès, or Segundo de Chomón, and which conceptually paved the way for all others, far from being mere adjuncts to "standard" filmmaking (i.e., default, "non-effect" cinematography), constitutively existed "within" it — perhaps even revealing the nature or essence of cinema (if there is such a thing!). For indeed, they make manifest what lies at the very heart of the cinematic apparatus itself and its operation. These earliest "trick" films exploit what makes the perception of moving photographs possible in the first place, namely those black, unexposed intervals that exist between the frames of a strip of celluloid, and whose marginal, invisible surface allowed for shooting

of special effects by distinguished between "visible," "invisible," and "imperceptible" effects. Several authors in this collection revisit Metz's taxonomy; see especially Jost, Lefebvre, Loew, and Odin. Christian Metz, "'*Trucage*' and the Film," trans. Françoise Meltzer, in *Critical Inquiry*, (3)4, 1977, pp. 657–675.

7 Present in several of Derrida's works, though especially in the analysis of Rousseau in the second part of *Of Grammatology*, and in "Plato's Pharmacy" (reprinted in *Dissemination*), the supplement is used to deconstruct the binaries that supposedly shape metaphysical thinking. According to Derrida, what could be seen as initially an "external" add-on, a mere addition or adjunct to something — supplementing it — as writing has long been conceived to be in relation to speech for instance, is in fact what makes that very thing what it is, i.e., what is otherwise missing or lacking from it in order for it to fulfill itself. This makes the supplement "internal" to that which it supplements (a sort of internal contradiction, as it were). In "Plato's Pharmacy," Derrida shows that only a blind reading of Plato could see him as simply condemning writing. For in writing against writing, Plato's writing (and his constant recourse to metaphors pertaining to writing and written signs) reveals writing as that which discloses (or fulfills) the essence of speech (considered to be the kernel of language and therefore the true object of linguistics) and thus as internally belonging to it. Jacques Derrida, *Of Grammatology*, trans. Gayatri Charkravorty Spivak (Baltimore: Johns Hopkins University Press, 1976); and "Plato's Pharmacy," in *Dissemination*, trans. Barbara Johnson (Chicago: University of Chicago Press, 1983): 61-171.

interruptions, for montage, and all the tricks of the famous substitution splice technique as well as the animating of inanimate objects. They also exploit another basic fact of photography (whose emulsion records only light), according to which shooting a black surface leaves the corresponding area of celluloid unexposed, thus making possible the production of composite images through multiple exposures. (Both these properties of cinema, which were so brilliantly exploited by Georges Méliès, are discussed in the essays of Donald Crafton, Martin Lefebvre, Benoît Turquety, and Marc Furstenau in this volume). But if special effects stand as some sort of supplement to "standard" cinema, could critical blindness or resistance to them — resistance to accounting for them and thus, ironically, resistance toward cinema itself — reflect that they threaten to deconstruct the cinema, at least the dominant realist cinema (and the conception we have of it) that they so often appear to serve?

2. ***Ideological Commitment to the Seamless Narrative Text* (*or, The Bad Object*).** There is an irony in our discipline's current interest in special effects. Scholarship is rendering them visible just as the technology is emerging to make them more imperceptible and more seamless than they have ever been. This is not to say that effects are always used imperceptibly nowadays — far from it, as any summer blockbuster will prove. Take the 2019 computer-generated "live action" version of *The Lion King* (Jon Favreau), a remake of the original 1994 Disney animated film (Rob Minkoff and Roger Allers): it can only be understood as a *tour-de-force* demonstration of the full capacities of current effects technologies, meant to be experienced *as* special effects.[8] But the fact is that a great deal of the effects work happening these days goes unnoticed (Brainstorm Digital's demo reel for Scorsese's *Wolf of Wall Street* [2014], which is easily found on the web, is a case in point),[9] which should make us wonder why effects that were more noticeable in previous eras did not elicit more scholarly interest.

There is no question that it was the film industry's embracing of digital technologies at almost every station of the workflow, although especially in production and post-production, that has led to special effects attracting

8 The computer-generated imagery of animals in the film is truly remarkable and, at first glance, is very difficult to distinguish from photographic imagery. Yet as an anthropomorphic fantasy, the effects become visible or perceptible *as* effects the moment the animals start to speak — this is, one might say, the specific special effect on display in the film.

9 As the headline of a 2014 article by Eric Limmer in the online technology magazine *Gizmodo* puts it, "It's Crazy How Much of *The Wolf of Wall Street* is Actually CGI." See https://gizmodo.com/its-crazy-how-much-of-the-wolf-of-wall-street-is-actua-1501402962.

so much attention of late. Moreover, as a result of digitization, the current degree and variety of image manipulation is probably unsurpassed in the history of cinema (save only for animated films). It was not, though, so much towards special effects *per se* that scholars turned to with the onset of cinema's digital age. Rather, it was to the "ontology" of the photographic image and the fate of cinematic "indexicality." An anxious discourse initially arose as film theorists became concerned that films might not be recordings of the world anymore, even though the culturally dominant form of cinema was based on using these recordings for the purposes of representing fictional worlds, a purpose for which special effects, past and present, have always been well-suited and widely used. While analog-era effects did not circumvent the recording nature of the medium as many digital effects now do (though it could be argued that the "save" function in computer software is a recording device of some sort, simply not a *photographic* one), they nonetheless exploited several of its limitations in order to introduce a rift between what is photographically recorded and what is represented.

The notion of "limitation" here is to be understood in the spirit of Rudolf Arnheim's consideration of the formative power of film and the ways in which the film image differs from reality — i.e., the world in itself as it exists and is perceived independently of its cinematic recording. Thus, close shots of the "life-size" prop hand of Kong holding Ann Darrow (Fay Wray) show real, *recorded* events, and yet they rely on a framing that only shows that part of the gorilla's body (the effect is furthered with process photography for the background views), hiding the fact that the hand is not actually part of a gigantic beast but rather a mechanical facsimile. An effect is created simply by using a prop and cropping reality through framing so that instead of seeing Fay Wray sitting on a large fake hand (as a production still with a wider framing might clearly disclose), the audience sees Ann Darrow held in Kong's grip. The same could be said for the use of miniatures which relies in part on the fact that the camera does not discriminate size. As for glass shots, rear projections, double exposures, substitution tricks, or optical printing, they all equally use the recording nature of film while exploiting the fact that photography, because it produces an image, does not distinguish between recording the world and recording an image (especially when the images appear photo-realistic on screen), making possible the blending of world and image.

In a sense, none of these traditional effects challenge the idea of cinema as a recording medium that captures aspects of the world on film. After all, that actors in a studio stood in front of a screen where images of the jungle are projected is a fact that the camera has recorded. What they do challenge, however, is any notion that the (fictional) world the film depicts photographically is an *honest* "transference of reality from the thing to its reproduction," as André Bazin famously wrote.[10] When Hitchcock uses a miniature train in the *dénouement* of *Secret Agent* (1936), the photographed train is not meant to depict a miniature but rather a full-size train. Yet it is *no more* a photographic record or "transference of reality" of that full-size train than a drawing would be as a likeness of one (hence the connection made earlier between noticeable effects and illustrations in novels). However, the film's rhetoric, the inclusion of this train in the narrative — the film's main cast of characters are on board — the continuity in eventhood, the use of montage and sound, all work together to make viewers understand and, following what was said above, perhaps even accept that the miniature (it is quite noticeable, after all) stands for the full-size train that, in the fiction, is carrying the film's principals. Through the alchemy of the fiction that the film constructs, what Bazin claimed was the nature of the photograph (in this case, the photographic images of the miniature train) starts adulterating: what the photographic images show and what they represent is different.[11] The process is completed when such effects become imperceptible (often with technological advances). At this point, the difference does not resolve itself in fiction, for it is now also perceptual in nature: once an effect becomes imperceptible, perception and cognition are unsuspectingly tricked. Thus, whereas in a non-effects fiction film what is seen on screen and what exists in reality actually correspond term to term (even if what we see is a street in Toronto passing for one in New York), the same cannot be said when effects become imperceptible: the worldly "reality" depicted,

10 Bazin, of course, is comparing photography with painting when he says that "Photography benefits from a transfer of reality from the thing to its reproduction" (trans. Martin Lefebvre). The two published translations, Hugh Gray's and Timothy Barnard's, are noticeably different: "Photography enjoys a certain advantage in virtue of this transference of reality from the thing to its reproduction." André Bazin, "The Ontology of the Photographic Image," in *What is Cinema?*, vol. 1, trans. Hugh Gray (Los Angeles: University of California Press, 1967), p. 14; "Photography transfers reality from the object depicted to its reproduction," André Bazin, "The Ontology of the Photographic Image," in *What is Cinema?*, trans. Timothy Barnard (Montréal: Caboose, 2009), p. 8.

11 Patrick Maynard has emphasized this aspect of photographic and cinematic representation, with his distinction between "detection" (or recording) and "depiction" (or fiction). He considers what he describes as the "endemic confusion about photography," which he argues is the result of "a failure to develop a simple terminological distinction between *a photograph of something* and *a photographic depiction of* something" (114; emphases in original). See Patrick Maynard, *The Engine of Visualization: Thinking Through Photography* (Ithaca and London: Cornell University Press, 1997).

which serves as the material substratum for the fictional world of the film (its diegesis), was never truly photographed *per se* (even if the effect in question involves photography, as in glass shots or the use of props), but only *appears* to have been photographed as depicted.

Could it be then that special effects were more or less willfully overlooked by post-war critics and theorists because they undermine the "ontological" specificity of photography — an aspect of filmmaking only acknowledged when cinema massively adopted photography-mimicking digital effects and computer imagery (CGI)? Yet special effects remained largely invisible to critics and theorists when the mood swung over to apparatus and ideology critique. Considering post-1968 debates, one wonders if special effects could have offered at the time a site from which to question the realism their use usually sought to serve — thus de-naturalizing cinematic realism, as it were — and to question the idea that cinema, through its tie to photography, possesses some deep, ontological predisposition for realism. The anti-idealist and anti-realist camp would instead elect montage — whose noble genealogy included the Soviet avant-garde — and the rest, as the saying goes, is history. Might it be, then, that special effects were a bad object for both realists and anti-realists alike?

We know that Bazin seems to have distinguished between two sorts of special effects: mechanical *trucages* or props that are homogeneous and continuous with the real world and whose effect can therefore be recorded (e.g., the hidden strings that dictate the balloon's movements in *Le ballon rouge* [*Red Balloon*, Albert Lamorisse, 1956]), and those that require manipulating the image. We might think of them as "special effects that express a belief in reality" and "special effects that express a belief in the image." Bazin shows himself willing to accept the first type but not the second one, which he likens to montage in their ability to cinematically create an event. Thus in "Montage interdit" (first translated as "Virtues and Limitations of Montage"), Bazin criticizes the use of composites and process shots:

> True, [...] techniques such as rear projection, make it possible for two objects [...] to be seen together [...]. The illusion here is closer to perfection [than it is by creating the effect through editing], but it is not undetectable. In any event, the important thing is not that the trick effect be invisible but whether or not they are used, just as the beauty of a fake Vermeer could never take precedence over its lack of authenticity.[12]

12 Bazin, "Editing Prohibited," in *What is Cinema?*, trans. Timothy Barnard, op. cit., p. 78.

The critique brings together two issues. The first concerns the noticeability of the effects; the second one their inauthenticity or dishonesty. The first issue might seem to be a technical matter — such that in several cases it could be solved by advances in effects technology and filmmaking know-how — which is why a moral issue is brought in to buttress the argument. In both instances, however, it is the photographic nature of film that is at stake, by which is meant its ability to record (and, for Bazin, to reveal) reality.

It should be obvious by now that our practical-minded convention-agreeing or game-playing account of spectatorship in the face of noticeable special effects, while it may serve as one explanation of their "cultural invisibility," contrasts with Bazin's more fine-grained view of realism. As complex as this notion may be in his writings, Bazin's realism, at least as it is implicitly defined in the quotation above, opts for a more discriminating understanding of "self-containment" and "obtrusiveness." Accordingly, anything that hinders the illusion of a world existing independently of the image — a world shown to viewers in such a way that what they see could have been shot in a single take (i.e., a photographable world) — risks appearing "unrealistic."[13] Theoretically speaking, then, detecting the presence

13 This is why Bazin's realism can accept editing and special effects, as long as they do not give the impression of creating the event depicted. This is a theoretical stance for Bazin. As a critic, however, he was sometimes more forgiving. In his 1952 review of *When Worlds Collide* (Rudolph Maté, 1951), he writes: "The end of the world special effects, the tidal wave that engulfs New York, the explosion of the earth attracted by a rogue planet, the construction and launch of the rocket ship, its landing on planet Zyra, all display a rather astonishing realism" ("*Le choc des mondes*: L'arche de Noé atomique" initially published in *Le Parisien libéré* 2424, 30 June 1952, and reprinted in André Bazin, *Écrits complets*, vol. I, Hervé Joubert-Laurencin [ed.], [Paris: Macula, 2018], p. 953). Most of the effects mentioned by Bazin here use miniatures. However, the view of New York submerged is a painting, and the landing of the spacecraft on Zyra shows a miniature with a very noticeable painted background. As soon as the ship lands, two of the film's principals go the deck to see whether the air is breathable (it is!) and take a look at humanity's (only white Americans!) new planet. The camera follows their off-screen gaze by panning to the valley below. This is a composite shot done on the optical printer that merges the actors with a painting that looks as if it were from an animated cartoon. It is difficult to think of these effects, including the miniatures, as anything but an equivalent to montage, in the sense that Bazin uses the term to critique Jean Tourane's *Une fée... pas comme les autres* (*The Secret of Magic Island*, 1956) in "Forbidden Montage." The claim here is that because Tourane is using animals that are not acting their parts but are mostly standing still or moving minimally in each shot, the entire film relies on montage and the accretion of shots: in short, the actions depicted *could not* have been filmed in a single take, for the animals are incapable of performing them continuously. But the same description equally applies to the special effects segments in Maté's film, since what is depicted in those moments only exists in miniatures and composites, i.e., in cinema. In the terms of "Forbidden Montage," only certain mechanical effects (like the strings of *Ballon rouge* or the life-size animatronic shark of *Jaws* [Steven Spielberg, 1975]) or else pyrotechnics,

of an effect could therefore jar the viewer into a skeptical stance regarding the reality of the world depicted. In an otherwise realist film, a noticeable painted backdrop or rear projection might give the impression that an *alien* or *foreign* (in German *fremd*) element has suddenly intruded in the film — an element that, like montage, belongs to cinema rather than to a world (that appears) independent from it. And while Bazin doesn't quite think of the matter in such terms, what he objects to can be construed as a potential *Verfremdungseffekt*, as indeed highly noticeable rear projection has sometimes been used by later modernist filmmakers (Straub-Huillet, for instance, in *Chronik der Anna Magdalena Bach* [*The Chronicle of Anna Magdalena Bach*], 1968).[14]

Yet post-1968 critics did not seize on special effects' adulteration of photography's tie to reality as a site of critique of realism and illusionism nor in their attempt to replace theoretical idealism with Marxist materialism and ideology critique, not even through deconstructionist readings of special effects films. Were these scholars as blind as their predecessors to the implications etched out by Bazin in "Montage interdit," blinded by filmmakers' attempts to seamlessly "suture" (to use a term from that era!) special effects shots to realistic non-effects cinematography regardless of their success in achieving it? But there might also have been something

wind machines, and stunts can be said to be sufficiently homogeneous and continuous with the real world so as not to "fall" into the realm of "montage." Yet the fact that Bazin found the special effects in Maté's film "realist" arguably attests to a certain willingness to overlook their obvious "unrealness."

14 As briefly discussed above, it seems that classical-era audiences and critics were inclined to avoid any "modernizing" or "political" readings of such noticeable effects, which, in any case, would have worked against the grain of the films themselves (not a commonly held notion at the time). Viewers instead appear to have turned a blind eye to any potential aesthetic or political implications, endeavoring to coherently, if not seamlessly, integrate such effects into the fictions they appeared to serve. Turning a "blind eye" does not mean not noticing; rather, it means minimizing as much as possible what can detract from the film's unity of style or from what is implicitly constructed as its intent. By contrast, in the segment of *Chronicle of Anna Magdalena Bach*, where Bach gives an outdoor recital at night, the filmmakers make no attempt to use rear projection to approximate a unified narrative space—quite the opposite in fact. There is a clear mismatching of perspective, angle, focus, and lighting between the foreground (Bach playing harpsichord) and the slanted projected building in the background. This is simply an additional strategy used by Straub-Huillet to estrange the viewer in a modernist film that eschews most conventions of classical realism and underlines the artifactuality of the film. Note that for viewers, the only difference between Straub-Huillet's use of rear projection and classical Hollywood's — both being noticeable — lies in their ability to infer a different intention in each instance, that is to say, whether or not viewers recognize that some effort is made either to mask or to emphasize the hiatus between what is photographically recorded and what is represented.

else at stake, something deeper still that made special effects somewhat unpalatable for ideology critique.

For one thing, of course, effects form a little industry of their own within the film industry (and now beyond). Every era has had its own state-of-the-art effects, and they often involve patents, require specialized technicians, and entail extra expenditures. This is why the more up-to-date (and expensive) effects usually only find their way into big-budget films (at the very best, films that could occasionally fit into the *Cahiers du cinéma*'s notorious "Category E"!). But more to the point might be the second issue Bazin raises in the quotation above, one concerning morality or ethics in representation.

With part of their lineage found in stage magic shows (the other source being conventional theater, as François Albera, and Frank Kessler and Sabine Lenk remind us in their essays), early special effects were more than attractions, they were also, in many cases, *trompe-l'œil*. However, there are two types of *trompe-l'œil*, and we need to distinguish between perceptual and cognitive deceit. In perceptual deceit alone, the effect is (or seeks to be) seamless, though we are aware nonetheless, because of the content, that an effect is being used; whereas in cognitive deceit the *trompe-l'œil* is (or seeks to be) both seamless *and* imperceptible: viewers are simply unaware that an effect has been used. One strives to be a *visual fiction*, the other a *visual lie*. But, if seamlessness is a condition of imperceptibility and of a successful lie, any *attempt* at seamlessness that does not involve an impossible content (viz., hiding a cut, attempts to blend foreground and background in rear projection, etc.), can be understood as an *attempted lie*, even if the effect is noticeable. As a result of this lineage, and somewhat ironically, special effects — whose images were neither factual nor truthful, strictly speaking — could only have been conceived of as a bad object by both Bazin *and* post-68 critics. For the latter, imperceptible special effects could only have been complicit with dominant realist ideology in their effort to mask their true nature, while most noticeable effects might appear as failed attempts at doing so rather than properly modernist gestures. Either way, as a result, special effects could appear as morally corrupt for both Bazin and the "anti-Bazinians" for whom, unlike montage, special effects seemed almost willy-nilly to serve the seamless narrative text rather than actively resist it.

3. ***Ideological Commitment to Art and Authorship* (*or, It Ain't What You Do...*).** There is, finally, a third strand in our account that can help explain the prior lack of scholarly attention given to special effects and which can

itself be broken down into several more specific factors as they all relate to the discipline's longstanding commitment to art and authorship.

a) ***Authorship/Craft.*** Unlike Méliès, most filmmakers do not shoot or create the special effects sequences for their films (although there have been exceptions and different directors have had varying degrees of involvement with the production of special effects).[15] Highly specialized skilled artists, craftspeople, and special technical consultants are usually hired for this purpose. (At one point during the classical era, all the major studios had a special-effects department.) Some effects, such as background plates for rear projections or computer-assisted compositing, have to be shot separately, which is usually undertaken by a second unit. In this regard, we might think of special effects shots as analogous to the landscape settings and backgrounds of large-scale Renaissance paintings, which most times were painted by apprentices, while the Master took charge of the figures in the foreground.

The use of second units or the need for specialized crews certainly furthered the sense that special effects are a marginal aspect of filmmaking, at least until the advent of a constant flow of blockbuster effects-dominated films where both narrative (plot and character) *and* special effects seem to equally vie for the spectator's attention (in these films, in fact, the narrative often becomes a mere pretext for a smorgasbord of effects). Yet the comparison with landscape painting falls short aesthetically. Having gained their complete independence from the depiction of characters and events, landscape came to thrive as a genre of its own starting in the seventeenth century.

Although the work of certain special effect artists has been celebrated, (from Willis O'Brien and Ray Harryhausen to Douglas Trumbull and John Dykstra), their overall social status is still closer to that of craftspeople than that of full-fledged *auteurs* or artists. Thus, as magnificent as Trumbull's effects for *2001: A Space Odyssey* were for 1969 (and still are), as an *artistic* achievement *2001* is first and foremost considered to be a Stanley Kubrick film. Although film studies as a discipline has taken its distance from criticism and aesthetic evaluation over the years — not to mention auteurism — thanks in good measure to film theory, cultural studies, and film history, the stigma of craftsmanship attached to special effects has certainly been a factor in their earlier neglect by the mainstream academic film

15 James Cameron, for instance, worked as a special-effects assistant and matte artist in the early stages of his career.

studies community, a community whose object of study, whether individual scholars care about it or not, has acquired the status of artistic medium. In the social hierarchy of filmmaking, special-effects units are thus seen as involving highly skilled workers whose task is to meet and fulfill the director's conception of their film.

b) ***Fragments.*** The fact that special effects are created by a specialized crew also points to what seems to be their fragmentary nature with regards to an entire film. Thus, filmmakers and viewers (when the effects are not imperceptible to them) often use the terms "special-effect shot" or "special-effect sequence" to distinguish these segments from the rest of a film. And though they may be an important attraction in a film (what would the success of the *Star Wars* saga be without special effects?), they seem to have a practical purpose in illustrating a story — or, better yet, displaying *moments* or *elements* belonging to a story. Thus, when the *only* admirable aspects of a narrative feature film are said to be its special effects, the film is usually considered a failure. Their fragmentary aspect might furthermore be reinforced when used as attractions (which by definition excludes imperceptible effects). Like any attraction, special effects have the potential to stand out in a film as quasi-independent fragments, spectacular moments that are almost experienced for their own sake, *all the while* serving the narrative. Think, for instance, of the use of "bullet-time" in *The Matrix* (Lana and Lilly Wachowski, 1999): while viewers understood the action depicted in the narrative, they often individuated these moments in their viewing experience ("Wow! *This* is so cool!") and could wonder how the effect was achieved all at once.

There is little doubt that the experience of film viewing builds upon integrating fragments of different sorts (not just spectacular special effects) in a constant back-and-forth between part and whole-in-the-making (this corresponds to what is also known as a hermeneutic circle), but it is the integrated result — wholeness as finality — that tends to be seen as the valued object, not the fragments in themselves. Thus, as long as special effects were conceived as fragments momentarily fulfilling the requirements of a story as best possible — a furthering of the idea of *supplement* mentioned above — they failed to receive sustained attention in film scholarship.

c) ***Technical discourse.*** Of course, one should not think that nothing of interest was written about special effects prior to the digital age — far from it. It is just that *academic* film scholars mostly stayed away from the topic, just as for a long time they shied away from many of the more technical aspects of filmmaking (including in-depth discussion of various cameras, lenses, editing devices, projectors, etc. and their impact on film style, form, and meaning). Thus, for a long time, special effects were mostly the province of technical discourse and discussed in the pages of trade publications like *American Cinematographer*. Film Studies, on the other hand, initially grew out of film criticism. If interest is currently rising for "Industry Studies" in our discipline today, in the wake of film history's interest for technical developments and patents, it is because many feel that a major gap in scholarship needs filling. Now it is fair to say that from the perspective of criticism or any phenomenological or viewer-based approach to film, it can be difficult if not impossible to identify what sort of special effects are being used in a given segment of a film. Moreover, for the longest time, film scholarship was conducted (it often still is) with only the most basic knowledge of the medium's technical means. Therefore, for previous generations of scholars, studying special effects might well have meant investigating the technical means of filmmaking (a secondary interest at most) regardless of the aesthetic, cultural, or ideological import of the medium. Today the tide has obviously turned. With the introduction of digital effects and with the realization of their widespread use, special effects have become an important site from which to deepen film theory and to investigate the production and the reception of films as well their aesthetic proclivities.

***

The essays in this collection represent a wide range of views on special effects from diverse perspectives, but all accept that the question of special effects can no longer be relegated to the margins of film studies. A complete account of the cinema needs to acknowledge special effects as a constituent part of the medium, indeed as a major technical but also aesthetic component of filmmaking and as an important part of the experience for the audience. Each of the essays builds upon and develops specific aspects of the burgeoning field of special effects studies. They have been divided into three sections that reflect what we, as editors, consider to be each chapter's primary focus. But as the reader will discover, these sections are anything but airtight, as several issues or concerns carry over between them. The opening section gathers essays that address special effects mostly through theoretical or philosophical questioning, considering some of the basic concepts necessary for their analysis. Essays in the second section trace the historical developments of the technologies and techniques of special effects and discourses around their use. Authors in the final section of the book turn to specific examples of film for analysis to investigate various questions raised by special effects from the late silent period to the digital age.

It is, then, as noted, around these three axes—*Concepts, Techniques, Films*—that this collection is organized. We conclude with an *Envoi* by Dudley Andrew, in the very traditional sense of the term, as a "letter of dispatch," setting the discipline of film studies upon a new course with special effects as a central topic but without abandoning or forgetting the accomplishments of the past, the enduring significance of, for instance, a figure like André Bazin, still so central to so much of contemporary film theory. In light of the significant advances made recently in the study of special effects, it is not the goal of this collection to argue for the legitimacy of their study but to quite simply proceed with both theoretical and historical analyses. Accepting that special effects are a basic aspect of cinematic representation, though, requires that the question of what such representation consist of be asked anew.

Concepts: It is just such a fundamental question that is addressed in the first section, in which certain basic questions are revisited in light of an acknowledgment of the integral nature of special effects. In this section, the authors find, through the consideration of special effects, openings into sundry perennial problems of film theory and suggest some solutions to various theoretical impasses. Several key terms of film theory are reconsidered — notably realism, simulation, imagination and the imaginary, indexicality — as well as more general philosophical concepts such as meaning, representation, and the phenomenological and pragmatic questions of experience, belief, interpretation, and even ethics. Less an exhaustive survey of the theoretical field from the perspective of special effects, this section provides some speculation on specific aspects of their theoretical significance.

Martin Lefebvre, in the opening essay, explores the notion of "gap" as a way of addressing the conceptual and constitutive nature of the special effect. The same question is raised by both Donald Crafton and Benoît Turquety in their essays. There is, of course, and most importantly, the real, material gap that existed originally between film frames, which allowed for the very first kinds of special effects, created for example by Georges Méliès, who used editing to produce effects of substitution and transformation. From this basic material fact, though, Lefebvre then considers the many ways in which our experience of the cinema is governed by some more general and significant sense of a "gap" that exists between film worlds and the real world of phenomenological experience, and the role that special effects have played in the construction of film worlds, which necessarily engage the audience, to some inevitable degree, eliciting some reflection on the constructed nature of those worlds. Building on earlier work he has done on the question, Lefebvre offers a careful analysis of the concept of "indexicality," which is at the center of the phenomenological account of film viewing that he outlines.

John Belton, too, is concerned with the basic phenomenological question of the viewer's experience when confronted with a necessarily constructed image. For Belton, contemporary visual effects raise the fundamental question of the very nature of the image, which is, he argues, and as revealed most explicitly by digital effects, necessarily "composite." Like Lefebvre, and like François Jost in his essay, Belton raises the issue of indexicality, providing a summary of recent debates on the matter. He challenges the traditional accounts, though, which tend to conceive of it as a constituent element of particular kinds of imagery, offering instead the more pragmatic argument that it is the spectator who determines whether "something is or is not indexical." Building on the work of Stephen Prince on contemporary visual effects, Belton describes the complex spectatorial experience of engaging with increasingly intricate audio-visual compositions. As a deliberate "manipulation of reality," the cinematic image necessarily involves us in an imaginative undertaking, as Belton emphasizes. He traces the links between the image and the imaginary — which he distinguishes, though, from the use of the term in the psychoanalytic work of Christian Metz. Less a potentially deceptive or illusory phenomenon, the composite image — characteristic of the cinema — is the site of imaginative engagement.

François Jost places the question of special effects in the cinema within the context of the longstanding concern in modern philosophy, at least since Descartes, of the relation between perception and reality, doubt and belief. This philosophical problem is given acute manifestation in the cinema, which, Jost argues, has developed an extensive repertoire for the creation of certain kinds of perceptual "tricks," or *trucages*. These are, Jost insists, as integral a part of cinematic representation as the more familiar elements of cinematography and *mise en scène*, as ordinary a part of filmmaking, we might say, and as François Albera also argues, as framing, lighting, montage, and color. While *trucages*, or special effects, may of course be used to create deceptive imagery — to "feign" — this must be distinguished from the more "ludic" use of such techniques in the creation of fictions. The relation between cinematic tricks and belief must be described according to a clear distinction between beliefs about the real world and beliefs about the worlds of fictions. Jost builds on the famous work by Christian Metz[16] on film and *trucages* to develop a "general theory of *trucages*" while critiquing some aspects of Metz's approach and developing some of his insights in a pragmatic direction instead. Concerned less with the problem of deception, which Metz seemed to emphasize, Jost describes instead the role of effects in relation to what the spectator believes to be watching (viz., fiction or documentary).

Marc Furstenau asks whether the use of special effects can be incorporated into a revised theory of cinematic realism. Special effects have often been thought to disrupt the essentially realistic nature of cinematic representation. Long described as illusory, or as the techniques for the creation of illusion, the main concern had been that special effects would create only a false or misleading realism. Alternatively, the specific illusions that special effects create have been understood to be in the service of the generally illusory nature of cinematic representation itself. Furstenau argues, though, that illusionism and realism have to be separated and clearly distinguished as concepts. Against an illusion theory of cinematic representation and of pictorial depiction generally, he argues that realism in the cinema is thoroughly a matter of style, and that it need not become entangled with apparently related but in fact tangential questions of perception, belief, and "reality" but focused on the stylistic and thematic uses to which special effects may be put according to a more deflationary account of realism.

Techniques: From such general and abstract speculations about special effects, the second part of the book moves on to some more specific considerations of the historical development of effects technologies and the critical debates about their use. Frank Kessler and Sabine Lenk trace the history of debates about special effects, specifically in Germany, back to those in the early nineteenth century about the proper staging of theatrical dramas, when there was quite acute concern about the proper use of so-called "mechanical artifices." Surprisingly, they discover that while there was considerable emphasis in early critical accounts of dramatic spectacle on the need for "discretion" when using such artifices, lest they overwhelm and obscure the "poetic and dramatic force of a text" when using mechanical means to visualize these on stage, there was far less concern about the use of tricks in early cinema. Given the inherently "mechanical" nature of the cinema emerging at the beginning of the twentieth century, it seemed

16 Metz, "*Trucage* and the Film," op. cit., pp. 657–675.

more obviously appropriate to employ the full force of cinematic artifice. Kessler and Lenk draw our attention to what are in fact often very different discursive contexts for different art forms and show that, while there are some important continuities between the theater and the cinema, there are also quite different critical concepts that have been applied to them. In the process, Kessler and Lenk also manage to distinguish two different schools of thinking about theatrical effects: the German and the French.

François Albera makes a related argument, tracing the origins of special effects back to early cinema, where they emerged as a constituent part of the new form, partly as an extension of techniques that had been developed in the theater but also as distinct methods unique to the cinema. Emphasizing, like Kessler and Lenk, the mechanical nature of the cinema, Albera describes the emergence — first in the theater but then in the new medium — of the "great machinist." This is who we would now call the effects technicians, those craftspeople with an increasingly specialist knowledge charged with the task of realizing the visions of the dramatist and the filmmaker. Many of those working in theater, and then in film, became famous in their own right, providing the indispensable means for the creation of technically complex representations. Albera traces the history of the term "special effect" — in French, "*effet spécial*" — which is, and remains, he says, perfectly "ordinary," understood as those techniques that were necessary, even inevitable, as the cinema developed aesthetically. Built into its representations is the very fact of the cinema's "machinic" nature. At the heart of the new medium, he argues, is the very question of modernity itself, of automation, industrial production, technical expertise — the basic question of the relation between the human, labor, and the machine.

Developing the notion of the "gap" raised by Lefebvre, Donald Crafton describes the very significant "spaces between the frames" through the exploitation of which special effects first become possible. He describes the cinema, intriguingly, as "a system for structurally omitting information," allowing, and again as an inherent capacity, for the creation of deliberately designed compositions. Like so many of the contributors to this volume, Crafton very carefully considers the material basis in the cinematic apparatus itself to trace the origins of special effects. He also returns to Méliès, and to Henri Bergson, whose description of the cinema as consisting of discrete units of stillness helps us to see that, as Crafton puts it, and as Benoît Turquety also suggests, "the cinema has always been digital."

Turquety considers in very precise detail the nature of the effects pioneered by Georges Méliès, specifically his famous "substitution trick" but also his use of other techniques, including superimpositions, multiple exposures, mattes, and so on. He, too, describes how Méliès exploited the "gap" in the film recording process — as well as black backgrounds as empty spaces on which to superimpose other imagery — to create his intricate composite compositions. In contrast to more usual accounts of photography and photographic processes, Turquety describes the medium's fantastic possibilities, inherent in the very material constitution of the cinema, deployed almost as soon as it was invented to create elaborate and fanciful compositions. Through a re-reading of the early-twentieth-century French philosopher Henri Bergson, Turquety argues that the sorts of manipulations thought to be characteristic of the digital era can be seen to have their origins in these early experiments, as Bergson seemed to have intuitively understood in his account of the cinema, despite his reservations about the new medium as consisting of discreet still images or temporal samplings.

Katharina Loew charts the history of the transition from the "tricks" of early cinema, created by amateur generalists, to special effects, understood as the domain of technical specialization. She argues that, starting around 1910, special effects have not, for the most part, been used for the creation of elaborate, "attention-grabbing" spectacle. Rather, they have typically been meant to go unnoticed, creating seamless but often complex composite representations. On the basis of a modified version of the taxonomy of effects provided by Metz, she looks back to the silent feature film, where we may see the emergence of "standardized composite techniques," and the increasing specialization necessary to create the look of mainstream cinema.

For Roger Odin, special effects must be understood as part of the means by which the cinema is used as a medium of communication; he contends that one needs, moreover, to distinguish in any analysis between different "spaces of communication." Like others in this collection, he begins with Metz's formative essay on *trucages* but revisits it through his own pragmatics of communicative spaces. Thus, the specific value of any effect is shown to be determined by the different "cinematic spaces of communication," which include, for instance, different genres. Special effects, though, are by no means exclusive to the cinema, according to Odin. Indeed, he seeks to expand the notion of "special effects" to all forms of image manipulation made possible in the digital age and by new modes of image production, including the use of cellphones to shoot still or moving images. Building, too, on the work of Jacques Aumont, Odin describes the production of effects as a fundamental technical craft of the modern digital world, which produces not "representations of reality" but a new kind of communicative utterance.

Finally, Phillipe Marion considers various recent films that use the contemporary technology of motion capture, which he describes as a "perpetual special effect." He posits that the cinema is no longer essentially a camera art, and he describes what he calls the "subjugation of the profilmic" to such techniques as motion capture. He notes that while effects still tend to be described as "special," they are in fact now an entirely routine part of filmmaking. As a routine element, he considers the question of how they are at once experienced both as effects by an increasingly knowledgeable and informed audience and as the means for the creation of a convincing diegesis. Special effects, he argues, are a part of the very "ontology" of the cinema, "part and parcel" of the film medium itself now. Marion also considers the effects that new digital effects techniques are having on traditional aspects of cinematic composition, like editing.

Films: In the third section of this volume, several specific case studies are undertaken. Janet Bergstrom offers an account of a particularly exemplary special effects film, F.W. Murnau's famous *Sunrise* (1927), his first Hollywood film after emigrating to the United States from Germany. Bergstrom describes the emergence of a cadre of specialized effects technicians, whose contribution to this film was indispensable to its success. The film was marketed at the time as a technical feat to be marveled at for its special effects. Bergstrom describes the various aspects of the filmmaking, which combined the in-camera effects of the cinematographer and set designers, but equally importantly the postproduction work of the effects technicians and the careful work of matte painters. The film is notable for its use of a newly patented process, the "Williams shot," and marks the beginning of an increasingly complex division of labor that continues to characterize filmmaking to this day.

Suzanne Liandrat-Guiges considers that most famous of movie monsters, *King Kong*, and the significance of the figure in the 1933 film by Cooper and Schoedsack. There is, she argues, an inherent paradox in the use of special effects in the film, so obvious as effects, which nevertheless create such a compelling fictional world. This is not, however, any kind of straightforward deception or illusion but a more subtle and complex spectatorial experience that weaves the special effects with the entire formal and rhetorical design of the film. The perceptual aspects of special effects — which, as she notes, are explicitly thematized in the film through a *mise-en-abyme* structure that incorporates a film within the film — are the basis for a more complex conceptual experience, according to the "open perspective" that she provides. Liandrat-Guiges shows how the film's visual design and plot offer a layered perspective that is at times created by, and at times literally mirrored by, the use of special effects, so that a "nesting or mirroring principle presides over the film's aesthetic." Her claim, in the end, is that proper understanding of special effects in a film cannot be limited to the aims of simply offering a "realistic" depiction of what cannot otherwise be depicted; rather, their meaning resides in the way they relate to the entire system of a film, including its plot.

This section concludes with two very careful and detailed historical analyses of particular effects in two important films that exemplified and promoted specific techniques. Kristen Whissel recounts the use of 3D effects in *House of Wax* (André de Toth, 1953), starring Vincent Price. Often dismissed as little more than a gimmick film, Whissel convincingly demonstrates how it very effectively exploits the specific effects of 3D, namely the "negative parallax" effect, by which the items of the film world seem to extend physically into the space of the theater, the world of the spectator. The especially "uncanny" nature of this effect is compared to that of waxworks themselves, whereby, and according to a Freudian account of the *Unheimlich*, basic boundaries are effaced: between the past and the present, the living and the dead, the animate and inanimate — and between the screen space and the spectatorial space. The resulting effect, Whissel argues, is "emblematic" of the cultural and political fluidity of modernity itself, which was felt acutely in the 1950s, as enormous technological and cultural changes were underway in the cinema and beyond.

Sean Cubitt, too, describes the effects of special effects on the medium itself. Considering the example of the sci-fi action film *Oblivion* (Joseph Kosinski, 2013), Cubitt describes what he calls the peculiar "lack of identity" characteristic of digital special effects. As thematized or dramatized (deliberately or not) in this film, Cubitt argues that such a lack of identity corresponds to a more general, cultural sense of ontological and epistemological instability. The claim is that special effects have always responded to the different needs and interests of film audiences, while belonging to art's perpetual repositioning of the ancient and constitutive instability of the category of the "human." Read allegorically, the effects in a film like *Oblivion* thus reveal an emerging "world view," which understands human existence itself to be subject to increasingly intricate mathematical calculations.

The range of essays in this volume suggests, we hope, the vitality and the importance of special effects as a central topic in film studies. Yet, and as Dudley Andrew argues in the concluding essay, it is important to see how the newly emerging question of special effects is continuous, in many respects, with earlier historical and theoretical speculations about the cinema. Well-known for his very subtle exegeses of the critical writing of

André Bazin, Andrew explores the question of special effects in relation to Bazin's still very influential account of realism. Bazin, of course, explicitly raised the question of special effects on only a few occasions in his writing, yet, as Andrew argues, it is at these points — when, for instance, he considers the historical fortunes of the technique of the "superimposition" in the cinema — that the subtlety of his account is revealed. In contrast to the usual sort of distinction assumed to be operative in film theory, between a strict normative emphasis on realism and a strident formalism, we can discover in Bazin a much more complex and nuanced account, what Andrew calls a "flexible realism." Bazin was, he says, fully aware that the cinema was first of all a technical apparatus and subject necessarily to historical change, the result of both new artistic approaches by filmmakers and the changing tastes of audiences. The "ontology" of the cinema, in this respect, lies precisely in the fact of material change, in the constant reinvention of the medium, the development of various technical means for the creation of compelling representations that will continue to grip the imaginations of audiences. Even in the case of the most enduring and influential of the "realists" in film theory, then, we in fact discover a quite subtle historical but also phenomenological account of the effects of special effects, which are an inherent part of the always changing technical and aesthetic apparatus of the cinema.

Andrew is concerned to "open" Bazin to the contemporary digital era — not to simply vindicate his original claims but rather to find in his responses to the technical developments of his era a guide to our responses to our own. "Unlike the pure cinephiles at *Cahiers*," Andrew writes, "Bazin frequently stepped back from films to examine cinema as a medium evolving from, and alongside, traditional artistic media (theater, painting, and especially the novel), and in relation to emerging technologies that amounted to the new media of the 1950s (television, 3D, Cinerama, 'Scope')." It is, Andrew suggests, a lesson for film scholars today to be sensitive to the complex historical, discursive, and material and technical contexts within which the cinema emerged and which continues, unceasingly, to develop and change, with special effects at its heart.

***

Finally, we want to acknowledge that the impetus for this book was an event organized in Montréal by André Gaudreault, Viva Paci, and Martin Lefebvre and held at the Cinémathèque québécoise in November 2013. This gathering, entitled *The Magic of Special Effects: Cinema, Technology,*

Reception was a resounding success with over 400 attendants, and has led to numerous publications. Several of the essays here gathered — though not all of them — started out as papers presented in Montréal.

Bibliography

Bazin, André. *What is Cinema?*, vol. 1, translated by Hugh Gray (Los Angeles: University of California Press, 1967).

———. *What is Cinema?*, translated by Timothy Barnard (Montréal: Caboose, 2009).

———. *Écrits complets*, vol. 1, edited by Hervé Joubert-Laurencin (Paris: Macula, 2018).

Bukatman, Scott. *Matters of Gravity: Special Effects and Supermen in the 20th Century* (Durham, NC: Duke University Press, 2003).

DeMille, Cecil B. "A Director Looks at 'Process-Shots'," *American Cinematographer* (November 1936): 458–459.

Derrida, Jacques. *Of Grammatology*, translated by Gayatri Charkravorty Spivak (Baltimore, MD: Johns Hopkins University Press, 1976).

———. "Plato's Pharmacy." In *Dissemination*, translated by Barbara Johnson (Chicago: University of Chicago Press, 1983).

Keil, Charlie, and Kristen Whissel, eds. *Editing and Special/Visual Effects* (New Brunswick, NJ: Rutgers University Press, 2016).

Limmer, Eric. "It's Crazy How Much of *The Wolf of Wall Street* is Actually CGI," *Gizmodo* (14 January 2014). Accessed 17 November 2021. https://gizmodo.com/its-crazy-how-much-of-the-wolf-of-wall-street-is-actua-1501402962.

McClean, Shilo T. *Digital Storytelling: The Narrative Power of Visual Effects* (Cambridge, MA: MIT Press, 2007).

Metz, Christian. "*Trucage* and the Film," translated by François Meltzer, *Critical Inquiry* 3, no. 4 (1977): 657–675.

North, Dan. *Performing Illusions: Cinema, Special Effects, and the Virtual Actor* (London: Wallflower, 2008).

North, Dan, Bob Rehak, and Michael S. Duffy, eds. *Special Effects: New Histories/Theories/Contexts* (London: British Film Institute, 2015).

Pierson, Michele. *Special Effects: Still in Search of Wonder* (New York: Columbia University Press, 2002).

Prince, Stephen. *Digital Visual Effects in Cinema: The Seduction of Reality* (New Brunswick, NJ: Rutgers University Press, 2012).

Turnock, Julie. *Plastic Reality: Special Effects, Technology, and the Emergence of 1970s Blockbuster Aesthetics* (New York: Columbia, 2015).

Whissel, Kristen. *Spectacular Digital Effects: CGI and Contemporary Cinema* (Durham, NC: Duke University Press, 2014).

About the Authors

Marc Furstenau is Associate Professor and Head of the Film Studies Program in the School for Studies in Art and Culture, Carleton University (Ottawa, Canada). He is the editor of *The Film Theory Reader: Debates and Arguments* (Routledge, 2010), co-editor of *Cinema and Technology: Cultures, Theories, Practices* (Palgrave, 2008), and author of the forthcoming book *The Aesthetics of Digital Montage* (Amsterdam University Press).

Martin Lefebvre is Professor and Chair of the Mel Hoppenheim School of Cinema at Concordia University (Montréal, Canada). He is Editor-in-Chief of *Recherches sémiotiques/Semiotic Inquiry* and has published widely on semiotics and film. He is the author of *Truffaut et ses doubles* (Vrin, 2013) and has edited several volumes including *Techniques et technologies du cinéma* (with A. Gaudreault; PUR, 2015); *Landscape and Film* (Routledge, 2007); and *Eisenstein: l'ancien et le nouveau* (with D. Chateau & F. Jost; Sorbonne, 2001).

Concepts

1. Mind(ing) the Gap

Martin Lefebvre

Abstract

The author's purpose in this chapter is twofold: first to define special effects, and second to exploit this definition to reconsider how Charles S. Peirce's conceptions of iconicity and especially indexicality can help us logically account for some of the issues that pertain to representation in the cinema. The author wishes to show how Peirce's concepts not only remain viable for thinking about cinema but also offer a powerful tool for critically analyzing images. The argument is weaved around the notion of 'gap,' of which various instances are taken into consideration — whether the space between film frames or the gap between 'film world' and real world, that between fiction and fact, or between filmmaking and film viewing, as well as that which distinguishes iconicity and indexicality.

Keywords: special effects, Charles S. Peirce, icon and iconicity, index and indexicality

Introduction

The aim of this paper is to consider special effects — and what they mean for film scholarship — through various declensions of the idea, or metaphor, of a *gap* (or *écart* in French). In other words, I wish to examine how cinematic special effects — their being and use, our experience of them, and even our thinking about them — are predicated by various forms of difference or else by a series of discontinuities best grasped through the notion or "image" of a gap, divide, or "in-between" space. The gap is an image that comes to mind from considering some of the earliest cinematic special effects (such as those used by Georges Méliès). In the process, I will come to explain why I prefer to retain the term "special effects" rather than replace it with the newer phrase "visual effects" — the use of which has been quite cogently

Lefebvre, M. & M. Furstenau (eds.), *Special Effects on the Screen: Faking the View from Méliès to Motion Capture*. Taylor & Francis Group, 2022
DOI 10.5117/9789462980730_CH01

defended by Stephen Prince, among others.[1] Both terms, in fact, ought to coexist within our critical metalanguage, though with different ends. For instance, while contemporary CGI animated films like *Wall-E* (Andrew Stanton, 2008) use many of the same digital visual-effects technologies as do special-effects films like *Rogue One* (Gareth Edwards, 2016), my contention is that it is important to hold on to the term "special effects" to distinguish between the two. The point is simple enough: regardless of the technology it uses, *Wall-E,* unlike *Rogue One,* comprises no special effects to speak of. To my mind, it would be as meaningless for viewers to speak of special effects in discussing the Disney/Pixar film as it would be when considering a novel or a painting. For neither literature nor painting can be said to use special effects, at least not in any way that could approximate what is usually meant when using the term with reference to cinema.[2]

At stake here is the fact that "special effects" form a highly differential or contrasting notion, one whose definition resists any positive account of its own. Indeed, the very term "*special*" that forms part of the compound expression "special effects" marks the presence of a distinguishing quality, a difference, and therefore what I consider to be a gap or discontinuity of some sort. More specifically, the presence of special effects in a film only makes sense with reference to their absence, that is to say, with reference to some sort of baseline or "zero degree" of filmmaking that corresponds to cinema's ability to record and reproduce more or less closely, more or less approximately, how the independently existing visible world appears to us:[3] reference, in short, to that which is *not* special (or, at least, is no longer so). Special effects rely on the cinema's aptitude to diverge from this baseline — either through photographic or non-photographic means — yet without jettisoning it altogether. To use a term that was popular in the heyday of structuralist criticism, special effects require the expression of a specific *paradigm*: a virtual set pregnant with possibilities. In the absence of it — in the absence of the gap, deviation, or difference that fully expressing this paradigm introduces in the order of representation, as is the case with literature, painting, as well as animated films (whether CGI

1 See Stephen Prince, *Digital Visual Effects in Cinema* (New Brunswick, NJ: Rutgers University Press, 2012).

2 To be sure, there are specialized technicians/artists who use visual-effects software on animated films — adding fake "camera" flares or rendering the animation (with shading, textures, bump-mapping, depth-of-field, motion-blur, etc.).

3 That the camera has the ability to approximate how we see the world does not mean that it necessarily does so. It might be more accurate to say that a camera records the world however it "sees" it. It just so happens that how it sees the world often approximates how we see the world.

or traditional cell animation) — the term "special effects" simply becomes otiose. For neither literature nor painting nor animation rely on the same representational paradigm (based on the recorded images of the world) as that which makes special effects possible in the cinema.[4] In short, it is because many (most!) "live-action" films still rely, in some fashion or other, on cinema's photographic recording capabilities that we can still expect to make some theoretical or conceptual mileage out of the term "special effects." What follows is thus meant to unpack (but also complicate somewhat) these few ideas.

1. The Interval Between Frames: The Inaugural Gap

The Edison catalogue is today credited with the first substitution edit/special (or trick) effect, which was used to show the beheading of Mary Stuart in *The Execution of Mary, Queen of Scots* (Alfred Clark, 1894). But it is to the pioneering work of Georges Méliès that such effects are most readily associated. Indeed, it was the French magician-turned-filmmaker who not only popularized the substitution trick — as early as 1896 — but gave it its "*lettres de noblesse*" by his skillful control and imaginative exploitation of the various possibilities it offered. Often described as an in-camera edit technique, the process nonetheless usually required some actual cutting-room editing to shave off any undesired frames inadvertently exposed with the sudden interruption and resuming of the camera's hand-cranked recording action, thereby providing a (relatively) smooth or "invisible" matching effect between the two takes, the aim of which was to conceal a temporal hiatus in the shooting. As a result, any modification in the profilmic visual field between both takes, such as moving an object from one location in the composition to another — or else substituting, removing, or adding people or props — is perceived on screen as a sudden appearance, transformation, or disappearance, provided that the other parameters of the image are kept more or less constant or still, such as to create an otherwise strong impression of identity and continuity between the end of the first shot and the beginning of the second one.

4 One might wonder what this claim implies for pictorial *trompe l'œil* so often evoked in discussions of special effects. The distinction is mainly between means and results. Though paintings may occasionally be used as *trompe l'œil* in the context of some specific architectural setting or else as part of an apparatus, wherein they may deceive the viewer (as when matte paintings are used in films), they themselves *qua* paintings do not resort to special effects in the process.

The substitution trick effect thus works by masking the gap or discontinuity in the shooting and by creating a background continuity *against which* a substitution or transformation can be measured and appear to happen. Whenever the effect was emphasized for the sake of spectacle, viewers typically experienced either *spatio-temporal continuity* based on a *concealed spatio-temporal discontinuity* (e.g., the above-mentioned beheading of Queen Mary) or *spatial discontinuity* set in (falsely) *continuous time* (sudden appearances, disappearances, or moving of objects). It is worth noting that in the first instance, the spatial discontinuity involved can be purely extra-diegetic (in the Edison film, the diegetic executioner — distinct from the non-diegetic actor who plays him — is not aware that he has beheaded a mannequin that was substituted for the actress between takes); whereas in the second one, the spatial discontinuity (appearances, disappearances, etc. — cf. the example of Méliès's 1898 film *Le magicien* [*The Magician*], mentioned below in the next paragraph) is experienced by the inhabitants of the diegetic world — though not as the result of a substitution splice but rather as a trait of their world (whether or not an explanation is offered, characters in such a film may become aware that they live in a world where objects can suddenly appear or disappear or be substituted, etc.). Hence the existence of at least two different usages of the trick with regards to the depicted world, depending on whether the characters can perceive or are by definition oblivious — barring any metalepsis[5] — to the effects of the discontinuity enacted by the substitution, i.e., whether or not the effects of the substitution (though usually not its cinematic *means*) are diegetized. (The issue of the viewer's perception of special effects will be discussed later.)

It follows that the various effects achieved by the substitution splice are made possible by taking advantage of a basic or fundamental fact of cinema — one uncovered by all those who successfully worked at inventing various devices for the production of moving images — namely that the recording and playback of apparently continuous motion requires showing, in succession, a series of *discrete, discontinuous* still images (i.e., frames or fields of pixels, depending on the technology). With the substitution trick, this basic fact of cinema is often used to alter the depiction of the "cine-photographable" world by seemingly showing as continuous events that did

5 Gérard Genette, in his narratology, defines metalepsis as any "intrusion by the exradiegetic narrator or narratee into the diegetic universe [...] or the inverse." See Gérard Genette, *Narrative Discourse: An Essay in Method*, trans. Jane E. Lewin (Ithaca: Cornell University Press, 1980) pp. 234–235.

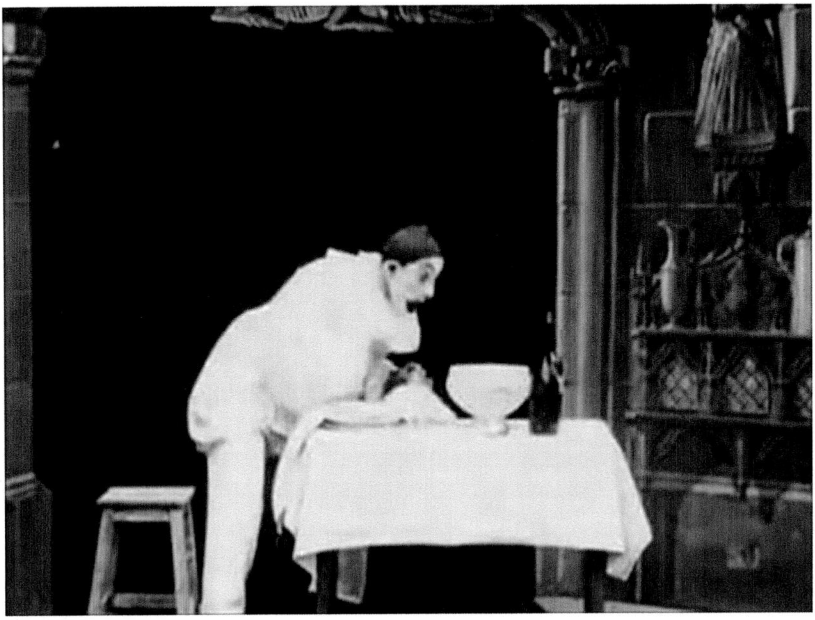

Fig. 1.1 *Le magicien* (Georges Méliès, 1898).

not actually unfold continuously in front of the camera. Discontinuity in *real world* events (i.e., halting the shooting of the film to move or remove an object, substitute a prop or add a character in the profilmic space, etc.) is thus covertly ushered into the film by exploiting the invisible discontinuity of the film strip — the gaps between the frames — yet all the while giving (or attempting to give) the impression of a temporally continuous *film world*[6] (which, depending on the situation, might be intended to pass for the real recorded world or else for a fictional one) on the ground that almost everything else seen in the succession of images appears identical or continuous. This is to say that the succession of images *appears* to respect the "normal" (though otherwise invisible) flow of intervals between frames on both sides of the cut. This is how, for instance, in *Le magicien*, Méliès was able to combine a match cut on movement with the sudden disappearance of a bench from under a Pierrôt as he is about to sit down (Fig. 1.1). The effect implies shooting the movement of sitting twice — once with the bench present and once without — and skillfully matching on action the

6 I use this term to mean the world as created by the film, whether this world gives itself to us as fictional or not. Using a term developed by the filmologists, we could also call it the *filmophanic* world. See Etienne Souriau's preface to *L'univers filmique* (Paris: Flammarion, 1953), p. 8.

succession of shots so that the character appears to fall to the ground in a single, continuous take.[7] In short, what Méliès exploited so well is the fact that, because continuity in the cinema (i.e., the impression normally created by uninterrupted takes) is achieved through discontinuity (i.e., through the mostly imperceptible succession of individual frames divided by short intervals), discontinuity in shooting can, under certain conditions, mask itself as a continuous single take. What *formally* distinguishes this process from a simple continuity match cut depicting a single action or movement across two different separate shots, and which may equally mask the temporal discontinuity of the shooting (as any indefinite amount of time may have lapsed between the takes — from minutes to hours, days, or more), is the attempt to hide the edit itself.[8]

Other early special effects taking advantage of the discontinuity or gap between film frames also included the frame-by-frame animation of inanimate objects (as popularized by Segundo de Chomón and Emil Cohl) and the use of multiple exposures. In the first instance, intermittent shots of an object in slightly varying positions in space give the viewer the impression that the otherwise inanimate object is moving in either temporal or spatio-temporal continuity (depending on whether

7 A much later but all the same famous use of the technique can be seen in the dream sequence of Buster Keaton's *Sherlock Jr.* (1929). Here, the impression of temporal continuity is achieved by match cuts of the character who either stands still or follows through on movement while the background world around him keeps changing. Unlike what goes on in *Le magicien* where one gets the impression that an object has disappeared from under the character, here we get the impression that it is the character who is constantly transported from one location to the other once his dream self has entered the (dream) film world of the movie screen and is subjected to the effects of its editing. This leads to a series of visual gags, including one that resembles *Le magicien*: Keaton is about to sit on a bench, but because he is the victim of a shot (and location) change, he ends up falling off a sidewalk into the middle of a street with passing cars. Though it is the same gag as the one in Méliès's film, and achieved through the same basic technique, its meaning is different: here it is the character who is made to unnaturally jump in space and is submitted to spatial discontinuity, not the furniture of the world he inhabits.

8 To be sure, match cutting belongs to classical cinema's arsenal of "invisible editing" and it might be argued that it therefore also "obscures" the cut. However, this suppression — which often mimics the spectator's mental "movement" of interest toward something — is not really a form of camouflage or *trompe l'œil* as is the substitution trick. Match cutting, which may depict a single action over a cut without creating an impression of temporal hiatus, usually unites frames that *vary* slightly from each other in shot size and/or viewing angle, with sometimes a very brief duplication of the action being matched. These variations betray the cut, even if the continuity in movement between the shots helps to visually ease the transition between them. However, by avoiding changing the shot size and camera angle, both *The Execution of Mary, Queen of Scots* and *Le magicien* attempt to completely conceal the cut, thus combining the technique of the match cut with that of the substitution trick.

the background space is continuous or changing; in the latter case, the impression is that of a single object moving through different *tableaux*. As for multiple exposures, the effect relies on exposing and processing two or more distinct shots "on-top" of each other as a single section of film is run through the camera more than once (more controlled layering effects were later achieved using an optical printer — which eventually became computer controlled — until most such layering came to be realized by digital means), giving the impression of spatial contiguity and temporal continuity out of discontinuous takes.

It might be said, then, that early effects of the sort described so far "inhabited" or at least operated from the space created by the gaps found to exist between film frames, those very same gaps that enable cinematic motion in both its recording and its restitution. Not surprisingly, manipulating the rate of succession of these gaps (and thus of consecutive still images) led to other early special effects, such as accelerated or slowed motion. By increasing or decreasing the near 1:1 standard ratio in the frame succession rate between camera and projector (usually achieved by under- or overcranking the camera — though the same effect could be produced by under- or overcranking the projector. With the development of optical printers such effects could also be realized in postproduction; and needless to say, they can now be obtained digitally by postproduction manipulation of the frame rate, although the use of specialized high-frame-rate cameras during the shooting will produce sharper slow motion images).

Now, I'm not proposing that the formal condition of possibility identified so far for some popular early trick effects — the exploitation of the physical gap separating film frames — stand for a general definition of special effects. For one thing, backward or reverse motion, which was also considered to be an early special effect, was not initially achieved by manipulating this gap. Furthermore, it should be noted that nothing in the cinematic *technique* of the substitution edit *necessarily* aligns it with what is usually meant by a "special effect," certainly in everyday parlance. For instance, consider a case of true "invisible" editing: a series of still shots of an unmoving and unchanging everyday household object filmed at great intervals in time but in the exact same, controlled shooting conditions. Once joined together and screened, the segments would give the impression of a single continuous take, much like time-lapse cinematography often does, but without movement and therefore without imparting an impression or awareness of any sleight of hand or any departure from a simple continuous recording of the cine-photographable world (there is no substitution to be seen anymore). As far as I know, no such *stand-alone* film actually exists, not because it is impossible to make — quite the contrary — but simply because it would be a pointless achievement.[9] Its very possibility, however, is instructive because it now forces us to consider a different kind of gap relative to special effects: it is a gap that cannot be accounted for solely in *quantitative* terms (such as the physical gap between frames) with regards to the recording of reality, but must be considered in *qualitative* terms with regards to the world depicted (a gap in the kind of world represented).

2. Worldly (or Qualitative) Gaps

In as much as the above-mentioned early special effects imply a departure from cinema's "zero degree" — all the while existing in a paradigmatic set with it — they also make possible a departure from reality itself (the cine-photographable, profilmic world), at first by altering the cinema's ability to continuously record our world according to the camera's own limitations (the point arguably applies to film editing as well, not merely to the special effects mentioned above); and secondly in terms of the being (or ontology) of that world.

The preceding section ended with the example of an improbable film using the technique of the substitution splice to falsely create the impression of uninterrupted continuity such as would otherwise be achieved by a continuous take. This raises the question as to why we might hesitate to call such impression of continuity a "special effect." The answer cannot lie in the technique itself, which is similar to that used by Méliès for some of his most famous special-effects films. We must therefore look elsewhere. In our example, notice that the technique is used *solely* to hide the ellipses in shooting and that the fake continuity it achieves does not interfere with the ontological and existential connection the camera otherwise establishes between the real world and the film world. To put it differently, the above hypothetical film would not have *looked* any different had it been shot in a single take. The obvious implication here is that the inanimate object thus depicted actually exists, continuously and independently of it being filmed. This leads us to suppose that, as far as the substitution splice is

9 The only potential reason for making such a stand-alone film would be to make an indefinitely long film. Theoretically, such a film could be much longer than permitted by the amount of celluloid mounted on a camera magazine or, in the digital era, by the space available on a hard drive. Imagine the impression of a single take lasting an indefinite number of lifetimes, though actually composed by several indistinguishable takes (and generations of filmmakers)!

concerned, the hiding of the edit[10] is really part of a more elaborate "deceit" that concerns the real world (the one in which you and I as well as every other existent being, past and present, live in) and the camera's theoretical ability to record it. It might well be for this reason also that regular match cuts — which are elliptical in terms of the real-world shooting of the film, but not in terms of the narrative fictional world depicted — do not register with viewers as special effects (i.e., no one, certainly today, calls them such). For unlike Méliès's substitution edits (which to this day we still think of as special effects), they do not show anything about the world that a single take, theoretically speaking, could not show (this comment concerns perceptual space alone, not the psychology of the action nor the rhythm of the *découpage,* which are different matters altogether).

But if this observation helps us distinguish between the fake continuity of the substitution edit and that of the match cut, it may also raise more questions than it answers. In particular, it forces us to ask whether it is a necessary condition — a marker — of special effects that they be used to depict qualitatively and ontologically alternate, impossible worlds. Take, for instance, the case of matte paintings: should the matte shots of Tara's exterior in *Gone with the Wind* (Victor Fleming, 1940) or those of Hindley Hall Mansion's interior in *The Paradine Case* (Alfred Hitchcock, 1947) be conceived differently from those of *Star Wars* (George Lucas, 1977) on such grounds? In both *Gone with the Wind* and *The Paradine Case,* the fake matte set extensions appear fully congruent, qualitatively speaking, with our existing world, merely standing in for otherwise "equivalent" existing worldly locations, as would more elaborate and costly sets. This means that although Tara and Hindley Hall are purely fictional locales and have no worldly existence, unlike *Star Wars*' Death Star they are nonetheless made to resemble real existing houses (though they aren't meant to stand for any particular existing one, except in fictional terms — unlike the mattes of the White House in *Dave* [Ivan Reitman, 1993]. I shall return to this distinction in section 4).

Let us consider the matter more closely before coming to an answer. One obvious paradox of cinema is that it can show us fictional film worlds by using recordings of the camera-independent existing world masquerading as imaginary. This is how most of Méliès's special effects of the sort discussed earlier distinguish themselves: what is depicted cannot be recorded with

10 See Donald Crafton's essay in this collection where he explains that Méliès would often stage the action in the lower part of the frame so as to direct the viewer's gaze away from the splice happening at the top of the frame.

a single uninterrupted shot, for it is no longer congruent with the cine-photographable independent world, nor is it understood (or appear) to be so qualitatively. Indeed, not only do these images show us something that *did not* actually occur in front of the camera; they seek to show us something that *could not* have happened in front of the camera in a continuous take — a set of events and actions to which we can now include reverse motion events such as spilled milk returning into a glass. Hence the notion of a qualitative gap operated by special effects between the real world and the film world. The implication here is that in recorded live-action cinema, the real world may somehow interfere with the creation of fictional worlds. Thus, the early special effects discussed above belong to a wide arsenal of techniques and practices developed to *loosen the grip* of the real world on the fictional film world.

Now, if one considers solely their goal of foiling or somehow altering or manipulating the camera's tie to our world in order relax the latter's dominion over it, then such "trick" effects can indiscriminately include: creative geography editing (by which the viewer is led to infer a seamless continuous space from shots taken at different locations, a form of editing that can be combined with either match cutting or a shot/counter-shot pattern in order to buttress the inference); matte shot composites (achieved on-set by glass paintings or Schüfftan mirror shots, or else in postproduction); rear and front projection shots; but also makeup, costumes, and prosthetics; fake studio sets (from Old West towns to the rotating room where Fred Astaire dances on the walls and ceiling in Stanley Donen's 1951 *Royal Wedding*); painted set backgrounds; stunt and body doubles; miniatures and models; puppets and animatronics; etc. To some extent, even acting could be considered a means to diminish the hold that the real world exerts over the film world, although no one would, of course, consider it a special effect per se (for reasons that will become clear in the next section).

All of the above practices can be used to depict worlds that are either *impossible* or else *difficult* or *impractical* to show with a camera, on the ground of three basic constraints. These include sundry practical constraints, such as legal, ethical, or security constraints: one shouldn't commit a crime to illustrate the beheading of Mary Stuart; shooting in the vicinity of a real tornado or storm may prove too hazardous — think of Ford's *The Hurricane* (1937), de Bont's *Twister* (1996), or Petersen's *The Perfect Storm* (2000); etc. There are also financial and technical constraints: it cost less to use mattes in *The Paradine Case* than it would have to build an entire mansion as a set, and it was technically more convenient in terms of "image control" to shoot the live action on the studio lot and to add mattes later than to shoot the entire Hindley Hall scene on location in some existing mansion. Finally, there are "ontological" constraints: *Star Wars*' Death Star only exists in fiction, as do King Kong (Merian C. Cooper and Ernest B. Schoedsack, 1933) and the invisible man (James Whale, 1933), and therefore cannot be captured by the camera without some form of deceit.[11] Moreover, ontological constraints also affect the film apparatus: no real existing camera can move in the way the "viewpoint" is seen to move in the famous CG-enhanced long take in *Panic Room* (David Fincher, 2002) as burglars break into a house.

The question before us, however, is whether or not all the practices mentioned above, which seek to evade or overturn the camera's subjection to the cine-photographable world, should all be considered special effects? On one hand, these would include various means and techniques used to manipulate and exploit either the workings of the apparatus that produces the image — diverting it from its baseline — or the image itself that has been produced by it: in-camera effects; the substitution splice and the other early effects mentioned above; post-production composites (which are a form of intra-image montage) and image manipulations (for instance, the full range of CG interventions); etc. Also included, on the other hand, would be all those processes used for manipulating or masquerading the world and/or exploiting the camera's optical limitations at producing an image during the shooting (such as its insensitivity to scale or to what lies outside the frame; its limited apprehension of depth or three-dimensionality; its sensitivity to light — meaning that of its film stock or of its light sensors; its image resolution, etc.). The latter comprise such diverse means and techniques as fake and painted sets, makeup and prosthetics, stunt and body doubles, puppets and animatronics, miniatures, glass paintings placed between the camera and the live action, the Schüfftan process, etc. And what of lens filters and gels or else bizarre effects created by lighting that can all alter how the reality that lies in front of the camera looks? Should they also be considered "special effects"? Now, because a number of these techniques came to require the use of ever more highly specialized production and postproduction teams, often working in concert with each other, and because they all share the common goal of *untightening*, or, in several cases, of simply *manipulating* or *diverting* the real world's claim on the film image, many of their practitioners ended up lumped together in large studios under the heading of "special-effects departments" working alongside "technical departments" (which often included the more "theatrical" means

11 Films set in the past may or may not fall into this category.

of fictional world building, such as set construction and costume),[12] before being eventually outsourced to independent effects companies.

Yet any such groupings are somewhat loose, *ad hoc*, and determined chiefly by the terms of a given technique and the industrial organization of filmmaking labor and craft. Furthermore, from the perspective of the matte painter, it makes little to no difference at all whether the effect sought is the composite integration of live action into a painting of Hindley Hall, the White House, or the Death Star. So what of the qualitative-ontological gap that marked the substitution trick as a special effect but is absent from match cutting and from the hypothetical fake continuity film mentioned at the end of the preceding section? Here the answer is simple enough: notice that the distinction in question was made on the grounds of the *viewer's experience* rather than that of the filmmaker. Hence the importance of considering a third form of gap, which overlaps with the qualitative-ontological one: namely that which divides filmmaking practices from the phenomenology of film viewing and of special effects.

3. The Gap Between Filmmaking and the Viewing Experience

It was Christian Metz who most famously paved the way for a phenomenology of special effects in his essay "*Trucage* and the Film."[13] There, he distinguished between "*trucages visibles*," "*trucages imperceptibles*," and "*trucages invisibles*."[14] For Metz, visible effects comprise such image manipulations as accelerated and slow motion (to which we could add time-lapse photography, reverse motion, and noticeable multiple exposures). Metz

12 David Bordwell, Janet Staiger, and Kristin Thompson give the examples of Warner Bros. which inaugurated its "Special Research Department" in 1929; RKO, which rounded up its special-effects technicians into a single department in 1932; and MGM, which in 1936 separated its special-effects technicians from the art department. The new unit "prepared miniatures, process work, and full-size composites while another department handled matte paintings and optical effects." See *The Classical Hollywood Cinema. Film Style and Mode of Production to 1960* (New York: Columbia University Press, 1985), p. 324.

13 Christian Metz, "*Trucage* and the Film," trans. Françoise Metzler, in *Critical Inquiry* (3)4 (1977), pp. 657–675.

14 The French word "*trucage*" (sometimes spelled "*truquage*") is difficult to translate into English. It is a noun derived from the substantive "*truc*" or from the verb "*truquer*," which means "trick" or "to trick." In the language of cinema, its use is often synonymous with "*effets spéciaux*" (special effects). The English "trick" itself comes from the Old French verb "*trichier*" (which is now "*tricher*," which means "to cheat"). Perhaps, then, it should come as no surprise that the first great "cheat" of cinema was a Frenchman!

seems to imply here that the effect can be "seen" because it so obviously affects the entire visual field of the image itself. Indeed, his examples, which also include fades, dissolves, wipes, and other such optical effects, concern the cine-photographic apparatus which is thus manipulated to produce certain kinds of images. Imperceptible effects are those forms of trickery that go entirely undetected by the viewer. Examples would include an actor wearing shoe lifts to make him appear taller; successful use of stunt or body doubles (perhaps we could add fully integrated and photo-realistic matte paintings — whether by hand or CGI— which fool the viewer into believing that what is shown is a photographic depiction of an actual existing location, as in the above examples from *Dave* and *The Paradine Case*. In fact, many of today's unobtrusive and realist CG image modifications fall into this category). Finally, somewhere between the two previous forms lie the invisible effects, which comprise effects like the "invisible man" or Méliès's substitution trick, that is to say, special effects that do not appear to affect the film image's ability to represent the world and yet are nonetheless perceptible in the sense that the viewer clearly realizes or else "senses" that some form of deceit or mystification —whose means may well be left undetermined — is happening.[15] This triad represents for Metz a spectrum of "perceptual regimes" ("*régimes perceptifs*"[16]) by which viewers already versed in the conventions of fiction filmmaking can experience special effects.

But as convenient as it may be as a starting point, Metz's taxonomy is nonetheless marred by the blind spot of his objectivist/immanentist structuralist epistemology. To put it more simply: his phenomenologically inspired scheme actually turns out to be lopsided and self-contradicting, weighing more heavily on the side of filmmaking than on that of the viewing experience, thus jeopardizing the very distinction it would seem useful to develop between filmmaking practices and the phenomenology of special effects. In other words, Metz wishes to map how viewers respond to special effects but not how this response can help us define or delineate what constitutes special effects in the experience of cinema — e.g., to explain why no one thinks of match cuts on action as special effects, even though they present as continuous (in the filmophanic or narrative world) an action

15 One way to help us think through the distinction that Metz seeks to make between visible and invisible (though perceptible) effects is to use Metz's semiological vocabulary: visible effects involve both the cinematic signifier and the signified, while invisible effects become perceptible at the level of the signified alone.

16 Ibid., p. 180.

that is not (in the real existing world). In so doing, Metz falls victim to what he himself criticizes: importing a filmmaker's craft conception of special effects into theoretical discourse.[17] Indeed, how else could one account for the category of the "imperceptible effect"? Only the filmmakers, who are in the know, can call an effect truly "imperceptible," *not the viewer for whom there is no effect to be perceived while they screen the film*. Therefore, to speak of "imperceptible effects" is to consider the viewer's response from the perspective of the filmmakers who trick them. However, this is *not* to deny the existence of imperceptible effects. Indeed, they exist as such for those who produce them but not, experientially speaking, for those who are unaware of them and for whom there is no perceptual, qualitative gap between what is seen and the camera's ability to record the world. In short, Metz's tripartite nomenclature should give way to a more basic, higher-level dualism: *the filmmakers' special effects* and the *viewer's special effects*.

The first category is based on craft or professional notions of special effects. It is often subdivided today into *optical effects* (effects that involve cinematography and are produced with the camera and the lighting, such as fish eye or wide angle or split dioptric lenses; exposure effects; filters; in-camera irises, fades and dissolves; multiple exposures; slow and accelerated motion; time-lapse photography; bipacking for still background matte composites and Williams process for travelling mattes; etc.); *special effects proper* (which are also achieved during the shooting but are interventions onto the real world being filmed: mechanical effects; puppets; animatronics; miniatures and models; prosthetics, body suits and makeup; set design and set painting; matted glass shots; rear and front projections; riggings, stunts and body doubles including those involving crashes, water tanks, fire, explosions, zero gravity or high g-force acceleration; weather effects: fake snow, sprinklers for rain, wind machines; squibs that simulate bullet impacts; fake blood, slime and other substances; etc. And although set design and props are usually relegated to other departments, it is hard to see why they should not be considered special effects in this scheme if not for the *ad hoc* organization of studio labor and other historical considerations); and *visual effects* (relative to postproduction work, most of which is now achieved digitally, through software-based enhancing and retouching, color

17 Metz's statement of intent comes up early in the article: "The concept of *trucage* presented here must not be confused with the 'special effects' of which studio technicians speak [...] [T]he notions of technicians — who sometimes have a *professional*, and therefore corporate, personality — cannot not automatically be considered as theoretical concepts." Metz, "*Trucage* and the Film," op. cit., p. 659.

grading, painting, compositing, and morphing — either from scratch or from material that has already been recorded, as in motion and performance capture-based characters. It also includes effects traditionally achieved in postproduction through lab film processing and color timing work; and through the use of optical printers whether to reproduce optical effects or to integrate principal photography into matte plates, etc.).

The second category is a different matter altogether: it is not defined by the historical divisions of studio filmmaking labor but rather qualitatively and ontologically, that is to say, by the viewer's ability to see and sense the distance or gap that separates the image screened from the cinema's native ability to both record the world that lies independently in front of the camera and show it as such.

This has many implications, the most obvious being that the set of terms that fall into this category need not parallel those of the previous group, i.e., the filmmakers' special effects. For instance, and for reasons I explain below, viewers do not tend to discuss sprinkler rain as a "special effect" — even when the effect is so clearly fake (for a discerning eye, at least) as to be "perceptible," as in the obvious studio setting for Gene Kelly's "Singin' in the rain" song and dance act in Stanley Donen's 1952 film. The same goes for much color grading as well as several effects that concern qualities of the images in themselves without affecting the ontology of the world depicted or else the idea that what is recorded is a camera-independent world: wipes, fades, dissolves, irises, split-screens, wide-angle or telephoto lenses, exposure, so-called "special effects" filters (as opposed to polarizing, ND, hard-edged or soft-edged GND, or white balance filters whose presence is mostly imperceptible). All seem superadded to whatever is depicted — which is why slow and accelerated as well as reverse motion and time-lapse photography ought to be considered differently, as *bona fide* special effects, for unlike fades or dissolves, they can be used to affect the qualities of the world they depict.

4. The Reality/Fiction Gap

At this point, two issues require a brief discussion: fiction and interpretation. We can begin with the first term. Surely, there is no debating that fiction is one way of loosening reality's grip on cinema. But this is an ambiguous statement given how most live action films are still made, namely by using a camera (even if this is no longer the *sole* source of images in live-action films — more on this later). For clearly the camera does not care whether it is recording fact or fiction, its *indexical* link is with whatever lies in front of it, whether or not this is how things normally appear in the world or how they are made-up to appear just for its sake by *mise en scène*. The studio sets, the prosthetics the actor wears, the rigged car crashes and stunts that are recorded are *real* — how else could they be recorded? — even though they stand for beings and events whose reality, most times, is purely fictional. However, if we allow for prosthetics and body doubles to fall under the heading of special effects *simply on the grounds that they masquerade the world*, then the implication would be that the entire domain of cinematic fiction should equally fall under this heading for the viewer. And this would lead to a rather strange conception whereby films made without the help of special effects (from the point of view of the filmmakers) would appear *entirely* made up of them (from the viewer's perspective) by the sheer fact that they are fictions! If we wish to avoid turning every fiction film (including those of Renoir or of the Italian Neo-Realists!) into an entire special effect, we therefore need to be more discriminating, and this is especially true for what Metz calls "invisible" special effects.

I have been putting a lot of emphasis so far on the camera and its recording function. The reason for this becomes clear when we consider the problem raised by fiction. What everyone (filmmakers and viewers alike) understands about special effects in narrative fiction films is that they pertain to fiction by first pertaining to perception. In other words, they are not produced for the (sole) sake of fiction, *but for the purpose of showing something that cannot or should not — for the various reasons outlined earlier — be shown otherwise* (*i.e., without some form of trickery or deceit*) *using a camera*. What comes first with special effects, therefore, is a relation to the camera (this is the representational paradigm mentioned in the introduction), not to the fiction; or to put it in a slightly different way, what is initially at stake is a *perceptual fiction*, not so much a *fiction of enactment*[18] — even though the former may, and indeed most often does, serve the latter. This distinction is akin to what demarcates magic from acting as different forms of deceit and fiction. This explains why the "make-believe" of acting as well as that which relates to the "everyday" or "ordinary"-looking furniture of the film world (by which I mean some of the more *theatrically inspired* tools of the trade: costumes, "basic" makeup, sprinkler rain, realistic built sets and props) usually do not stand out as special effects for most viewers even when seen as "fake."

18 The case for motion and performance capture is slightly different and related in this regard to rotoscopy.

Of course, for the viewer, the boundary between perceptual and enactment fiction is not always obvious to delineate, and it is better to think of it as a spectrum, with acting at one end, the "invisible man" and his ilk at the other, and filmed prosthetics, death-defying stunts, and mechanical effects often lying somewhere in the middle for many viewers, as long as they are identified or believed to be gerrymandered so as to trick our perception. (As we shall see below, digital cinema is having an impact on the ability of viewers to *correctly* place an effect along this spectrum). For example, theater can rig its actors so that they may appear to fly or float, yet completely hiding the rig may prove difficult. In a film, on the other hand, full baring of such a device without any apparent intention to conceal it — imagine a Brechtian superhero film! — might not be considered a special effect at all, whereas erasing all traces of the rig — as in *Green Lantern* (Martin Campbell, 2011) will make it so. Concealing the effect, even though it is still perceptible due to its fantastic content, is what ensures the perceptual fiction. It is also where the pleasure of much special effects can be found: in seeing something we *know* is false while it appears perceptually true (or "camera-true").

If we should not accept turning entire films into special-effects movies on the ground of their fictionality alone, adopting the viewer's perspective on special effects while in the act of watching a film nonetheless implies that there will be occasions for discrepancy between *the filmmakers'* and the *viewer's special effects*. For instance, many traditional matte paintings and digital composites simply pass under the radar of even the most discriminating viewers who just do not perceive them. If there is no reason to suspect deceit, and if the effects used are truly imperceptible and reproduce something (or a possible instantiation of a type of thing) that exists in our world and for which there are no obvious constraints that restrict its recording by a camera, there will be a mismatch between the filmmakers (who are aware that effects of some sort were used) and the viewer (who is not aware and believes by default that what is seen was actually recorded by the camera without deceit — all the while knowing that the film is a fiction).

Such discrepancies are perhaps even more frequent now that we have entered the digital age, as filmmakers use various software to touch up details and imperceptibly alter — tweaking, changing, adding, deleting — sundry elements in the image. As mentioned above, this can impact viewers' ability to correctly situate an effect along the perceptual/enactment fiction spectrum. For instance, director Terry Gilliam explained that actor Tobey Maguire's fake hair in *Fear and Loathing in Las Vegas* (1998) ended up requiring digital postproduction work to hide the line on his forehead formed by the bald cap on which fake hair was attached. Here, the filmmakers used a combination of special effects (filmed prosthetic) and visual effect (digital retouching). For the viewer who has reason to believe that what is seen is not the actor's real head of hair, the overall effect is certainly perceptible, although its digital aspect is not: such a viewer will most likely believe the effect to be achieved solely by the wearing of a wig and therefore will have it stand on the side of enactment fiction rather than perceptual fiction or, if you will, on the side of theater rather than magic. The same viewer might well consider that this is not a special effect at all and that it is akin instead to the makeup the actor is wearing or to his "costume," all of which appear to be recorded without any perceptual deceit.

Inversely, in the digital age, viewers may erroneously believe digital manipulation is used to create an effect when in fact there is no perceptual sleight of hand involved, the camera simply recording what lies in front of it. Watching *Apollo 13* (Ron Howard, 1995), many viewers believed (and still do) that the weightlessness of the actors and props was achieved digitally — as is the case with *Gravity* (Alfonso Cuarón, 2013) where wires were digitally removed and where most of the sets and floating props used CGI and digital compositing — when instead they were done aboard a NASA aircraft (nicknamed the "vomit comet"!) capable of producing zero gravity for short periods of time.

In all these mismatches, we see how what the viewers know and believe — both about the world and about cinema — plays an important part in interpreting and even appreciating special effects. Knowledge of the real world and of constraints on shooting conditions can be used hermeneutically to conjecture and interpret the presence of special effects that cannot otherwise be seen or identified. This is especially obvious when there is a qualitative gap between the film world and our world such that a camera *could* not have really recorded an event for it has no real-world counterpart: the T-1000 liquid metal terminator of *Terminator 2: Judgment Day* (James Cameron, 1991), the autobots of *Transformers* (Michael Bay, 2007), the folding of the Parisian cityscape of *Inception* (Christopher Nolan, 2010), the astral voyages of *Dr. Strange* (Scott Derrickson, 2016), are just a few obvious examples.

Of course, not all special or visual effects are used to create fantastic worlds that have no real-world counterparts or else entirely forgo to record settings (or actors) that possess what philosophers name *haecceity* (the *hic et nunc* quality of actuality and existence). Take, for instance, the case of miniatures, or "bigatures" as director Peter Jackson likes to call them, the use of which is still widespread in the effects industry today. Using miniatures involves *existents*, namely scaled model sets that are *really* captured and recorded by the camera, relying in part on the latter's lack of any sense of scale. Today the use of miniature sets is usually accompanied with other effects — mostly postproduction CG — to fine-tune and correct the overall impression that the filmmakers seek to convey. The same can be said of motion and performance capture, which rely on recordings of actors' movements and facial expressions that "drive" the effect as they are digitally "painted" over and mapped onto otherwise non-existent and non-photographed beings like the Na'vi of *Avatar* (James Cameron, 2009) and the very expressive and talkative apes of *War for the Planet of the Apes* (Matt Reeves, 2017), or grafted onto a real, photographed, existing body that is different from the actor to whom the facial features belong, as was the case for Brad Pitt's face in *The Curious Case of Benjamin Button* (David Fincher, 2008).[19] Examples of modern-day usage of miniatures include, among many others, the explosion of the White House in Emmerich's 1996 *Independence Day* (it was digitally composited in postproduction with a shot of an alien spaceship shooting a light ray at it from above) or the tsunami scene from *The Impossible* (Juan Antonio Bayona, 2012) as the giant wave hits and destroys the cottages of a Thai holiday resort where the Bennet family are staying (some of the scale model shots were digitally composited into other views so that the entire scene uses different types of effects, including CGI and stunts). However, *how* the effects are achieved, while it matters very much to the filmmakers, is of little relevance to a phenomenology of special effects. What matters instead — and in this they are no different from the fantastic events of *Inception* or *Dr. Strange* — is that they become perceptible by how they appear or else are understood to diverge from recorded views of the real, existing world.

But could it not be argued that in the use of miniatures or motion-capture, there is after all something of the real world that is recorded? As is the case with standard film acting, we have here some *existent* that is photographed and that stands, as a fiction, for something else. True. However, this can't

19 Facial replacement and motion capture would certainly warrant a separate study. They offer intriguing forms of hybridity. Motion capture's use of an actor's movements to "drive" an animated being is not entirely unlike using a human being to equally "drive" a body suit such as the ones worn by the two actors who played the monster in *Creature From the Black Lagoon*. Though the effect is created by recording the actual movements of an actor, it seems unlikely that the movements themselves are being recorded for their own sake or that they are identifiable in their "*haecceity*" by the viewer. It is worth noting, however, that facial replacement shares a certain family resemblance to the practice dubbing an actor's voice "over" another actor's body. Indeed, the controversy that erupted around *Black Swan* (Darren Aronofsky, 2010) and Natalie Portman's dancing (her face was placed over that of professional ballet dancer Sarah Lane whose body is seen) is not without reminding us of the situation depicted in the plot of *Singin' in the Rain*. I'll return to motion capture in my final remarks at the end of the paper.

be the whole story. First of all, acting is usually a very visible form of deceit that requires the viewer's willingness to abide by the convention whereby the actor is made to stand for a character. Secondly, actors in films most often portray human beings — there are exceptions, of course, which may involve deception and effects of various sorts, from costumes and masks (*Creature from the Black Lagoon*, Jack Arnold, 1954) to motion-capture and CGI (Andrew Serkis as Caesar in the recent *Planet of the Apes* franchise). However, this means that in the canonical situation at least, actors are used to stand in for something, i.e., characters, that belong to the same general type they themselves belong to: human beings. Miniature models like the ones mentioned above, on the other hand, serve to depict a *different embodied type of object than that to which they belong*, namely full-size buildings (such as they would appear if filmed), ones that are *not* miniature models. Therein lies the perceptual deceit that is not normally found in acting.[20] Of course, as is the case with other invisible effects, viewers only become aware of the effect — or infer it, rightly or wrongly — when the depiction is such that it is incompatible with what they know or else believe to be the case about our world or when they strongly suspect (rightly or wrongly) it would otherwise be subject to the above-mentioned constraints they know can affect filmmaking. Thus, at stake in our examples is the impossibility of making a film by destroying the real White House or a real hotel resort with a tidal wave; or, considering the cost, the unlikeliness of destroying sets made to a 1:1 scale with these buildings — which, if it were the case and if this became known to the viewer or else was perceptible, would push the effect toward enactment fiction, next to stunts and the crashing of real cars.

Finally, if true imperceptibility has little value for a phenomenology of special effects (at least until they become known through sources other than film viewing alone), room should be made for effects that may be conceived as *seeking invisibility*, yet whose perceptibility is not limited solely to the impossible or improbable nature of the content they depict, or to impossible camera work[21] — given the viewer's knowledge of film, of the world, and of the basic restrictions mentioned earlier — but also include visible "failures" to deceive perception. Here I would include effects that "show" (or "betray") themselves by their limitations to "falsify" perception — i.e., to falsify how reality looks when it is filmed: fringing of mattes; the mismatching of color,

20 With the exception perhaps of when characters are severely injured or dismembered. Not surprisingly, this is where special effects often come into play.

21 For instance, think of the digitally tweaked and "impossible" long takes of Alfonso Cuarón's *Children of Men* (2006) or *Gravity* (2016).

of movement, and of visual texture in rear projections; the lack of gravity and the unlikely physics of early miniatures and models; the stiffness of mannequins thrown down cliffs or inside volcanoes; the unnatural choppiness of stop-motion animation; the uncanny bouncy motion of early CG characters like *The Phantom Menace*'s Jar Jar Binks (George Lucas, 1999) or Spider-Man (Sam Raimi, 2002), etc. Such effects are perceptible even though the *means* used may seek imperceptibility. It is here, furthermore, that special effects become especially prone to the effects of time. What changes historically, as refinements and newer means for creating effects gain in imperceptibility, is not so much the viewers' perception itself but rather their willingness to accept the conventions of an effect whenever the means used to create it betrays it, thus introducing yet another gap that would need attending to: a gap in perception between how we see the world, how we normally see at the movies (when no effects can be detected), and how the film world appears once the special effect itself become visible. For example, reviewers of the original *King Kong* clearly identified the perceptual gap involved in the stop-motion action. Joe Bigelow of *Variety* wrote:

> It takes a couple of reels for "Kong" to be believed, and until then it doesn't grip. But after the audience becomes used to the machine-like movements and other mechanical flaws in the gigantic animals on view, and become accustomed to the phony atmosphere, they may commence to feel the power. As the story background is constantly implausible, the mechanical end must fight its own battle for audience confidence [...] While not believing it, audiences will wonder how it's done.[22]

Regardless of the old-fashion charm it possesses for a number of cinephiles and film scholars, it may be more difficult for many audience members today to become "used" to the jerky movements of the original *King Kong* after screening Peter Jackson's *King Kong* (2005) or Jordan Vogt-Roberts's *Kong: Skull Island* (2017).

5. Enter Indexicality

It is a truism to say that digital cinema has revolutionized how films are made. The entire workflow has been transformed in important ways, and

22 Published 6 March 1933. Accessed online July 2016: http://variety.com/1933/film/reviews/king-kong-2-1200410783/

visual effects have now become ubiquitous. There's hardly a film — or even a single shot! — made today that does not use them for minor touch-ups or else to create entire virtual sets composed from hundreds of elements, either photographic or CGI, into which actors and props, initially shot against a green screen or a video wall, are digitally composited. In many ways, the principles of pre-digital effects are still in place — composites, whether digital or employing glass paintings, still seek to create an impression of spatial contiguity by combining heterogeneous source elements — yet they now possess an unprecedented flexibility. The result is that filmmakers have gained an almost complete plastic control over the image. And because many of these digital effects are seamless, they are oftentimes completely invisible as long as they do not challenge the obvious constraints mentioned earlier according to which visual effects can turn, for the viewer, into special effects.

As far as film scholarship is concerned, the digital turn, and the exponential growth of visual effects that has accompanied it, has offered a new opportunity to revisit classical film theory as well as some foundational ideas about cinematic representation. Among the issues that have resurfaced is a concern for cinema's indexicality, although most will now readily concede that digital cameras do not rob images of their indexical value — since like their predecessors, they offer a record or trace of whatever stood in front of the lens. Visual and special effects, however, offer a different kind of challenge to indexicality — though not necessarily a new one (except for their sheer volume and the flexibility they entail, as mentioned above).

What I acknowledged at the outset of this paper as cinema's "zero degree" or baseline is probably best identified by the earliest films of the Lumière catalogue: here, the real world — the world of *existents* — is captured by the camera in a single take and there is no distance, no gap,[23] between it (as it exists independently) and the film world (as it appears on screen). With fiction, however, a wedge gets driven between the real world recorded by the camera and the film world we are shown. I called it a paradox earlier that cinema can show us fictional film worlds by using recordings of the camera-independent existing world masquerading as imaginary. The reason for calling it so is that two distinct universes of reference — the real world that lies in front of the camera and the fictional world it is meant to depict — share the image in such a way that they may be understood to overlap or be superimposed upon each other. For example, when Lauren Bacall and Humphrey Bogart embrace in *To Have*

23 Except, of course, for those that are medium-specific: framing, black and white, absence of sound, absence of "real" depth, etc.

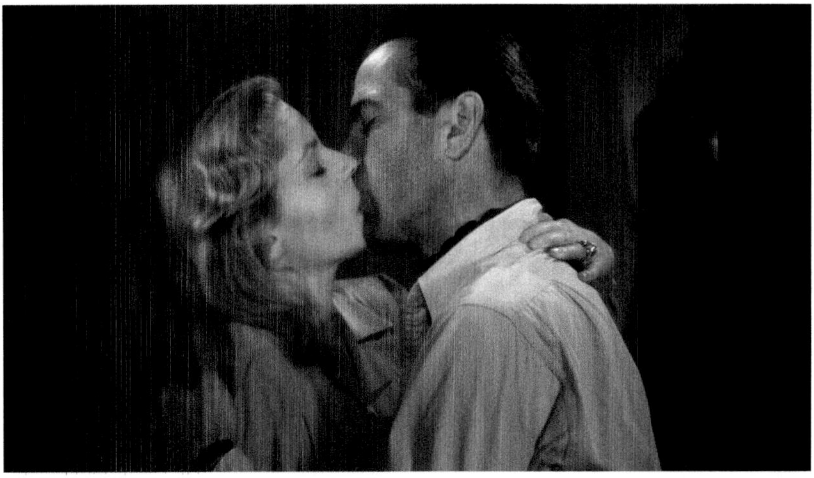

Fig. 1.2 *To Have and Have Not* (Howard Hawks, 1944).

and Have Not (Howard Hawks, 1944), these two worlds are intertwined as both actors *and* characters kiss on screen (Fig. 1.2).[24] To say that there are two distinct universes of reference here means that two statements such as "Bogart and Bacall are kissing" and "Harry Morgan and Marie Browning are kissing" can both strike us as true descriptions of the image, though only in as much as they direct our attention or point to two sets of "objects" that do not share the same worlds, and whose "existential-status" are different: the names "Bogart" and "Bacall" are used here as indexical signs of individuals who exist (or have existed) in the world in which we live, while those of "Harry Morgan" and "Marie Browning" index fictional individuals belonging to a fictional world.[25] Visual and special effects are usually understood as further dividing these two worlds of reference, to the point where it would seem to make little sense to speak of indexicality when the image we see — save perhaps for a few details — has been completely constructed, digitally painted, and composited in postproduction. In fact, the claim made that digital cinema is more like painting or drawing than photography seems to only find its full realization when visual effects are taken into consideration. So where does this leave us with regards to indexicality?

24 In itself, of course, this phenomenon is nothing new: theater also juxtaposes real and fictional worlds, only it does so without the mediation of a recording device.

25 Of course, there are many hybrid cases where a fictional world employs as its characters, events, or settings people, events and places that have (or have had) a corresponding real-world existence.

Tom Gunning has argued that indexicality — by which he means the "diminished way" this concept has been used in film studies — "may not be the best way, and certainly should not be the only way, to approach the issue of cinematic realism."[26] This is absolutely true, especially since realism can be found in various other means of representation — such as painting or literature — that do not rely on anything like a direct trace or imprint of what they depict the way that photography does. And yet, regardless of realism, both can represent using indices. The fact is that realism and indexicality should be understood as entirely separate issues. In light of this, it seems to me that what is sorely needed, what visual and special effects require of film studies, is that we come to clarify as much as possible the questions raised by these terms.

We need first to go back to the source of indexicality, to the work of American philosopher Charles S. Peirce and consider the basic role that he saw indexicality as serving in representation. To put it simply, the index is a concept developed by Peirce to answer the problem raised by the representation of particulars or individual *existents*, that is to say, objects such as they can be experienced or observed. A problem itself arising in the context of Peirce's development of a realist epistemology. An individual existent is something that manifests its being in its resistance or its reactivity, it possesses *haecceity*, a "thisness," that Peirce saw as expressing itself as a kind of dualism — in the sense that resistance requires a force against it, and vice versa. This, then, is the nature of experience itself. Peirce writes:

> Whenever we come to know a fact, it is by its resisting us. A man may walk down Wall Street debating within himself the existence of an external world; but if in his brown study he jostles up against somebody who angrily draws off and knocks him down, the sceptic is unlikely to carry his scepticism so far as to doubt whether anything beside the ego was concerned in that phenomenon. The resistance shows him that something independent of him is there. When anything strikes upon the senses, the mind's train of thought is always interrupted; for if it were not, nothing would distinguish the new observation from a fancy.[27]

26 Tom Gunning, "Moving Away from the Index: Cinema and the Impression of Reality," in *Differences*, (18)1, 2007, p. 21.

27 Charles S. Peirce, CP 1.431. In quoting from Peirce, I will use the standard reference format: CP, followed by volume number and paragraph number for the *Collected Papers of Charles Sanders Peirce*, vols. 1–6, Charles Hartshorne and Paul Weiss, (eds.) and vols. 7–8, Arthur W. Burks, ed., (Cambridge, MA: Harvard University Press, 1931–5 and 1958); EP2, followed by page number for *The Essential Peirce*, vol. 2, Peirce Edition Project, ed., (Bloomington: Indiana University Press,

Now, if the only way we can become aware of existents is when they strike us, this raises the question of their representation. How can a sign, or representation, come to stand for — and convey — the object of an experience or observation, an individual existent? The answer would seem simple enough: it must itself reproduce, or come to embody in its very functioning, the dualism of experience. That is to say, it must be dyadically affected by, or connected to, its experienceable object in such a way that awareness of the sign can then point in the direction of it or else lead its interpreter to it. "The index," writes Peirce, "represents its object by virtue of a character which it could not possess did the object not exist."[28] It is a sign that is determined by the *existence* of its object— and it is in this sense that photography is usually thought of as indexical. However, pointing fingers, the *index rerum* and *nominum* of a book, proper names, and pronouns are also indices.

In the above example from *To Have and Have Not*, two sets of proper names were used as indices. We noticed, however, that one set references fictional or imaginary beings. How can we claim these to be existents such that they can be referred to by indices? This is where the notion of a "universe of discourse" (a term Peirce borrowed from Augustus De Morgan — though he also came to use the term "sheet of assertion") becomes useful. A universe of discourse contains all the objects of a given domain of reference that are taken to be observable and on the basis of which assertions can be made concerning them by using indices. Peirce therefore recognized that fictional beings can be referenced by an index, even though, unlike what happens with non-fictional entities, it is impossible to experience or observe them in any other way than by imagining them through the fiction wherein they appear: "When the universe of discourse relates to a common experience, but this experience is of something imaginary, as when we discuss the world of Shakespeare's creation in the play of *Hamlet*, we find individual distinction existing so far as the work of imagination has carried it, while beyond that point there is vagueness and generality."[29]

Once this is established, it is important to see the gap that fiction can introduce in the overlapping of worlds mentioned above. For as soon as one says: "Harry Morgan and Marie Browning are kissing in a hotel in Port de France" or "Harry Morgan takes Johnson on fishing trip on the high seas," these statements cease to be true of the visual contents of Hawk's film when

1998); and Ms. followed by manuscript number and page number for Peirce's manuscripts held at the Houghton Library of Harvard University.

28 Charles S. Peirce, Ms. 142: 3

29 Charles S. Peirce, CP 4.172.

we replace the name "Harry Morgan" with the name "Humphrey Bogart" (the hotel is a movie set in Hollywood[30] and rear projections are used for the fishing scenes whenever we see the actors). This should not be a surprise since, after all, the film's narrative is *about* Harry Morgan, not Humphrey Bogart. We could try to counter the objection by saying instead: "Humphrey Bogart and Lauren Bacall are kissing while pretending they are in a hotel in Port de France," but even though this may well be the case, the image itself gives no evidence of this: it merely shows us the couple embracing in a room whose location is *fictionally indexed* (in good measure through dialogue) as being in a hotel in the capital of Martinique.

This is not to say, however, that the film's ability to reference individual existents from our world — as opposed to fictional ones — should be neglected. When we appreciate Bogart's acting it is clearly because the film — not the narrative — records it and thus makes it possible for us to study it by our electing to see *this* (i.e., individual, particular) performance as one of the objects represented indexically on the screen (though iconicity is also involved through which we apprehend the qualities that this particular performance embodies), and whose independent concrete existence we see as affecting the image. It might be argued that as long as the Academy Awards will continue to present actors with Oscars for their performances, indexicality will continue to be a stake in the film experience![31] But the fact remains that the possibility of semiotically using film images for any of the purposes an index may serve by being existentially affected by its real-world object (in this instance: what is recorded by the camera), is actually part of the experience of watching live-action films (including narrative fiction); and that it has been so ever since early cinema whether or not we engage in it at every moment of watching a film and in full awareness of doing so (for we also engage with the film as an iconic and a symbolic sign, thus serving a host of other semiotic purposes). Indeed, this happens every time the independent existence of the real world captured by the film draws our attention and makes us turn toward it or attend to it: when in

30 The real-world location of this room is nowhere indicated in the film, of course. However, anyone investigating the whereabouts of Humphrey Bogart at the time of shooting his scenes would discover that he was at Warner Brothers Studios' Stage 28 in Burbank.

31 This is merely a way of emphasizing that without photographic indexicality, it is difficult to imagine how cinema could have developed the star system that has been so vital for its success. Here too, however, things are changing as actor's performances may be enhanced and transformed through digital alterations with the body/face split of facial replacement or the altering of facial features or movements — for instance, in *Dark Shadows* (Tim Burton, 2012), Johnny Depp's blinking eyelids were eliminated from his face.

a crime movie shot on location I suddenly recognize a street I lived on, or in a melodrama the Paris hotel I once stayed in, or when I search for one of Hitchcock's cameo appearances, or else when I read the end credits to see what locations were used because I wish to visit the landscapes which I found so arresting while watching the film.[32]

It is true, of course, that the current omnipresence of visual effects is having a toll on the viewer's ability to do this *accurately* and *confidently*. We may find a landscape especially striking, only to discover it was not really photographed and that none of it actually exists in our world. For although most films continue to shoot material, it is becoming increasingly difficult — impossible actually — for viewers to differentiate what has been recorded by a camera from what is shown without having been filmed, and to distinguish what has been composited together into a shot to create the impression of a unified contiguous space from several heterogeneous elements (both photographic and non-photographic), or what has been digitally manipulated (how many of us saw the facial replacement of Armie Hammer on someone else's body at a first screening of *The Social Network* [David Fincher, 2010]?) — at least, as long as what is seen doesn't appear outright as a special effect for the viewer (as is the case with Brad Pitt's facial replacement in *The Curious Case of Benjamin Button*). In short, viewers may know that existents from our world are still there to be seen in films, but they are not quite sure where exactly anymore — since not even the actor's body is immune to compositing. Does this mean we have no cause to discuss indexicality anymore? In answering, let us begin by taking on the issue from a slightly different perspective and consider this time the filmmaker's task.

32 Of course, these are not the only indices I can use in my experiencing a film. There is an indefinite number of them, which may point to their objects in as many directions. For example, looking at the image, I can see it as an index that a wide angle or a telephoto lens is being used, or that the cameraperson is operating a Steadicam. Certain stylistic features may also point to the filmmakers who use them, much like a signature points to its owner. However, such indices are related to the apparatus and its use, not to the photographic nature of the image as imprint of the profilmic world that lies in front of the camera. Choosing one index over the other is relative to one's purpose as an interpreter of signs. Note that in all those cases where the index points to what is seen as the occurrence of a type or a habit (as may be the case with a filmmaker's style), what is really at stake is an instantiation, *token*, or *replica* of what Peirce calls a symbol, i.e., a general type related to its object — this object being itself of the nature of a type — on the grounds of a rule, a convention, or a habit (i.e., a general principle capable of an indefinite number of occurrences — which is what an artistic style is). In such a case, the sign and the object as well as the interpretation of their relation are all tokens of a type — individual instantiations of something abstract and general — as is the singular, existential *relation* that obtains between the sign and its object.

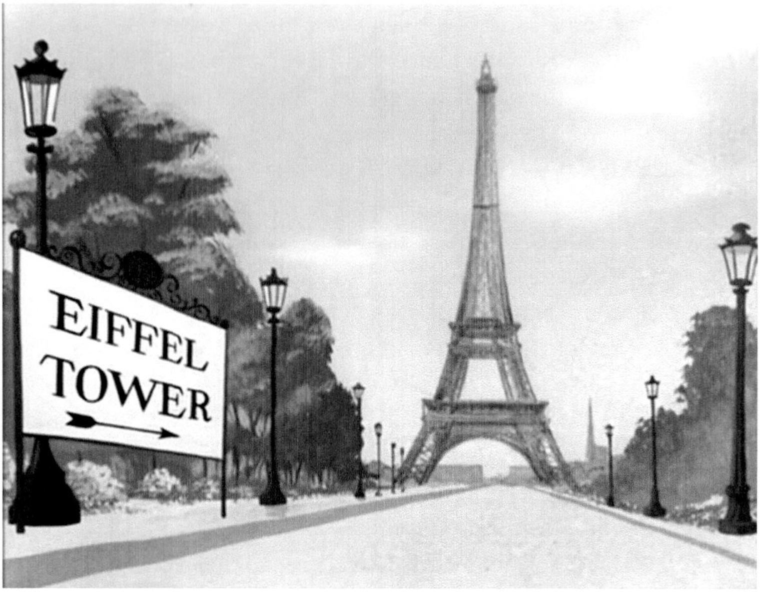

Fig. 1.3 *French Rarebit* (Robert McKimson, 1950).

Indices, as I mentioned earlier, are used to represent individual existents. They are the only type of sign that can do this, since pure icons merely stand for self-identical qualities, and symbols stand for types (on the ground of laws or habits), neither of which can be said to *exist* and therefore directly act upon and determine the sign in representing them except as being utterly indeterminate, that is to say, vague (e.g., some such possible object) or general (e.g., a type of object). The implication, therefore, is that without indices, filmmakers cannot under any consideration represent individual existents. Now this might not be entirely a hindrance for fiction filmmaking should we wish to rule out indices standing for individuals that belong to our real-world universe of discourse. For the sake of intelligibility, we could even let the characters in such a fiction continue to use indices in exchanging amongst themselves (e.g., "come here," "where is Peter?," etc.), without it affecting our "no index" rule. The simplest way to achieve this might be to imagine forgoing recording anything with a camera and therefore to make an animated film that avoids visually representing any individual existent. A good example of this, among others, might be Chuck Jones's 1942 Bugs Bunny short *Case of the Missing Hare*, where the friendly cottontail battles it out with a pompous magician in some non-descript, vague space made up of some exterior decor with a hollow tree (where Bugs resides) and, later, the stage of a theater (where the magician performs and suffers

Bugs's wrath). Now compare the film to another Bugs Bunny cartoon, *French Rarebit* (Robert McKimson, 1950), where our *lapin* friend finds himself in Paris as two chefs seek to turn him into a rabbit stew — *sacrebleu!* — to their great detriment, of course! Here, however, a series of landmarks are presented verbally and visually, including an unmistakable view of the Eiffel Tower (Fig. 1.3). Obviously, this is a drawn likeness of the famous Paris landmark — what Peirce would call an iconic sign. But it is more than that as well: it is also an index of it — namely, the representation of an individual existent. The fact that it is drawn rather than photographed simply implies that it is not a *photographic index*, but to be sure it is an indexical sign just the same and one that refers to our universe of real-life existents, even though it is integrated into a fiction. And so too, it follows, is the digitally created Times Square circa 1922 of Baz Lurhmann's *The Great Gatsby* (2013). To see through this more clearly, we now need to return to Peirce and properly distinguish icons and indices.

6. The Gap Between Iconic and Indexical Signs

A pure icon is a *self-representing sign* in that its connection to its object is one of complete qualitative sameness or self-identity: "A pure icon," writes Peirce, "does not draw any distinction between itself and its object. It represents whatever it may represent, and whatever it is like, it in so far is. It is an affair of suchness only."[33] And elsewhere he writes: "No pure Icons represent anything but Forms; no pure Forms are represented by anything but Icons."[34] By "form," what is meant here is a quality quite regardless of its *hic et nunc* manifestation or embodiment in some existent. In other words, it is a mere *possibility* and, as such, cannot give any indication at all regarding the existence of its object. The upshot is that a pure icon *is* its object, for the latter is the quality that it itself is as a *qualitative essence* or *continuum* (differentiating them as "sign" and "object" fulfills merely logical and functional requirements, not real-world existential/numerical distinctions). The pure icon has no means of signifying its object other than its qualitative self-identity with it. As indicated earlier, only indices can denote particulars — objects that are existentially and thus numerically distinct from the signs that stand for them — by virtue of an existential tie to them, whereas icons have no such purchase on particulars. A pure

33 Charles S. Peirce, EP2: 163.

34 Charles S. Peirce, CP 4.544.

icon, then, is a very abstract, almost ethereal kind of representation: it has no concrete existence proper and no "being" other than that of being a mere (self-signifying) possibility, or, to put it differently, it is a possible (as opposed to some actual, embodied) sign of its own self as a possibility. The pure icon is still a sign, but it is so in *potentia*, not in *actu,* for only embodied signs *act* concretely as signs. But what of actual images like paintings, drawings, even photographs and films which we usually think of as iconic because they represent their object on the grounds of some likeness they share with them? Surely these are *existents* in their own right and as such can be materially distinguished from what they represent. The answer lies in the distinction Peirce makes between pure icons and *iconic* signs, that is to say, embodied signs that can iconically (i.e., qualitatively, on the ground of a likeness) represent something numerically other than themselves as a possible (as opposed to an existing) object.

When Peirce himself, or commentators of his work, seek to explain iconicity by offering concrete examples of icons — a figurative painting, a diagram, a map, a sample paint chip, i.e., *actual* artefacts that signify on the grounds of a likeness — what is referred to are not pure icons but iconic signs or *hypoicons*. Whereas the pure icon is a mere possibility standing for itself as a possibility, a figurative painting is an actually existing thing that while also standing for itself (for its own *sui generis* qualities as a painting) may as well stand for *other* formally identical, though numerically distinct potential manifestations of the form it embodies (it is worth repeating that iconicity, on its own, cannot offer any assurance as to the *existence* of its object, whereby the latter is always represented as a mere qualitative possibility, or a form). In Peirce's phenomenological language, this means that a hypoicon injects an element of Secondness (viz., the duality or otherness of concrete existence) into the semiotic process of iconicity. What this implies is that hypoicons can both self-represent and alio-represent (that is, represent something other than themselves).

Let's make this more tangible. Imagine a painting of a horse. On one hand, the work may be used to signify the qualities it embodies in its depiction of the animal, though quite regardless of the embodiment itself on canvas. This means that the painting can self-signify, or auto-represent itself, not as an existent but as a *type* of thing of which it is an embodiment. This could include the *form* exhibited by the painted horse, but it can also comprise the style of the work. For instance, the art teacher might say to her students: "Just look at the shape of this horse!" to (indexically) bring to their attention and make them appreciate the quality of *design-type* being embodied on the canvas (for this shape is something general which could also be reprised any number of times). She could also say: “This is so *typically* Remington,” in which case she is again using the painting as qualitatively self-signifying a type, namely the painter’s style. In both cases, the painting is used to self-signify, namely to stand for qualities it possesses and embodies (but doing so regardless of this particular embodiment). On the other hand, it is also possible to look at the painting as iconically signifying something *numerically other than itself*, an object with which it nonetheless shares a quality. This is part of what distinguishes the hypoicon from the pure icon. From the perspective of the quality that they share, this other (possible) object — a possible horse — is also a (possible, not an actual) replica or token of a type that the painting embodies, although the painting’s concrete embodiment of it (through paint, brush strokes, lines, etc.) is not shared by the object (otherwise this possible object would not be numerically distinct from the painting itself). This is a complicated way of saying that now the painting can be seen as iconically representing a (possible) horse, even though an actual painting and a possible horse are two different sorts of things. In this instance, the hypoicon is *alio*-representing, since it stands for something other than itself.

If the painting does not represent an actual horse, it is because nothing about it points toward an existent. It is not an index, and the horse it depicts could be either an imaginary one or an actual one: the painting offers no information to this effect. Now, the visual depiction of a merely possible real horse is also the depiction of a possible horse-type. That is to say, it may be a black horse or a grey horse, a big or a small horse, a stallion or a mare, or else an Arabian, an Appaloosa, or a halter-type Quarter horse, or else it could be one of, say, Tom’s horses or one of the neighbor’s horses, etc. This means that without an index, the object of our painting is a *vague* generality (what Peirce calls a *negative* generality), representing *some* possible horse that belongs to some possible horse-type. Yet, the possible horse of the painting is less vague than that of the following sentence: “Tomorrow I will buy myself a horse” (what will it look like? Will it be male or female? Big or small?), and less positively general than that of the following proposition: “The horse is a noble animal,” by which is meant the entire species and therefore all the individual occurrences of *Equus ferus caballus*.[35] The reason for this is that the painted horse may have color, spots, a given shape, and

35 In CP 1.427, Peirce distinguishes negative and positive generality in the following terms: “Generality is either of that negative sort which belongs to the merely potential, as such, and this is peculiar to the category of quality; or it is of that positive kind which belongs to conditional necessity, and this is peculiar to the category of law.”

so on. In other words, it possesses certain qualities that both curtail the realm of possibilities of various aspects of the horse-type (even though any number of such horses may be embodied, e.g., in other paintings, in reproductions and copies, in real life, etc.) and enable one to refer to it indexically (even though it is not a real horse that one can mount, it can obviously be pointed to: see how, for instance, I can refer to the "saddled brown horse" of Frederick Remington's *Cow Pony*). We might add that our ability to recognize and identify the *alio*-represented object of a hypoicon in a figurative medium — "this is an X-type" — is relative to other embodiments of its object-type, either as mere possibilities (this is what happens with the depictions of unicorns or dragons) or as real-worldly existents (as in the case of our painted horse, where the possibility depicted by the painting is congruent with existents from our world, i.e., with real existing horses that concretely embody the horse-type), all of which form a continuum of qualitative possibilities that may be represented through positive generality once named or conceptualized.

In the end, therefore, what distinguishes the pure icon from the hypoicon is the *mode of being* of the sign. As Peirce writes:

> A possibility alone is an Icon purely by virtue of its quality; and its object can only be a Firstness. But a sign may be iconic, that is, may represent its object mainly by its similarity, no matter what its mode of being. If a substantive be wanted, an iconic representamen [a term Peirce often uses as a synonym for "sign"] may be termed a *hypoicon*. Any material image, as a painting, is largely conventional in its mode of representation; but in itself, without legend or label it may be called a *hypoicon*.[36]

Thus, where a pure icon is a form that is self-representing, an iconic sign is a concrete, embodied — and therefore *impure* — form that may be used as a self-representation (regardless of its own concrete embodiment) or else to *alio*-represent a possible object with which it shares this same *form*. It follows that the *alio*-representing hypoicon is *formally*, though not materially, *identical* with its object. In the following passage, Peirce describes the experience of iconicity:

> I call a sign which stands for something merely because it resembles it, an *icon*. Icons are so completely substituted for their objects as hardly to be distinguished from them. Such are the diagrams of geometry. A diagram,

36 Charles S. Peirce, EP 2: 273.

> indeed, so far as it has a general signification, is not a pure icon; but in the middle part of our reasonings we forget that abstractness in great measure, and the diagram *is for us the very thing.* So in contemplating a painting, *there is a moment when we lose the consciousness that it is not the thing, the distinction of the real and the copy disappears*, and it is for the moment a pure dream — not any particular existence, and yet not general [by which he means not *positively* general]. At that moment we are contemplating an *icon*.[37]

Taking in the painting, the interpreting mind sees (*feels* might be a better term to use) a quality, which is to say that the image shows merely the possibility of "something," i.e., some possible and unknown "X" of such or such a description: "X is like this." If this "X" is identified as a possible (as opposed to an actual) embodiment of the horse-type because the qualitative "like this" is identified as qualities otherwise known to belong to horses, then this interpreting mind says to itself something like (while considering the painting): "This would be a horse if it were real existent.[38]" However, if a caption is added that reads: "Trigger, the famous movie horse," then the situation is transformed, and the composite of painting *and* caption can now be used as an index. What is represented is no longer a possible manifestation of the horse-type but an actual existing embodiment of it: Roy Rogers' legendary horse. The sign is now such that it can point its user toward the existing object of experience that has affected or determined it, so that she *can* also experience it (at least theoretically — or through other indices of it, if it is an object from the past). This is how we should understand Peirce's claim that a proposition "represents an image with a label or pointer on it."[39]

Let us now complete our understanding of indices before returning to special effects. The index points in the direction of its object by being really affected by it, yet it doesn't actually substitute for the object. In the previous example of painting and caption, the move from hypoicon to index must be explained. To a child who does not know of the horse called "Trigger," the painting along with its caption merely says: "something called Trigger is like this," without pointing to the universe of real-world observable existents (indeed, the same could be done with the image and caption of a

37 Charles S. Peirce, CP 3.362; emphasis added.

38 By which we mean a real existing or *living* horse as opposed to a real existing *depiction* of a horse.

39 Charles S. Peirce, CP 5.543.

unicorn or a dragon). What is required for this sign to function indexically, therefore, is a certain amount of collateral knowledge concerning its object: one must independently know of Trigger. The surest way to achieve this collateral knowledge is through observation or experience, though in some cases — especially when this is impossible — other signs determined by the object, other indices, may help approximate this observation. In a manuscript from 1907 Peirce writes:

> [A] sign is something which functions triadically. [...] Every sign, in functioning as such, produces a mental effect. How shall we name the entire mental effect which a sign by itself is calculated, in its proper significative function, to produce? The word *signification* is somewhat too narrow, since, as examples will soon show, this mental effect may [also] be of the nature of an emotion or of that an effort. Permit me to call this total proper effect of the sign taken by itself the *interpretant* of the sign. But merely producing a mental effect is not sufficient to constitute the object of a sign; for a thunderclap or an avalanche may do that without conveying any meaning at all. In order that a thing may be a true sign its proper significate mental effect must be *conveyed* from another object which the sign is concerned in indicating and which is by this conveyance the ultimate cause of the mental effect. In order to be the cause of an effect, or efficient cause, as the old phrase was, it must either be an existent thing or an actual event. Now such things are only known by observation. It cannot be itself any part of the mental effect, and therefore can only be known by collateral observation of the context or circumstances of utterance, or putting forth, of the sign. But the sign may describe the kind of observation that is appropriate and even indicate how the right object is to be recognized. But although the full realization of the meaning requires the actual observation, direct or indirect, of the object, yet a close approach to this may be made by imagining the observation. If the sign is not a *true*, but only a *fictitious* sign, it is the mere semblance of a sign. If, however, it be so true as to profess to be in certain respects fictitious, the conditions of a true sign hold in slackened force.[40]

The main idea of this passage is that a sign is a *mediation* between an *object* and some consequent effect of that object being the case, what Peirce calls the *interpretant*. This is what distinguishes the avalanche in itself from a written account or image of it bringing to our mind the idea and horror of the

40 Charles S. Peirce, Ms. 318, pp. 39–43.

avalanche. (Of course, the avalanche itself may also be a sign of something, say, global warming, in which case it mediates between an object and an interpretant of its own — although this is a different matter.) But which avalanche? An actual existing avalanche can only be known by experiencing or observing it *hic et nunc*. However, a sign may describe or indicate the kind of collateral observation to be made in order to realize its meaning[41] or else it can *point* toward the individual existent represented so that it might be observed. As we know, the only way a sign can point in the direction of a particular being or event is through an index. In canonical cases, an index works by referencing a given object set within a universe of discourse which corresponds to the aggregate of individuals that exist (or are experienceable) for the users of the sign. In the above quote, Peirce is explaining that although no index can, by itself, substitute for the role of collateral observation in making known an independently existing individual object, indices may be able to approximate this by combining with an icon (or several of them) so as to help us *imagine* such observation. This is especially important in cases where collateral observation has become impossible. (For instance, if the object lies in the past or is not otherwise available for observation: how else can we *know* of Napoleon today barring opening up his tomb? Moreover, any such observation would be very limited in terms of the knowledge it would yield of the man!). Of course, what we thusly imagine — and think to be true (unlike fiction, which does not share this pretense) — may turn out to be false or erroneous should no such object actually exist (or have ever existed), in which case says Peirce, we have the mere *semblance* of an

41 In a different section of the same manuscript (Ms. 318 contains substantially different drafts of an unpublished article), Peirce writes:

> It should be mentioned that though a sign cannot express its Object, it may describe, or otherwise indicate, the kind of collateral observation by which that Object is to be found. Thus, a proposition whose subject is distributively universal (not plural or otherwise collectively universal), such as 'Any man will die,' allows the interpreter, after collateral observation has disclosed what single universe is meant, to take any individual of that universe as the Object of the proposition, giving, in the above example, the equivalent 'If you take any individual you please of the universe of existent things, and if that individual is a man, it will die.' If the proposition had been, 'Some Old Testament character was translated,' the indication would have been that the individual must be suitably selected; while the interpreter would have been left to his own devices to identify the individual. (pp. 37–38)

Notice that in such cases, Peirce is discussing signs (symbols, actually) whose objects are initially represented as either general or vague and require additional indexical determination to fulfill their meaning. This section of the manuscript was published, with an introduction by Helmut Pape, under the title "Charles S. Peirce on Objects of Thought and Representation," in *Noûs*, (24)3, 1990, pp. 375–395.

indexical sign (think of circumstantial evidence "pointing" the detective in the wrong direction and leading him to falsely accuse someone; or else fake "clues" left by a cunning thief at the scene of a crime which, when "properly" interpreted by the detective, will lead to the wrongful arrest of an innocent person). As for fiction, as we saw earlier, Peirce recognized that fictional or imaginary beings may be indexed. Thus, a painting of a man with the caption: "the famous Sherlock Holmes, as he appears in *A Study in Scarlet* by Arthur Conan Doyle" would be a true indexical sign — one professing its fictionality — though it would be so only in a "slackened way" because it is impossible to experience its object in any other way than to imagine it in relation to Doyle's fictional account. In fact, we might think of fiction as offering a situation not entirely unlike what is the case with the historical past, with the important caveat that with history there is always the possibility of unearthing new evidence to help us approximate collateral observation made through imagination.

There is a final distinction we need to make about indices. It concerns the fact that not every index points to its object in quite the same way or, better yet, relies on quite the same kind of collateral knowledge in so doing. On these grounds, Peirce saw the importance of distinguishing between two classes of indices which he called respectively *reagents* and *designations*. He writes:

> An index represents an object by virtue of its connection with it. It makes no difference whether the connection is natural, or artificial, or merely mental. There is, however, an important distinction between two classes of indices. Namely, some merely stand for things or individual quasi-things with which the interpreting mind is already acquainted, while others may be used to ascertain facts. Of the former class, which may be termed designations, personal, demonstrative, and relative pronouns, proper names, the letters attached to a geometrical figure, and the ordinary letters of algebra are examples. They act to force the attention to the thing intended. Designations are absolutely indispensable both to communication and to thought. No assertion has any meaning unless there is some designation to show whether the universe of reality or what universe of fiction is referred to. The other class of indices may be called reagents. Thus water placed in a vessel with a shaving of camphor thrown upon it will show whether the vessel is clean or not. If I say that I live two and a half miles from Milford, I mean that a rigid bar that would just reach from one line to another upon a certain bar in Westminster, might be successively laid down on the road from my house to Milford, 13200 times, and so laid

> down on my reader's road would give him a knowledge of the distance between my house and Milford. Thus, the expression "two miles and a half" is, not exactly a reagent, but a description of a reagent. A scream for help is not only intended to force upon the mind the knowledge that help is wanted, but also to force the will to accord it. It is, therefore, a reagent used rhetorically. Just as a designation can denote nothing unless the interpreting mind is already acquainted with the thing it denotes, so a reagent can indicate nothing unless the mind is already acquainted with its connection with the phenomenon it indicates.[42]

It follows, then, that whenever true, undoctored photographs are used indexically to stand for what they depict, they are clearly reagents. Their semiotic potential in this regard lies in the fact that we know how they are produced or, in Peirce's words, that we are "acquainted with [their] connection with the phenomenon [they] indicate." In short, there is no need here for sign users to have prior knowledge of the object signified, since independent knowledge of how the sign reacts to it is sufficient for it to function indexically. Equally important for us, however, is the fact that Peirce's comment suggests that representations such as drawings and paintings can also be truly indexical — *even without a caption* — as long as we are "already acquainted with the thing [denoted]," which is to say, on the basis of our knowledge that the object thusly drawn to our attention is a particular existent. And this brings us back to the Eiffel Tower of *French Rarebit* or, for that matter, Times Square or the Queensboro Bridge in *The Great Gatsby*.

In both films, these well-known landmarks are represented as specific individual existents, not as vague or general objects. The fact that neither are photographic — at least in the sense of a true, undoctored photograph — does not alter their indexical status. Furthermore, we need to understand that referring to an existent does not imply that the sign will necessarily offer an accurate representation of it in all of its particular qualities (no *alio*-representing hypoicon sign ever does). Indices only portray qualities when they are accompanied by iconic signs whose function it is to depict them by way of exemplification. In a case such as *French Rarebit*, a rough sketching of the Parisian landmark is sufficient for us to recognize it (in the film, to be sure, our recognition is sustained by other indices as well, including text, dialogue, the recognition of other landmarks, etc.). The same is true for drawn caricatures of well-known people or politicians: they are indexical and iconic signs whose iconicity does not pretend to

42 Charles S. Peirce, CP 8.368, n. 23.

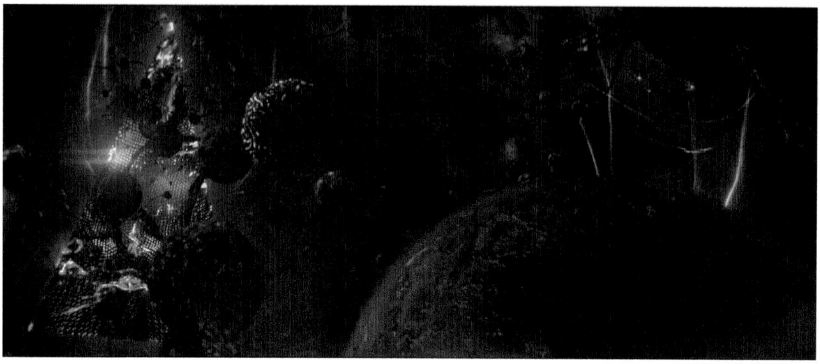

Fig. 1.4 *Dr. Strange!* (Scott Derrickson, 2016).

offer lifelike renderings of their object (as far as iconicity is concerned, the object is always but a mere possibility, not an existent). In other words, a sign may be a true index without the iconic sign accompanying it being a "complete" or "true" (in the sense of totally faithful) likeness of its object, which is why, as I mentioned at the start of this section, indexicality and "realism" should be treated separately. Of course, iconicity can play an important role in designations (especially in the case of designation by images), since activating our knowledge of the existence of the object may rely on our ability to recognize it through its likeness. Yet, even in such cases, the functions of iconicity and indexicality remain distinct. Consequently, liberties taken in drawing the Eiffel Tower or in depicting New York c. 1922 do not impede their function as indices as long as we can recognize them in their *haecceity*. (For instance, there are a couple of websites devoted to historical inaccuracies in the depiction of New York and its buildings in *The Great Gatsby* — these efforts merely prove that the film is indexical in this respect. It would make no sense to do this with the realm of Dormammu in *Dr. Strange* [Scott Derrickson, 2016]!) (Fig. 1.4).

Photographs are indices that involve, for the most part, fairly accurate likenesses of their real-world objects. As indices, however, they may also, in certain circumstances, combine the two classes of index outlined above. A photograph of a famous landmark will stand for its object both on the basis of our independent acquaintance with this existent *and* on the grounds of our knowledge of the workings of photography. In most situations, however, this doubling of the semiotic function goes easily unnoticed, merely offering redundancy in the way we use the photograph to represent. It might only become obvious when what we otherwise know to be a photograph of something or someone we are acquainted with is so dark or else so out of focus that we cannot make out the object, in which case our ability to use it as a designation based on our recognition of its likeness is jeopardized; or else, when we learn that what we thought was a photographic record is only a *semblance* of a reagent while still being a designation, a situation that certain imperceptible special effects puts us in.

7. Back to Special Effects

To make all of this more concrete, let us consider traditional matte paintings. If you recall the earlier discussion (in section 3), a distinction was made between imperceptible and invisible mattes on the basis of their congruence with existents from our world. As hypoicons, the mattes of the White House in *Dave* (Fig. 1.8), of Hindley Hall Mansion in *The Paradine Case* (Figs 1.5 and 1.6), and of the Death Star in *Star Wars* (Fig. 1.7), represent their objects along a spectrum of possibilities whereby only the Death Star stands as the pictorial embodiment (all hypoicons are embodied forms) of a possible type of object not congruent with existents of our real-world universe of discourse (i.e., not otherwise embodied in this universe — and therefore not *actively* a type determining its occurrences but only a possible type to which the Death Star, as a possible object, would belong if it were real). As a result, matte paintings of the Death Star — though they show an object that can be indexically referenced in the fiction — are not *themselves* indices of an actual existing space station, *nor can they pass for one (at least for those of us who know that the Death Star is not part of the furniture of our real-world universe of discourse) even though they may look "real" to the film's viewers thanks to the photo-realism of the paintings and the way the cinematographers smoothly integrate them into the movie.* The situation is different for the other two sets of mattes. Hindley Hall's interiors pictorially embody a type that is congruent with existents from our universe of discourse: they manifest the same architectural type as do existing eighteenth or nineteenth-century British country mansions. Thus, if the effect goes undetected,[43] the matte paintings are mistaken for photographic indices pointing toward an existent

43 In this case, discerning eyes might notice the eerie atmospheric effect — soft focus and grain characteristic of the era's matte shots using an optical printer — in the long shots of the manor's rooms, which used matte paintings, distinguishing them from the closer, sharp focus shots made on sets with props. Some inconsistent floor shadows can also be noticed as the lawyer Keane (Gregory Peck) is about to exit Mrs. Paradine's room. Given the diegetic source of light, his shadow should extend forward in the foreground where other shadows appear on the painted matte — not to mention some continuity issues: a piano seen on set in a close shot seems to be absent from the matte painting used for long shots of Mrs. Paradine's room.

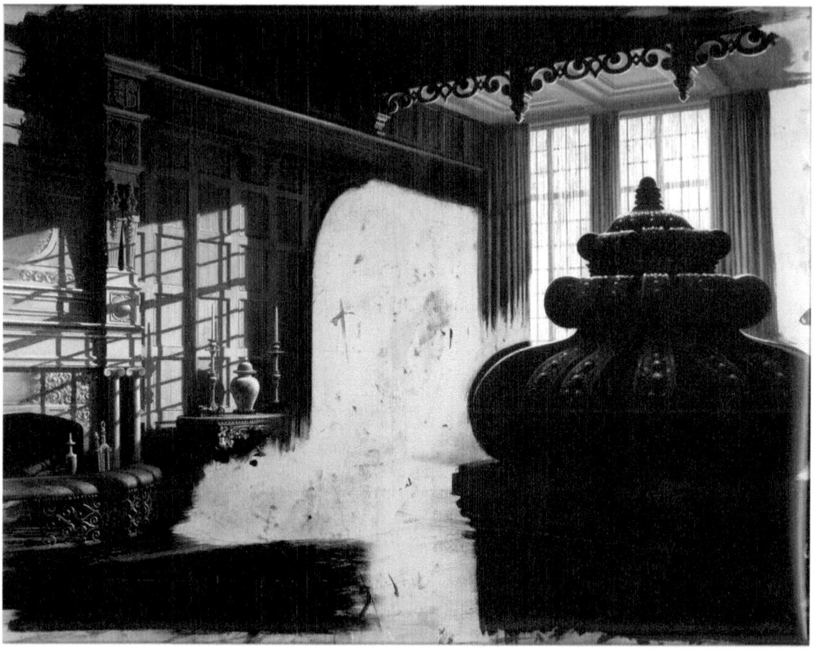

Fig. 1.5 Matte painting by Spencer Bagdatopoulos for *The Paradine Case* (Alfred Hitchcock, 1947).

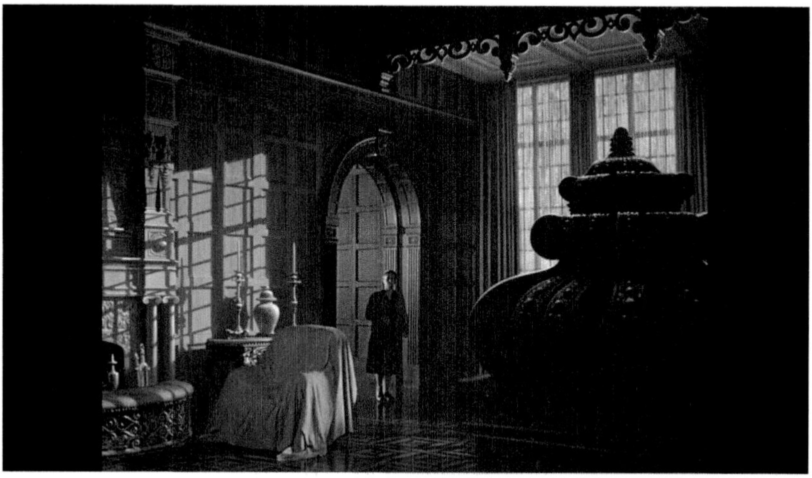

Fig. 1.6 Composite with matte painting in *The Paradine Case* (Alfred Hitchcock, 1947).

manor which the unsuspecting viewer believes was experienced by the camera and is (or was once) available for collateral observation — which is the *raison d'être* of the effect, after all (i.e., to be *believable in the way undoctored photographs are*). For the viewer, then, the imperceptible effect is, to use Peirce's term, the *semblance* of an index (a reagent), which is precisely

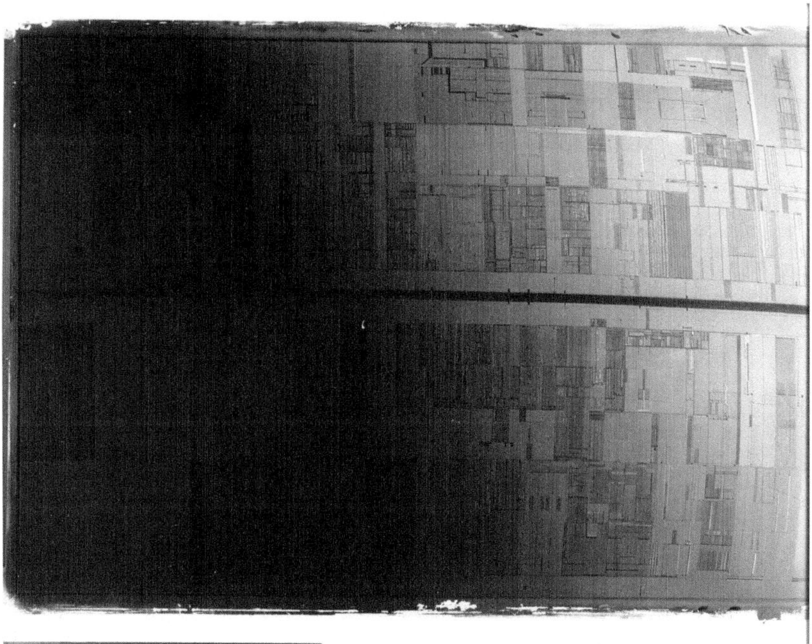

Fig. 1.7 Matte painting by Ralph McQuarrie for *Star Wars* (George Lucas, 1977).

where the deceit lies. This is why I claimed earlier that the deception created by the matte effects in the case of Hindley Hall is of a different nature than that of the Death Star. For although both work hard at deceiving perception, the difference lies in their "embodiment" of an object-type, i.e., whether it is a possible object-type — although the viewers might well think it is an *actual* photographic embodiment — about which we know that replicas exist (some such things as Hindley Hall do exist) or merely a possible object-type which, viewers know, is without any existing replicas (no such things as the Death Star exist, except in the fictional world of *Star Wars*) — a distinction that belongs to what I called earlier an ontological gap in representation.

The situation of *Dave* is different, however. Here, most of the views of the White House use matte paintings as set extensions. In what follows, I will refer only to the view from the North Portico as the First Couple return from an evening "out on the town." It is a brief shot where the entire building is shown. Technically speaking, what is going on here is fairly simple: a glass painting is used to stand in for a "normal" shot taken by a camera. In semiotic terms, the filmmakers are using the *semblance of a reagent* to imperceptibly imitate a photographic view of the White House. However, because the matte painting shows a likeness of a well-known landmark that we recognize, it offers viewers a real index, namely a *designation*. If we look

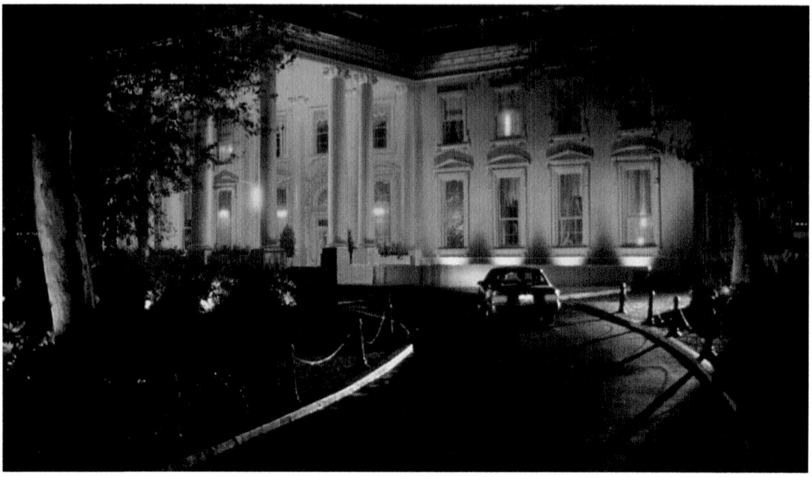

Fig. 1.8 Composite with matte painting of the White House in *Dave* (Ivan Reitman, 1993).

at this more closely, we will see that it is in the relation between these two classes of indices that the deception of the special effect lies.

We know that what viewers are *actually* looking at, although most are not aware of this, is a photographic image of a painting of the White House (the shot that we see in the film is a composite, but let us leave this out until later). Now, a condition for the success of the effect is that the photographic depiction of the painting be such that its object — i.e., the White House as it is made immediately available by way of a pictorial representation of it — comes to substitute itself for the actual object of the photographic image that appears on the film screen. In other words, viewers must not be made aware that they are looking at the *photographic depiction of a painting*. Rather, they must see through these layers to "access" the White House. Put simply, the effect requires of the photographic image that it act as nothing more than a "dumb," transparent window, appropriating as its own the object of the painting, even though in so doing the viewer takes the shot as depicting the real White House instead of depicting a depiction of it by way of a painting. What viewers are looking at and what they *see* or believe they are looking at are therefore two distinct matters. Because the glass matte is quite imperceptible and photo-realistic, viewers who do not know that using the grounds of the White House to shoot a narrative film is forbidden (except perhaps for its occupants!) may thus be inclined to believe that they are *really* seeing the building — that is, really seeing a photographic image of it — rather than a painted image of it used as *trompe l'œil*.[44]

44 Of course, someone might also think the building is a set, a facade, built on a studio lot — which is how it might have been done prior to the use of glass shots or Schüfftan process shots.

We also know that a painted likeness of the White House can be used as an index of it for those who recognize the building. It turns out that matte artist Paul Lasaine worked from photographs of the White House to get the likeness of the building right, with those photos standing for the real building as both index and hypoicon of it. Yet however much photographs may be affected by — and resemble — what they depict, they also possess their own *sui generis* qualities, namely through lens distortion, stock grain, or in how they react to light, etc., and a good matte artist will take this into consideration. The matte painter's task, in other words, is to replicate how an object looks when filmed, not so much how it looks to the naked eye. The north entrance of the White House in *Dave* is shown with some slight lens tilt distortion and light glare that reproduce how the building might look when photographed at night. Whereas such qualities of the image can normally be used as indices of the functioning of a camera (and not of the existent object depicted by the image), because this is a painting, they are in this case as much a part of what is being depicted as the White House itself. For Lasaine and the filmmakers, the pictorial rendering of glare and lens distortion are used as hypoicons (they embody a type of "apparatus effect" through likeness), not indices — and certainly not as indices of the photographic nature of the image.[45] For the unsuspecting viewer fooled by the *trompe l'œil*, however, the lens effect and glare are merely a *semblance* of an indexical sign of the apparatus — and thus part of the overall deception. As for the image of the White House, while it is clearly indexical, it is only so — though unbeknownst to the viewer — on the basis of object recognition and thus as a *designation*. It is therefore unrelated to photography's connection to visible existents making up our real universe of discourse: *the index, in other words, is not where the viewer thinks it is*.

What is interesting to note here is how, regardless of whether or not Lasaine copied an actual photograph, the indexical status of the painting literally crosses over to the film for the unsuspecting viewers who recognize the building. For it carries over not because the film image is a photographic index of Lasaine's painting of the White House (perhaps this is what Lasaine sees when watching the film!) *but because it is an iconic sign of it*, which *ipso facto* makes it an iconic sign of the White House *as well as an index of it*. We

However, I surmise most people who know this much about how films are made would quickly dismiss the idea on the basis of cost in relation to how briefly the North Portico appears in the film.

45 Of course, indexicality could also be involved: should these "photographic" effects be determined by an actual photograph embodying those same qualities, anyone aware of this relation (such as the production team) could theoretically use the matte painting as an indexical sign pointing to this photograph.

can put it this way: as a photographic record of Lasaine's painting, the film image is framed in such a way as to capture and reproduce (one is tempted to say: appropriate) some of its visual qualities, including the likeness of the White House which, for viewers with minimal worldly knowledge, is immediately recognized. On the basis of this recognition, the image can be used as an index of its object, thus further determining what is otherwise (erroneously) believed to be a photographic index of the famous landmark.

The upshot is that the view of the North Portico is a designation; but the deceit is that, for viewers convinced by the matte painting into believing they are seeing a photographic view of the building, the reagent that superadds itself to the designation is merely a *semblance*. And because the trickery concerns the way the object appears — how it is represented — and not its actual existence, the only way to discover the deception is to gain knowledge about the production of the sign (i.e., find out how the shot was made), since no actual collateral observation of the object (the White House) would reveal it to be a semblance of a reagent (i.e., a "faked" photographic image).

Compare what is going on in *Dave* to the discussion regarding Bogart and Bacall at the outset of section 5. Again, two universes of reference are superimposed: the White House as the official residence and workplace of the real-life President of the United States and the White House as the residence and workplace of the film's fictional President.[46] Both buildings, the existent and the fictional one, can be referred to indexically with true propositions as long as their universes of discourse are carefully distinguished. In a sense, perhaps, this would be a bit more like Bogart "playing" himself in a fictional film. But there is another, deeper, difference, for those two universes are not superimposed in exactly the same fashion. In *To Have and Have Not*, there is a way of describing all that which we see the characters doing as being *necessarily* true of both the characters and the actors: for instance, although actor Walter Sande does not die when the character of Johnson is killed during a shootout, the visible bodily "performances" of actor and character are both identical and existentially tied — with those of the latter being determined by the former — in such a way that it is true to claim that both the actor and the character can be seen lying on the

46 The fact that the film's titular character, Dave Kovic, is merely a stand-in for the comatose and dying elected President — in the spirit of Chaplin's *The Great Dictator* (1940) — could be seen as a fun thematic duplication (though most likely by fluke, since this is an age-old theme of narrative fiction and theater) of what is going on perceptually with the matte shots: deception based on likeness.

floor in their respective worlds. The overlap can be explained by the fact that, as long as one plays along with the convention of cinematic fiction, a number of visible qualities that belong to the actor are transferred to the character through their hypoiconic embodiment on the screen. Moreover, the photographic nature of the process implies that the image, which offers a likeness of the character-embodying actor, also serves as a trace of him. In *Dave*, on the other hand, the image of the White House, while it is a *bona fide* index pointing toward its object, is not a *photographic* index of the building (although this is not perceptually apprehended by the viewer). As we can see, it isn't that indexicality has disappeared, but *one class of index has been replaced by another and the photographic layering of iconicity and indexicality has been disturbed.* As a result, fiction can assert its rightful course, which is precisely what takes place when the film partially merges the two universes of discourse (the real and the fictional) in a composite shot that shows an actual car driving up to the building. The composite result, therefore, is really a complex hypoicon that is a likeness typifying a situation, but one that involves an indexical element referencing our world — for instance, an actual/fictional car of such and such a make driving up to the White House some evening (the actual evening can only be referenced in the fictional world). Yet it is a hypoicon that seeks to pass for a photographic index, for a reagent. And it does so by using in turn a designative index (Lasaine's painting of the White House which is photographed and rendered indexically, though invisibly so for the unwary viewer), the semblance of a reagent (the photo-realism of the painting and its hypoiconic rendering of photographic qualities), a true photographic reagent (the moving car, which also involves a photographic hypoicon) all for the purpose of tricking the viewer into believing that what is shown is a single contiguous space belonging to both universes of discourse (reality and fiction), when it only belongs — as far as the complete composite is concerned — to that of fiction (no such actual car really drove up to the White House on the day the scene was shot).

When successfully fulfilling its intended purpose, an "effect shot" such as the matte composite of *Dave* is not a special effect for the viewer, though it is a postproduction visual effect for the filmmakers. Special effects — in the way I have been using the term here — are not deceitful in quite the same way — though, of course, they are related. The difference is especially striking for the class of effects we call "invisible," as they involve this time the viewer's *disbelief* at the *semblance of indexicality* (with regards to the real world and to CG-simulated photographic images of it) that special effect films offer. Depending on the type of effect, its photo-realism, and its ability to blend with the truly indexical aspect of the film (or else truly imperceptible visual effects), this disbelief *collides* with what is perceived and with how we have come to habitually interpret photographic images (and what today may pass for photographic images). The blowing-up of the White House in *Independence Day* — which used a miniature whose real blowing-up was recorded — is a case in point. No reasonable viewer believes this to be a photographic record of the actual White House being destroyed in order to make the film, even if this is what it *looks* like. Here, what needs to be considered are the conditions that explain why a statement such as "This is the White House" can stand as a true description of the image in both the real and the fictional universes of discourse, while "This is the White House exploding" may only do so with regards to fiction, and "This is a miniature of the White House (exploding)" is suppressed even though it is not entirely so (which is why the effect is not imperceptible, although it is invisible). It is here wherein resides the perceptual fiction of (perceptible) special effects. The ontological gap either calls on a fiction that indexes our real-world universe of discourse while designations perceptually impersonate reagents (*Independence Day*'s image of a miniature — itself a reagent — posing as an image of the real White House) or else uses pictorial embodiments of object-types that are merely congruent with existents from our universe of discourse yet — and herein lies the special effect — in a way that breaks the practical, financial/technical, or ontological constraints that otherwise restrict the use of a camera in recording a situation or event (an example out of many: the destruction of Tony Stark's mansion in *Iron Man 3* [Shane Black, 2013]).

As we can see, a film's ability to refer to our world of existents is not limited to its use of photographic means. We could add that when Albert Dieudonné (*Napoléon*, Abel Gance, 1927) or Marlon Brando (*Désirée*, Henry Koster, 1954) play Napoléon, they are also used by the filmmakers as indices of the well-known *empereur des Français* — though not through photographic means (i.e., not as photographic indices of Napoléon). The same is true of any reconstructed worldly event: from the storming of the Winter Palace in *Oktyabr': Desyat' dney kotorye potryasli mir* (*October*, Sergei M. Eisenstein, 1928) to the D-Day landing on Omaha Beach in *Saving Private Ryan* (Steven Spielberg, 1998). Such indices are more like names, designations, than reagents. In each case, the particulars they signify are integrated to a fiction, yet the dual or hybrid status of what they stand for — indexed reality (presented with the help of icons) *and* iconic fiction (indexed with regards to the universe of discourse of the fiction) — is part of the experience they offer viewers.

Although the above examples of matte shots and miniatures far from cover the entire range of visual (imperceptible) and special (invisible) effects, it seems clear that filmmakers' aspirations to fashion images that break free from, modify, or divert the camera's attachment to the real, cine-photographable world still rely in most cases on the idea of a camera's indexical potential in order to create perceptual fictions and forms of deceit whose experience[47] and meaningfulness require this idea. Whether or not they actually involve the representation of an existent from our world with which spectators are already acquainted, depicting possible occurrences of object-types consistent with our world, or else showing impossible objects or events, effects typically offer semblances of photographic reagents — feeding off our acquaintance of the photographic process — whenever they mimic the texture and qualities of camera-produced images, that is to say, whenever they offer viewers an icon of live-action cinema. Effects in this instance — from Méliès to CGI — may be seen to exist in the gap that forms between the idea of a camera's indexical potential and efforts to curtail its deployment, all the while exploiting it.

Visual and special effects want it both ways: they seek to free the cinema from the camera's close dependence on reality, and yet they also aspire to give viewers the impression of being realized through the recording action of a camera — even when it is obvious that this cannot be the case. What is especially interesting about this fact is that, in doing so, visual and special effects help us become aware of how difficult it actually is to come to terms with the role or function of photographic indexicality in cinema, including in what I have referred to as "zero degree" fiction filmmaking.

On one hand, and as intimated earlier, it seems obvious that without photographic *reagency*, the cinema would not have developed its star system. Characters and stars, however, are merely the most visible manifestations of the separate tracks that meet in the image: reality and fiction. Indeed, whenever cinema is used according to its baseline, we should think of the real world of particulars captured by the camera as an underlying layer always ready to pierce or tear through the fabric(ation) of the fiction as soon as viewers — for whatever reason or purpose — "retrieve" their awareness that in looking at the film they are also looking at a representation of the real world and its particulars. Hence the idea that the film's photographic images may be used indexically. On the other hand, it is also clear that filmmakers — especially in classical cinema — make efforts in the opposite direction by offering viewers the depiction of fictional worlds. In

47 For viewers, of course, the experience of imperceptible effects is basically a non-experience.

the celluloid era, most images in live-action fiction films were recorded with a camera (and, notwithstanding a few exceptions, visual and special effects occupied but a marginal amount of screen time in any given film). But what exactly were filmmakers seeking to represent in the process? Were they seeking to represent the particulars of our world or were they instead trying to create fictional worlds filled with their own particulars (except for overlaps like the White House, Times Square, or the Eiffel Tower) by way of qualities made manifest in our world? As long as films relied mostly on photographic images, the question perhaps was merely "academic": reality and fiction cohabited in both Hollywood musicals and in Italian Neo-Realist films. The difference in these films was stylistic and artistic, but it had nothing to do with photographic indexicality. Neo-Realist films, of course, sought to represent reality, the lives of Italians in the aftermath of WWII. The films sometimes have an almost documentary feel to them, and yet they are still fictions whose characters, situations, and environment are meant to stand, for the viewer, *not for actual existents but for exemplary types of existents*. The reality indexed by the camera can thus fade next to that which the fiction constructs — what is at stake here is *realism*, not photographically indexed reality (even if both are very close). Though the viewer may use images of *Roma città aperta* (*Rome Open City*, Roberto Rossellini, 1945) or *Ladri di biciclette* (*Bicycle Thieves*, Vittorio De Sica, 1948) indexically, shooting mostly on location, using the real world as its material, also gives the images in these films a *flavor*, a *quality of worldly presence* relative to those locations such as only photography can provide, and perhaps it is *this quality* that was sought out by a Rossellini or a De Sica, rather than the representation of each real-world particular recorded by the camera (need I emphasize that I think this to be the case?). The fact is, however, that this quality — this *icon* — accompanies all instances of photographic indexicality. Wherein lies what I twice before referred to as the paradox of using images of real particulars to create fictional worlds. In this regard alone (I am not discussing here aspects of *mise en scène* or editing, which, as we know, have an important role to play in the stylistic identity of Neo-Realism), the key difference between an MGM musical and a Neo-Realist film lies in choosing *what* is being recorded.

Now, if visual and special effects throw this paradox into relief, it is because they interfere with the way fiction cinema meshes the two tracks of reality and fiction as the camera knots together indexicality and iconicity. Our thinking about cinema is often confused in this regard because we have difficulty distinguishing the indexical from the iconic. This is most obvious in Bazin's famous "Ontology" essay, where he writes of the photographic image that it offers us "the object itself" and that "it is the model" (2009: 8).[48] The idea has dumbfounded many a film theorist over the years.[49] However, from Peirce we know that iconic signs are indeed indistinguishable from their objects in the quality that they share (as mentioned above, hypoicons are *formally* indistinguishable from what they signify). Therefore, like all icons and iconic signs, the photograph and its object do share the same identity (formal, not numerical). Bazin's mistake was to think that such is the case *only* or *especially* for photographic images and that this is in particular due to their nature as "automatic imprints" — that is, as indices rather than icons. However, unlike "zero degree" cinema, visual and special effects' perceptual fictions can more readily appear as offering a model of cinematic "semioticity," where iconicity dominates over photographic indexicality.

This becomes especially obvious when we consider motion capture (MOCAP). It has been argued that MOCAP is indexical since the technology for it requires recording the actual actions and movements of an individual.[50] But the fact of the matter is that MOCAP makes it very difficult (though not entirely impossible, mind you) to use the end result seen on screen in such a way as to point toward that individual existent in its *haecceity*, except in a very vague fashion.[51] Given this difficulty, might it not be a more accurate description of MOCAP to see it as *extracting* general qualities from an individual existent and transferring them to an inexistant fictional being? In which case, MOCAP would "iconize" rather than "index": its concern would lie not in representing the actual fact of qualitative embodiment in an existent but in representing the qualities an existent manifests *regardless*

48 André Bazin, "The Ontology of the Photographic Image," in *What is Cinema?*, trans. T. Barnard (Montreal: Caboose 2009), p. 8.

49 See Daniel Morgan's "Rethinking Bazin: Ontology and Realist Aesthetics" in *Critical Inquiry*, (32)3, 2006.

50 See for instance Marco Grosoli's "Motion/Performance Capture and the Afterlife of the Index. A Reconsideration of André Bazin's 'Myth of Total Cinema'" in *Recherches sémiotiques/ Semiotic Inquiry*, (31)1-2-3, 2011, pp. 152–173.

51 One would first have to recognize that what is shown is produced through motion capture and know independently that the technology requires the recording of a body's actual movements and actions. The same, of course, is true of photography. However, by not providing us with a likeness with which to identify the individual whose movements are recorded and shown in the image, motion capture also keeps us in the dark as to its identity. Perhaps like a photograph of a person whose face and body is covered by shadows, we see a vague existing something whose qualities in this case lie only in the forms of the movements captured. In certain instances, of course, the actual existent can still be identified, especially when facial movements are captured: one can get recognizable glimpses of Andrew Serkis in installments of the *Planet of the Apes* series or of Zoe Saldana in *Avatar*.

of its existence by transferring them to the representation of an imaginary being. Perceived from this angle, MOCAP can be seen to loosen the knot that otherwise binds indexicality and iconicity in photography — separating qualities from an existent — even if the end effect is to offer a semblance of index by mimicking the feel of motion photography as much as possible. In this regard, MOCAP can be used to remind us of how much live-action fiction cinema — from Neo-Realism to *Avatar* — thrives on likeness, more so perhaps than on the representation of actual, worldly, individual existents, even though it cannot seem to entirely do away with them. Could it be then that "iconizing" — the transferring (the stealing?) of qualities from the real world over to the world of fiction — has always been the goal of live-action fiction cinema, even when filmmakers use their camera to stick as close as possible to worldly existents so as to capture as many aspects of reality as possible? Is photographic indexicality, notwithstanding what I wrote above regarding the star system and the different ways it can manifest itself in the experience of viewing, merely an epiphenomenon in the way we use fiction film? These questions are not meant to argue that all fiction cinema since *L'arroseur arrosé* (*The Sprinkler Sprinkled*, Louis Lumière, 1895) — perhaps one of the earliest fiction films made with a camera — neatly folds into visual and special effects and has always been indifferent to reality. After all, since the whole point of much visual and special effects, especially today, lies in their ability to masquerade as photographic and therefore build onto photography's association with the index and its representation of individual existents, the perceptual fiction they set up seems to ask of us that we entertain the notion (either as a consciously held false/fictional belief, as in the case of *Star Wars*; or as mistaken belief in the case of *Dave*) that what we are seeing was directly present to the camera in the way that we now perceive and come to understand it on the screen. What this points to in the end is the need for untangling a number of questions concerning cinema and our use(s) of it, and the fact that our thinking about film can no longer avoid taking into consideration the entire domain of visual and special effects.

Bibliography

Bazin, André. "The Ontology of the Photographic Image." In *What is Cinema?*, translated by Timothy Barnard (Montréal: Caboose, 2009).

Bigelow, Joe. "King Kong," *Variety* (6 March 1933). http://variety.com/1933/film/reviews/king-kong-2-1200410783.

Bordwell, David, Janet Staiger, and Kristin Thompson. *The Classical Hollywood Cinema. Film Style and Mode of Production to 1960* (New York: Columbia Press, 1985).

Genette, Gérard. *Narrative Discourse: An Essay in Method*, translated by Jane E. Lewin (Ithaca: Cornell University Press, 1980).

Grosoli, Marco. "Motion/Performance Capture and the Afterlife of the Index. A Reconsideration of André Bazin's 'Myth of Total Cinema'," *Recherches sémiotiques/Semiotic Inquiry* 31, nos. 1-2-3 (2011): 152–173.

Gunning, Tom. "Moving Away from the Index: Cinema and the Impression of Reality," *Differences* 18, no. 1 (2007): 29–52.

Metz, Christian. "*Trucages* and the Film," translated by François Metzler, *Critical Inquiry* 3, no. 4 (1977): 657–675.

Morgan, Daniel. "Rethinking Bazin: Ontology and Realist Aesthetics," *Critical Inquiry* 32, no. 3 (2006): 443–481.

Pape, Helmut. "Charles S. Peirce on Objects of Thought and Representation," *Noûs* 24, no. 3 (1990): 375–395.

Peirce, Charles S. *Collected Papers of Charles Sanders Peirce*, vols. 1-8, edited by Charles Hartshorne and Paul Weiss (vols. 1-6); and Arthur Burks (vols. 7-8) (Cambridge, MA: Harvard University Press, 1931-1958).

———. *The Essential Peirce*, vol. 2, edited by Peirce Edition Project (Bloomington: Indiana University Press, 1998).

Prince, Stephen. *Digital Visual Effects in Cinema: The Seduction of Reality* (New Brunswick, NJ and London: Rutgers University Press, 2012).

Souriau, Étienne. *L'Univers filmique* (Paris: Flammarion, 1953).

About the Author

Martin Lefebvre is Professor and Chair of the Mel Hoppenheim School of Cinema at Concordia University (Montréal, Canada). He is Editor-in-Chief of *Recherches sémiotiques/Semiotic Inquiry* and has published widely on semiotics and film. He is the author of *Truffaut et ses doubles* (Vrin, 2013) and has edited several volumes including *Techniques et technologies du cinéma* (with A. Gaudreault; PUR, 2015); *Landscape and Film* (Routledge, 2007); and *Eisenstein: l'ancien et le nouveau* (with D. Chateau & F. Jost; Sorbonne, 2001).

2. Images as Visual Effects

John Belton

Abstract

The advent of digital imaging has prompted historians and theorists such as Stephen Prince to re-theorize the nature of the image, using special effects as a lens through which to view contemporary motion imaging. This paper draws on Prince's work to argue that the photographic image has always been a special effect. If, as Bazin suggested, the term "image" refers to "everything that the representation on the screen adds to the object there represented ... as opposed to what it reveals of it," every image is a construction, a composite of sorts. From this perspective, there is no radical break between the nature of the analog and the digital image; the latter is merely an extension of the other. Both analog and digital imaging are products of addition. The additive nature of digital imaging (and projection) provides a perspective from which it is possible to view digital imaging as a form of special effects.

Keywords: special effects, digital imaging, composite imaging, analog imaging, Stephen Prince

Dedicated to Stephen Prince (1955–2020)

The advent of digital imaging has prompted historians and theorists of the image to return to a basic question in film theory — that of the nature of the image. It has prompted us to explore ways in which we need to re-think what an image is and what it once was from the perspective of digital imaging techniques. Those who engage in this enterprise tend to fall into one of three camps — each one defining itself in terms of the problematic notion of "indexicality". In Charles S. Peirce's trichotomies of signs, an indexical sign "points to" its referent and shares an existential connection with it, in the

Lefebvre, M. & M. Furstenau (eds.), *Special Effects on the Screen: Faking the View from Méliès to Motion Capture*. Taylor & Francis Group, 2022
DOI 10.5117/9789462980730_CH02

manner of a photograph.[1] One group of scholars locates a radical break in the nature of the image, arguing that digital imaging lacks the indexicality of analog imaging associated with photography and the traditional cinema. A second group traces a continuity that bridges this so-called "break." They argue that digital imaging possesses a "kind" of indexicality, an indexicality that is mediated through numbers. A third group suggests that analog and digital imaging are *both* indexical but that there is a difference between the nature of analog and digital indexicality. For them, indexicality is relative rather than absolute.

Stephen Prince, in his book *Digital Visual Effects in Cinema: The Seduction of Reality*, provides a valuable overview of the first group. For Prince, David Rodowick, Lev Manovich, Jonathan Crary, and Sean Cubitt argue that digital imaging marks a radical break with analog cinema because it lacks the latter's indexicality.[2] Prince himself belongs to the second group: he rejects the notion of a radical break between analog and digital cinema, arguing for a larger continuity that connects them. Digital technology is seen as an extension of the tools of analog cinematography, a continuity that he then proceeds to demonstrate in his review of the history of visual effects.

But perhaps Tom Gunning is the most notable proponent of the continuity camp. In "What's the Point of an Index? Or, Faking Photographs," Gunning argues that there is little or no difference between analog and digital imaging. Digital imaging supposedly has a weak "truth claim" because it lacks indexicality. Gunning argues that both analog and digital cameras record light; they just do it in different ways. Citing Peirce, Gunning notes that traditional analog photographs are both indexical and iconic. He also notes that most of Peirce's examples of indices do not share an iconic resemblance with their referents. In fact, a number of them record or register indexical data in the form of *numbers* (thermometers, speedometers, medical instruments for reading blood pressure, etc.).[3] Gunning notes that that is exactly what digital imaging does. What Gunning ignores, however, is that what is indexical in, say, the thermometer is the column of mercury in the glass tube, not the numbers on the side of that tube. In fact, those numbers, added for the convenience of the user of the instrument in measuring the column of mercury, share no existential bond with the mercury; the relationship

1 Philip Rosen, *Change Mummified: Cinema, Historicity, Theory*. (Minneapolis: University of Minnesota Press, 2001), p. 18.

2 Stephen Prince, *Digital Visual Effects in Cinema: The Seduction of Reality* (New Brunswick, NJ: Rutgers University Press, 2012), p. 50.

3 Tom Gunning, *Nordicom Review* (25)1–2 (Sept. 2004), p. 40.

of those numbers to the column of mercury is conventional (or "symbolic" in Peirce's parlance) in that they represent one of several possible scales of measurement; they are in either Fahrenheit, Centigrade, or Celsius, not to mention Kelvin. At any rate, for Gunning, the fact that digital technology converts analog, indexical information into numbers does not necessarily make those numbers non-indexical. In fact, Gunning takes the argument one step further, insisting that digital imaging *is* indexical. In both analog and digital photography, original information — light waves — undergoes processing. In analog photography, the processing is photochemical; this processing figures as a stage of mediation between image and referent. In other words, analog photography is not a direct but a mediated process, mediated by lens, film stock, exposure rate, type of shutter/speed, and type of processing.[4] In digital photography, the processing is numerical. The numbers into which the profilmic event is translated are, for Gunning, not arbitrary, but determined by that profilmic event.

A much more obscure form of this argument was made by Laura Marks years earlier in her *Millennium Film Journal* essay, "How Electrons Remember." Marks argues that in analog imaging techniques, particle-waves (consisting of either light or electrons) establish a physical connection between image and referent. "In photography, film, and analog video," she explains, "it is possible to trace a physical path from the object represented, to the light that reflects off it, to the resulting image."[5] In digital imaging, this connection is broken at two different points along the chain from referent to image. The first break occurs when analog light waves from the referent "are translated into symbols," that is, into "strings of numbers."[6] For Peirce, nothing could be clearer — an indexical sign has been transformed into a symbolic sign. Numbers function as a conventionalized system for encoding original data. The second break occurs when certain intermediate signals representing states between 0 and 1 are assigned whole number values of 0 and 1. Original data is "rounded off" in order to conform it for digital processing.[7] Marks, however, concludes her argument by establishing a middle ground between the indexical and the symbolic. Rather than emphasizing the two indexical breaks she has identified, she opts to reconnect the digital image to electrons to which they owe their being.

4 Ibid., p. 40.

5 Laura Marks, *Millennium Film Journal* 34 (Fall 1999): The Digital. Reprinted in *Touch: Sensuous Theory and Multisensory Media* (Minneapolis: University of Minnesota Press, 2002), p. 162. Page references are to *Touch*.

6 Ibid., p. 171.

7 Ibid., p. 172.

"Although it no longer bears an analog relationship to its initial object, the digital image relies for its existence on the fundamental interconnectedness of subatomic particles."[8] Having untied digital images from one chain of material existence, Marks then re-secures them to another. The result is that digital images *do have* an indexical relationship to the subatomic particles that generate them. In asking if an image is indexical or not, Marks suggests that we must also consider the following question: "Of what is it indexical?" Images that are no longer indexical of some profilmic scene may nonetheless still be indexical of some other moment in the chain of procedures that results in image-formation. All signs, in fact, have indexical components. A painting, for example, consists of brushstrokes that index the movements of the painter's hand in the execution of the painting.

With a bit of imagination, is it not possible to see that an existential bond exists in computer graphic imaging (CGI) between the images that are produced and the computer keystrokes that produced them?[9] At one end of the indexical equation stands a series of operations involving specially written programs that are used to create images; at the other end are those images. The former has a physical connection with the latter. Indexicality is a term that can describe the series of relations within this equation.

The work of Gunning and Marks points in the direction of the third group, represented by Mark Wolf, who argues that analog and digital imaging processes are both indexical, but the nature of their indexicality differs.[10] Analog indexicality is a product of a physical connection established by light waves, while digital indexicality is a product of a physical connection mediated by binary numbers (data and code). In effect, digital imaging, which simulates photochemical imaging, introduces a new category of indexicality, much as, for André Bazin, the cinema introduced a new category of presence vis-à-vis the theater. From this perspective, then, there is both analog and digital indexicality — two forms of referentiality defined by a series of differences that are relative rather than absolute. Each indexicality is indexical in varying degrees and, more importantly, is indexical in different kinds of ways to its referent. For Wolf, the analog photograph emerges as the "most" indexical, more or less directly referencing a profilmic event. But even it possesses different degrees of indexicality determined by the

8 Ibid., p. 174.

9 I am thinking here of Anton Braxton Soderman's 2006 SCMS paper, "The Index and the Algorithm," in *Differences* (18)1, pp. 153–186.

10 Mark Wolf, *Abstracting Reality: Art, Communication and Cognition in the Digital Age* (Lanham, MD: University Press of America, 2000), pp. 261–262.

resolution of the image (the greater the resolution, the more indexical it is).[11] The digital photograph is "less" indexical — on a sliding scale of indexicality — depending on the number of pixels, i.e., the resolution inherent in the image. Its relation to its referent is mediated by algorithms, quantization, and/or fractal compression. Computer-generated imagery — including simulations — are the least indexical; there is no real-world referent or profilmic event to which they refer. Instead, they reference data sets. As Wolf explains, "the referents are not objects [in the real world] but laws of physics [abstracted from the behavior of objects in the real world]."[12] In effect, Wolf's discussion of indexicality shifts the focus of discourses about indexicality from whether or not an image is indexical to *what it is indexical of*. Wolf's argument returns us to Laura Marks's question, "Indexical of what?"

Braxton Soderman takes up and extends the tentative gestures made by Wolf to develop a more thoughtful account of digital indexicality where there is no real-world referent. Though he acknowledges the differences that exist between digital photography and CGI/computer simulation, Soderman sees a deeper similarity between them. Both are indexical of the algorithms that underlie their existence. Shifting "the 'existential bond' implicit within the indexical sign of the image from a 'law of optics' (as Peirce calls it) to a *law of computational execution*," Soderman argues that algorithms "force" a digital image to appear much in same way that photographic film is forced to receive an image.[13] In other words, though the physical contact between image and referent may be broken (or non-existent), there remains a physical connection between algorithm and image. The fact that an image resembles a real world or simulated referent is a matter of iconicity — equations within the program that instruct the data to simulate this or that. Indexicality lies elsewhere — between the image and the algorithm. Soderman's insight is extremely important for understanding certain major works of digital art that actually depict the work of algorithms. The avant-garde works by Jim Campbell (*Illuminated Average #1, Accumulating Psycho*) that he cites as examples involve the use of computers and clearly leave a record of that use in the image. But even more important, indexicality becomes an issue of iconicity. In that indexicality is the result of physical contact between a referent and a sign that establishes a chain that moves in one direction (from referent to sign); the completion of this chain takes place with a "certification" of indexicality by someone for whom the sign is indexical. At

11 Ibid., p. 261.

12 Ibid., p. 276.

13 Soderman, op. cit., pp. 10–11.

this point in the chain, the direction of movement goes from viewer to sign and then on to referent. The crucial link in this chain is clearly the viewer for whom something is or is not indexical. It is the viewer who invests the sign with indexicality — whether that indexicality is real or simulated. From this subjective perspective, if a digital image looks indexical, it is indexical. Indexicality, in other words, is ultimately a matter of *for whom* a sign is indexical. The question of indexicality is actually a series of questions. The first question, as we have seen, is "Indexical of what?". Another question might be "Indexical for whom?".

As I have argued elsewhere, it is the spectator who creates indexicality:

> In its simulation of analog imaging, digital imaging actually engages in a form of iconicity that confounds any simple understanding of indexicality, because it is indexicality itself that digital imaging mimics. In the realm of computer-generated imagery, digital artists carefully construct an apparently profilmic scene out of bits and bytes. The nonexistent referent acquires an existence thanks to the iconic abilities of digital artists and their software. In the particular instance of CGI, indexicality exists, if at all, only as an *effect* of the digitally produced image. It is only indexical in a hallucinatory sense — as a consequence of its resemblance to an imagined referent.[14]

Unlike Wolf, Soderman does not address the issue of analog digital indexicality, i.e., digital photography for which there is a pro-filmic event. As a result, he leaves an important question unanswered — if you cannot see the work of the algorithm, how is that algorithm indexed? Is our knowledge that an image is digital enough to give it a different indexicality from that of an analog photograph?

If Soderman can connect indexicality to the algorithm, the usefulness of the term "indexicality" becomes more and more problematic. If that which is indexical is no longer a profilmic referent, then we are now talking about indexicalities (as Wolf did) instead of indexicality. And we do not mean the same thing anymore. As Tom Gunning suggested, perhaps the term "indexicality" is no longer adequate for understanding the differences between analog and digital imaging?[15] And as Martin Lefebvre reminds us, discussions of the image that focus exclusively on its indexical elements at the expense of its iconic and semiotic components not only misuse Peirce's

14 John Belton, "If Film is Dead, What is Cinema?" *Screen* (55)4 (Winter 2014), pp. 465–466.
15 Gunning, op. cit., pp. 46–48.

theory of the sign but also fail to account for the role of the image in its function as an image, i.e., in terms of what it conveys about its referent.[16]

If one traces the dominant discourses about digital imaging, they often return to the notion of the manipulability of the digital image and the consequences of that in terms of the image's "truth claims." Within these discourses, digital imaging is often introduced through an example of its inherent mutability and manipulability — that is, through its ability to lie. The ability of digital imaging to lie then prompts a review of the history of analog photography and a review of its manipulations of the truth. Often, these discourses will conclude with a reference to postmodernism, noting that traditional truth claims were essentially ideological in nature — that our notion of the real was only a reality effect — and that digital imaging actually helps us understand these ideological operations better than analog photography does — replaced by a new paradigm based on digital technology.

Typical of such discourses on digital imaging in the popular media is Andy Grundberg's 1990 *New York Times* article, "Ask It No Questions: The Camera Can Lie."[17] Grundberg, photography editor of the *Times,* characterizes photography "as a medium of truth and unassailable accuracy," noting that it has functioned as legal and scientific evidence for over 150 years. But "today," he notes, "the veracity of photographic reality is being radically challenged" by computer imaging that "allows anyone to alter a photographic image at will." In other words, digital photography lacks the truth-claims of analog photography.

One controversial instance of image-doctoring had prompted public debate a few years before Grundberg wrote his article. In the February 1982 issue of *National Geographic* magazine, the cover photograph had been digitally manipulated in order to "adapt" a horizontal view of the pyramids of Giza for the vertical format of the cover page. The editors of *National Geographic* moved one of the pyramids closer to the other to fit them both within the magazine's famous yellow borders.[18] Four years after Grundberg's piece, another controversial manipulation occurred. In the last week of June 1994, both *Newsweek* and *Time* magazines published the *same* police

16 Martin Lefebvre, "The Art of Pointing. On Peirce, Indexicality, and Photographic Images," in *Photography Theory* (*The Art Seminar, II*), James Elkins, ed., (New York: Routledge, 2007), pp. 1–2 (page citations come from an unpaginated copy of the text provided to me by the author).

17 Andy Grundberg, "Ask It No Questions: The Camera Can Lie," *New York Times*, 12 August (1990), pp. H 1, 29.

18 Hubertus Amelunxen, Stefan Ilghaut, Florian Rotzer, eds., *Photography after Photography: Memory and Representation in the Digital Age* (Amsterdam: G + B Arts, 1996), p. 41.

headshot of O.J. Simpson who had been accused of murdering his wife and one of her friends. While *Newsweek* ran the photo more or less as is, *Time* extensively retouched it, darkening Simpson's skin tone to make him blacker and, according to some observers, more sinister.[19] One can expect faked photos on the cover of tabloid publications such as the *National Enquirer* but not on the covers of *National Geographic* or *Time*.

For Grundberg, the credibility of a photographic image is determined in large part by "the social perception of the medium." "In the nineteenth century," he writes, "photography's hold on the public imagination was so great that even blatantly faked 'spirit' photographs, in which ghosts seemed to hover over the living, thanks to the now commonplace technique of double exposure, were taken at face value."[20] For Grundberg, the current era of computer-manipulated imaging coincides with a different social perception of the medium — that of postmodernism in which "reality is not a fixed subject but a constantly evolving construction of culture." Citing Jean Baudrillard, Grundberg explains that photography is no longer "a mirror of reality but merely a hall of mirrors."[21] Grundberg's alarmist concern for "the veracity of photographic reality" with which he begins his piece gradually segues into a mellow meditation on the cultural construction of reality.

A similar discourse informs the argument that structures one of the major academic texts on photography in the digital age, William J. Mitchell's *The Reconfigured Eye: Visual Truth in the Post-Photographic Era* (1994). As Mitchell's subtitle suggests, "digital imaging dramatically changes the rules of the game," calling into question the photograph's traditional association with "visual truth."[22] Quoting Susan Sontag, Mitchell acknowledges that an analog "photograph passes for incontrovertible proof that a given thing happened."[23] But the traditional truth claims of analog photography are "subverted by the emergence of digital imaging."[24] Notions of "subversion" soon give way to detailed discussion of the differences between analog and digital photography. The fact that analog photography has a long history of image manipulation provides him with a platform to distinguish between photographs as things in themselves (as ontological entities) and the ways in which they are used (as ideological instruments). Thus, photographs can

19 Ibid., p. 74.

20 Grundberg, op. cit., p. 1.

21 Ibid., p. 29.

22 William, J. Mitchell, *The Reconfigured Eye: Visual Truth in the Post-Photographic Era* (Cambridge, MA: MIT Press, 1994), p. 31.

23 Ibid., p. 24.

24 Ibid.

be used to deceive or to prove the existence of things that do not exist, such as fairies or the spirits of the dead.

As an example, he cites the Cottingley fairy photographs. Elsie Wright and Frances Griffiths faked a photograph of "Alice and the Fairies" to hoodwink Sir Arthur Conan Doyle who reportedly viewed the photograph as evidence of the existence of fairies.[25] Doyle also believed that photographs were capable of recording the spirits of the dead. The genre of spirit photography, as Grundberg notes, exploits the evidentiary nature of photographic imagery in an attempt to ground belief in fact.

The ideological nature of photography's special relationship with the real is no better illustrated than in Stalinist-era photographic revisionism. This original image of Lenin haranguing a crowd of followers on 5 May 1920 includes the figure of Leon Trotsky standing on the lecture platform's steps to Lenin's right (Fig. 2.1). After Lenin's death in 1924, Trotsky fell into political disfavor and was systematically removed from documents recording the history of the revolution, as seen in this revised version of the original (Fig. 2.2). In each instance, the photographic image has been manipulated to serve a narrative purpose — that of a fiction deliberately concocted to exploit the notion of photographic credibility. Mitchell's discussion of analog and digital photography follows discursive formulae when he concludes his book by viewing digital imaging as an exemplary form of postmodern representation. Much as postmodern literary theory has "shaken our faith in the ultimate grounding of written texts on external reference," "the emergence of digital imaging has irrevocably subverted the [...] certainties" of traditional photography.[26]

For Grundberg and Mitchell, digital imaging is both a symptom and a product of postmodern epistemology. It *exemplifies* a loss of faith in the image and, at the same time, *produces* that loss of faith by undermining our perception of traditional technologies, such as photography, in terms of their relationship with the so-called "real." For both Grundberg and Mitchell, digital imaging's subversive characteristics force us to revisit analog photography and to see it from a postmodern perspective — as a medium fully capable of convincing fakery. The implication is that the credibility with which we have invested it is the product of an ideological need — our desire to establish an existential bond between photograph and referent. In effect, digital imaging becomes a wedge that enables us to see that reality is an ideological construct. The notion of "credibility" is

25 Ibid., p. 196.
26 Ibid., p. 225.

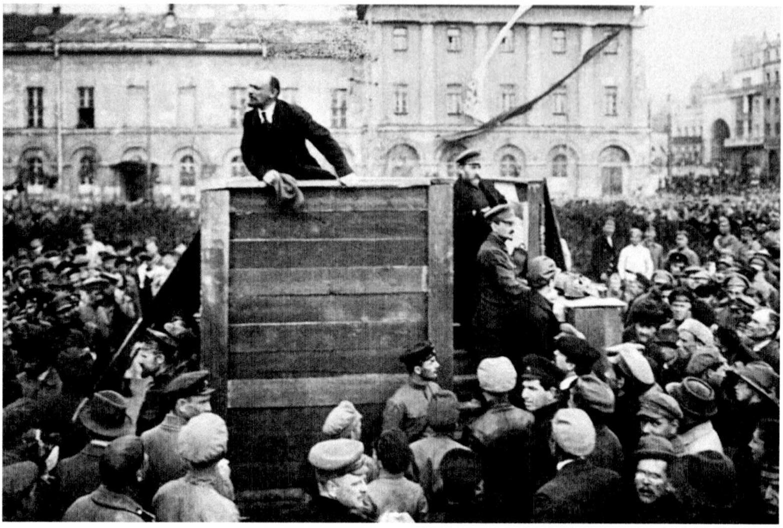

Fig. 2.1 Photograph of Lenin with Trotsky.

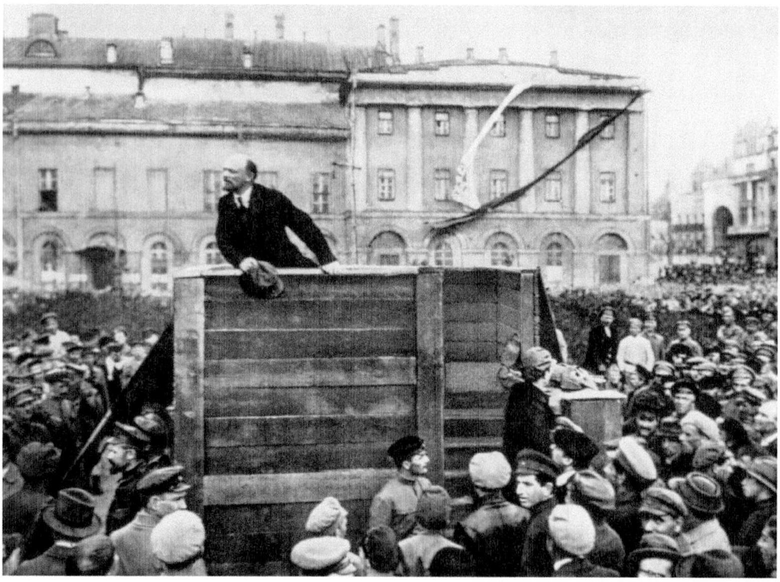

Fig. 2.2 Photograph of Lenin with Trotsky erased.

less a matter of what we know about how technology works than of what we know about how ideology works.

In terms of the issue of truth claims, Gunning, like Grundberg and Mitchell, points out that digital imaging did not invent fakery; with the advent of film negatives, analog photography could fake images; he illustrates this with a review of the genre of spirit photography. Gunning then argues that indexicality is not absolute but is subject to external validation of one kind or another. An observer must recognize the correspondence between image and referent. If an image is to be used as legal evidence, it must be supported by other forms of documentation (verbal discourses, written testimony) that establish its validity.[27]

In his book *Digital Visual Effects in Cinema*, Prince makes a case for the larger continuity that links digital and pre-digital visual effects. The main thrust of his argument is that digital imaging has not significantly changed the nature of cinema, contrary to what some critics — Steven Shaviro among them — have suggested.[28] In general, Prince admits that "there have been changes, but these mainly involve expanded imaging toolsets — a larger range of creative options for manipulating image properties than existed in the era of photo-chemical film."[29] For Prince, one of the culprits responsible for pronouncements of a radical break between photographic and digital imaging is film theory itself. Prince says that

> theories of cinema privilege live action staged before the camera, homogenous and organic visual spaces, and a suspicion of visual illusion. A customary way of thinking about cinema sees it as a medium that is photographically-based and oriented toward live action — that is, toward filming actors and events arranged before the camera during production. Cinema has never existed without visual effects, and yet they largely are non-existent in our theories and aesthetics of the medium.[30]

Prince proposes a revision of conventional, photographically based theories of film realism in an attempt to acknowledge the constructed nature of the image using visual effects as a site for re-defining the image as a visual effect. He concludes his book with a thought-provoking expansion of the definition of visual effects to include any manipulation of the image. Citing digital matte artist Craig Barron, Prince suggests that cinema, as the by-product of various optical toys, *originated* as a visual effect. For Prince, "this perception stands film theory on its head. Theory has prized what we might call organic images, ones reflecting a minimal amount of post-production processing, and this bias is characteristic of theory's rather puritanical conception of

27 Gunning, op. cit., p. 43.
28 Prince, op. cit., p. 149.
29 Stephen Prince, Columbia Film Seminar talk, 24 October 2013, p. 1.
30 Ibid.

the medium."[31] Then Prince moves beyond this historical observation to propose a point that follows through on his complaint about the "bias" of film theory. He asks,

> what if cinema's essence, indeed, is composed of visual effects? Editing is a visual effect. So is lighting and deep-focus cinematography. These are normative indexes of film style and yet they are constructions in ways no less artificial than the use of matte paintings or hanging miniatures. The costumes worn by actors are visual effects, and in light of cinema's composited nature, it makes no sense to differ with the logic that motion capture provides actors with the means of wearing digital costuming and make-up. The digital era compels us to really think about the implications of narrative cinema's amalgamated nature — and to recognize that digital imaging builds upon the aesthetic traditions established in analog filmmaking. Digital imaging conserves the past even while opening doors onto new optical domains.[32]

In terms of editing, surely fades and dissolves are visual effects since they are, for the most part, the product of an optical printer. But does that make a straight cut a visual effect?

To Prince's list of "effects," one might add the crucial illusion of motion created by still frames projected at 24 frames per second, not to mention the hyper-reality of high frame rates such as the 48 frames per second used in filming Peter Jackson's *The Hobbit* (2012). To frame rate, we might also add sound, which not only features sound *effects* but dialogue and music that are often added to the image in post-production. And while we are at it, should we not consider 3D a visual effect? It is certainly considered an effect within the effects industry[33] when flat films are converted to 3D, as was the case with *Man of Steel* (Zack Snyder, 2013), *The Avengers* (Joss Whedon, 2012), *Alice in Wonderland* (Tim Burton, 2010), the last few *Harry Potter* films, and the re-release of *Titanic* (James Cameron, [1997] 2012), not to mention the re-purposing of *The Wizard of Oz* (Victor Fleming, [1939] 2014).

However, it must be said that Prince's expansion of the notion of visual effects is a bit disconcerting in that the operations he is describing are

31 Prince, op. cit., p. 227.

32 From a talk Stephen Prince gave at the Columbia Film Seminars in New York City in 2003, p. 13. A slightly shorter version of the passage can be found in *Digital Visual Effects in Cinema,* op. cit., p. 227.

33 Christopher Cram, "Digital Cinema: the Role of the Visual Effects Supervision," *Film History* (24) 2 (2012), p. 185. See also Cram's email to the author, 3 December 2011.

not generally included among what the film industry refers to as "visual effects." In fact, my one contact within the visual effects industry — a former student — remains rather skeptical about the notion that every image today is a visual effect.[34] If the terms "visual effects" or "simple effects" function within the film industry in a much narrower way, then we can argue that Prince's expansion of that term runs the risk of rendering that term problematic for theorizing the nature of the image because it no longer has the precision it once had. But Prince's polemic does call for us to think about both analog and digital imaging in ways that bypass the conventional indexical/non-indexical opposition that has tended to become a *cul de sac* in film theory and that point to a more promising notion of the constructedness of the image.

Prince argues that digital technology extends photographic realism in what he calls "a seduction of reality." The phrase "seduction of reality" comes from James Cameron, who used it to characterize the role of visual effects in *Avatar* (2009). The notion is that visual effects "seduce" reality. In Prince's words, "they beguile it so as to draw close, study and emulate it, even transcend it."[35] For Prince, visual effects extend traditional notions of cinematic realism. Visual effects do not so much transcribe or record reality as translate it; reality thus undergoes a process of mediation or remediation, a term introduced to the study of new media by Jay David Bolter and Richard Gruson (2000). From this perspective, reality bridges the break between analog and digital imaging, but it is refashioned in the process.

Contemporary digital cinema is, for the most part, a composite of different kinds of images. Original photography frequently combines 35mm analog images with digital images. Special-effects sequences combine practical (i.e., analog) and digital effects with CGI (computer-generated imagery). The final film is then put through a digital intermediate process involving conforming the negative, importing and integrating visual-effects elements, incorporating an edit decision list, performing color correction and timing, and outputting the finished digital file to film (prior to 2014) or to a digital cinema package (DCP). In order to produce a seamlessly credible diegetic world, the inherent heterogeneity of these different kinds of images must be concealed. For example, film grain might be added to digital images and removed from 35mm film images to achieve a homogeneity in their "looks." I want to argue that this manipulation of the image constitutes an expansion of visual effects beyond the spectacular special-effects sequence

34 Christopher Cram, email to the author, 4 October 2013.

35 Prince, *Digital Visual Effects*, op. cit., p. 9.

to the film as a whole. The image today is, in effect, a processed form of representation (along the lines of Manovich's argument about cinema as a form of painting).[36] It is neither analog nor digital but a simulation of something in between. If, strictly speaking, "special effects" is a term used to describe any kind of effect that is *added* after an image has been captured, then contemporary imaging in the cinema is a special effect.[37]

In 2012, I edited an issue of *Film History* on digital cinema that included two essays on visual effects and, in 2013, I responded to a paper that Prince gave on visual effects at the Columbia Film Seminars in New York City. What I have learned from Prince and those essays on visual effects is, first of all, that the term "special effects" is obsolete. "Special effects" are more properly known as "visual effects." Terms used to describe effects in the cinema have changed from generation to generation. In 1928, the Academy recognized certain technical contributions to the cinema through the category of "Best Engineering Effects." From 1939 to 1962, the Academy gave awards for "Best Special Effects." This category included both visual and sound effects. From 1964 to 1971, the category was known as "Best Special Visual Effects." Sound effects became a separate category in 1964. From 1977 (*Star Wars,* George Lucas,) to the present, the category has been called "Best Visual Effects." Effects include practical effects that are non-computer-generated effects achieved during filming that can be seen on camera such as fire, fog, rain, car crashes, explosions, and visual effects that are created in post-production through CGI, digital matte painting, or other techniques. In the pre-digital era, these effects were accomplished by using matte paintings, miniatures, image compositing, optical printers, or other devices; now they are produced digitally.

What I have also learned from Prince and others is that visual effects or VFX is a remarkably productive site for thinking about the nature of the contemporary cinematic image. If, in what follows, I seem to shadow Prince's project to some extent, it is because his call to rethink our theorization of the image from the perspective of digital imaging is so compelling. But I think the thrust of our arguments differ significantly enough to justify the intervention I hope to make in the discourse about visual effects. I, for one, am not interested in visual effects as the "seduction of reality" or in the relation of visual effects to theories of realism. If both Prince and I look at

36 See Lev Manovich, "What is Digital Cinema?" http://manovich.net/index.php/projects/what-is-digital-cinema.

37 For the definition of a "special effect," see Christopher Cram's email to the author, 4 October 2013 at endnote #40.

the constructed nature of the image — in part, at the composited nature of the image — I want to argue that contemporary digital imaging is less a matter of realism than of simulation. Digital imaging does not strive after an impression of reality as after a simulation of the characteristics of analog imaging that involves a processing of the image that renders it "imaginary." My use of the term "imaginary" is not exactly the imaginary of Christian Metz's imaginary signifier, which finds its identity in Lacan's mirror phase. It does rely on the notion of the image as a composite of sorts — not of presence and absence but of multiple layers of image. The image is imaginary in that its chief concern is not with the reproduction of a profilmic reality or scene but with its internal coherence as an image — with the creation of an image that effaces its actual constructedness and composite nature in an attempt to create a homogeneous "look" that satisfies the needs of the diegesis for credibility. It is an image that seeks internal coherence among the elements out of which it is composited more so than any relation it might enjoy with reality.

At the same time, the image is imaginary in a Bazinian sense. In part, what I mean by "image" and "imaginary" can be understood in terms of the differences between the image and reality — with the image as something that is a manipulation of reality. The distinction comes from Bazin's "The Evolution of the Language of the Cinema." There he distinguishes between two opposing trends in the realm of film aesthetics — that of those directors who put their faith in the image and that of those who put their faith in reality. Within this contrast, the term "image" refers to "everything that the representation on the screen adds to the object there represented [...]" "as opposed to what it reveals of it."[38] Bazin's term "image" encompasses what he calls "the plastics of the image" and "the resources of montage."[39] If the image is the sum total of what has been "added to" the object represented, then the image is a visual effect — for one of the basic definitions of visual effects is "something that has been added to live action footage."[40] This sort of manipulation of the image is precisely what VFX engages in.

Another way of looking at the constructedness of the image is to see that in every image there is a tension between the object represented and the representation of that object — between that part of the image that resists meaning and that which imposes it, between whatever existential

38 André Bazin. *What Is Cinema?*, vol. 1, trans. Hugh Gray (Berkeley: University of California Press, 1967), pp. 24, 28.

39 Ibid., p. 24.

40 Email from Christopher Cram to the author, 4 October 2013.

bond it may have with its referent and its status as a sign (i.e., between its indexical nature and its iconic and symbolic nature). It is this tension that defines analog imaging and that the tools of digital imaging can now, if they so choose, eliminate.

From yet another perspective, the image today is a construction much as the soundtrack has always been a construction. If sound editing and mixing involves the combination of heterogeneous materials — dialogue, music, and sound effects — into a whole, is that not what digital compositing, the digital intermediate, and other digital tools do to the contemporary image?

I want to look at the way digital technology has — to use Prince's term — "seduced" not reality but the image. I want to see Prince's "seduction of reality" as a new reality — a "processed" version of reality. And though it may be an extension of analog cinema, I think it also needs to be distinguished from analog cinema. I propose to use, as Prince has done, the field of visual effects as a site for exploring the differences that inform this continuity.

The compositing of live action footage and matte paintings has long been a staple of visual effects. The challenge is to combine the look of photography with that of an image that has been painted. In analog cinema, this composite image was frequently created with a glass shot in which a painting on glass was placed between the camera and a partial set. A similar effect can be achieved by matting out part of the image during filming, rewinding the footage, and then filming a matte painting. This sort of matte work can also be done in the lab with two photographic images. The same effect can be accomplished with a hanging miniature. Or images can be combined through the Schüfftan process in which a miniature and full-sized subjects are filmed simultaneously using a silvered mirror in which some of the silver has been removed. In some instances, matte paintings can achieve a remarkable photorealistic effect, as in Albert Whitlock's *trompe-l'oeil* painting of Covent Garden in Alfred Hitchcock's *Frenzy* (1972) (Fig. 2.4). I am told that the only live-action element in this shot is one of the trucks parked in the street (Fig. 2.3). Can you tell which one? All of these instances of visual effects involve the internal coherence of the image — they involve the production of a composite that does not betray its hybrid nature.

Compositing in digital cinema is similar to that in analog cinema, but rather than trying to manipulate one (or more) image to match the dominant image element (whether that be live action, miniature, or painting), all the composited elements of a digital image are manipulated to produce a homogeneous, synthetic image that is ultimately different from any one of the "original" images out of which it is made. In short, while in analog-effects work it is most often the live action element that the visual effects attempt

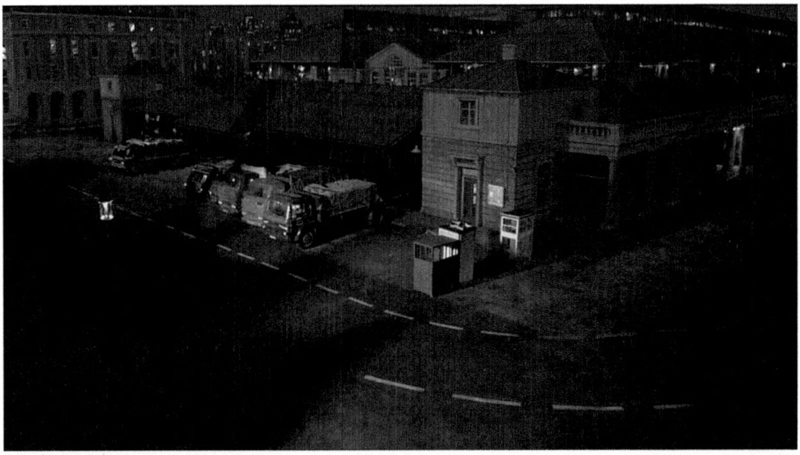

Fig. 2.3 Still from *Frenzy* (Alfred Hitchcock, 1972).

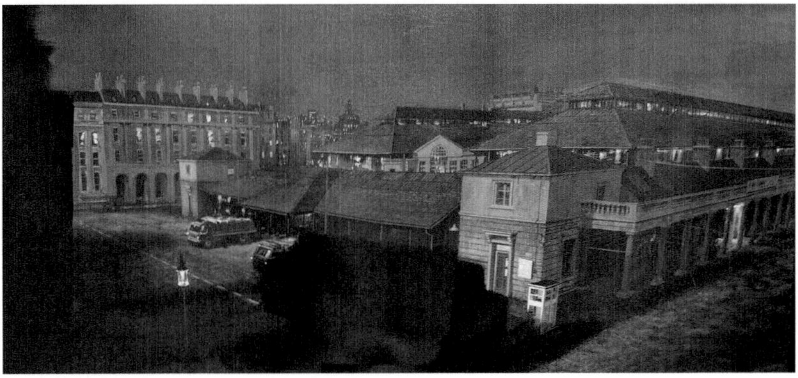

Fig. 2.4 Albert Whitlock matte painting for *Frenzy*.

to match, in digital-effects work each element of the whole image undergoes processing to produce the match. There is a roughly four-minute clip online from *The Great Gatsby* (Baz Luhrmann, 2013) that provides a fairly good example of digital compositing of live action footage shot in front of a blue or green screen and computer-generated digital mattes.[41] The film's VFX supervisor, Chris Godfrey, put together this before-and-after reel of some of the compositing work that was done by the Sydney-based effects house, Animal Logic, which was one of the seven effects houses that worked on the film.

Luhrmann shot the film entirely in Australia, using CGI to reconstruct New York City and Long Island ca. 1922. The clip — which consists primarily

41 See https://vimeo.com/68451324.

of wipes from live action, green-screen footage to the finished shot with its digital additions — provides a remarkable example of digital compositing and the constructed nature of the image.

Over 1500 of the film's shots involved visual effects of the sort seen in this clip, not to mention the color grading that occurred in the digital intermediate phase. Animal Logic's Ingrid Johnson notes that Luhrmann "wanted to create a photorealistic version of New York City and Long Island that was historically accurate but also a bit more modernized. For example, the cars were replicates of the cars of the time but with a modern edge. The costumes, everything is a bit amped up." Her colleague, Andy Brown, adds that "There was a big emphasis on historical reference photos to make sure there was authenticity. [But] Our brief was to make everything bigger, brighter, more colorful. Baz didn't want it to be a romanticized New York with sepia tones or brown; he wanted a broader and more saturated color palette, which was almost Technicolor in look."[42] The Technicolor look can be found in the highly saturated colors of the party sequence.

The look of the film has its origins less in actual photos of New York from the 1920s than in an artist's rendering of that period. According to Adriene Hurst, "Look development continued from the start through to the end. Adrian (Hauser), the film's senior digital intermediate colorist, worked on the grade with the director Baz Luhrmann, Catherine Martin and the DP, and they needed to spend a lot of time developing and enhancing the final look of the film. 'I find that digitally acquired footage can too easily look "digital," that is, bland and homogenous,' he said." Hauser then traces the film's look back to artists' sketches. He confesses,

> I was fascinated with the colour rendering used in the original hand-drawn storyboards. They used a luxurious colour palette, so we created LUTs (Look Up Tables) and colour processes that steered away from the digital, clean look and brought us into that world. Simon Duggan's original photography was beautiful to start with, so the process focused more on polishing and embellishing.[43]

Luhrmann's images are perhaps some of the most extreme examples of what I am calling "imaginary." They reek of aesthetic artifice. They look processed. Luhrmann's Times Square, though based on period photographs,

42 http://library.creativecow.net/kaufman_debra/VFX-Great-Gatsby-Animal-Logic/1.
43 http://www.digitalmedia-world.com/Features/cutting-edge-transforms-gatsby-into-living-colour (accessed 9 October 2017).

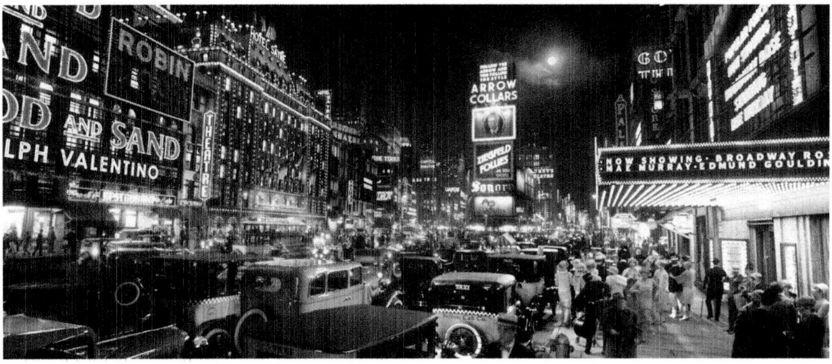

Fig. 2.5 Still from *The Great Gatsby* (Baz Luhrmann, 2013).

is an imaginary place — a place that only exists in the imagination (Fig. 2.5). Gatsby's yellow Duesenberg, for example, is an overly saturated yellow — something like Kyle Hadley's yellow sports car in Douglas Sirk's *Written on the Wind* (1957). In his review of the film, Tom Doherty found that the look of the images effectively conveyed the illusory nature of Fitzgerald's vision. Doherty writes that "the age of the computer has allowed the re-creation of a city of dreams with no material foundation. The environs of digital space and the multi-plane perspective of 3D seem suited to the mists of a Jazz Age that never was, an apt correlative for Gatsby's own vision of an imagined past."[44]

Few contemporary films look as processed as *The Great Gatsby*, but more and more of them have been tweaked in post-production by going through the digital intermediate process. As *Variety* commented: "The DI basically does for movies what PhotoShop and similar software can do to enhance still photos, though on a more sophisticated level."[45] Though it includes virtually all post-production operations, the DI process has become primarily associated with color grading. Digital technology has been used for years for film editing, computer-generated imagery, and special effects. What is notable about the DI process is that it involves the digitization of all or nearly all of the film's images, rather than the partial digitization involved in these earlier features of the post-production workflow.

Contemporary films regularly contain CGI and images filmed digitally or photographically. The "look" of these three kinds of images is visibly different: CGI looks somewhat synthetic; digital images are extremely (i.e., too) sharp

44 Tom Doherty, "The Great Gatsby," in *Cineaste* (38)4, 2013, p. 47.

45 Jack Egan, "Color Schemes," online *Daily Variety*, (6 January 2005). See http://variety.com/2005/film/awards/color-schemes-1117915901/.

and cold, while images obtained with film are softer and warmer.[46] One of the tasks of the digital intermediate process in post-production is to reconcile these different looks into a homogeneous hybrid that combines the features of all three into a new category of image whose "processed" look lends itself to the filming of fantasy.

In 2007, almost 70 percent of Hollywood films went through a DI.[47] For some of those films, the DI is nothing more than a digital version of the color grading that took place in the days of analog postproduction. But for many films, the digital intermediate process opens the doors to a manipulation of the image that was never possible with analog imaging in that it permits the manipulation of individual *parts* of the image without manipulating the image as a whole. As I noted elsewhere in a discussion of the digital intermediate in films such as *Pleasantville* (Gary Ross, 1998) or *Sin City* (Robert Rodriguez and Frank Miller, 2005), the hybrid nature of the image is revealed in ways that it is not in most digital filmmaking.[48] The mixture of color and black and white in these films forces us to perceive the image in terms of picture elements, not whole pictures. To cast this in the language of motion-picture technology, photochemical imaging practices necessarily treat the image as a whole; digital-imaging practices necessarily treat the image differently — as an array of discrete picture elements. With photochemical color grading, it is impossible to adjust one color or part of the image without adjusting it all. DI permits more focused adjustments — changes on the level of minute picture elements. The manipulation of individual pixels in DI is one of the most compelling examples of the constructed nature of the digital image, even when that manipulation effaces its own intervention in the construction of the image.

The transition of the industry from film to digital that took place in the wake of *Avatar* (2009) has rendered the notion of a "digital intermediate" somewhat obsolete. Only one or two major filmmakers (Quentin Tarantino, Christopher Nolan, Paul Thomas Anderson) still shoot on film; everyone else shoots digitally. Thus, there is no need to render film into the intermediate

46 For a discussion of the differences between the film and video "look," see Alan Roberts, "The Film Look: It's not Just Jerky Motion…", BBC R&D White Paper (December 2002), p. 15. The paper can be found at http://downloads.bbc.co.uk/rd/pubs/whp/whp-pdf-files/WHP053.pdf. See also André Gaudreault and Philippe Marion, *The End of Cinema?: A Medium in Crisis in the Digital Age* (New York: Columbia University Press, 2015), p. 75.

47 Robert Hoffman, email to the author, 1 August 2007.

48 John Belton, "Painting by the Numbers: The Digital Intermediate," *Film Quarterly* (61)3 (Spring 2008), pp. 58–65.

stage of a data file that will subsequently be rendered back into film. As Robert Hoffman of Technicolor reports,

> in the very recent past (the last year actually) we've started using the term "color finishing" rather than DI given that with today's digital workflows there's nothing "intermediate" about it. Before it was sandwiched between production photography on negative and negative film out. And although we continue to do some film outs, creating either an archival negative master, or print master, most of the time the films are simply staying digital all the way downstream.[49]

But the function of the DI remains. It enables technicians in post-production to manipulate the image pixel by pixel. As Ben Murray, head of creative services at Technicolor-PostWorks NY, says, "I think the spirit of the digital intermediate carries on in that the entire film is an effect. In that spirit, to this day and even more so, every shot in a film is a visual effect, and is an opportunity to minimize distractions and help storytellers focus their story and fix things until the last second. That wasn't possible before."[50]

The projection of those images in the movie theater provides yet another argument for treating the image as a visual effect — as a composite, as something created by means of the addition of images. Digital projection is additive; analog projection is subtractive. With film projection, light from the projector passes through the image which "filters" the light, rendering the results of that filtration as a whole image on the screen. In digital projection, a digital light processor, outfitted with millions of micromirrors, projects individual pixels onto the screen where those pixels combine to form an image. If the image is not constructed in production or postproduction, it most certainly is during projection.

In conclusion, digital imaging brings with it the potential to reveal to us the composite nature of the image. For the most part, it has chosen not to do this. In a film such as *The Great Gatsby*, digital visual effects that were featured in 1500 shots have dictated the look of the film as a whole — a look that is bound up with fantasy and the imaginary. The less fantastic look of other works of digital cinema merely illustrates the ability of digital imaging to simulate the non-digital look of traditional film. In both cases, digital

49 Hoffman, email to the author, 25 August 2015.

50 Linda Romanello and Mark Loftus, "The Evolution of the Digital Intermediate," *Post Magazine*, (2 April 2015), http://lightiron.com/about/news/evolution-digital-intermediate/.

imaging performs the work of addition upon the object of representation, a work that closely resembles that performed by visual effects.

Bibliography

Amelunxen, Hubertus, Stefan Ilghaut, and Florian Rotzer, eds. *Photography After Photography: Memory and Representation in the Digital Age* (Amsterdam: G+B Arts, 1996).

Bazin, André. *What is Cinema?*, vol. 1, translated by Hugh Gray (Berkeley: University of California Press, 1967).

Belton, John. "Painting by the Numbers: The Digital Intermediate," *Film Quarterly* 61, no. 3 (Spring 2008): 58–65.

———. "If Film is Dead, What is Cinema?" *Screen* 55, no. 4 (Winter 2014): 460–470.

Braxton Soderman, Anton. "The Index and the Algorithm," *Differences* 18, no. 1 (2007): 153–186.

Cram, Christopher. "Digital Cinema: The Role of the Visual Effects Supervision," *Film History* 24, no. 2 (January 2012): 169–186.

Doherty, Tom. "*The Great Gatsby*," *Cineaste* 38, no. 4 (2013): 45–47.

Egan, Jack. "Color Schemes," *Daily Variety*, (6 January 2005). Uploaded 15 June 2013. http://variety.com/2005/film/awards/color-schemes-1117915901/.

Gaudreault, André, and Philippe Marion. *The End of Cinema? A Medium in Crisis in the Digital Age* (New York: Columbia University Press, 2015).

Godfrey, Chris. Vimeo. "The Great Gatsby VFX." Uploaded 15 June 2013. https://vimeo.com/68451324.

Grundberg. Andy. "Ask It No Questions: The Camera Can Lie," *The New York Times*, (12 August 1990).

Gunning, Tom. "What's the Point of an Index? Or, Faking Photographs," *Nordicom Review* 25, no. 1–2 (September 2004): 39–49.

Hurst, Adrienne. Digital Media World. "Cutting Edge Transforms Gastby into Living Color." Accessed 9 November 2021. https://www.digitalmediaworld.tv/in-depth/423-the-great-gatsby-in-living-colour.

Lefebvre, Martin. "The Art of Pointing. On Peirce, Indexicality, and Photographic Images." In *Photography Theory* (*The Art Seminar, II*), edited by James Elkins (New York : Routledge, 2007): 220–244.

Manovich, Lev. "What is Digital Cinema?" Accessed 7 November 2021. http://manovich.net/index.php/projects/what-is-digital-cinema.

Marks, Laura U. "How Electrons Remember." In *Touch: Sensuous Theory and Multisensory Media* (Minneapolis: University of Minnesota Press, 2002): 162–178.

Mitchell, William, J. *The Reconfigured Eye: Visual Truth in the Post-Photographic Era* (Cambridge, MA: MIT Press, 1994).
Prince, Stephen. *Digital Visual Effects in Cinema: The Seduction of Reality* (New Brunswick, NJ: Rutgers University Press, 2012).
Roberts, Alan. "The Film Look: It's not Just Jerky Motion..." *BBC R&D White Paper* (December 2002). http://downloads.bbc.co.uk/rd/pubs/whp/whp-pdf-files/WHP053.pdf.
Romanello, Linda, and Marc Loftus. "The Evolution of the Digital Intermediate," *Post Magazine* (2 April 2015). http://lightiron.com/about/news/evolution-digital-intermediate/.
Rosen, Philip. *Change Mummified Cinema: Historicity, Theory* (Minneapolis: University of Minnesota Press, 2001).
Wolf, Mark. *Abstracting Reality: Art, Communication and Cognition in the Digital Age* (Lanham, MD: University Press of America, 2000).

About the Author

John Belton is Professor Emeritus of English and Film at Rutgers University, Chair of the Board of Editors of the Society of Motion Picture and Television Engineers, and editor of the Film and Culture series at Columbia University Press. He is the author of five books, including *Widescreen Cinema* and *American Cinema/American Culture.*

3. The Pragmatics of *Trucage*: Between Feigning and Fiction

François Jost

Abstract

Christian Metz once famously defined special effects (or *trucages*) in terms of a hiatus with regards to the analogical reproduction of reality. A pragmatic theory, however, must take into consideration the context wherein special effects appear as well as the role of the spectator. Therefore, the goal of this essay is not so much to describe special effects but rather to grasp *when* something on the screen is experienced by the spectator as a special effect. Special effects are not necessarily self-defining. Hence, determining whether something is a special effect depends in great measure on the knowledge and beliefs that spectators possess not so much regarding cinema but relative to the world. In this essay, the author distinguishes three types of counterfeits. First, one finds counterfeits of reality whereby a *trucage* falsely turns an icon into an index. Next, one finds counterfeits of *lived* reality which open up the vast domain of *feigning*, that of the *as if* — i.e., of that which could well have happened. Finally, there also exists a counterfeiting of documents that proceeds by modifying or touching up a pre-existent representation.

Keywords: feigning, falsification, spectator, pragmatism

In his second *Meditation*, Descartes examines the illusions of the senses and ponders the degree of trust he can put in what he sees from his window: "If by chance I were to see from a window men passing in the street, I should not fail to say upon this sight that I see men [...] and yet what do I see from this window if not hats and coats that could cover-up specters or feigned men whose movements are activated only by springs?"[1]

1 This translation is closer to the French text than most known English translations (either Elizabeth Heldane's and G.R.T. Ross's for the 1911 Cambridge University Press edition; or John

Lefebvre, M. & M. Furstenau (eds.), *Special Effects on the Screen: Faking the View from Méliès to Motion Capture*. Taylor & Francis Group, 2022
DOI 10.5117/9789462980730_CH03

A similar situation is commonly encountered by spectators who, while watching certain films, are not always sure whether what they see refers to reality, or rather to a feigned reality, or what is called quite simply a *trucage* or trick whose supposed realism is identified above all with an illusion of the senses, an illusion that Descartes contrasts with truth. Indeed, it was in relation to illusion rather than to representation that the "trick" was defined in the nineteenth-century theater. It is one thing to admire an academic painting in which an attempt has been made to render a mythological scene as realistically as possible, but quite another to allow oneself to fall into the trap proffered by a *trompe-l'oeil* prolonging the columns of a church. In the second case, the spectator confuses, at first sight, a painting that is by definition a two-dimensional image with the three dimensionality of a real architectural space. The French term *trompe-l'oeil* elegantly expresses the spectator's ephemeral feeling of having been fooled.

If color, framing, lighting, and montage are notions that can be studied in themselves as components of the image or the film, *trucages* have a status of their own. They are, so to speak, intrinsically *pragmatic* in the sense that they are informed by the idea of a communicational effect based on the intention of a transmitter who seeks to deceive a receiver by means of an object, namely the film. Like the "trick," the French term "*trucage*" has an ambivalent meaning in that it refers to a technical procedure (there are specialists of "*trucage*" in the film industry) while at the same time connoting that the spectator has been deceived. "Therefore, *trucage* cannot exist without deception," as Metz comments, very convincingly, in a famous article that offers the first semiological analysis of *trucage*.[2] Described as a "stratagem" (the primary meaning of the word "trick") designed to mislead the spectator, *trucage* should not be understood as a breach in reality or, in other words, as something connected to the idea of *representation* but as an *illusion*, an erroneous perception.

Cottingham's more recent 1986 version, also for Cambridge). Here is the French text: "[…] *si par hasard je ne regardais d'une fenêtre des hommes qui passaient dans la rue, à la vue desquels je ne manque pas de dire que je vois des hommes* […]; *et cependant que vois-je de cette fenêtre, sinon des chapeaux et des manteaux, qui peuvent couvrir des spectres ou des hommes feints qui ne se remuent que par ressorts*?" René Descartes, *Méditations métaphysiques* (Paris: GF Flammarion, 1992), p. 89.

2 See Christian Metz, "*Trucage* and the Film," trans. Françoise Metzler, in *Critical Inquiry* (3)4 (1977), pp. 657–675. I use here the term "semiology" rather than "semiotics," since it is the one that Metz preferred to use in accordance with the Saussurian tradition. It is only in 1977 that the term "semiotics" appears in the title of one of his books: *Essais sémiotiques* (Paris: Klincksieck,1977).

Error, trickery, deception… The semantic field covered by *trucage* is associated with falsity and its opposite: truth. What counts in *trucage* is, therefore, not so much the relationship between image and reality but the way in which we conceive of that relationship.

But what is the truth of an image? Our conception of *trucage* depends on how we answer this question. The most widespread attitude since semiology was first developed has consisted in gauging truth by means of a judgment concerning the *quality* of its correspondence to reality. All procedures that widen the gap vis-à-vis the analogical reproduction of reality can thus be considered to belong to the realm of *trucage*. This is more or less the position taken by Metz who, nonetheless, seems very hesitant on this point.

Towards the end of his essay, Metz affirms that "montage itself, which is at the basis of all cinema, is already a perpetual *trucage*."[3] It should be noted in passing that this assertion conforms perfectly to Descartes's hypothesis of some evil genie who tricks him into confusing illusion with reality: "I shall think," he says "that the sky, the air, the earth, colors, shapes, sounds and all external things are merely the delusions of dreams which he [the evil genie] has devised to ensnare my judgement."[4] For Metz, the evil genie is the *cinéaste*, who plays on what Samuel Coleridge called our willing suspension of disbelief in order to convince us that a matte painting or other forms of composite showing a landscape are the result of real photographic shots. Yet, a few lines after having identified montage with the notion of *trucage*, Metz nuances his argument relative to montage: "but, at the same time, [the spectator] will not be unaware that he has seen several photographs: he will not have been fooled."[5] This hesitation on Metz's part reveals the ambivalence of the word *trucage* as both a technical procedure modifying reality and a deception. By conceiving of it as something that is distanced from reality but that does not trick or fool the spectator, Metz implies that one does not entail the other. Montage is thus characterized as a technical procedure that "tricks" reality, yet without necessarily deceiving the spectator.

While the truth of the image is to be found in a judgment pertaining to the quality of the relation between the *representamen* (or sign) and its object,[6] a general theory of *trucage* and deception should not be reduced to the problem of analogy alone, which was a central concern for early

3 Metz, "*Trucage* and the Film," op. cit., p. 672.

4 René Descartes, *The Philosophical Writings of Descartes*, vol. II, trans. John Cottingham et al., (Cambridge: Cambridge University Press, 1984).

5 Metz "*Trucage* and the Film," op. cit., p. 672.

6 I am using these terms in reference to C.S. Peirce's semiotic theory. For Peirce, a sign is a representamen that stands for the object it denotes by way of another sign called the interpretant.

semiology. Therefore, while I am basing part of my argument on Metz's essay, I shall nonetheless propose a different approach, one that takes into account both the context in which *trucage* is used and the role of the spectator, understood equally here as the viewer of fiction films and that of various types of television programs. My approach will be at once pragmatic and philosophical. In fact, I will focus less on describing what *trucage* is and more on attempting to define – to parody Nelson Goodman – *when there is* trucage *for the spectator.* For *trucage* is not an *index sui.* It depends to a large extent on the knowledge and beliefs of the spectator, not only about the cinema but about the world in general.

Three Relations to Truth

But before developing this point, I should start by defining the semiotic framework that makes it possible to think about illusion and deception in film. Although the most popular conception of the filmic image is that it is a sign of reality, it is not the only one. In the Peircean perspective, the image is thought of as a *representamen* that stands for an object. However, we must go beyond the idea of plain denotation by which the image is confined to its role of referring to a "motif" by way of analogy. For the illusion produced by *trucage* belongs to a sphere I call *counterfeited reality* or, more exactly, that of the counterfeiting of *sensible* reality, since what is at stake is the counterfeiting of the visible world by an act that breaks the semiotic relation – an indexical relation – between the sign (the moving image) and its object (the visual "motif" shown). The duplicity of *trucage* thus consists in presenting an icon as though it were an index. For although it is produced independently of reality by means of an operation involving a *resemblance* to — rather than an *impregnation* of — reality, it functions *as if* it were a simple imprint of the real. Consequently, while *trucage* is based on a false relationship with reality, the truth of the image is identified with what *appears* to be the case, not with what is actually the case.

However, when the object belongs to the human realm rather than that of the inanimate world — for example the backdrop or some ecosystem providing a background for a character — the notion of deception is displaced and becomes more complicated. Take the everyday simple case offered by the setting up of a TV interview. The interviewee responds to a number of questions after which, almost invariably, the cameraman asks them to carry out a few simple actions, e.g., to sit at a desk, to leaf through the pages of a book, to walk into the studio, etc. These actions are recorded for the edit, or, more frequently, to provide the interview with some context. None of the actions would have been carried out by the interviewee had they not been asked. Sometimes, for the needs of the shoot, the interviewee will sit at a desk just to add to the illusion. The interviewee is thus part of a *trucage* of reality that consists in convincing the spectator that the segment has been shot spontaneously. As it happens, this kind of situation is a sign that refers not to lived reality but rather to a stereotype: writers write, media analysts watch TV, intellectuals read... What we have here is thus a *counterfeit of the lived*, which opens onto the vast field of *feigning*, a field that, to be brief, can be characterized as the province of the *"as if"*: the television director will ask the interviewee to do "as if" they were reading a book, etc. To the degree that the *"spectrum"* — the filmed subject, as Barthes might have put it[7] — is an accomplice of the deception, truth is to be identified here with *sincerity*.

A further semiotic relationship should be envisaged, namely that in which the sign relates or corresponds to an object that is, simultaneously, its source. This is a relationship that has traditionally been very important for theories of enunciation. This source, for instance, can be an author. I remember a filmmaker friend, Erwin Huppert, who once publicly presented a short film he (falsely) claimed had been made by S.M. Eisenstein and allegedly first shown at the International Congress of Independent Cinema at La Sarraz.[8] Using chemical products in a highly skilled way, and leaving the film in the oven for a few hours, the result was rendered even more convincing by the addition of Cyrillic inter-titles; indeed, many people — including, I believe, famed Swiss film critic Freddy Buache — were taken in! But the source can also be an era, a style, or a way of filming. In 1995, a documentary showing the dissection of an extraterrestrial discovered in Roswell, New Mexico was broadcast by television channels around the world (I will return to this example later). All such *documentary counterfeits* constitute the domain of *fakery*. In this context, truth will be measured in terms of *authenticity*.[9]

7 See Roland Barthes, *Camera Lucida* (New York: Hill and Wang, 1981).

8 The event was held 3–7 September 1929.

9 In this essay, I will not address further the counterfeiting of lived experience in moving images, for it would take us too far astray with regards to the subject of this book. (For a fuller discussion, see: François Jost, *La télévision du quotidien. Entre réalité et fiction*, Bruxelles and Paris : de Boeck Université/INA (2001). [2e édition January 2004]; see also François Jost, *The Promise of Genre*: https://www.academia.edu/220307/The_promise_of_genre. (2004); and François Jost, "*Loft Story: Big Brother* and the Migration of Genres," in Ernest Mathis and Janet Jones, eds., *Big Brother International* (London: Wallflower Press, 2004), pp. 105–122.

Metz and Trucage

Let us now return to that foundational essay by Metz, for whom the truth of the image is measured essentially in terms of its conformity with sensible reality. His system of categorization is based entirely on what is perceived by the spectator. According to Metz, there are three categories of *trucage*:

· *Imperceptible trucage*, or in other words the *trucage* that the spectator does not see: tricks designed to disguise the fact that the leading man is shorter than the leading lady; the use of a stunt double when a scene is too difficult or dangerous for the actor; a night scene shot in the daytime using a "day-for-night" procedure, etc. The kind of *trompe-l'oeil* technique mentioned above is often used in backdrops to create an illusion aimed at the spectator: a few planks of plywood are transformed into the heavy bars of a prison cell thanks to the addition — in postproduction — of a sinister grating sound; a few expertly applied strokes of paint and synthetic materials are transformed into a luxurious marble wall; a few hours of make-up and the young leading lady appears ten years older… For Metz, these *trucage* are essentially associated with the "profilmic," or in other words with the manipulation of what appears in front of the camera, the effects of which are therefore visible. It should be noted in passing that Metz does not take sound into account, as I did above.

· *Visible trucage*, which is not associated with the "profilmic" but rather with the materiality of film, either at the time of the shoot or in postproduction. Metz is thinking here about slow motion, dissolves, wipes, etc. It is reasonable to imagine that he would, if he were alive today, add a whole range of special effects made possible by digital technology, including such techniques as deformations, shrinkage, screen rotation, amongst a host of others. All these tricks are *visible* for the spectator as excrescences or additions that belong to audio-visual enunciation rather than to the world of fiction.

· *Invisible trucage*, "invisible because we do not know where it is, because we do not *see* it; but it is perceptible because we perceive its presence, because we 'sense' it."[10] This is the case, for example, of a number of films by Méliès and of various versions of *The Invisible Man* (James Whale, 1933; and more recently Leigh Whannell, 2020), in which furniture moves of its own accord, doors open as if by magic, and where the power of the gaze alone seems sufficient to make glasses fall from tables. Anxious to remain within the

10 Metz, "*Trucage* and the Film," op. cit., p. 664.

perceptive paradigm, Metz curiously fails to recognize *trucage* as a kind of manipulation, as if he wanted to avoid considering knowledge in his theory, i.e., basic knowledge of the world that the spectator is deemed to possess. Indeed, if we think that *trucage* must have been used when a piece of furniture is shown moving on its own, is it not first of all because we *know* that furniture does not move around without some form of intervention (human action or otherwise)? Clearly, since Metz's framework is so resolutely immanentist, this consideration does not cross his mind. As we shall see, it would have called his entire system into question.

The distinction between the invisible (what we do not see) and the perceptible (what we do not see but sense) is too vague to be applicable. I therefore propose that visible *trucages* be referred to instead as "ostensible" *trucages* in that they differ overtly from the laws governing our worldly existence.

Critique of the Metzian Paradigm

Although the field of *trucage* has expanded considerably since Metz's article was first published, with the number of possibilities for manipulating images having increased exponentially now that the digital age is upon us, my critique is not aimed at such historical conditions (though they will be factored in later). Rather, it concerns the very paradigm on which Metz's essay is based, namely the paradigm of the visible, or, more generally, that of the perceptible. For if we take the perceptible as our starting point, we obviously run the risk of being pulled, almost mechanically, towards the logical consequence that the cinema itself — all of it — is but a gigantic "ostensible" *trucage*. Indeed, Metz went as far as to write that "cinema in its entirety is, in a sense, a vast trucage [...]."[11] Yet at the same time, he declined to envisage postproduction processes such as color timing or sound mixing as *trucages*, though it would have been coherent with his position. In fact, he hesitated between the two senses of the word "ostensible" *trucages* to which I referred in the introduction: "ostensible" *trucages* as a procedure in itself, and "ostensible" *trucages* as an action onto others, in this case the spectator. When he asserts that the cinema is a *trucage*, he is taking an ontological position, applying what is, in the final analysis, a Platonic point of view — "ostensible" *trucages* as a degraded form of the representation of reality. On the other hand, when he talks

11 Ibid., p. 670.

about the visible or the "sensed," he adopts the phenomenological point of view characteristic of his work as a whole;[12] for it should not be forgotten that his underlying idea is that the image is analogous to reality. And, just as with Barthes, this iconicity sweeps all before it, and in particular what concerns enunciation. It is with sole consideration for the visible that he judges "ostensible" *trucages*, which is why he fails to see the limitations of his model. In his taxonomic scheme, enunciation is only applied to optical effects (dissolves, wipes, etc.).

I remember what first prompted me to work on the notion of "enunciation." I was seven years old and I did not understand why, when the hero turned off his bedside lamp, we still saw the room as if it were (almost) in full daylight. I only wrote about this feeling 25 years later… But, joking aside, the problem with a method that focuses solely on the image in order to isolate the "imperceptible," the "visible," and "the invisible because we do not know where it is […] because we do not *sense* it"[13] is that it inevitably leads us to an aporia, one that has become all the more evident in the era of the digital image.

Jean-François Leroy, the director of the "Visa pour l'image" photo festival, often complains about photojournalists' uses and abuses of Photoshop. Of the 100,000 photos presented each year to the World Press Awards, the "Nobel Prize" of photography, 20 percent are rejected out of hand on the grounds that their authors have made too great a use of software. In such photos, details regarded as aesthetically substandard are erased, contrasts are heightened to highlight particular characters or figures, and, above all, colors are saturated to draw attention to the sky or to render it more strange or forbidding. In this way, a grey slum can be transformed into a colorful, poetic landscape or cityscape.

In such cases, it would be a truly impressive feat if one were capable of determining whether such manipulations are quite simply "invisible" because we do not know "where" they occur, because we cannot see them, or else, if the *trucage* is "perceptible" "because we perceive its presence, because we 'sense' it," as Metz intimates. This kind of modification in the visual representation of reality is not new, and one would have to be either very naïve or very optimistic to imagine that it is possible to decide which photographs truly reflect bare reality and which do not.

12 Cf. François Jost, "Préface à l'édition espagnole des *Essais sur la signification au cinéma, tome I et II*", (Barcelona: Paidos, 2002); and Dominique Chateau and Martin Lefebvre, "Dance and Fetish: Phenomenology and Metz's Epistemological Shift," in *October* 148 (2014), pp. 101–130.

13 Metz, op. cit., p. 664.

For one thing, no photo can reproduce the nuances perceived by our eyes; just ask any amateur photographer looking at a disappointing silver print that was supposed to capture a glorious image of a backlit landscape bathed in sunlight! In this regard, the question of how we are to understand the meaning and scope of the concept of *trucage* acquires a certain urgency. But, at the same time, it is illusory to think that we can provide a definitive answer for it. As long as we take as our measuring yardstick conformity with the real, we can only arrive at an aporia. In the end, the problem simply becomes a matter of aesthetics (does beauty reside with nature or with art/artifice?) or ethics (should we fabricate beauty from poverty?).

The second difficulty, which I touched on above, is the highly confused status of what Metz calls the "sensed." When a door opens on its own in *The Invisible Man*, do I "perceive the *trucage*," as he says; do I "sense" it? Would it not be more accurate to say that I perceive it only if I have some *knowledge* of it (if only vaguely so). After all, I cannot exclude the idea that the spectator (say, a child) believes that it is possible for people to be invisible in our world. Many adults believe in ghosts, including those who reject the idea, as Paul Veyne commented while classifying the papers of a recently deceased friend.[14]

Let us consider a document that seems to have been recorded with a camcorder.[15] It is a video that features a child with extraordinary powers, including the ability to knot the sheets of his bed around a bed post without physical intervention and to make a cup fall from a table simply by looking at it. In other words, he seems to have the power of telekinesis. For those who believe in this kind of phenomenon, a document such as this one that has all the features of having been recorded using an analog video-recorder can serve as a testimonial or an element of proof regarding what is shown. In other words, the video works as an index for these viewers. On the other

14 Paul Veyne, *Les Grecs ont-il cru à leurs mythes? Essai sur l'imagination constituante*, (Paris: Seuil, 1983).

15 This is a short film (6'55") that was shown on the Franco-German channel Arte along with 12 other short films in a program entitled *Documents interdits* (known in English as *The Forbidden Files* and *Banned Reels*). No other information was given and, at the end of the show only, French scriptwriter Jean-Claude Carrière appeared and explained that the various documents that were shown which had perhaps fooled viewers into believing that what they depicted was true were faked and produced to test their gullibility. The document I am referring to is known as *L'Enfant* (*The Child*). It shows the strange life of Peter, as seen through videos shot by his father. *Documents interdits* is distributed by MBA Films and Arte France/La Sept. Today, many of these short films can be found on the Internet where they are sometimes presented as authentic video records of paranormal phenomena.

hand, spectators of a more rational — or more circumspect — nature will see the video as no more than a clever *trucage*. While there are many such examples in the field of images shot as "signs of the world" (there are plenty to be seen on YouTube), there also exists a (possibly smaller) number of examples of productions explicitly designated as "documentary" signs.[16] I am thinking in particular of the aforementioned sequence, shown by television channels throughout the world in 1995, of the dissection of an extraterrestrial found in Roswell. In order to authenticate the sequence, the producer had it examined, in France, by a surgeon, a cameraman, and a special effects technician, all of whom claimed that they would have been incapable of producing such a result. Their conclusion? It was not a *trucage*!

My point is the following: while it is true to claim that, in *ontological terms*, the creation for the camera of an object associated with the representation of an extraterrestrial creature is either a *trucage* or a fake, it only functions as such *if* it corresponds to the beliefs held by the spectator. From a pragmatic perspective, it is therefore less the analysis of the visible that reveals the *trucage* than the knowledge and beliefs of the individuals who see it.

That said, there is, of course, a major difference between the *trucages* of *The Invisible Man* and those used in the above-mentioned examples of telekinesis or extraterrestrial portrayals. *The Invisible Man* only aims to fool us in order to provoke astonishment and pleasure, while the other two films are intended to deceive credulous individuals into believing in the existence of paranormal phenomena. This dichotomy traces a line between, on the one hand, *trucages* associated with feigning, by which I mean *trucages* that function *as if* what they show derived from reality — semiotically speaking, *trucages* that imitate an 'utterance' of reality (that imitate a visual statement of fact); and, on the other, *trucages* that "avow" their fictional status.

In overtly fictional works, the reception of *trucage* functions in a very different way. For example, at the end of Tarkovsky's *Stalker* (1979), there is a scene that thematically is very similar to the one described above in which a child makes a number of glasses move down a table and finally fall to the floor just by staring at them. The *trucage* is not designed to make us believe that telekinesis exists in *our* world but rather that it exists, for our pleasure, in the diegetic world of the film.

Finally, it might be worth noting that Metz also omits to take into account the aesthetic aspect of *trucage* "for itself" (to use a Hegelian term), that is to say, *trucage* that is used to create a universe that has nothing to do with

16 François Jost, *Realtà/finzione. L'Impero del falso* (Milan: Castoro editrice, 2003).

our own sensible world and therefore seeks not to deceive the spectator but instead to construct an original world.

The Interpretants of Trucage

But let us leave aside the last remark above and draw the conclusion the analysis of our two examples requires of us: *trucage* only acquires meaning in relation to another sign, an "interpretant sign" as Peirce would say, namely *reality* or *fiction* in this case. In recent years, I have endeavored to account for the importance of this pragmatic interpretive process in our constant, everyday dealings with moving images by developing a system that takes into consideration three distinct worlds: the *real world*, the *fictive world*, and the *ludic world*. To avoid a lengthy digression, I will not fully explain this system (I have written about it at length elsewhere),[17] but I can sum up its central proposition as follows: "every film and television genre can be categorized with reference to the three worlds mentioned above which play the role of interpretant signs, therefore giving meaning to whatever is seen to be going on in these worlds."

In short, when the real world is the interpretant, then the images are interpreted in reference to the world we live in (though not necessarily by analogy, I might add); when it is the fictive world that is the interpretant, they are interpreted in reference to a mental world; finally, if the ludic world is the interpretant, the images will be interpreted reflexively. In the ludic world, representation (the sign) is understood to refer to itself at the same time as it is referring to an object. With regards to the cinema, this "transparency-cum-opacity,"[18] as linguists say, emerges from the field of cinematic enunciation and sometimes engenders a feeling of *gratuitousness*, in the sense that the latter is a component in the definition of play as "a non-imposed, *gratuitous* [my italics] physical or intellectual activity in which one takes part for *fun* and *pleasure*" (*Larousse* Dictionary).

Film and television genres are therefore interpreted with regards to their respective positions on the following diagram (Fig. 3.1):

17 See Jost, *La Télévision du quotidien*, op. cit., "Les Mondes de l'image: entre fiction et réalité," in *Recherches sémiotiques/Semiotic Inquiry*, (23) 1-2-3, (2003), pp. 165–184. (In English: *The Promise of genre*: https://www.academia.edu/220307/The_promise_of_genre, "*Loft Story*: *Big Brother* and the Migration of Genres," in *Big Brother International*, pp. 105–122.

18 François Récanati, *La transparence et l'énonciation* (Paris: Seuil, 1979).

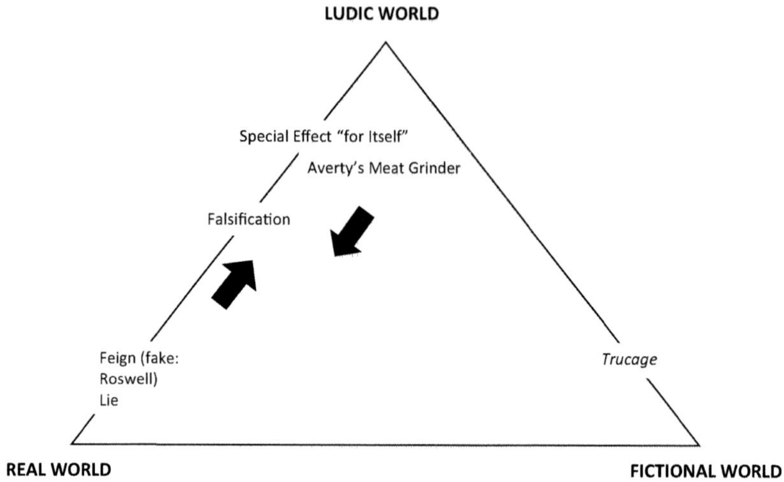

Fig. 3.1 The three worlds of cinema and television.

It follows that *trucage* will function, both semiotically and aesthetically, in quite radically different ways whether it is considered in relation to one or the other of the three worlds of reference posited here. For in each case, as we shall see, a different conception of truth is called into play.

Trucage and Reality. The Real World

Among the visual documents that take their meaning from a direct reference to the real world, archives occupy a prime position. Although these documents, once categorized as archival, do not exist in raw form anymore but are immediately reorganized by the discourses they serve to support, it is important that their integrity not be called into question. Yet, as the following two examples show, this is not always the case. The first example is supplied by the film *Un spécialiste* (*The Specialist*, Eyal Sivan and Rony Brauman, 1999), which deals with the Eichmann trial. Let us leave aside the fact that the soundtrack comes from Israeli television, which, strictly speaking, is itself a *trucage*. What really deceives the spectator are other modifications carried out digitally: for example, the reflection of the faces of the judges on the glass cabin in which the Nazi is held, and the travelling shot inside the courthouse, which was built from scratch for the purposes of the film. Whether they are linked to the "visible" or to enunciation, the *trucages* are invisible, imperceptible, and undetectable. And what are we to think of the sound effect accompanying Eichmann's quill pen in order to render the scene more readable?

The second example, *Apocalypse* (Isabelle Clarke and Danielle Costelle, 2009) is a series of archival documentary films about the Second World War broadcast by the TV channel France 2. To make it more accessible to young people, the filmmakers colorized the archival footage, no doubt thinking that war is prettier when it is in color.

In both cases, *trucage* can only be identified by means of some collateral information (e.g., a knowledge based on interviews with the filmmakers, or on the history of techniques, etc.). Without such information, archival material, even after it has been subject to iconic manipulations, appears primarily as an index. The film, in other words, functions *as if* it were a simple duplication of the real, though the finished product is separated from reality by a number of operations and is better characterized by resemblance (iconicity) than direct contact (indexicality). It is not a *fake* in the sense of an entirely fabricated document passed off as genuine, but a series of *falsifications* based on remodeling and modifying a pre-existing document. If examples such as the first one tie in to the notion of feigning (*trompe-l'oeil*), the second case seems more like an outright lie, since it deceives the spectator about the semiotic nature of the images themselves.

The Fictive World

But what about *trucages* that modify our vision of history? One example is provided by *Citizen Kane* (1941) in which Orson Welles showed his press baron-protagonist in the company of Adolf Hitler. Another is to be found in *Zelig* (Woody Allen, 1983), where this time it is the character played by Woody Allen who is seen with the Nazi dictator. More recently, in *Forrest Gump* (Robert Zemeckis, 1994), the hero is seen meeting Presidents Kennedy, Johnson, and Nixon. These manipulations have, however, raised some questions. Speaking of *Forrest Gump,* Olivier Mongin, the editor of the French journal *Esprit* spoke quite despairingly of "electronic revisionism," intimating that such effects could be used to give young people a false view of history. However, unlike the archival manipulations described above, which are actively deceptive in terms of the degree of indexicality and iconicity of the documents featured, in fiction, *trucage* is used in what can be described as a largely iconic context and can in no manner be considered as an existential proof of something, which would be the case if a historical figure had been made to disappear, suffering a fate similar to that of Lavrentiy Pavlovich Beria in the days of the old Soviet Union. This approach, which necessarily involves the fictionalization of existing archival footage of the Nazi dictator or of various US Presidents, is linked to a phenomenon of enunciative feigning (or "pretending") that renders the two documents compatible. Playing with codes and bringing the enunciation of a film up to date are just two of the more or less subtle pleasures associated with reflexivity in all its forms. Therefore, we are grateful to these filmmakers for showing us that they possess a specialized knowledge that concerns cinema in the way they can imitate the look and feel of a given genre or period in film history, all the while providing us with the feeling that we belong to the *happy few,* an elite that shares with them a specific form of cinephilic knowledge. This use of *trucage* is, to a great extent, ludic.

Trucage for the Sake of Amusement and Art

When it is applied to the realm of the ludic, falsification is by no means scandalous. Examples include faking archival footage in which contemporary actors imitate long-dead politicians, the use of digital techniques to age current news items in order to create an anachronistic effect, or else creating the illusion that a live singer is performing side by side with a long-deceased one in the form of a hologram.

In cases such as these, manipulation of sound and image is not necessarily intended to provide an illusion of reality; it is unshackled by the rules of traditional representation. This includes the entire gamut of films and television programs in which audio-visual manipulations are used for either fun or art. At this point, it is less a question of deceiving the spectator than of providing amusement. We are dealing here with *special effects rather than trucage*: procedures are used to distort sound and image for its own sake rather than for the sake of sensory illusion and deception.

Nevertheless, the distinction I am drawing here is not airtight. And there is nothing to guarantee that an image designed to be playful will not be interpreted in reference to the real world. I am thinking of the well-known TV variety show by Jean-Christophe Averty (*Les raisins verts*, 1963) where we see a newborn ground into minced meat by a meat grinder. Although it was very clearly a plastic baby, and although the *trucage* is fairly basic, the inferences required to recognize the ontological status of the image were, it seems, ineffective on some viewers, and the broadcaster ended up receiving an avalanche of complaints! Similarly, there is nothing to stop us from seeing the Roswell extraterrestrial as a highly successful pastiche and to categorize it as belonging to the ludic world.

But let us return to the curious paradox of a *trucage* that does not seek to deceive, a kind of trick that I prefer to call a "special effect." Averty's TV program could doubtless be viewed from this perspective. But, to focus on something a little more contemporaneous, I would like to consider a recent film, Steven Spielberg's *The Adventures of Tintin* (2011).

First, it is important that we dispense with some common beliefs that have accompanied the emergence of synthetic images. Many people considered it worthwhile to warn us against these images which, having been transformed into mere icons, no longer have any link to reality. The critiques offered up by these denigrators of digital technology belong to the same family as those warning against electronic revisionism. Of course, they are both characterized by a basic error of analysis. In the case of the synthetic images of characters in Spielberg's film, the truth of the matter is that they are not "pure" products of the imagination, as it was when Tintin's inventor Hergé *drew* them. Rather, they are based on "performance *capture*" which, as we know, is distinct from traditional drawn animation in that it consists in recording not only the movements of live actors but their most intimate facial expressions as well. While creators of cartoons not only had to think of how to represent movement but also take *timing* into account, performance capture captures both. Traditional animation was a *cosa mentale* in which the slightest error assumed gigantic proportions: a mistaken appraisal of timing could render a comic effect insipid; to correct the error, the entire sequence had to be re-done. Leaving aside the skills of the illustrator — which could be considered a somatic index — animation was tied, broadly speaking, to the notion of the icon. Performance capture, on the other hand, is at the origin of a style of animation based on an indexical contact with reality, with all subsequent digital manipulations (which distinguish it from "traditional" footage) being performed intentionally.

Moreover, it is the same intentionality that accounts for the other aspects of the film, such as the clothes the characters wear or the various backdrops which are all computer generated: it could be argued that *everything* in the image was planned. If the term "profilmic" refers to everything that has been laid out in front of the camera, then the synthetic images are, without doubt, the most "intentional" of all images. However, it is clear that the intention of the filmmakers is not to fool the spectator into believing that the characters are real human beings but rather to recover, in the use of color and the rendering of traits and texture — or, in other words, in the plastic aspect of the film — something of the original comic strip, even if the result, as Belgian *bande dessinée* artist François Schuiten notes, is different from Hergé's famous "clear line" design.

It is important to observe that these three components of the image — the semiotic, the iconic, and the plastic — are not in phase with one another. While performance capture delivers "realistic" movement in the sense that it conforms to what would otherwise have been provided by live action shots, the costumes and backdrops situate the film in a comic strip/cartoon universe, creating for the spectator a kind of semiotic chimera in which the sensible is mixed with the mental, and real movement with pictorial imaginings. Or, to express the idea in another way: recognition of real movement becomes mixed with the pleasure of invention, the illusion of reality with enunciation. As a special effect, Spielberg's use of performance capture breaks with the logic of *trucage* as used, for instance, in *The Ghost and the Darkness* (Stephen Hopkins, 1996), in which (nearly) the same technique was applied to create the illusion of a lion attacking a man. While in Hopkins's film the technique was used to ensure that the scene did not end with the recording of a real bloodbath, in *The Adventures of Tintin* the point is to exploit digital technology to create imagery rather than to fool the spectator.

In order to avoid thinking of *trucages* as a set of technical procedures the efficiency of which is subordinated exclusively to visibility, we need to first trace a boundary between *trucages* that pretend to be real (feigning reality as a *trompe-l'oeil*) and those associated with fiction. Only the first ones imitate an uttering of reality. And, from this point of view, how they are created matters little. In every era of film and television, we find procedures that, to borrow Descartes's famous maxim, "advance [as if] masked." Only time succeeds in unmasking them. For instance, for the contemporary spectator, changes in the backdrop behind the characters are more obvious in Méliès than, later, in back projections, which themselves are more "visible" than video inserts, which are themselves more apparent than digital inserts. But none of this is very important. Beyond the inevitable variations in degrees of "realism" and reality effects, these procedures all aim to create a perceptive illusion and to foster a belief in spatial or geographical unity.

On the other hand, *trucages* associated with fiction are not necessarily designed to deceive. They testify to their enunciative status and are indifferent to mimetism; the special effects in *The Adventures of Tintin* provide a good example. After the era of *trucages* designed to produce images of dinosaurs and similar creatures in naturalistic contexts, contemporary special effects now present us with impossible worlds, worlds that sometimes defy Euclidian space or that construct "irrealist" points of view. Long gone is the time when Hitchcock refused to put the camera in a fridge! Today, the camera routinely films around objects, scenes, and characters in a continual whirl that Laurent Jullier compares to a merry-go-round ride.[19] And this takes us back to the ludic world from which our journey started.

My point, in the end, is that discussions of *trucages* should emphasize both their changing status with regards to a given world of reference (say, reality or fiction) and the problems this gives rise to in any given audio-visual document. For instance, we have established that in fiction *trucages* are used to arouse pleasure in the spectator. However, in genres associated with the real, their use raises issues that are at once aesthetic and ethical. Not only do they call into question authenticity, which is the ontological "promise" of archives and *reportages*, but their use is also criticized for what, in fiction, is appreciated, namely aestheticizing or "prettifying" the real. A filter can be used to emphasize a feeling of mystery in a fiction film, but in photojournalism or documentaries its use runs the risk of inappropriately rendering misery supportable. Thus, the Brazilian photographer, Sebastião Salgado, known for his *reportages* on people working in poor conditions, has often been accused of using his subjects to create beautiful images. Not surprisingly, Susan Sontag, a filmmaker in her own right and a specialist of photography, has attacked Saldago's work for the "inauthenticity of the beautiful" in it.[20] In this regard, aesthetics (using a filter, photoshopping to make the image beautiful, etc.) ushers in ethical questions that, contrary to what the adversaries of "electronic revisionism" believe, simply do not have the same meaning in the different contexts of feigned utterances of reality and fiction.

Bibliography

Barthes, Roland. *Camera Lucida* (New York: Hill and Wang, 1981).

Chateau, Dominique, and Martin Lefebvre. "Dance and Fetish: Phenomenology and Metz' Epistemological Shift," *October* 148 (2014): 101–130.

Descartes, René. *The Philosophical Writings of Descartes*, vol. 2, translated by John Cottingham et al. (Cambridge: Cambridge University Press, 1984).

———. *Méditations métaphysiques* (Paris: GF Flammarion, 1992).

Jost, François. "The Promise of Genres," *Réseaux* 6, no. 1 (1998): 99–121. https://www.academia.edu/220307/The_promise_of_genre.

19 Laurent Jullier, *L'écran postmoderne. Un cinéma de l'allusion et du feu d'artifice* (Paris: L'harmattan, 1997).

20 Susan Sontag, *Regarding the Pain of Others* (New York: Farrar, Straus, & Giroux, 2003).

———. *La télévision du quotidien. Entre réalité et fiction* (Bruxelles and Paris: De Boeck Université/INA, 2001).
———. *Realtà/finzione. L'impero del falso* (Milan: Castoro editrice, 2003).
———. "Les Mondes de l'image: entre fiction et réalité," *Recherches sémiotiques/Semiotic Inquiry* 23, no. 1-2-3 (2003): 165–184.
———. "*Loft Story: Big Brother* and the Migration." In *Big Brother International*, edited by Ernest Mathis and Janet Jones (London: Wallflower Press, 2004): 105–1130.
Jullier, Laurent. *L'écran postmoderne. Un cinéma de l'allusion et du feu d'artifice* (Paris: L'harmattan, 1997).
Metz, Christian. "*Trucages* and the Film," translated by François Metzler, *Critical Inquiry* 3, no. 4 (1977): 657–675.
Récanati, François. *La transparence et l'énonciation* (Paris: Seuil, 1979).
Sontag, Susan. *Regarding the Pain of Others* (New York: Farrar, Straus & Giroux, 2003).
Veyne, Paul. *Les Grecs ont-ils cru à leurs mythes? Essai sur l'imagination constituante* (Paris: Seuil, 1983).

About the Author

François Jost is Professor Emeritus at Université Paris 3 — Sorbonne-Nouvelle where he created the Centre d'Études sur l'Image et le Son Médiatique (CEISME). He has written and edited more than 30 volumes on cinema and television and has published some 270 articles and book chapters. He is also the author of a novel, *Les Thermes de Stabies,* and has written scripts and directed several works of fiction and documents, including *La mort du révolutionnaire, hallucinée*, which won the Critics' Prize at the Festival du jeune cinéma of Hyères in 1979.

4. Realism, Illusion, and Special Effects in the Cinema

Marc Furstenau

Abstract
Special effects have long been described as illusory or as the techniques for the creation of illusion. In this respect, special effects have been understood to be in the service of the generally illusory nature of cinematic representation itself, with realistic depiction reduced to the status of illusion. Against an illusion theory of cinematic realism and of realistic pictorial depiction generally, the argument here is that realism in the cinema is thoroughly a matter of style and that it need not become entangled with the apparently related but in fact tangential question of illusion and the related questions of perception, belief, and "reality." Considering a particular film by George Méliès, *The Vanishing Lady*, which realistically represents a magic trick, an "illusion," a more limited and specific account of realism is proposed.

Keywords: special effects, realism, illusion, perception, depiction, George Méliès

Special effects have been put to many different uses in the cinema, but one of the most common and most important has been to aid in the creation of realistic depictions. A consideration of special effects, that is, must be able to account for the role that they have played in the history of cinematic realism. To say so, though, is to raise perhaps the thorniest question about film, if not about art more generally. As John Hyman has said, "The use of the term 'realism' in art history is so messy and unclear that perhaps it should simply be abandoned."[1] It is arguably as messy and unclear in film studies.

1 John Hyman, *The Objective Eye: Color, Form, and Reality in the Theory of Art* (Chicago and London: University of Chicago Press, 2006), p. 181.

Lefebvre, M. & M. Furstenau (eds.), *Special Effects on the Screen: Faking the View from Méliès to Motion Capture*. Taylor & Francis Group, 2022
DOI 10.5117/9789462980730_CH04

It is certainly the case that the question of realism in the cinema has been the source of some considerable theoretical puzzlement. It is at the heart of perhaps the most enduring and seemingly intractable debate in film theory. Hyman does not, in the end, advocate abandoning the term "realism" but tries instead to clarify its meaning. I will also try to do so for cinematic realism, which poses some particular problems when compared to drawing, painting, and sculpture, which are Hyman's concern — although my solution will be to emphasize the basic similarities between film and these other arts, to think of special effects that is, in relation to the more general issue of depiction and of film as an art of depiction, of picture-making. To say this, though, is to directly confront another problem. "A picture," as Hyman has also said, "is certainly an unusual object of vision."[2] A film, understood as an elaborate picture, is just as unusual, but what makes it so is not that it is an illusion, something false and misleading, as is very often argued. The most common mode of depiction in film is, arguably, realistic, and it is important to understand the role that special effects have played in the elaboration of cinematic realism, as a particular depictive mode. A principled distinction must be established between realism and illusion, though, in order to provide a basis for such an account of special effects in the cinema.

I.

Tom Gunning has made perhaps the strongest plea recently for a theory of realism in the cinema that could incorporate special effects, and especially digital special effects, which have been such a hard problem for film theorists. Gunning very helpfully points out what the source of the difficulty has been. The problem is partly the result of the apparently irreconcilable views on the matter of realism. As he says:

> While cinema has often been described as the most realistic of the arts, cinematic realism has been understood in a variety of ways: from an aspect of a sinister ideological process of psychological regression to infantile states of primal delusion, to providing a basis for evidentiary status for films as historical and even legal documents.[3]

2 John Hyman, *The Imitation of Nature* (Oxford: Basil Blackwell, 1989), p. 42.

3 Tom Gunning, "Moving Away from the Index: Cinema and the Impression of Reality," *Differences: A Journal of Feminist Cultural Studies*, (18)1 (2006), p. 29.

It is difficult, he implies, to conceive of an account of realism that would reconcile these very different conceptions. Indeed, each of these seems to be offered as a caricature of any sensible account of realism, representing the extremes to which theorists are led on the matter rather than reasonable proposals for an understanding of what it is to create realistic depictions in the cinema. Worse, as Gunning argues, is that claims about realism tend to take the form either of a panegyric, a polemic, or a plea. "Cinematic realism has been praised as a cornerstone of film aesthetics, denounced as a major ploy in ideological indoctrination, and envied as a standard for new media."[4]

More importantly, Gunning argues that a theory of cinematic realism would have to overcome the longstanding bias for the *photographic*, thought to be the constitutive material basis of the cinema as a medium, understood as fundamentally "indexical," in the now very familiar term of art, as the primary source, that is, of the cinema's capacity to represent realistically. This is indeed the most valuable part of Gunning's argument and is precisely the challenge for a viable theory of cinematic realism in the digital era. If "realism" is to be a useful concept for talking about a kind of depiction, it should be applicable to any particular form or medium and not limited to some privileged material thought to have a greater innate capacity for realistic depiction. It should be applicable, in a consistent way, across the digital divide and be able to explain a certain use of special effects, across the range of possible uses, in both the so-called analog and digital cinemas.

This is the problem that Gunning addresses, and he proposes a solution by first of all addressing this central tenet in most discussions of realism in the cinema. The specific notion of "indexicality" has tended to be the guiding assumption, with the unfortunate consequence, Gunning says, importantly, that what would otherwise be thought of as a legitimate mode of cinematic representation — namely the hand-drawn animated film — has largely been set aside, separated from the cinema understood as a photographic medium and given, therefore, theoretical short shrift. Indeed, the animated film has largely been on the margins of theoretical speculation about the cinema — not merely set aside but often contrasted with what is supposed to be the more genuine "photographic" cinema.

Animated films are thought to have more to do with painting and drawing, forms that the cinema was supposed to have overcome as a new and inherently more realistic mode of depiction. Like drawing and painting, though, animated films must be understood as capable of creating realistic representations, the source of any such realism having necessarily to be

4 Ibid.

found somewhere other than in an "indexical" link between the representation and what it is a representation of, between depiction and the depicted.[5] As the cinema in the digital era is becoming more and more like animated film, though, it is thought to be more and more like drawing and painting, which is to say *limited* in its capacity for realistic representation. The advent of digital special effects and computer-generated imagery (CGI), which, if not hand-drawn, is more "animation" than photo-cinematic representation and which constitutes an ever-greater part of the depiction itself, can, in this respect, only be understood to have wholly transformed cinema into what it had traditionally been compared against. Theories of realistic depiction in the cinema, based on the assumption that the cinema is photographic, or "indexical," seem to have come to a dead end, unable to account for the fact that the marginal form of animation has now apparently become definitive. As film becomes (computer-)animated film, it seems to have become wholly special effect. This is the problem according to Lev Manovich, for example. Asked in an interview with Dan North about how to "define and delimit the subject" of special and visual effects, Manovich says that to answer that question means "addressing the identity of the moving image for the last twenty years. This is the central question: do we have *cinema*, or is it *all* special effects. And what is 'special effects'?"[6]

An account of special effects, though, should not have to depend on the assumption that, if they were to become so extensive, as they seem to have in the digital era, they would effectively alter the very identity of the cinema itself, that we would no longer even have anything recognizable as "cinema" but something new that consists largely — or only — of "special effects." Manovich seems concerned (or is concerned to show) that the cinema has now assumed the identity of animation, that is, a form of depiction consisting largely or entirely of special effects — defined as *intentional* depiction as opposed to the *causal* generation of photographic imagery. But this suggests that it is either one or the other, that cinema is either (mainly) photographic

5 The term "indexicality" is used in a rather restricted and limiting fashion in much of film theory, to mean simply a physical or "existential" link between depiction and depicted, to mean, that is, representations that are created through some direct causal process, so that photographs are thought of as comparable to fingerprints or bullet holes. This is not the place to enter into this debate, but for a more detailed discussion and an alternative account of the index in film theory, see Martin Lefebvre and Marc Furstenau, "Digital Editing and Montage: The Vanishing Celluloid and Beyond," in *Cinémas*, (13)1-2 (2002), pp. 90ff. See also Martin Lefebvre's essay in this collection.

6 Dan North, "Afterword: A Conversation with Lev Manovich," in Dan North, Bob Rehak, Michael S. Duffy, eds., *Special Effects: New Histories/Theories/ Contexts* (London: Palgrave, 2015), p. 267.

and special effects are a marginal aspect, or that it is a form of animated film, a generalized form of special effects. Special effects have always been a part of cinematic depiction, though, and if that is so, it should not matter that there are more of them now, or even too many. Indeed, when would "more" become "too many?"

II.

The array of techniques that have been grouped under the category of special effects has always been an integral aspect of the cinema. If there has not yet been a satisfactory explanation of the role that they have played in the history of cinematic realism, if the fact of special effects in an otherwise "realistic" medium has been a difficult one to account for, it should not mean that, now that there seem to be more of them, we solve the problem by simply declaring that the "identity" of the cinema has changed as a result of their supposed increase (having presumably reached some sort of "tipping point"). In fact, it is important to note, Manovich is a champion of the animated film and welcomes computer-generated imagery. Like Gunning, he rightly notes that, historically, animation has been marginalized in film studies and especially in film theory in its preference for questions about the supposedly "photographic" nature of the cinema. Yet Manovich reproduces, and merely reverses, the traditional dichotomy of film theory, simply preferring one kind of cinema (a fully animated, special-effects cinema) over another (having, by fiat, posited the two different cinematic "identities" in the first place) and suggesting that simply because there are more animated films now, or that more films contain elements that could be described as animated, one kind of cinema has been replaced by another kind of cinema, or by something that is no longer even cinema as traditionally defined. He does not explain how an account of the cinema could in fact include both its photographic and animated forms, or any other forms that might be imagined, and how these might relate to the question of realistic depiction.

We need an account of the cinema, and of realistic depiction in the cinema, that does not depend on the assumption that the cinema once had one identity that has been replaced by another, such change understood to be the effect somehow of special effects themselves. If special effects are in fact integral to the history of cinematic depiction, as Stephen Prince has persuasively argued, then the fact of their relative ubiquity over the course of the history of the cinema should not change our conception of the cinema as a mode of depiction. It is simply the case that special effects, along with various other techniques, have always been used to create realistic depictions, and they continue to do so. It is, moreover, difficult to determine if it is indeed the case that the use of special effects is any more extensive now than at any other time in the history of the cinema. It is not even clear, really, how one would make such a determination in any case. The cinema must be understood as a mode of depiction that has used many different media of depiction in order to create both realistic and non-realistic representations. The latest media — digital media — are in a fundamental sense no different. This is Gunning's point, and it is worth pursuing and developing, establishing, as he says we should, some sense of historical continuity.

But what is it that can be said to have remained the same across the digital divide? On what basis can a continuity be established? Gunning, unlike Manovich, wants to say that there is a specific feature or constitutive component of cinematic depiction that can be found in both photo-generated cinema and in animated film, including its recent manifestation in the digital cinema. Its identity, that is, has remained the same. Manovich argues that the cinema's original identity has been restored, that there has been a photographic interlude in the cinema, which has ended, and the cinema is now what it was at the beginning — animation. What had been obscured, during that interlude, was the fact of "artifice," so that realism could not be seen for the illusory effect that it was, thought to have been generated spontaneously rather than being the result of formal effort, of stylization. He charts the initial phase in the history of the cinema, which he says was from "animation to cinema," a fact obscured in most histories:

> Once the cinema was stabilized as a technology, it cut all references to its origins in artifice. Everything that characterized moving pictures before the twentieth century — the manual construction of images, loop actions, the discrete nature of space and movement — was delegated to cinema's bastard relative, its supplement and shadow — animation.[7]

The importance of this for Manovich is that the specific value of animation is reasserted as the primary value, which is to undermine a realist aesthetic and to reincorporate depictive or representational strategies from the avant-garde. What he describes as the "language of new media" undermines what he presents as its opposite, a "totalizing realism" that can only be understood

7 Lev Manovich, *The Language of New Media* (Cambridge, MA and London: MIT Press, 2001), p. 298.

as illusion. He compares the new media to "traditional cinema or realist theater, which aims at all costs to maintain the continuity of the illusion for the duration of the performance. In contrast to such totalizing realism, new media aesthetics has a surprising affinity to twentieth-century avant-garde aesthetics."[8] As a major representative of the avant-garde, Manovich approvingly names Bertolt Brecht, whose "strategy of revealing the conditions of an illusion's production, echoed by countless other leftist artists, has become embedded in hardware and software themselves."[9]

This is, though, effectively no different than the realist's claim that the means of producing the illusion, or realism, is "embedded" in the original "hardware and software" of the cinema, in the camera and the photosensitive film. In both cases, realism is conceived of as illusion. The choice, then, is to decide whether that is good or bad, whether it should be revealed or whether it is what we should take pleasure in. In most other accounts of special effects, the concern is not with the ideological unmasking, through avant-garde techniques (now understood to be somehow "embedded" in new digital media), but rather to accommodate ourselves to the fact that illusion is just what the cinema has always offered and to show how that can be created either photographically or through the use of special, or "visual," effects. This is the argument that Stephen Prince makes. Special effects, or what Prince calls "visual effects," have, he insists, been a constitutive part of the cinema since its inception. Insofar as the cinema is used to create fictions (he distinguishes between fictional and documentary cinema, only the latter depending on what he calls "photographic truth"), it

> marries a compelling presentation of sound and moving images to the depiction of what are often worlds of the imagination. The more perceptually convincing these imaginary worlds can be made to seem, the more virtual and immersive the spaces of story and image become [...]. [D]igital tools enhance the perceptual realism of effects sequences.[10]

8 Ibid., p. 207.

9 It is not at all clear how "avant-garde strategies" could be "embedded" in software and hardware. It is partly against just such an assumption — that styles, or "strategies," are something that can be manifested in the material or physical medium itself — that I am arguing. For a more comprehensive argument against such "materialist" accounts of style as embedded in a medium, see Joseph Margolis, "Mechanical Reproduction and Cinematic Humanism," *What, After All, Is a Work of Art? Lectures in the Philosophy of Art* (University Park, PA: Pennsylvania State University Press 1999), pp. 101–127.

10 Stephen Prince, *Digital Visual Effects in Cinema: The Seduction of Reality* (New Brunswick, NJ and London: Rutgers University Press, 2012), p. 183.

As such enhancements, he argues, these "digital tools," or visual effects, "help to fulfill one of cinema's historical objectives."[11] While I largely accept Prince's historical argument — being doubtful though that an art like the cinema can have anything like "historical objectives" — my objections are to his notion of "perceptual realism" and to his insistence that the pleasures of the cinema, enhanced by visual effects, are, as he says, the "pleasures of illusion."[12]

This is the most common way in which the cinema is understood, though, and special effects are now, for the most part, understood as what can create, or most powerfully contribute to, the "illusion" of realism. It is very often said that, when we watch a film with special effects, what we see are illusions, and that they, in turn, and necessarily, draw our attention to the illusory nature of cinematic representation itself. Special effects have been defined as "artificially produced illusions."[13] Michelle Pierson describes special effects as "techniques of cinematic illusion."[14] A special effect, says Norman Klein, is a "technological marvel [that] controls an illusionistic environment."[15] Specific genres, especially science fiction films, are said to exploit the "illusionistic potential of special effects."[16] Prince concludes his account of special effects, specifically digital visual effects, with a chapter entitled "The Pleasures of Illusion." The most recent

11 Ibid.

12 This is the title of the conclusion of his book.

13 Dan North, Bob Rehak, and Michael S. Duffy, "Introduction," in Dan North et al., eds., *Special Effects: New Histories/Theories/ Contexts* (London: Palgrave, 2015), p. 1. By "artificially produced illusions," the authors seem to be positing a distinction between those produced through the use of special effects and the illusions created (naturally?) by the basic photographic processes of cinematic representation. They argue that "film has always been a mechanical medium whose major attraction was the technological reconstitution of fictional worlds using any available resources," p. 4. The main resource, of course, has been photography, and they note that much of the debate about special effects is a debate about the consequences of their use for what they call the "truth claims of the medium," insofar as it has depended on photography and thereby taken advantage of the "privileges ... conferred upon the photographic image by its indexical relationship to its subject," p. 7. But, they add, and here they seem to be establishing the distinction between artificial and other modes of producing illusions, "we should not imagine that photography and film were ever available in a pure, honest state free from the spectres of illusion and deception," p. 7.

14 Michelle Pierson, *Special Effects: Still in Search of Wonder* (New York: Columbia University Press, 2002), p. 49.

15 Norman M. Klein, *The Vatican to Vegas: A History of Special Effects* (New York: New Press, 2004), p. 13.

16 Angela Ndalianis, "Special Effects, Morphing Magic, and the 1990s Cinema of Attractions," in Vivian Sobchak, ed., *Meta-Morphing: Visual Transformation and the Culture of Quick Change* (Minneapolis: University of Minnesota Press, 2000), p. 255.

digital special effects are understood to increase the power of illusion, to increasingly encompass all aspects of cinematic representation. Joel Black argues that "while special effects once allowed filmmakers to present glimpses of the unreal world of dreams (*Un chien andalou*, Luis Buñuel, 1929; *The Wizard of Oz*, Victor Fleming, 1939; *Spellbound*, Alfred Hitchcock, 1945), today's sophisticated effects are increasingly used to create a heightened illusion of reality itself" — what he calls the "reality effect."[17]

There is an astonishing degree of consensus on the matter. Illusionism is the most common basis upon which special effects are analyzed, with the corollary that realism itself is an illusion. In these and other similar accounts of special effects in the cinema, the concept of illusion is only occasionally argued for, or carefully presented as a concept. It is indicative, though, of the degree to which our understanding of visual or pictorial depiction is based on a concept of illusion that such accounts merely begin with that assumption and then go on for the most part to describe, in various but often quite similar ways, what it is to experience such illusion. Most notable about recent accounts is the degree to which they seek to revise earlier more suspicious or dismissive claims about the illusory nature of the cinema, presenting it now as a "pleasure" rather than ideological threat, while continuing to accept the assumption that it is indeed an illusion. Several recent works on special effects explicitly contrast their approach to the sort of ideological critique of the illusory nature of cinematic representation that had once been common. But they do not (or cannot) dislodge the basic assumption, given the more comprehensive philosophical perspective deeply embedded in such critiques, which tends still to be a widely accepted one.

From this perspective, illusion is understood as a triggering of perceptual recognition capacities, and realism as the effect of our feeling as though we are having, to some degree or other, a visual experience comparable to that of perception itself. Even Gregory Currie, one of the most vociferous opponents of the illusion theory of cinema — what he calls the "myth of illusion" — cannot in the end escape the notion of a link between realism and perception and so cannot, despite explicitly saying that he will, offer a clear alternative to illusionism. He describes and rejects "the claim that film is realistic in its capacity to engender in the viewer an illusion of the reality and presentness of characters and events portrayed." This view is, he says, "widely believed. Studio publicity writers, Marxist critics of Hollywood film,

17 Joel Black, *The Reality Effect: Film Culture and the Graphic Imperative* (New York and London: Routledge, 2002), p. 8.

as well as more conventional figures like Arnheim, constantly assert it."[18] Yet Currie, wedded as he is to a strict cognitivism, characterizes our experience of likeness or similarity between depiction and the object depicted as the result of what he calls, after Flint Schier, "triggering recognitional capacities."[19] He argues that the "recognition capacity [is] triggered at a lower, sub-personal, level of functioning; it is something that happens within me."[20] Currie rejects a strictly Cartesian theory of mind but offers a homuncular explanation in its stead (which is, I think, no better). He argues for the concept of "person as constituted by a hierarchy (or complex of hierarchies) of intelligent creatures or homunculi [...]. The person or agent himself occupies the top level of the hierarchy."[21] Ultimately, he admits: "Perhaps this view of picture recognition constitutes a sort of illusionism, about pictures [...]. But this is an illusionism we can live with."[22]

I argue, however, that in our accounts of realistic pictures, we need not live with any sort of illusionism. Realism can be described in such a way that does not require the positing of such novel entities as "homunculi," which operate within us at a "sub-personal" level. Realistic depiction is what we deal with directly, as a matter of depiction, as a style, which we may recognize as such without being fooled, deceived, or "triggered" in any way, with the often-suggested corollary that what we experience is confused with or somehow intimately linked with our experience of reality itself. The pleasures of realism are not the pleasures of illusion. But illusion is also not the risk of realism. We need not think of realism either as what must be repudiated or embraced *as* illusion. Manovich argues that animation, as the basis of the cinema, achieves an anti-realist aesthetic but in the same way that the realists argue that the photographic basis of cinema ensures realism. Manovich argues that it has always been illusion. Gunning also argues that it has always been illusion. Manovich never clarifies what he means by illusion. Gunning does. For Gunning, the illusion is what is created by the fact that cinematic representation effectively confronts us with what can only be experienced as real movement, and that this is what underpins an "impression of reality." Gunning is one of the few film theorists to try to work out a coherent account of illusion, which he effectively offers as the

18 Gregory Currie, *Image and Mind: Film, Philosophy, and Cognitive Science* (Cambridge: Cambridge University Press, 1995), p. 20.

19 Ibid., p. 80, note 1.

20 Ibid., p. 83. See Flint Schier, *Deeper Into Pictures: An Essay on Pictorial Representation* (Cambridge: Cambridge University Press, 1986), p. 61.

21 Ibid., p. 84.

22 Ibid., p. 86.

basis of any coherent theory of realism. He does not subscribe to the view that cinema produces a "totalizing realism" (by which he seems to mean a *complete* illusion), nor that animation performs an avant-garde, anti-realist function. Rather, he tries to find that aspect of the depiction that creates enough of an illusion to create the "impression of reality." Insofar as he does so, though, he is still committed, it seems to me, to a theory of realism as an effect created in our minds, causally, by a depiction — a depiction that seems to *cause* the same, or some similar, effects as that of reality itself. Is the cinema an illusion? Are special-effects techniques in the service of the overall depictive goal, which is to create an illusion, the "impression of reality," even if only a partial one? I argue that realism is not best defined as a means for the creation of the "impression of reality" but is rather a matter of depiction. This distinction has to be clarified, and a more careful consideration of Gunning's argument provides the opportunity for such clarification.

III.

The challenge, Gunning says, for a newly articulated theory of cinematic realism, is to account for the changes to the cinema, but also for what has historically been marginalized in past accounts of the cinema. It would, he argues, have to account for not only photographically generated imagery but animated film as well, and even for the special effects created by CGI. Yet this is just the gap that seems more pronounced in film theory now than ever. As Gunning says:

> Given that as a technical innovation cinema was first understood as "animated pictures" and that computer-generated animation techniques are now omnipresent in most feature films, shouldn't this lacuna disturb us? Rather than being absorbed in the larger categories of cultural studies or cognitive theory, shouldn't the classical issues of film theory be reopened?[23]

Indeed. The most basic question of such "classical" theory is, of course, the question of realism. Gunning's solution is to make the very fact of "animation" — meant literally as imbuing the image with movement or *animating* it — the central defining feature of realistic depiction in the cinema. The one constant, he argues, among all the various kinds of films

23 Gunning, op. cit., p. 34.

— photo-generated, hand-drawn, or digitally synthesized — is movement, and it is on this basis that he argues for a theory of cinematic realism based on an analysis of the depiction of motion. Yet, for Gunning, movement seems like something more than mere depiction, and his "realism" is based on an attempt to identify what seems most manifestly "real" in the cinema, establishing what I will call a "relation of identity" as the basic criterion of realistic depiction. By contrast, I will argue that realism is characterized by other kinds of relations, of depictive resemblance or correspondence, of likeness or similarity — recognized as such. For Gunning, though, realism has to be based on some stronger form of relation, in a more direct connection to reality itself. Central to his argument for a realism of movement, of literal "animation," is this claim: "We do not just *see* motion and we are not simply affected emotionally by its role within a plot; we *feel* it in our guts or throughout our bodies."[24] The movement in film is not, that is, just "like" movement, it does not merely "resemble" movement, it does not only "correspond" to the movement as we see it in reality. Rather, it seems as though it *is* movement, that it is effectively *identical* to real movement.[25]

Gunning builds upon the argument that Christian Metz made, in what Gunning calls one of his "pre-semiotic" essays, from his early "phenomenological" phase. While usually associated with "postclassical film theory,"

24 Ibid.; emphasis in original.

25 The issue of movement and its representation has arisen in art theory more generally, around the question of the depiction of movement in drawing and painting in otherwise still imagery. In *Art and Illusion*, in a chapter entitled "Conditions of Illusion," Ernst Gombrich argues that Diego Velázquez solved the problem of representing motion in painting in his work *Las Hilanderas* (or *The Fable of Arachne*), c. 1657–60. The picture depicts a woman at a spinning wheel that appears to be turning. Gombrich says, about the history of the puzzle of representing such scenes with moving objects in them, that it "needed the imagination and skill of a Velázquez to invent a means of suggesting that 'uncertain glimpse' in the spinning wheel of the *Hilanderas*, which appears to catch the so-called 'stroboscopic effect,' the streaking after-image that trails its path across the field of vision when an object is whizzing past" (Ernst Gombrich, *Art and Illusion: A Study in the Psychology of Pictorial Representation* [Oxford: Phaidon, 1960], p. 192). Any other of the techniques used in the past, he says, "destroys the illusion of movement" (ibid.), which is what, he implies, Velázquez has managed to create. John Hyman has rejected this claim, noting that, in Gombrich's argument, "the word 'illusion' is being used carelessly... For there is no illusion: nothing appears to be other than it is... If the word 'illusion' is used to describe this painting, this is really a kind of hyperbole, which is quite misleading. What is remarkable about the painting is that the motion of the wheel can actually be seen and not be inferred from any other cues (Hyman, *The Objective Eye*, op. cit., p. 204). I will argue that even in film, where one can indeed see movement, what one sees is quite simply the depiction of movement, as Hyman insists. Even if it is arguably depicted more easily and effectively in film, it is no more an "illusion" than it is in the painting. I return to Gombrich, to Hyman's critique of Gombrich's illusion theory, later in this essay.

Gunning says, Metz in this essay is concerned with a question that is at the heart of classical theory, the question of the "impression of reality" in the cinema. His answer, while distinct from, say, Bazin's concern with "the possibility of grasping the mysteries of Being,"[26] as Gunning puts it, is similarly phenomenological. Against the view of the "later apparatus theorists," a version of which Metz himself would in fact become and who saw realism as "a dangerous ideological illusion,"[27] Metz is concerned in this earlier work only, or otherwise, with the task of describing the simple phenomenological fact of the impression of reality. Gunning quotes Metz: "Films release a mechanism of affective and perceptual participation in the spectator [...] films have the appeal of a presence and of a proximity."[28] Metz sets out, then, Gunning says, "to give this physiological effect a phenomenological basis."[29] Of all the aspects of cinematic representation thus conceived, for Metz, it is "*movement* [. . .] that produces the strong impression of reality."[30] Gunning summarizes Metz's view: "Spectator participation in the moving image depends, Metz claims, on perceiving motion and the perceptual, cognitive, and physiological effects this triggers. The nature of cinematic motion, its continuous progress, its unfolding nature, would seem to demand the participation of the perceiver."[31] For Metz, Gunning says, "the motion we see in film is real, not a representation,"[32] a claim that Gunning accepts, with a slight hedge. "Ultimately," he says, "I think there is little question that phenomenologically we see movement on the screen, not a 'portrayal' of movement."[33] Calling it "real" in this respect, does not, however, "commit us," Gunning insists, "to the nonsensical position that we take the cinema image for reality, that we are involved in a hallucination or 'illusion' of reality that could cause us to contemplate walking into the screen, or interacting physically with the fictional event we see portrayed."[34]

There is, though, a confusion at the heart of this account, manifested first of all in the phrase "the impression of reality" as the basic description of realism, which I would rather define as the "depiction of reality." The

26 Gunning, op. cit., p. 25.

27 Ibid., p. 41.

28 Ibid., pp. 40–41. See also Christian Metz, "On the Impression of Reality in the Cinema," in *Film Language. A Semiotics of Cinema* (Chicago: University of Chicago Press), 1974, pp. 3–15.

29 Ibid., p. 41.

30 Ibid.

31 Ibid., p. 42.

32 Ibid., p. 43. This is, I should emphasize, Gunning's summary of Metz's argument.

33 Ibid., p. 44.

34 Ibid.

distinction is an important one. To suggest that the cinema strongly — and more strongly than other modes of depiction — creates such an *impression*, and especially that it does so insofar as one aspect of that depiction is in fact not itself a depiction, to suggest that it somehow, if only partially, transcends the very condition of depiction, is to confuse basic categories. By contrast, to say that realism is the "depiction of reality" is to commit oneself to the principle that a depiction is a depiction all the way through, that there are no elements of a representation that can be described as *not* a representation, no elements of a portrayal that can be described, as Gunning says, as "not a portrayal." Something cannot simultaneously be and not be something. This is the basic principle of non-contradiction, which can be traced back to Aristotle, who argues, in the *Metaphysics*, that "It is impossible to hold (suppose) the same thing to be and not to be."[35] Moreover, and despite Gunning's insistence that this does not commit us to an illusionism, it cannot be understood as anything else. To say that realistic depiction in the cinema is not a *complete* illusion, to say that we are not absolutely convinced by the depiction to the point that we might try to physically engage with it, that we might "contemplate walking into the screen," is to hold out a wholly unlikely, indeed "nonsensical" view, if it is meant to describe all experiences of realistic depiction. This is in fact, though, another failure to avoid contradiction if one insists, as I do, that something either is or is not an illusion. Either we are entirely convinced that a depiction that we see is not a depiction, and we are then experiencing an illusion, or we see that it is a depiction, and therefore not an illusion. Illusion cannot be partial.[36]

35 Aristotle, (*Metaphysics* IV 3 1005b24 &1005b29–30). I hesitate to add that, in *Hamlet*, as one significant dramatic account of the Aristotelian principle, Shakespeare insists on the only logically consistent conjunction "or" when Hamlet ponders the question, "To be or not to be." The presence of the ghost in the play, though, suggests that one may both "be" and "not be," but the question is whether this is a hallucination, an illusion that Hamlet is experiencing in his fevered state of mind. And, of course, in stagings of the play, the ghost illusion, that is, the specific theatrical technique (or special effect) used to create the representation of a ghost, is created in ways that either suggest that it is an illusion, in Hamlet's mind, *or* not. This is an effect that should have a key place in the history of special effects.

36 It is Rudolf Arnheim, of course, who made the most famous and influential argument that the cinema creates a "partial illusion." In *Film as Art*, having described the state of cinema at the time, in the early 1930s, as lacking certain realistic features — it was "silent" and in black and white but also flat, etc. — still, it was powerfully realistic, insofar as it was convincing, or at least more convincing than any prior form of visual representation. Despite its limitation, he argues, everyone who goes to see a film nevertheless accepts the screen world as being true to nature. This is due, he argues, to the phenomenon of 'partial illusion.'" See Rudolf Arnheim, *Film as Art* (Berkeley: University of California Press, 1957), p. 15. Nowhere, though, does he give

Illusions are possible, of course, as in the rare cases of *trompe-l'oeil*, which can fool a viewer, as I in fact was when visiting the Galleria Spada in Rome, which features in the courtyard a magnificent instance of a forced perspective optical illusion, an architectural *trompe-l'oeil*. It is a colonnade, a corridor with columns on either side, at the end of which is a sculpture that appears life-sized (at least two meters tall) at its apparent distance of about 40 meters. The statue is in fact only about eight meters away, and the statue only about 60 centimeters tall — the effect created by shortening the last columns and raising the floor. The effect was wholly convincing, until I decided to walk down the colonnade. Having seen what I thought (mistakenly) was a large sculpture at some considerable distance and having discovered upon a moment's further inspection that it was rather a much smaller statue that was much closer, I was now regarding it as what it really is (and realizing the mistake I had made). Such a *trompe-l'oeil* is an illusion until it is suddenly not an illusion — or, more precisely, it is understood as a particular kind of depiction, one that can create illusion. An illusion, that is, strictly speaking, is that particular phenomenon that can be prompted by a (very particular kind of) depiction but is not identical to it, nor is it a general feature of all realistic depictions, which tend in almost all cases not to be realistic by causing false belief — as this architectural trick had done — but through some other means.

The "nonsensical" possibility that Gunning dismisses, then, is indeed just the one that I experienced, believing that I saw a large statue, mistaking a fictional depiction for what is real. The moment that I realize it is only a depiction, though, I do not begin to seek those aspects of it that are "not depiction" in order to account for the illusory effect it first had on me. When I see the shorter columns, I understand what is true. I treat my experience as the brief moment of false belief that it indeed was (what an illusion, strictly defined, creates) and which has been replaced now by a correct belief, which is that what I see is a depiction, indeed, that what I had seen all along was a very realistic depiction, of the sort that we can call an illusion, or that we should say more accurately is the sort of depiction that can create the possibility of illusion, understood strictly as the source of a false belief. The depiction that I saw is in fact carefully designed to create a false belief through the application of a variety of very specialized techniques that do not even really offer much guarantee of success. Even if achieved, it is necessarily fleeting and easily overcome. Moreover, such success depends

anything like a persuasive account of this supposed psychological "phenomenon," which he derives from the research of the Gestalt psychologists in Germany with whom he trained.

on the artist having made a great deal of effort for what is often a very slight return. Illusions are vanishingly rare occurrences, very difficult to achieve and, when successful, usually very brief (they can inevitably be "seen through"). Most works of depictive art, the vast majority, do not make such an effort and do not seek to produce anything like an illusion. Illusion is the rarest, indeed the most aberrant, form of depiction (or, more precisely, a rare possible *effect* of depiction), the least likely to succeed on its own terms and, in the critical judgement of most, the least aesthetically satisfying form of depiction. It is, in this respect, an unlikely candidate as the exemplary instance of all (realistic) depiction, even if thought of as illusory only to some degree.

Most accounts of (realistic) depiction, though, like Gunning's, depend on the assumption that all depictions are, to some extent, even if usually a small extent, illusory, a view that depends on what I insist is the false notion that illusion can be a matter of degree, or isolated and distinguished from other non-illusory elements of a depiction. Gunning's claim is that the one part of the depiction that is illusory is the depiction of movement, that at least in this respect we are fooled, or, as he says, it is the one part of the depiction that exceeds or transcends the status of depiction or "portrayal," existing as real movement. The only way that this can be understood, though, as I am suggesting, is to say that this is the part of the depiction that succeeds in creating — and in fact maintains — an illusion. According to Gunning, in the case of the illusion at Galleria Spada, I would have to accept that what I see when I am in the grip of the illusion is the "real" length — the length of 40 meters — rather than acknowledge that only about eight meters really separate me and the small statue.

Illusion is, though, an either/or wager. The colonnade at Galleria Spada, the "special effect" designed to create the possibility of illusion, realizes its design goal only when I am fully gripped by a false belief. Looking at it now, no longer in thrall, having overcome my false belief, knowing the truth, but still marveling at its achievement as a realistic depiction (it is a complex realistic depiction of a life-sized statue) and admiring the tricks that it uses, I am now marveling at it, correctly, *as* a depiction, which is what it always was, regardless of what I may have (falsely) believed about it at any moment — but that rare kind of depiction intended to cause illusion. To say that something like this is an illusion, then, is only to say that it is a depiction that *may* be experienced — but only *incorrectly* — as *not* a depiction. An illusion, in this respect, is in fact a demonstration of the principle of non-contradiction in that, while I falsely believe that it is real, I am in fact confusing something for something that it is not. The contradiction is resolved, as it must be, certainly in cases like this where it is so easy to overcome one's false belief, the moment I realize what it is and therefore what it cannot be.[37]

Another way to put the matter is to say that a depiction is a material thing, while illusion is non-material. It is a belief, a *mental phenomenon* (putting aside for the moment the difficulties with the term) — to experience an illusion is to have a (false) belief. A picture can create a false belief, but it itself is not that belief. That some (very few) pictures (and other forms of representation) can, indeed, produce that experience of false belief that we call "illusion" does not tell us anything important about what pictures, even realistic pictures, are or how they normally (usually) are experienced, even if we reduce the illusory aspect to a partial degree or to an isolated component, like movement. Richard Wollheim has made this point while suggesting what is so attractive about the illusion argument, especially in the form that has it that all pictures, all depictions, are to some degree illusory:

> That representation is a kind of partial or inhibited illusion, working only for one sense or from one point of view, and that to see something as a representation is to enter into this illusion so far as is practicable, is a view that has obvious attractions: even if only because it offers to explain a very puzzling phenomenon in terms of one that is easy and accessible.[38]

Gunning's account of movement as that one part of the cinematic depiction that is "not depiction," that is itself "real," and that thereby gives cinematic realism its strong charge, is an instance of effectively giving into the temptation to explain the puzzling nature of depiction itself by recourse to the notion of partial illusion, if he is indeed asking us to accept the claim that, somehow or other, when I see movement in a film, I actually see it, or have

37 My interpretation and argument here is based loosely on what is still the most powerful refutation of the so-called "argument from illusion" in philosophy, which J.L. Austin offers in *Sense and Sensibilia*. Rejecting philosophical accounts of perception itself as fundamentally illusory, Austin asks if illusion applies at least to depiction, or if certain modes of depiction are more powerfully illusory, considering *trompe-l'oeil* paintings but also magic tricks, photographs, television, and, significantly, the cinema: "Is the cinema a case of illusion? Well, just possibly the first man [*sic*] who ever saw moving pictures may have felt inclined to say that here was a case of illusion. But in fact it is pretty unlikely that even he, even momentarily, was actually taken in… One might as well ask whether producing a photograph is producing an illusion—which plainly would be just silly." See J.L. Austin, *Sense and Sensibilia* (Oxford: Oxford University Press, 1962), pp. 26–27.

38 Richard Wollheim, "On Drawing an Object," in *On Art and the Mind: Essays and Lectures* (London: Allen Lane, 1973), p. 23.

the "impression" that — or the false belief or feeling that — I am really seeing movement. Despite its attraction, though, and as Wollheim says, such a view is "misguided. For, in the first place, it does not fit our experience."[39]

Gunning's argument is exemplary insofar as it does what most arguments about depiction do, which is to try to find what distinguishes pictures from other kinds of representations, e.g., literary descriptions, and to isolate what makes them unique, on the assumption that they are unique. Pictures are thought uniquely to "trigger" a kind of response in us that is comparable to — if not identical with — perceptual response. If this were so, it would be odd, and indeed pictures are often thought of as odd things, as I noted in my introduction — "unusual," as John Hyman says. Yet what makes a picture seem so unusual is often the result of some conceptual confusions in our descriptions — our theoretical accounts — of pictures rather than anything about the pictures themselves, and certainly not as a result of pictures somehow transcending their very status as pictures, even if only partially, offering something that is "not picture." Hyman is responding to another argument by Wollheim, a famous and influential argument that he proposed as an alternative to the notion of a picture as an illusion but which in fact does not escape from the logic of the theory of illusion. Pictures do not fool or deceive us, Wollheim argues, but they do depend on, or generate, what he describes as a "special perceptual capacity."[40] This is contrasted with our ordinary perceptual capacity or "straightforward perception," which is "the capacity we humans and other animals have of perceiving things present to the senses."[41] By contrast, "the special perceptual capacity is something that animals may share with us but almost certainly do not, and it allows us to have perceptual experiences of things not present to the senses."[42] By this he means an experience of what is depicted as though it is what we are seeing ("present to the senses") and not merely a picture of it.

For Hyman, this distinction is "unintelligible unless taken in conjunction with the causal theory of perception."[43] It is a causal theory of percep-

39 Ibid. "To enter into an illusion (as opposed to seeing through it) depends by and large on a subversion of our ordinary beliefs; whereas to look at something as a representation seems not to necessitate either denial or erroneous belief *vis-à-vis* reality," (Wollheim, "On Drawing an Object," op. cit., p. 23).

40 Richard Wollheim, "Seeing-as, Seeing-in, and Pictorial Representation," *Art and its Objects: An Introduction to Aesthetics* (New York: Harper and Row, 1968), p. 217. The passages I quote are those cited by Hyman, in his argument against Wollheim, on which I depend here.

41 Wollheim, p. 217.

42 Ibid.

43 Hyman, *The Imitation of Nature*, op. cit., pp. 20–21.

tion — and the causal theory of depiction that is derived from it — that Hyman rejects. Wollheim, he says, despite his reservations about illusion, is "the most influential exponent of the causal theory of depiction."[44] As a theory, Hyman says, "it proposes that to see something is to have a visual experience which was caused by what it is an experience of," and he argues that "the theory is not coherent."[45] It is not coherent insofar as it derives from an incoherent theory of vision according to which pictures create a very particular (not to say peculiar) experience in our minds, which is, namely, to perceive something "not present to our senses." Hyman insists that this cannot be:

> For "present to the senses" ordinarily means "perceptible." A sound is present to the senses if it is within earshot, and a picture is present to the senses if it is within view. Thus Wollheim would appear to be comparing straightforward perception (of the perceptible) with a special and, on the whole, distinctively human capacity for the perception of the imperceptible, which is absurd.[46]

What Wollheim is arguing is that, as he himself puts it, pictures afford us "visions of things not present."[47] Hyman offers a summary of such a position:

> A vision in this sense is not a revelation, but a visual experience in the causal theorist's sense, that is a constituent of vision, a mental state which visible objects cause in the sighted. Human beings are blessed with the capacity for unusual visual experiences that have not been caused by what they are experiences of, and a picture will generally cause a sighted human being to have an experience of just this sort. The task of the painter, on this view, is to produce something that will cause in the spectator a particular mental state of this sort.[48]

This is, Hyman says, ultimately a theory of pictures as illusion, or of pictures defined as what causes in us a kind of illusion, to some degree or other.

Hyman compares Wollheim's account to the arguably even more famous and influential account offered by Ernst Gombrich, which Hyman

44 Ibid., p. 20.
45 Ibid., p. 41.
46 Ibid., p. 21.
47 Wollheim, Richard, *Art and its Objects* (Cambridge: Cambridge University Press, 1980), p. 218.
48 Hyman, *The Imitation of Nature*, op. cit., p. 21.

summarizes: "The psychological effect that Gombrich has in mind is certainly the visual experience which would ordinarily be caused by seeing whatever the artist depicted."[49] As Gombrich explains, rejecting a "totalizing" or complete illusion for some version of partial illusions, "what may make a painting like a distant view through a window is not the fact that the two can be as indistinguishable as is a facsimile from the original: it is in the similarity between the mental activities both can arouse."[50] Gombrich acknowledges that there are more and less illusory pictures, that there are techniques available to create representation that are merely recognizable as what they depict and those that function to create some "triggering" effect, so that we experience a mental event comparable to that triggered by objects of perception themselves. The history of art, for Gombrich, is to an important degree a search for such techniques as distinct from other techniques. The history of art, Gombrich says, in a passage quoted by Hyman, "may be described as the forging of master keys for opening the mysterious locks of our senses to which only nature herself originally held the key."[51] Gombrich is not, as Hyman notes, offering a general theory of depiction but only of illusory pictures. When pictures are "illusory," though, in such a view, they are so on the basis of their relation with — or their sharing the identity of — perception itself, in contrast to the view I outlined above. Hyman concludes that "Gombrich's theory is of deliberately inexact scope: his topic is illusion, and illusion, he contends, is 'always a matter of degree'."[52]

The failure of such accounts, even in partial form, for Hyman, is the failure to explain how a natural phenomenon like perception can in fact be reproduced in such a compelling way by a picture. Moreover, the theory

49 Ibid., p. 22.

50 Gombrich, E.H., "Illusion in Art," in E.H. Gombrich and R.L. Gregory, eds., *Illusion in Nature and Art* (London: Duckworth, 1973), p. 240. Quoted in Hyman, *The Imitation of Nature*, op. cit., pp. 22–23.

51 Gombrich, ibid, p. 201; quoted in Hyman, ibid., p. 23. For a particularly effective account of such a developmental and progressive history, which sees it as a movement towards what Gombrich calls the "forging of the master keys" to perception, through ever more convincing realistic depiction, see Arthur Danto's essay, "The End of Art." Danto argues that history has ended, and that while it may once have been the goal of (some) art, it is no longer, and he gives a reason why, presenting it as his "historical thesis" (derived from an idiosyncratic reading of Hegel's *Aesthetic*): "that the task of art to produce equivalences to perceptual experiences passed, in the late nineteenth and early twentieth centuries, from the activities of painting and sculpture to those of cinematography — in the fact that painters and sculptors began conspicuously to abandon this goal at just about the time that all the basic strategies for narrative cinema were put in place." See Arthur Danto, "The End of Art," in *The Philosophical Disenfranchisement of Art* (Columbia University Press, 2005 [1986]), p. 99.

52 Hyman, *The Imitation of Nature*, op. cit., p. 23. See Gombrich, "Illusion in Art," op. cit., p. 196.

of perception itself is a flawed one, and, as Hyman says, neither Wollheim nor Gombrich "has actually defended the causal theory of perception."[53] That is to say, they advance a causal theory of depiction, which implies a causal theory of perception, which they do not in fact propose or defend. The assumption, on the basis of which depiction is described as the depiction of the illusory nature of perception, is never in fact argued for but is implicit, merely presented as a possibility in the examples of those pictures that can in fact create the effect of illusion. Those who do explicitly advocate a causal theory of perception, the most significant being H.P. Grice and P.F. Strawson, do so on the basis of the fact that we are susceptible to error or confusion in our experiences of perception. That we may be fooled, that we may misperceive, that we may be susceptible to illusion is offered as proof that perception itself is a causal phenomenon, since such errors must be mental errors. So, perception itself is what causes correct and possibly incorrect "accounts" in our minds of what we see, with "what we see" being only and necessarily indirectly experienced. Perception itself, in this respect, following this line of thinking, is thought of *as a kind of illusion*, and this is what is supposedly reproduced or depicted in realistic pictures — they are representations of the illusory nature of perception itself — or, pictures recreate or simulate perception in this informative respect (but on an erroneous theory of perception).[54]

53 Hyman, ibid.

54 While this may seem to be a rather arid philosophical distinction being made here, it is a question that is at the heart of the contemporary technological culture of audio-visual representation, where the promise continues to be that a final and complete coincidence of depiction and perception will be achieved — comparable to the vaunted "singularity" that is promised with "AI," when computer operations would coincide perfectly with thinking or consciousness. Both claims are incoherent, mistaking and combining one thing with another. Ever since the marketing campaigns for the "Cinématographe," from which we have inherited the apocryphal stories of the "panicking audiences" confusing the image of a train with a real train, the media industry has sought to convince us that a "perfect" representation is always on the horizon, with the next television or projection system, or with Virtual Reality (VR) or Augmented Reality (AR) or IMAX-3D. At the heart of such claims, dutifully pursued by film and media theorists, is an incoherent causal theory of perception and of depiction. Patrick Maynard is another philosopher who has sought to dispense with such theories to account for visual representation and who makes this link with the discourse around media technologies. "Instead of thinking," Maynard says, "like many theorists, of depictions primarily as triggers for environmental seeing routines, we shall need to think of environmental scene, feature, object, and so forth [terms that he introduces for pictorial analysis] as recognitional capacities put to work, often through what I am calling transfer, in a certain kind of activity: the perception of depictive pictures *as* pictures" (2003, p. 59). This may seem paradoxical, he notes, but having our perceptual capacities "put to work" or "stimulated" by a picture is not to say that we think, under the spell of illusion, that we are really seeing what is depicted. Moreover, that is not even the most

This is why recourse is so often made to that instance of depiction, to *trompe-l'oeil*, that seems so clearly to create a genuine instance of illusion, which is then generalized as a basic feature of all pictures, despite, as I say, its being so rare.[55] As Hyman says, though, and in contrast to what is effectively claimed in such accounts, "illusions are not simply misperceptions, for illusions typically drive a wedge between perception and judgment."[56] As I have explained above, such a wedge can be removed, and a correct judgment can be made. That correct judgment, though, is the judgment that we in almost all cases make, correctly, when we see a picture — namely,

significant part of our experience of pictures, he argues, insisting that "the idea that interpreting pictures has as its main point the stimulation of those processes is wrong," and instead he will "argue for the autonomy of pictorial perception against a tradition of treating it as environmental perception under constrained conditions, against a powerful techno-economic movement towards reducing those 'constraints.'" See Patrick Maynard, "Drawings as Drawn: An Approach to Creativity in an Art," in *The Creation of Art*, Berys Gaut & Paisley Livingston, eds. (New York: Cambridge University Press, 2003), p. 84 n. 9. If VR or AR, that is, are conceived technologically as reducing the technical constraints on depiction that are thought to be preventing them from becoming indistinguishable from that which it is thought to be producing (as an illusion), such a goal is impossible, based as it is on an incoherent account of both perception and depiction.

55 Lev Manovich, for instance, in his chapter in *The Language of New Media* entitled "The Illusions," begins: "Zeuxis was a legendary Greek painter who lived in the fifth century B.C. The story of his competition with Parrhasius exemplifies the concern with what was to occupy Western art throughout much of its history. According to the story, Zeuxis painted grapes with such skill that birds began to fly down to eat from the painted vine." Lev Manovich, *The Language of New Media*, (Cambridge, MA and London: The MIT Press, 2001), p. 177. He compares the latest 3D-graphics technology, "RealityEngine," with such (mythical) painterly skill. "In terms of the images that it can generate, RealityEngine may not be superior to Zeuxis. Yet it can do other tricks unavailable to the Greek painter," p. 177. Such historical comparisons, I argue, are uninformative and indeed perpetuate erroneous theories of both depiction and perception by positing that illusion has been the main goal in the history of art, even if only "Western" art. Manovich cites his source for the story: Stephen Bann, *The True Vine: On Western Representation and the Western Tradition* (Cambridge: Cambridge University Press, 1989). The story was first recounted by Pliny in his *Natural History*, in which Parrhasius is the winner of the contest. While Zeuxis fools the birds, in his turn "Parrhasius himself produced such a realistic picture of a curtain that Zeuxis, proud of the verdict of the birds, requested that the curtain should now be drawn and the picture displayed." Zeuxis, realizing his mistake, "yielded up the prize, saying that whereas he had deceived birds Parrhasius had deceived him, an artist" (Pliny, *Natural History*, XXXV, pp. 65–66; quoted in Hyman, *The Imitation of Nature*, op. cit., p. 24). Hyman: "If it seemed to Zeuxis just as if he saw a curtain only until he had realized his mistake, then Pliny's story is not, after all, about a trompe l'oeil. Hence, the causal theorist (of perception) is committed to the view that a trompe l'oeil painting of, say, a broken pane of glass is one that is apt to cause a visual experience such as it would be correct to report thus: 'It seems to me just as if I see a broken pane of glass'. In other words, the causal theorist (of perception) is committed to a causal theory of depiction of the least ambitious sort, viz., one that is restricted to trompe l'oeil." (p. 25).

56 Hyman, *The Imitation of Nature*, op. cit., p. 24.

that it is a picture and not some mitigated form of misperception, or still to some degree, in some part, a misperception, on par with the misperception posited (implausibly) as an irreducible part of perception itself. As Hyman says, "the causal theorist (of perception) is committed to a causal theory of depiction of the least ambitious sort, viz. one that is restricted to trompe l'oeil."[57] He adds: "The causal theory of depiction proposes that pictures and hallucinations are visual experiences of things not present to the senses; and the causal theory of perception proposes that to see something is to have a visual experience which was caused by what it is an experience of. These theories are obviously entangled."[58]

A causal theory of depiction is only intelligible, he insists, if the causal theory of perception is. Such a theory is not plausible, though, to the extent that it depends so heavily on what can in fact be distinguished from normal perception, from what Hyman calls "aberrant perceptual phenomena,"[59] like hallucinations or illusions. These can be accounted for in terms other than those provided by a causal theory of perception. "A picture is an unusual object of vision," then, Hyman explains, "not because it allows us to see what is not there to be seen, which is absurd, but because when looking at a painting, the natural answer to the question 'What do you see?' is a description of the depicted scene, and not the disposition of pigments."[60] The "disposition of pigments" in a painting, that is, is not the basis upon which one may claim that what we see "is not there," in the manner of a hallucination. It *is* there. We *can* see it. *How* we can (also) see a depicted scene is indeed a complex matter, but it is not to be explained by saying we are having something like a hallucination, that I am in a state of false belief. I believe that I am looking at a picture, and that looking at a picture, a painting, for example, means that if those pigments are arranged in particular ways, I will see the depiction of a scene. That is what makes the picture realistic. And I can admire it just on such terms.

Rather than a relation of identity, that is, which is what Gunning, specifically on his reading of Metz, is in fact offering, and which in fact is based on just such confusions, the relation between a representation, a portrayal, a depiction, and what it is a depiction of is of the kind I have suggested above. To resemble, to be like or similar to, or to correspond to is what in fact defines depiction. Accounts of special effects largely argue that they

57 Ibid., p. 25.
58 Ibid.
59 Ibid.
60 Ibid., p. 42.

reveal the degree to which the cinema is an illusion. Rather than trying to accommodate them to a theory of realism, special effects have, for the most part, been understood as the means for the creation of illusion; or, theories of realism or realistic pictorial depiction and perhaps especially cinematic realism have been based on, or reduced to, a theory of illusionism, which depends, usually implicitly, on an unconvincing causal theory of perception. So I argue that a different basis is required for a viable theory of realism that would be able to accommodate both traditional and digital special effects, that does not depend on any degree of illusion, and that is not based on a theory of physiological triggering.

Gunning is right when he says that a theory of realism would have to be able to explain what have been thought of as marginal instances of the cinema, such as animated film, or supposedly marginal aspects of cinematic production such as special effects, computer-generated imagery, and so on. He is wrong, though, to say that such a theory of realism must be based on the fact of movement as what is truly, phenomenologically "real" about cinematic depiction. A theory of realistic depiction needs to be based on the principle that any instance of such (realistic) depiction is depiction, as I say, *all the way through*. The relation between a realistic depiction and reality is not a relation of identity, or causally generated, as it is so often claimed, implicitly or explicitly. This is what has made theorizing about realism, about realistic depiction, so difficult, what has been the source of so much of the puzzlement, if not conceptual confusion: the notion that *realism* depends on there being some real relation with "reality," a relation of identity. There need not be. Indeed, real*ism* is precisely defined by another kind of relation, by being *about* reality, not by sharing any of the properties *of*, or somehow blending with or blurring into, reality. For special effects to be fully integrated into any account of cinematic depiction, then, and especially realistic depiction, a clearer account of realism is required.

IV.

In the final section of this essay, I will consider the historical origins of the notion that special effects in cinema are an illusion and that they contribute in some fashion, in this respect, to realistic depiction. My example will be the cinematic representation of what may well be an actual illusion. There is, in this example, often thought to be a supposed coincidence of subject matter and form — that the illusion depicted is depicted itself through illusion. I will argue, though, that while what is depicted may be described as an illusion, the manner of its depiction is not. It is realistic, but in the sense in which I have been proposing. I will, that is, consider an example that is often offered up in accounts of special effects in the cinema, the origins of which are traced to a particular moment in the early history of the medium, when Georges Méliès began to film the magic tricks, or "illusions," that he had been performing for some years, mainly at the Théâtre Robert-Houdin in Paris, which he had purchased in 1888 from the widow of the famous and very influential conjurer, Jean-Eugène Robert-Houdin.[61]

Among the many tricks that Méliès performed there, one was the very familiar illusion of "The Vanishing Lady," achieved on stage in the traditional manner, with the use, that is, of a trap door, through which the "lady," Méliès's assistant, would escape, after he had draped her with a large cloth, obscuring the contrivance, producing the illusion that she had "vanished." The trick is thought by Paul Hammond to have been originally designed by another magician, Bautier de Kolta.[62] John Frazer, in his history of Méliès's films, does not mention de Kolta. He does, though, describe the films' origins in Méliès's stage magic performances and his inventive updating of this familiar stage illusion in particular, which had been performed by many magicians and which audiences in Méliès's theater would likely have known well, likely expecting the use of a trap door. Apparently hoping to confound audience expectations, though, Méliès (and/or de Kolta) added as a part of the trick's *mise en scène*, a large, unfolded page of a newspaper, laid in front of or beneath the chair on which the lady was seated. As Frazer explains: "The paper was used to persuade the audience that the woman could not simply drop through a trap in the floor. Actually, the trick on the stage was

61 On the history of his purchase and the refurbishing of the theater, with some detailed descriptions of the conjuring shows and other entertainments that he offered there, see Paul Hammond, *Marvellous Méliès* (London: Gordon Fraser, 1974), p. 19ff; and John Frazer, *Artificially Arranged Scenes: The Films of Georges Méliès* (Boston: G.K. Hall and Co, 1979), p. 28ff.

62 While he does not discuss the trick as performed on stage, Hammond says that the filmed version "was inspired by one of Bautier de Kolta's illusions" (p. 30). It is not clear from Hammond's description if Méliès himself performed the trick on stage at his theater or if it was de Kolta, who also appeared regularly at the Robert-Houdin. Hammond also describes the influence on Méliès of the English magician John Nevil Maskelyne, who, with George Alfred Cooke, ran the Egyptian Hall in London which Méliès had visited, seeing the two perform. Having acquired his own theater, Méliès developed his own program, building upon others he had seen, as Hammond describes. "From 1888 to 1907 Méliès invented thirty or so 'theatrical compositions,' many of them variations of older illusions (Maskelyne's), a number of which had been put on the market by his friend, the theatrical supplier Voisin, who had helped him make his start in conjuring" (Hammond, p. 22). As Hammond notes, describing this trade in illusion, in the apparatuses themselves, or their designs, audiences would have been quite familiar with them from fairgrounds and music-halls and in other theaters before seeing them represented in films.

accomplished by a combination of a stage trap, a flap cut in the newspaper and a break-away chair seat."[63]

It is the particular nature of this instance of *persuasion*, as Frazer describes it, that I would like to consider, insofar as it shows the possible formal complexity of stage illusions themselves. It is, moreover, as I will argue, such complexity that is in fact the very subject of the realistic cinematic representation of this particular illusion, as created by Méliès. It is important, in the case of these different forms of representation, that we ask what precisely is or is not persuasive about them, and what we think we as an audience are being persuaded of.

In his filmed version of the trick, *Escamotage d'une dame chez Robert-Houdin* (*The Vanishing Lady,* 1896), we see Méliès enter the stage from a door in the backdrop, to our right. He greets us, the audience (or the imagined audience in the theater), and then opens the door again for the lady who joins him on stage. He then unfolds a large page of a newspaper, places it on the floor, brings the chair forward, setting it on top of the unfolded sheet. The lady takes the seat, is covered by the cloth, and, after a slight gesture by Méliès, a magical flourishing of the hand, she suddenly disappears, the chair revealed to be empty when he pulls the cloth away. She is then replaced by a skeleton, which in turn itself vanishes, and finally she reappears. Standing up from the chair, she smiles to the audience, she and Méliès bow, and they leave the stage through the same door, returning briefly for a second bow, before the film ends. The trick is achieved in the film not through the use of a trap door and cut paper, but rather with a "stop-substitution," a form of in-camera editing, whereby the filming is halted, allowing the lady to leave the set, to have the skeleton placed on the chair, removed, and for her to return again. At each point, the filming is stopped and resumed after the substitution has taken place, giving the impression of a magical process of disappearance and reappearance.

The film, his seventieth in his first year as a filmmaker, includes the first known use by Méliès of the stop-substitution technique.[64] The technique

63 Frazer, op. cit., p. 60.

64 See Frazer, p. 60. Méliès began making films in 1896, the year this trick was filmed. In that year, he is known to have made a total of 78 films, only two of which were extant at the time Frazer was writing, one being *Escamotage d'une dame*. He may, that is, have used the technique before, but this is the first still existing example of its use. The other film from 1896 listed by Frazer is *Une nuit terrible* (*A Terrible Night*). An updated filmography in Elizabeth Ezra's *Georges Méliès: The Birth of the Auteur* (Manchester: Manchester University Press, 2000) adds a third from that year, which had presumably since been discovered, *Le manoir du diable* (*The House of the Devil*).

itself had been used before, about a year earlier, in Alfred E. Clark's *The Execution of Mary Queen of Scots* (1894), produced by the Edison company in New Jersey to create the effect of the Queen's beheading. Frazer argues, though, that "*The Vanishing Lady* (1896) goes beyond the device of *The Execution of Mary Queen of Scots*. In the latter, the replacement of the actor by a dummy is a pragmatic necessity (for the actor certainly), but this procedure does not create a new reality so much as it facilitates the seemingly impossible."[65] It is not quite clear what Frazer means here by "a new reality," which is what Méliès is supposed to have created, in contrast to what Clark had done, or what precisely the distinction is that he is proposing, but he argues that it is Méliès's use of the technique that is the more significant, offering a version of a common claim, describing cinema itself as a sort of conjuring trick. "Méliès has transformed a stage conception into the kind of magic possible through the camera. With one trick he introduces the alternate possibility to the factual recording of the visible world, a conscious attempt 'to overcome reproduction and to arrive at the free use of the means of cinema for graphic expression.'"[66]

Yet the difference between the two films, both of which use the same editing technique, the same basic, rudimentary special effect, is less significant than Frazer suggests. Both, in fact, "overcome reproduction." Both present an alternative to "factual recording." Neither is limited to what exists in the "visual world." In both cases, what we are presented with are events that have to be created through "graphic expression," to use the term that Frazer offers. They cannot be performed in what we could call their "integrity" and simply recorded on film, in a single take. In his re-enactment of the execution of Queen Mary, Clark has necessarily to fabricate it (as Frazer somewhat wittily notes), to create, with the means available to him, a realistic representation of the historical event, an imagined depiction, without having to actually sever anyone's head from their body. The means he found to do so are those that are made readily available by the cinematic apparatus. While both he and Méliès were clearly inventive and should be acknowledged for their discovery and their undeniably creative use of this technique, of the apparatus, the technique itself is an obvious one, and it is not surprising that it was found so early in the history of the cinema and then used so often in subsequent years. It is simple, but it is an especially efficient means for the creation of just such representations, for the depiction of just such events that are either

65 Frazer, op. cit. p. 61.

66 Ibid., p. 61; Frazer is quoting Hans Richter, "Film as an Original Art Form," in Lewis Jacobs, ed., *Introduction to the Art of the Movies* (New York: Noonday, 1960), p. 282.

impossible (and/or morally indefensible), in the case of the beheading, or very difficult, in the case of the stage illusion, to reproduce "in reality" but which one wants to represent realistically. The question, though, to which I will try to attend in this concluding section with some conceptual care, is what precisely it means to represent something realistically, and what role special effects can play (and have historically played) in the history of cinematic realism.

Both films, by Clark and Méliès, are, I would insist, realistic representations of events that are otherwise difficult or impossible to represent through simpler cinematic means. Each would, that is, be even more convincingly realistic if they had in fact been filmed in one continuous take, but in neither case could this be done, for different but related reasons. In order, then, to create these respective realistic representations, recourse had to be made to some other means, still "cinematic" means, that would allow the filmmakers to achieve their goals.

A closer description of the Méliès film will provide a basis for a further consideration of these questions. In the filmed version of "The Vanishing Lady," the stage illusion of which Méliès wanted to produce a cinematic representation, an element of that stage version appears but is now superfluous in strictly cinematic terms, serving no specific technical or any other obvious function. Like on the stage, Méliès first of all lays the unfolded newspaper page on the floor, before setting the chair on it, even though there is no stage door to be covered. In the film, the trick is achieved through (or, as I prefer to put it, is represented or depicted by) editing, to which the paper bears no relation, like the relation it does bear in the stage version to the assumed means for achieving the trick (to creating the "representation," a "depiction," of a "vanishing lady"). It is only there, then, as part of Méliès's specifically cinematic effort, namely, to create a realistic depiction of the trick on film in the form with which audiences would be familiar, endeavoring to make it resemble, in important details, the stage performance of it. This is just what Frazer in fact implies when he says about the filmed version, without any further speculation: "The paper was spread in front of the chair [in fact it is placed under it] because this was the way the trick was traditionally done on the stage."[67] We have, though, to ask why Méliès was concerned in the film to present the trick as it was "traditionally done on the stage" when he is creating a new cinematic version of it.

The newspaper was included on stage, as I have suggested, partly as an acknowledgement of the fact that audiences were quite sophisticated on

67 Frazer, op. cit., p. 60.

matters of conjuring, a very popular form of entertainment. Even in that context, however, it makes no "narrative" sense, if one can say that such stage illusions are often presented as little stories, where the various formal elements — waves of the hand, magic coverings, and so on — all function as part of the performance, as the "fictional" devices with which the supposed magic is accomplished, even when they may not in fact be functioning to produce the illusion itself. One way of judging a magician's creativity, in this respect, in both the design and the execution of an illusion, is by how well they make elements of the technical contrivance itself a part of a narrative in this way, thereby, in one sense, obscuring their actual function and making them part of the little tale being told, the fiction being created, which is that a lady is being made to vanish. The newspaper, apparently in the stage performance and clearly in the filmed version, is given no obvious narrative significance in this respect. On stage, it seems to have a different kind of function, what we may call a meta-discursive one, and in this respect this version of this familiar illusion takes a more complex and not merely narrative form. By placing the chair on top of the large, unfolded newspaper sheet, Méliès is effectively challenging his audience in what could even be described as a self-reflexive gesture, asking them to imagine how it is that he will accomplish the trick, which is likely familiar to them, if he covered the very space where they know there is supposed to be a trap door. He is effectively saying to them, presenting the matter self-reflexively, "I will pull this off, even though I seem to have made it impossible to do so, according to what I assume you know about such tricks."

It is likely that this aspect of the trick, as it had been staged, and presumably for which it had become famous — hence the informative detail in the film's title, "*chez Robert-Houdin*" — is included in this *realistic* cinematic representation of it in order that it be recognized as such, as the version of the trick famous from its theatrical performances, in the theater owned by and programmed by Méliès himself. In the case of such a familiar trick, with such a long history and performed by so many different magicians, it is important that one make one's version of it distinctive, either by developing it in some novel narrative form or by self-reflexively incorporating some element that would confound an audience's knowledge about it, explicitly addressing the fact that they may well know (or think they know) how the illusion is created. To this end, we could imagine the filming of the trick, on the stage at the Théâtre Robert-Houdin, in its original form, with the use of the trap door, in a single camera take. In one respect, this would make it more convincing or be a more convincing representation of the trick, in the form in which it was presumably widely known. With the camera properly positioned, from the right angle with suitable lighting, and with the important addition of the newspaper, a quite compelling filmed representation of the illusion could thus be created, really using a trap door beneath the chair, without, that is, having to simulate the trap-door effect through editing.[68] As such, we would be in the same position of the actual stage audience, being asked explicitly to try to guess how the trick is done given the presence of the newspaper, which is performing no other function except to call attention to the use of trap doors in versions of this illusion, and apparently preventing its use in this version.[69]

In fact, though, the cinematic technology of the time would not allow this. For the film, as Frazer notes, the trick "was performed out of doors at Montreuil on a little improvised stage"[70] — standing in for the Théâtre Robert-Houdin. Most films of the time, including Clark's, were shot outdoors, or in studios with no roof, or a retractable roof, so that natural lighting could be used, being more effective than any stage lighting that was available. It was, that is, very difficult, if not impossible, as the beheading was for Clark, for Méliès to recreate the stage illusion in reality, yet, and as the very important clue of the newspaper suggests to us, it is the self-reflexive version of this trick that he is realistically representing for us, drawing our attention specifically to its self-reflexive aspect in order that we may ask the very question that the theater audience was being prompted to ask. It is, moreover, an integral part of any realistic representation, as I am defining it, that we be prompted to ask such questions about the depiction, about what is represented in the depiction.

68 It is worth noting that most magicians, when performing tricks in their own name on film or more commonly on television, tend to announce, as Penn and Teller do in the introduction to their TV series *Fool Us*, that "there is no editing!"

69 I should add that this puts the other part of the film — the appearance and disappearance of the skeleton — in possibly a new light. If understood as a representation of the version of the trick that defies the audience to explain it as an effect of the use of a trap door, which also requires the large cloth covering the lady (as well as the chair, with its break-away seat), obscuring her passage through the door, then the fact that (or the illusion that) there is no trap door is further made manifest by the fact that the skeleton appears and reappears without at any point being covered by the large cloth, which performs this specific function. It is as though, having put the use of the trap door into doubt in his stage version, Méliès is emphasizing the extent of his formal accomplishment by representing, in now strictly and perhaps equally self-reflexively cinematic terms, a trick that could not even be achieved with a trap door but which would require some other contrivance that the audience may well be less familiar with. It is another but related question how familiar they were with the technique used to create both parts of the trick on film, namely stop-substitution, and how self-reflexive Méliès was in fact being in his use of it here.

70 Frazer, op. cit., p. 60.

Realistic depiction is characterized, I argue, following John Hyman, by what he calls "modality," which he says is a "general measure of realism" and which he defines as "the extent of the range of questions we can ask about a depicted scene."[71] His is a deliberately limited account of realism, which he hopes avoids the more intractable issues that are usually a part of more expansive and extensive accounts of realistic depiction and about realism in general, which tend to become tangled up in apparently related but in fact tangential questions of reality, perception, belief, and illusion. The matter of realism in depiction is complicated enough, as I have tried to show in my account of this brief film by Méliès, a realistic representation of an illusion, about which an extraordinarily wide range of questions may be asked. The film is realistic just in its very modality, in its realistic depiction of an illusion, not by being *itself* an illusion.

Bibliography

Aristotle. *Metaphysics*. In *The Basic Works of Aristotle*, edited by Richard McKeon, translated by W.D. Ross (New York: The Modern Library, 2001): 681–986.

Arnheim, Rudolf. *Film as Art* (Berkeley, CA: University of California Press, 1957).

Austin, J.L. *Sense and Sensibilia* (Oxford: Oxford University Press, 1962).

Bann, Stephen. *The True Vine: On Western Representation and the Western Tradition* (Cambridge: Cambridge University Press, 1989).

Black, Joel. *The Reality Effect: Film Culture and the Graphic Imperative* (New York and London: Routledge, 2002).

Currie, Gregory. *Image and Mind: Film, Philosophy, and Cognitive Science* (Cambridge: Cambridge University Press, 1995).

Danto, Arthur. "The End of Art." In *The Philosophical Disenfranchisement of Art* (New York: Columbia University Press, 2005): 81–115.

Ezra, Elizabeth. *Georges Méliès: The Birth of the Auteur* (Manchester: Manchester University Press, 2000).

Frazer, John. *Artificially Arranged Scenes: The Films of Georges Méliès* (Boston: G.K. Hall and co., 1979).

Gombrich, Ernst. *Art and Illusion: A Study in the Psychology of Pictorial Representation* (Oxford: Phaidon, 1960).

Gunning, Tom. "Moving Away from the Index: Cinema and the Impression of Reality," *Differences* 18, no. 1 (2007): 29–52.

Hammond, Paul. *Marvellous Méliès* (London: Gordon Fraser, 1974).

71 Hyman, *The Objective Eye*, op. cit., p. 200.

Hyman, John. *The Imitation of Nature* (Oxford: Basil Blackwell, 1989).
———. *The Objective Eye: Color, Form, and Reality in the Theory of Art* (Chicago and London: University of Chicago Press, 2006).
Klein, Norman M. *The Vatican to Vegas: A History of Special Effects* (New York: New Press, 2004).
Lefebvre, Martin, and Marc Furstenau. "Digital Editing and Montage: The Vanishing Celluloid and Beyond," *Cinémas* 13, no. 1–2 (2002): 69–107.
Lopes, Dominic. *Understanding Pictures* (Oxford: Clarendon Press, 1996).
Manovich, Lev. *The Language of New Media* (Cambridge, MA: MIT Press, 2001).
Margolis, Joseph. "Mechanical Reproduction and Cinematic Humanism." In *What, After All, is a Work of Art? Lectures in the Philosophy of Art* (University Park, PA: The Pennsylvania State University Press, 1999):101–127.
Maynard, Patrick. "Drawings as Drawn: An Approach to Creativity in an Art." In *The Creation of Art : New Essays in Philosophical Aesthetics*, edited by Berys Gaut and Paisley Livingston (New York: Cambridge University Press, 2003): 53–88.
Metz, Christian. "On the Impression of Reality in the Cinema." In *Film Language. A Semiotics of Cinema*, translated by Michael Taylor (Chicago: University of Chicago Press, 1974): 3–15.
Ndalianis, Angela. "Special Effects, Morphing Magic, and the 1990s Cinema of Attractions." In *Meta-Morphing: Visual Transformation and the Culture of Quick Change*, edited by Vivian Sobchak (Minneapolis, MN: University of Minnesota Press, 2000): 251–272.
North, Dan, Bob Rehak, and Michael S. Duffy, eds. *Special Effects: New Histories/ Theories/Contexts* (London: British Film Institute, 2015).
North, Dan. "Afterword: A Conversation with Lev Manovich." In *Special Effects: New Histories, Theories, Contexts*, edited by Dan North, Bob Rehab, and Michael S. Duffy (London: Palgrave, 2015): 267–275.
Pierson, Michelle. *Special Effects: Still in Search of Wonder* (New York: Columbia University Press, 2002).
Pliny. *Natural History*, Vol 10. Translated by H. Rackham. Loeb Classical Library (Cambridge, MA: Harvard University Press, 1961).
Prince, Stephen. *Digital Visual Effects in Cinema: The Seduction of Reality* (New Brunswick, NJ: Rutgers University Press, 2012).
Richter, Hans. "Film as an Original Art Form." In *Introduction to the Art of the Movies*, edited by Lewis Jacobs (New York: Octagon, 1970):282–288.
Schier, Flint. *Deeper Into Pictures* (Cambridge: Cambridge University Press, 1986).
Wollheim, Richard. "Seeing-as, Seeing-in, and Pictorial Representation." In *Art and its Objects: An Introduction to Aesthetics* (New York: Harper and Row, 1968): 137–151.
———. "On Drawing an Object." In *On Art and the Mind: Essays and Lectures* (London: Allen Lane, 1973): 3–30.

About the Author

Marc Furstenau is Associate Professor and Head of the Film Studies Program in the School for Studies in Art and Culture, Carleton University (Ottawa, Canada). He is the editor of *The Film Theory Reader: Debates and Arguments* (Routledge, 2010), co-editor of *Cinema and Technology: Cultures, Theories, Practices* (Palgrave, 2008), and author of the forthcoming book *The Aesthetics of Digital Montage* (Amsterdam University Press).

Techniques

5. Trick-o-logics 1810/1910: The Magic of Tricks and Special Effects

Between the Stage and the Screen

Frank Kessler and Sabine Lenk

Abstract
This chapter discusses the ambivalent attitude of German critics concerning tricks on the stage and on the screen. Starting from a transformation scene in Goethe's *Faust*, it discusses the difference between an early French performance of the play in 1828, which makes use of all the possibilities it offers to recur to spectacular effects, and later German productions that shun what critics at the time called "mechanical artifices." With respect to cinema, however, German critics rather praised its capacity to transcend the laws of causality and to depict the fantastic events in fairy tales as one of the medium's specific qualities.

Keywords: tricks, theater, early cinema, Goethe, *Faust*

In the first part of Johann Wolfgang von Goethe's tragedy *Faust*, at the beginning of the play, the tormented hero returns home after having been on an Easter Sunday walk in the countryside, accompanied by his assistant Wagner. A stray dog, a poodle to be precise, which they had encountered outside the city walls, follows him right into the house. Inside, in Faust's study, the animal lies down near the oven but then all of a sudden starts to become bigger and bigger, until none other than Mephistopheles appears amidst a lot of smoke.

Goethe thus wrote a regular transformation scene for his *Faust*. But in doing so, did the author wish to create the possibility for a spectacular stage effect? That is not at all certain. For quite some time, in fact, the play as a whole was considered impossible to stage, and according to the theater

Lefebvre, M. & M. Furstenau (eds.), *Special Effects on the Screen: Faking the View from Méliès to Motion Capture*. Taylor & Francis Group, 2022
DOI 10.5117/9789462980730_CH05

historian Herbert A. Frenzel, it was generally felt that the story was best left to take shape exclusively in the reader's imagination. Goethe himself, in spite of a few isolated remarks about staging a performance of *Faust*, never undertook actual steps to realize such a project. And even though in 1829 the first public representation of the play in Germany finally did come to pass, the issue continued to be a controversial one for at least half a century.[1]

Interestingly, one of the reasons for having doubts about the appropriateness of a performance of the tragedy in front of an audience had to do with its spectacular qualities. The writer Ludwig Bechstein, who later became famous for his collection of German fairy tales, wrote in 1831:

> A second demonic apparition that deserves consideration is the poodle. There is no beauty in him, and one would be well advised not to take him lightly: he may have fiery eyes, but the expansion is a different matter. A dense haze that rapidly envelops the monster and makes it disappear from the eyes of the audience is the best solution. It is not necessary to show how it swells to the size of an elephant; such mechanical artifices only detract from the essential matters and do not contribute anything to a perfect representation, which is based on entirely different things.[2]

Bechstein's critical remarks with respect to an all-too-literal representation of the written text on stage point in the same direction as Goethe's own comments on a French performance of his play in 1828 at the Théâtre de la Porte Saint-Martin. The tragedy was rather freely adapted but seems to have enjoyed considerable commercial success. "The French had to transform it and add some strong spices and other ingredients to the sauce. From what we have come to know about it, one can understand why that concoction did achieve such powerful effects," Goethe wrote in a letter to a friend, the playwright and art critic Johann Friedrich Rochlitz.[3]

The choice of words of both authors — Bechstein's "mechanical artifices" as well as Goethe's "concoction" (which partly refers to the reworking of his

1 The first published version of *Faust* by Johann Wolfgang von Goethe (Tubingen: Cotta, 1808). In 1819 and 1820, the Count Anton Radziwill organized private performances for his guests in his Berlin residence. The first public staging took place in Brunswick in 1829. See Herbert A. Frenzel, *Geschichte des Theaters. Daten und Dokumente 1470-1890* (Munich: DTV, 1984), p. 342. See also Wilhelm W. Creitzenbach, *Die Bühnengeschichte des Goethe'schen Faust* (Frankfurt am Main: Rütten & Loening, 1881).

2 Ludwig Bechstein, *Die Darstellung der Tragödie Faust auf der Bühne* (Stuttgart: Brodhag'sche Buchhandlung, 1831), p. 45. All translations from German or French sources are ours.

3 Quoted in Creitzenbach, op. cit., p. 28.

play by the French translator, or rather adaptor and editor, but undoubtedly also to the effects introduced by the *mise en scène*) — reveals quite some skepticism on their part with respect to the appropriateness of spectacular stage effects for a tragedy such as *Faust*. This viewpoint is shared by others as well. In 1907, the stage director Eugen Kilian recommended presenting the scene with the poodle in such a way that the transformation takes place off-stage, hidden from the eyes of the audience. He suggests starting the scene at a point when Faust has already entered the room and the poodle is sitting behind the oven. So there is no need to have a dog on stage and to figure out how to represent its metamorphosis. Kilian adds: "In the scene when the devil is summoned and which is full of rather unattractive abracadabra, many elements can be left out."[4]

In these discussions on Goethe's *Faust* in the nineteenth and early twentieth century, it seems there was a tendency to perceive a tension between the poetic and dramatic force of the text and the way in which some of its central scenes could be visualized on a stage. The authors clearly were in favor of a rather cautious approach when it came to the potentially quite spectacular moments in the play. In a study on the staging of *Faust* in Germany since the nineteenth century, published in 1929 in commemoration of the centenary of the first public staging of the play, Julius Petersen, a professor of literary studies, argued that one had to limit the visualization in order to enhance the imagination ("*Gesteigerte Bildhaftigkeit wird durch die Beschränkung des Bildhaften gewonnen*") and proclaimed that "everything that is simply theatrical in the drama has to be omitted so that it can become theater."[5]

This, however, does not mean that all kinds of effects were to be excluded. Kilian, for instance, suggests that the Easter Sunday walk could be represented using a "*Wandeldekoration*," that is, literally, a "changing background set." He was apparently referring to some sort of moving panorama and wonders why nobody seems to have thought of such a solution, as the scene is indeed very difficult to stage because it involves several distinct locations.[6] Goethe himself inquired in 1828 in a letter to the painter Wilhelm Zahn about the manufacturer of the magic lantern that was used for the apparition of the Earth Spirit in the performances of scenes from *Faust* organized by Prince Anton Radziwill in 1819 and 1820: "The Prince Radziwill, who made possible

4 Eugen Kilian, *Goethes Faust auf der Bühne* (München, Leipzig: Georg Müller, 1907), pp. 39–40. The "rather unattractive abracadabra" that Kilian wanted to be left out had, however, been written into the play by Goethe. So the author's words were not necessarily inviolable when the aesthetic norms had to be fulfilled.

5 Julius Petersen, *Goethes Faust auf der deutschen Bühne* (Leipzig: Quelle & Meyer, 1929), p. 48.

6 Kilian, op. cit., pp. 33–34.

the private performance of some scenes from my *Faust*, had the apparition of the Earth Spirit in the first scene presented in a phantasmagorical manner, in such a way that in the darkened theater a small, continuously enlarged illuminated head was projected onto a screen placed in the background, and so the head seemed to approach and move forward. This feat was apparently achieved through some sort of *lanterna magica*."[7] So it appears that Radziwill must have employed a device similar to the one used by "Robertson" (Étienne-Gaspard Robert) for his phantasmagorias.

Realist Sobriety or Illusionist Simulation

When the first public performances of Goethe's *Faust* were staged, German theaters were looking, it seems, for some form of realist sobriety that apparently distinguished them from their French counterparts. The Parisian audiences wanted to be thrilled by a "monumental simulation of reality" produced with the help of elaborate light and sound effects as well as optical tricks. This illusionist style, introduced in 1822 at the Paris Opera by Louis-Jacques Mandé Daguerre and his collaborators, rapidly conquered all of France.[8] Conversely, the German stage was dominated by Friedrich Schiller's ideas of the "*Schaubühne als eine moralische Anstalt*" (the theater as a moral institution), proclaimed by the dramatist in 1802. Audiences were to be taught the highest moral principles, while only the strict minimum of sensual excitement should be provided.[9] The main task of stage machinery was to make possible a rapid and efficient change of scenery.

At the risk of oversimplifying the contrast between the two countries, one might say that what German spectators expected from the theater in the first instance was an aesthetic and intellectual as well as moral education, while their French counterparts rather looked for entertainment and wanted to be impressed by spectacular effects. Frenzel sees a link between the political restoration after the end of the Revolution and the way French theater functioned as an "illusionist drug" (*eine illusionsfördernde Droge*), as he calls it.[10] In Germany, by contrast, the class system had remained intact

7 Letter from Goethe to Zahn quoted in Adolph Enslin, *Die ersten Theateraufführungen des Goethe'schen Faust* (Berlin: Verlag von Gebrueder Paetel, 1880), pp. 14–15.

8 See Frenzel, op. cit., pp. 293–294.

9 Ibid., p. 334.

10 Ibid., p. 273. The fact that Frenzel is a *German* theater historian should perhaps be taken into account here, as he seems to share his nineteenth-century predecessors' misgivings concerning spectacular stage effects.

and "*Bildung*" was the one distinctive marker the German bourgeoisie, who made up a large part of the theater audiences, could mobilize when creating their own hierarchy of values against the aristocracy, opposing in particular the "seriousness" of high art and the "shallowness" of idle amusements.[11]

Clearly, a *mise en scène* using spectacular effects and maybe even devices borrowed from Robertson's phantasmagorias was bound to provoke skepticism among bourgeois critics such as Wilhelm Creitzenbach:

> It has often been suggested that nothing is impossible for modern stage machinery, and many among the genuine friends of the theatre complain about the fact that impresarios spare no costs and engage in the most scandalous waste of money when putting on stage the fantastic elaborations of an inferior versemonger in a *féerie* or other extravaganzas, while they shy away from presenting Goethe's poetic intentions in his greatest play. But also the practitioners, who in recent times have competed with each other with their suggestions how to stage these scenes [the Easter Sunday walk, the Walpurgis Night], did not succeed in resolving the problems, neither Devrient with his tripartite stage, nor Dingelstedt with his moving panoramas.
> With all their craft the stage technicians can never create the images that our imagination can conjure up when fired by the author's words [...].[12]

We thus can distinguish two strategies of staging, both of which attempt to achieve realism but have a very different understanding of this term. This also has important consequences with respect to the status of stage tricks. On the one hand, there is a tendency towards a realist sobriety, favored in Germany, which prefers verbal evocation to visualization (no poodle, no transformation, everything is in the words of the dramatist). On the other hand, there is an "illusionist realism," as it were, favored in France, which Arthur Pougin, author of an encyclopedia of theatrical terms published in 1889, sums up as follows:

> [...] all the sounds, all of the various other effects have to be produced in the most precise and most natural way, exactly at the right moment, in

11 One might actually say that even today, these values continue to dominate the motivation of a majority of theatergoers, even though nowadays they probably seek intellectual elevation rather than moral education.

12 Creitzenbach, op. cit., pp. 15–16.

> order to affect the ear, the eye and the imagination of the spectator in the most vivid manner, to convey to him a feeling of reality and to provide him with a total illusion.[13]

From the viewpoint of the first approach, tricks have to be used 'with discretion' (if at all), and effects must not be foregrounded and should not become attractions in their own right, as they might distract the spectators' attention from the words of the dramatist. The trick is seen as being in opposition to the imagination, the technician in opposition to the author of the text. For the second approach, the possibilities offered by stage technology should be used to the fullest in order to achieve a maximum of realism, meaning here the convincing representation of extraordinary, sensational, magical, or supernatural actions or events. The technician is seen here as a translator of the dramatist's words, and the trick is one effect among many that can be used to affect the spectators' imagination.

In addition to the historical, cultural, and aesthetical factors that can account for the differences between the German and the French attitude towards theater, there may be yet another aspect to take into account. Most of the German theaters were financed by the king, a duke, a count, or other reigning aristocrat, or else by the citizens of wealthy cities such as Hamburg or Frankfurt. In France, in particular in Paris, there were numerous commercially run stages that needed to draw large audiences to survive economically. Offering spectacular entertainment was thus an important element in their commercial strategy, whereas German theaters were more focused on providing cultural capital.

Interestingly, in *Der Komet*, a German trade journal addressing the entire range of show people from fairground exhibitors to managers of variety theaters, one can find numerous reports around the turn of the century on spectacular effects presented on stage in Paris or London. These appeared in the "Miscellaneous" section of the journal and described special effects or tricks used in the performances of all sorts of theater plays. So even though the readers of *Der Komet* mostly worked in a different domain, they seem to have been interested in such developments of stage craft and technology. This is yet another symptom of the divide that in Germany separated the cultural institution of the so-called legitimate theater from the various forms of popular entertainment.

13 Arthur Pougin, *Dictionnaire du théâtre et des arts qui s'y rattachent*, Vol. 2 (Paris: Firmin-Didot, 1885) [facsimilé Plan de la Tour: Eds. d'Aujourd'hui, 1985].

The Three Logics of the Trick Effect

It is important to be aware of these ideas (which, as we have seen, reach back far into the nineteenth century at the least) because they have shaped the critical debates about theater as an art form. The role and status attributed to the trick in theater aesthetics is largely dependent on them. We can even find traces of these ideas in film criticism, in particular in the early period.[14] They also represent different cultural attitudes towards theater and later cinema, which are to some extent at least rooted in national traditions concerning criticism and aesthetical theory as well as the institutions and organizational forms of theater, including commercial interests. If we follow Christian Metz and consider tricks as "avowed machinations,"[15] their status indeed depends upon the general conception one has of theater as an art form. For the German critics, as we have seen, the very fact that there is a machination is problematic because it draws the audience's attention away from the poetic force of the text. For Pougin, but also for an author such as J. Moynet,[16] tricks are an essential ingredient to *mise en scène*, allowing for the creation of a "feeling of reality" and "total illusion" that is needed to affect the spectators. In that case, the effect one wants to produce and the means to achieve this must be in a balance. Thus, in line with Metz's analysis of how tricks function in classical narrative cinema, we find that in theatre the institution equally "wins twofold": the effect is attributed both to the diegesis — thus reinforcing the spectators' absorption in the dramatic action — and to the discourse, that is the powers of the theater (or cinema) to produce such thrilling effects.[17]

There is, in fact, a third approach that can be found in *féeries* and other forms of extravaganzas or spectacular plays but also, as far as cinema around 1900 is concerned, the films of Georges Méliès, Gaston Velle, Segundo de Chomón, and several others. Here the machination is not only avowed but flaunted. The tricks become an attraction in and of themselves; they function as *clous*, as such highlights were called at the time.[18] They were considered

14 Yet these debates continue even today. See, for instance, Réjanne Hamus-Vallée, *Les Effets spéciaux* (Paris: Cahiers du Cinéma, 2004), pp. 82–83, where she reproduces one article by Olivier Assayas and one by Pascal Bonitzer, who take diametrically opposed views on special effects.

15 See Christian Metz, "*Trucage* and the Film," in *Critical Inquiry*, (3)4 (1977), pp. 657–675, in particular pp. 664–665.

16 J. Moynet, *L'Envers du théâtre* [1873] (Paris: Librairie Hachette, [1873] 1888).

17 See Metz, op. cit., pp. 667–668.

18 See, for instance, Edmond Floury, "La cuisine théâtrale," in *La Revue théâtrale* (Nouvelle Série), 54 (1906), pp. 1387–1388. Méliès, too, mentions the importance of such *clous* in his films. See Georges Méliès, "Importance du scénario," reproduced in *Georges Méliès*, ed. Georges Sadoul (Paris: Seghers, 1970), pp. 115–117.

of vital importance for the success of a play or a film, whereas the diegesis was clearly subordinate to the spectacular effects and the story that was told merely provided some sort of narrative motivation for them. Méliès as well as the impresarios staging *féeries* in their theaters tried to outclass their competitors thanks to the originality and quality of the tricks they presented.[19] The case of Méliès is particularly relevant here because his tricks also bear witness to the capacities of the cinematographic apparatus, enabling him to "achieve the most impossible and improbable things in kinematography," as he phrased it.[20] J. Moynet demonstrates in his book that, throughout the nineteenth century, stage trick specialists had been capable of creating the most complicated stunning effects. As for the cinematographic tricks, they helped indeed to promote the new technology of animated photography. Méliès played a central role in this process, as did Segundo de Chomón, who explored the possibilities of stop motion animation, yet another technique that allowed the visualization of magic or supernatural events.

Negotiations: Tricks and the "Nature" of Cinema

Curiously, cinematographic tricks (or at least some of them) appear to have been less problematic to German intellectuals around 1900 than tricks performed on the stage. One can explain this on the one hand by the fact that the stakes were considerably lower with regards to cinema, for at the time it was generally not considered an art form. On the other hand, tricks belonged to a realm in which films did not compete with the established arts such as, in particular, theater and literature. However, not all types of tricks were appreciated. The writer Hanns Heinz Ewers, for instance, who in 1913 wrote the script for *Der Student von Prag* (*The Student of Prague,* Stellan Rye), declared in 1908 that he loathed magical trick scenes and qualified them as "repulsive."[21] The journalist and writer Alfons Paquet derided in 1912 a film, "apparently Parisian [produced] by one of the big companies," showing the performance of a prestidigitator and condemned it as "cheap

19 See also Frank Kessler, "The *Féerie* Between Stage and Screen," in André Gaudreault, Nicolas Dulac, and Santiago Hidalgo, eds., *A Companion to Early Cinema*. (Malden, MA and Oxford: Wiley-Blackwell, 2012), pp. 64–79; and "La féerie: un spectacle paradoxal," in *Lendemains*, (152) (2013), pp. 71–80.

20 Georges Méliès, "Kinematographic Views: A Discussion," reprinted in André Gaudreault, *Film and Attraction. From Kinematography to Cinema* (Urbana, Chicago, Springfield: University of Illinois Press, 2011), as Appendix B, pp. 133–152, quote on page 148.

21 Hans Heinz Ewers, "Der Kientop," *Der Kinematograph*, 61 (26.2.1908) [no pagination].

tricks presented by the worst kind of ham actors."[22] In both cases, magical acts produced in a stage-like fashion were rejected as being of little interest — or even, in the case of Ewers, as outright repulsive, maybe because, as Paquet put it, "one knows how easy it is to arrange the most paradoxical things for the cinematograph."[23] Both authors quite violently rejected the simple simulation of a stage act performed in front of a camera, maybe because they experienced it as a hybrid form of spectacle.

However, in the same article, Ewers also expressed his enthusiasm for the capacity of cinema to escape the laws of causality. The examples he gives are scenes in which the film is reversed, with the action running backwards, but for Ewers in these cases "the history of the world is reversed, cause becomes effect and effect becomes cause."[24] For the same reasons, the critic Julius Bab declared that fairy tales are a realm that is particularly suitable for cinema: "the abolishment of the laws of causality [...], the extraordinary leaps beyond the limits of time and space are the very essence of fairy tale imagination."[25] Conversely, according to Bab, the theater and specifically the *Sprechbühne* (the "talking stage") is bound to the laws of causality, which means for him that fairy plays somehow clash with the specific requirements of theater as an art form.[26] This opinion is shared by another critic, Max Schumann, who pointed out in an article published in 1911 that one should not represent magical effects on the stage because it is better to stimulate the spectators' imagination. Using all the possibilities offered by modern stage technology, one runs the risk, according to him, of turning the stage into a cabinet of curiosities.[27] So with respect to the representation of fantastic or supernatural events on the stage, the attitude of critics around 1910 still largely followed the precepts proclaimed in the first half of the nineteenth century, while cinema, on the other hand, was seen as being particularly suited for this task. As the philosopher Georg Lukács put it: "[...] the basic law of connection for stage and drama is inexorable necessity; for the 'cinema', it is possibility restricted by nothing."[28]

22 Alfons Paquet, Answer to a survey organized by the *Frankfurter Zeitung* in 1912 entitled "Vom Werte und Unwerte des Kinos," reprinted in Anton Kaes, ed., *Kino-Debatte* (München: dtv, 1978), pp. 62–66. The quoted comment is on page 64.

23 Ibid.

24 Ewers, op. cit., (no pagination).

25 Julius Bab, "Die Kinematographen-Frage," in *Die Rheinlande*, vol. 22 (1912), p. 314.

26 Ibid.

27 Max Schumann, "Die Zauberposse auf dem modernen Theater," in *Masken. Wochenschrift des Düsseldorfer Schauspielhauses*, (6)40 (1911), p. 634.

28 Georg Lukács, "Thoughts on an Aesthetics of Cinema," in Richard W. McCormick, Alison Guenther-Pal, ed., *German Essays on Film* (New York and London: Continuum, 2004), p. 13.

These German intellectuals thus rejected the use of tricks and effects on the stage, mainly because of their rather rigid aesthetical conceptions of theater as an art form, which is in line with the opinions of their nineteenth-century predecessors in this matter. Yet they appreciate and even celebrate certain uses of cinematic tricks. It is, one might say, the very same normative and essentialist approach to art that leads them to conceive of a realm that is proper to cinema, i.e., the realm of the fantastic. In this way, films, according to Ewers, Bab, Lukács, and others, can offer something other than a simple reproduction of nature without competing with (and commercially threatening) the legitimate stage.[29]

Cinematic Tricks, ca. 1910

The positive attitude of German intellectuals regarding tricks used to enhance cinema's potential to represent the fantastic is apparently linked to a shift concerning the use of cinematic tricks in a move away from the cinema of attractions, that is the reproduction of stage acts criticized by Alfons Paquet. Let us return once more to his remark that "one knows how easy it is to arrange the most paradoxical things for the cinematograph." This comment was published in 1912. Around that time, indeed, general knowledge about how cinematic tricks functioned had been disseminated quite broadly in the popular magazines. In 1908 already, Gustave Babin's series of articles in *L'Illustration* (1908) had revealed a number of trick effects, and Georges Méliès had reacted full of indignation.[30] Around 1913, some of the illustrations used by Babin circulated in France and in Germany as Liebig advertising trade cards. In 1909, the trade journal *Erste Internationale Filmzeitung* published an article signed by Fred Hood that presented various examples of trick effects, some probably taken from Babin's articles. He also offered a detailed description of the Vitagraph film *Princess Nicotine*

29 For an overview of the various aspects of the debates concerning the relationship between stage and screen in France and a comparison with the situation in Germany, see Sabine Lenk, *Théâtre contre cinéma. Die Diskussion um Kino und Theater vor dem Ersten Weltkrieg in Frankreich* (Münster: MAkS Publikationen, 1989).

30 Gustave Babin, "Les coulisses du cinématographe," in *L'Illustration*, 3396 (28 March1908), pp. 209 and 211–215; "Les coulisses du cinématographe," in *L'Illustration*, 3397 (4 April 1908), pp. 238–242; "Le théâtre cinématographique," in *L'Illustration*, 3427 (31 October 1908), pp. 281 and 286–289. All reprinted in *Les Grands Dossiers de* L'Illustration: *Le Cinéma* (Paris: SEFAG, 1987), pp. 22–33. George Méliès, "Les coulisses de cinématographie: Doit-on le dire?," in *Phono-Cinéma-Revue*, April 1908, pp. 2–3. See also Roland Cosandey, "Cinéma 1908, films à trucs et Film d'Art: Une campagne de *l'Illustration*," in *Cinémathéque*, (3) (1993), pp. 58–71.

(J. Stuart Blackton, 1909) that had been released the same year, his source being an article that had appeared in *Scientific American*.[31] As the articles by Lukács and others show evidence of at least a minimal knowledge about the way cinematic effects are achieved, one can presume that information about these matters did circulate quite widely in the years around 1910.

According to the logic of tricks Méliès subscribed to, revealing the secrets of the profession was simply inadmissible because this meant not only giving up a competitive advantage but also risking that audiences will no longer be impressed by the performance once they know how an effect can be achieved. Within the logic of the "avowed machination," however, this is less of a problem because the knowledge the audience possesses can even increase their appreciation. This is indeed the function of the "making of" features on today's DVDs and Blu-Rays In 1908, Gustave Babin put it as follows:

> When one has come to know the main [devices], those that are used most frequently, it appears that an attentive spectator is capable of explaining and describing even the most mysterious ones and to reconstruct what the professionals call the "shooting of the scene." The cinematograph, however, will lose nothing of its attraction for them, quite on the contrary. The marionettes that were the joy of our childhood did not cease to entertain us when one day someone with sharp eyes exclaimed from the parterre: "You can see the strings!"[32]

According to Babin, the knowledge the audience has with respect to the "avowed machination" thus allows for an even greater appreciation, as Metz put it, of the powers of the medium. The cinematic trick, then, no longer functions as the equivalent of a performance of a magical act or a stage trick but as a special effect that is judged according to its efficiency and quality of execution in the service of the diegesis.

Conclusion

As we have seen in these debates that extend over a century, the status of the trick in productions for the stage and for the screen depends largely on

31 Fred Hood, "Interessante Tricks in der Kinematographie," in *Erste Internationale Filmzeitung*, (46) [11.11.1909] (no pagination).
32 Babin, op. cit., 1987, p. 22.

the aesthetic norms that constitute the frame of reference for the critics. In Germany, the art of the theater was seen to reside mainly in the dramatic text, which had to be performed in such a way that the spectators' imagination was stimulated and the audience was thus culturally and morally elevated. An all too explicit visualization was thought to compete with and consequently block the imaginative force of the viewers. In France, on the contrary, the illusionist power of stage effects was supposed to bring forth a visible equivalent of what the writer's words wanted to evoke. When convincing, the prowess of the technicians was lauded, as it contributed to the overall effect and aesthetic appeal of the play as well as the pleasure of the audience.

As for cinema, German critics distinguished between the reproduction on screen of magicians' stage performances, which they seem to have loathed, and the possibilities that the cinematograph offered to overcome the laws of causality and strict necessity, which in turn were said to govern the art of theater. While this did not necessarily turn them into lovers of cinema, they considered these cinematic tricks to be specific to cinema and thus, almost paradoxically, saw them as the equivalent of fairy tale imagination and thus as aesthetically justified.

These debates thus reveal the complexity of tricks — not as a technical feat (even though they are this, too) but as an element that sometimes can appear foreign — even opposed — to the characteristics of an art form and sometimes as a means to fulfil them to the highest degree. In order to understand their status, we need to study the historical and cultural context in which they function. And this is, as Faust says after the transformation has taken place, "*des Pudels Kern.*"

Bibliography

Bab, Julius. "Die Kinematographen-Frage," *Die Rheinlande* 22 (1912): 311-314.

Babin, Gustave. "Les coulisses du cinématographe," *L'Illustration* 3396 (28 March 1908); "Les coulisses du cinématographe," *L'Illustration* 3397 (4 April 1908); "Le théâtre cinématographique," *L'Illustration* 3427 (31 October 1908). Reproduced in *Les Grands Dossiers de l'Illustration: Le Cinéma* (Paris: SEFAG, 1987): 22–33.

Bechstein, Ludwig. *Die Darstellung der Tragödie Faust auf der Bühne* (Stuttgart: Brodhag'sche Buchhandlung, 1831).

Cosandey, Roland. "Cinéma 1908, films à trucs et Film d'Art: Une campagne de *l'Illustration*," *Cinémathèque* 3 (1993): 58–71.

Creitzenbach, Wilheilm W. *Die Bühnengeschichte des Goethe'schen Faust* (Frankfurt am Main: Rütten & Loening, 1881).

Enslin, Adolph. *Die ersten Theateraufführungen des Goethe'schen Faust* (Berlin: Verlag von Gebdrueder Paetel, 1880).

Floury, Edmond. "La cuisine théâtrale," *La Revue théâtrale* 54 (1906): 1387–1388.

Frenzel, Herbert A. *Geschichte des Theaters. Daten und Dokumente 1470-1890* (Munich: DTV, 1984).

Hamus-Vallée, Réjeane. *Les effets spéciaux* (Paris: Cahiers du Cinéma, 2004).

Heinz Ewers, Hans. "Der Kientop," *Der Kinematograph* 61 (6 February 1908).

Hood, Fred. "Interessante Tricks in der Kinematographie," *Erste Internationale Filmzeitung* 46 (11 November 1909), n.p.

Kessler, Frank. "The *Féérie* Between Stage and Screen." In *A Companion to Early Cinema*, edited by André Gaudreault, Nicolas Dulac, and Santiago Hidalgo (Oxford: Wiley-Blackwell, 2012): 64–79.

———. "La féerie: un spectacle paradoxal," *Lendemains* 152 (2013): 71–80.

Kilian, Eugen. *Goethes Faust auf der Bühne* (Munich and Leipzig: Georg Müller, 1907).

Lenk, Sabine. *Théâtre contre cinéma. Die Diskussion um Kino und Theater vor dem Ersten Weltkrieg in Frankreich* (Münster: MAKS Publikationen, 1989).

Lukács, Georg "Thoughts on an Aesthetics of Cinema." In *German Essays on Film*, edited by Richard W. McCormick and Alison Guenther-Pal (New York and London: Continuum, 2004): 11–16.

Méliès, Georges. "Les coulisses de la cinématographie: Doit-on le dire?," *Phono-Cinéma-Revue* (April 1908).

———. "Kinematographic Views: A Discussion." Reproduced in *Film and Attraction: From Kinematography to Cinema*, André Gaudreault (Urbana: University of Illinois Press, 2011): 133–152.

Metz, Christian. "*Trucage* and the Film," translated by François Melzer, *Critical Inquiry* 3, no. 4 (Summer 1977): 657–675.

Moynet, J. *L'envers du théâtre* (Paris: Librarie Hachette, [1873] 1888).

Paquet, Alfons. "Vom Werte und Unwerte des Kinos." In *Kino-Debatte*, edited Anton Kaes (Munich: DTV, 1978): 62–66.

Pougin, Arthur. *Dictionnaire historique du théâtre et des arts qui s'y rattachent. Poétique, musique, danse, pantomime, décor, costume, machinerie, acrobatisme; jeux antiques, spectacles forains, divertissements scéniques, fêtes publiques, réjouissances populaires, carrousels, courses, tournois, etc., etc., etc.* (Paris: Firmin-Didot, 1885).

Sadoul, Georges. *Georges Méliès* (Paris: Seghers, 1970).

Schumann, Max. "Die Zauberposse auf dem modernen Theater," *Masken. Wochenschrift des Düsseldorfer Schauspielhauses* 6, no. 40 (1911): 632–637.

About the Authors

Frank Kessler is a professor of Media History at Utrecht University and currently directs the Research Institute for Cultural Inquiry (ICON). His main research interests lie in the field of early cinema and the history of film theory. He is a co-founder and co-editor of the *KINtop* yearbook and the *KINtop. Studies in Early Cinema* series.

Sabine Lenk is an archivist and film historian. She was a researcher at the University of Antwerp in the research project "A Million Pictures" before co-writing the EOS research project "B-Magic," in which she works as postdoc at Antwerp University (research group Visual Poetics) and the Université libre de Bruxelles (CiASp).

6. Those Ordinary "Special Effects"

François Albera

Abstract

This essay defends the hypothesis according to which either there are no "special effects" in cinema or else everything counts as a special effect as soon as our attention turns to the machinery that produces those effects rather than to their reception on audiences or else to the impressions of reality or of surprise that they arouse. Using for sole reference writings and films produced in 1908 and the notions of trick and *trucage*, it is demonstrated that the trivial concomitance so often found today between "special effects" and digital images already belonged to early cinema.

Keywords: trick, *trucage*, machinery, theater, early cinema, work

In seeking to develop this oxymoron of the *ordinary* nature of "special effects" in cinema — we might also say their banality — I will start from the hypothesis that *there are no* "special effects" in cinema; or, and this is the same thing in the end, that in cinema *everything* is a special effect. Laurent Jullier, writing just before techniques for the digital manipulation of images became widespread, described the "making commonplace" of the very notion of special effects and wondered whether it would not be possible to assert that "the very notion of special effects [is] consubstantial *par excellence* with the digital image."[1] Here, however, I would like to argue that this "making commonplace" and this "consubstantiality" of special effects were already a part of early cinema, which is to say a part of cinema itself.[2] In order to do this, I will examine principally the year 1908, because of a conjunction

1 Laurent Jullier, "L'ère numérique: vers l'évanescence du trucage," in *1895: Revue de l'Association française de recherche sur l'histoire du cinéma*, 27 (September 1999), p. 113.

2 I will not adopt here the break posited between the cinematograph, or the "kinematograph," and "cinema," which has seen various chronological and conceptual re-interpretations, from Coissac to Gaudreault.

Lefebvre, M. & M. Furstenau (eds.), *Special Effects on the Screen: Faking the View from Méliès to Motion Capture*. Taylor & Francis Group, 2022
DOI 10.5117/9789462980730_CH06

of texts and films that reveal, like a detail being "magnified" under a lens, an apt *conjuncture* for making the argument I wish to make.

This argument hinges on cinema's association with *machinism*. Equally celebrated by some and condemned by others, the machinic nature of cinema nonetheless distinguished it from all other forms of spectacles that existed at the time of its invention. Indeed, prior to any sort of symbolic use of it, prior to its use to document or to entertain, the cinema presented itself as a machine, as an assemblage of devices calling on users of a peculiar ilk: individuals compliant with the cinema's machinery. The most noticeable — and visible — aspect of this rapport was undoubtedly the handle that both camera operator and projectionist were required to crank. Yet compliancy to machines was also required for optical settings on the set and for chemical processing in the lab. Thus, the fact that cinema is a machine enjoins us to approach the issue of special effects from a medium specific angle, even though their lineage also connects them to the *théâtre des féeries* and stage magic shows. Yet such lineage only pertains to the finality of the effects, not to the means by which they are achieved. Importantly, differences in *how* special effects are achieved in cinema induce a different positioning from spectators and from commentators who developed new strategies in writing about them, sometimes leading to disputes on the topic.

One particularity of the cinema-machine, which really amounts to a social-technical dispositive, resides in its self-avowed nature and its ability to include the viewer in the production of special effects. As a result, it shows itself to be free from the illusions of magic and from the occult. The spectator, knowing that any effect is due to mechanical and technical means, thus theoretically possesses the ability to understand them and to turn this understanding into a pleasurable experience. Yet, the cinema's self-exhibition as an all-powerful machine also establishes it as a modeling device: modeling spectacular effects but also modeling bodies with regards to social behavior and modeling the social and physical environment. As the cinema-machine and its special effects shape the world seen on the screen, they also shape the world where humans live. As Leroi-Gouran once said, speaking of humanity, "we cannot completely dismiss the thought that some species change takes place whenever humankind replaces both its tools and its institutions."[3]

The expression "special effect" (or "special effects") does *not* appear in texts on the cinema at the time, even though it appeared — in French — in a theatrical lexicon in 1847, in a definition of the word "*truc*" (trick): "a

3 André Leroi-Gouran, *Speech and Gesture*, trans., Anna Bostock Berger (Cambridge, MA: MIT Press, 1993 [1964]).

technique achieved through machinery or the stage set to create an illusion, a special effect."[4]

In his *Lexique français du cinéma: des origins à 1930*,[5] Jean Giraud (1958) traces the use of the term "special effect" ("*effet spécial*") to 1912, and in particular in a relatively restrictive sense: in the singular (today we generally prefer the plural) and to describe only slow motion and fast motion: "a special effect obtained by intentionally varying the speed of the take while maintaining the normal speed when projected" (the quotation is from Léopold Löbel).[6]

Giraud is obviously not a definitive source. His work, which is too often used without further thought and from whom general dictionaries (such as the *Robert*) draw all their information on film vocabulary, dates from 1958: what sources were available to him at the time, and, especially, from what perspective was his investigation conducted? These questions need to be asked.

In addition, one need not adhere to nominalism. The word can come after the thing; various previous names can be incorporated into a new, all-encompassing term that elucidates them. I will thus not argue for the ordinary nature of "special effects" because of the absence of a term for them.

The question is rather: what were the names at the time for the phenomena, techniques, technologies, and effects that would later be categorized under the name "special effects"? And especially, what was the status at the time of these phenomena, techniques, etc.?

For the period in question, we can consider that "special effects" were part of what were called at the time "tricks" ("*trucs*") to describe a genre, "trick scenes," which included not only scenes with particular ("special") "effects," such as the "transformation scenes" with which they were associated in the vocabulary of Pathé, but also animated drawings and the animation of objects or other inanimate elements using frame-by-frame photography.[7]

4 William Duckett, ed., *Dictionnaire de la conversation et de la lecture*, Vol. 61 (supplement to vol. 9), (Paris: Firmin-Didot, 1847).

5 Jean Giraud, *Le Lexique français du cinéma: des origines à 1930* (Paris: Centre national de la recherche scientifique, 1958).

6 Léopold Löbel, *La Technique cinématographique: projection, fabrication des films* (Paris: Dunod et Pinat, 1912), p. 149. This volume was later republished several times (1922, 1927, and 1934). Löbel was a chemical engineer who initially worked for Pathé and then became technical director of the Lux laboratory (a company of photographers and cinematographers founded by Henri Joly in 1906, which closed in 1913).

7 See most of the volumes describing film technology, such as Jacques Ducom, *Le Cinématographe scientifique et industriel: Traité pratique de cinématographie* (Paris: Geisler, 1911), which was reprinted in his *Cinéma muet, sonore et parlant* (Paris: Albin Michel, 1931), and G.-Michel Coissac, *Les Coulisses du cinéma* (Paris: Paul Duval, 1929), among many others.

"Trick" was also a stage term, coming from fairy plays in which it described "each incident" in the play.[8]

The word "special," when it describes an "effect" obtained with the cinematograph, means "particular" or, even better, "proper to the cinematograph." In descriptions of trick effects or "trick scenes" – in particular those of Georges Méliès – we see a blend of techniques and procedures, many of which came from theater and illusionism and a few of which were "proper to the cinematograph" because they were tied to the capabilities of the device or to specific ways in which the medium was manipulated. A distinction was made, however, between the former and the latter. In 1908, Gustave Babin, for example, spoke of "resources used exclusively by *the device itself*, the Cinematograph, [which] will make it possible to obtain the most marvelous *effects*, completely unknown and impossible to create until now."[9]

The word "effect," associated with "resources proper to the device," proposes an initial formulation coupling these resources of the cinema device (the machine) with effects hitherto unseen. This is an initial formulation of the "special" (unique, proper to) nature of these effects obtained by the cinematic machine.

We might see in the advent of this word, in the meaning it took on ("specificity"), an indication of the *caesura*, the passage from the "cinematograph" to the "cinema," an idea to which I do not subscribe. This is undoubtedly the case if one were to hold to a certain level of prescriptive discourse being constructed in the process of cinema's legitimation-institutionalization. But I will argue for the recognition of a more essential continuity, one neglected by the discourses that support such differentiation. Of course, we will be attentive to these discourses but with the aim of corralling them back to our approach to the phenomena we propose to examine. This is all the more important since the claims made in the primary sources will prove, upon examination, to be contradictory, given that they were run through with multiple perspectives and various and divergent interests.

The Great Stage Machinist

To begin, let us look at the term "effect" itself, which the French *Littré* dictionary describes (definition no. 7) as "a term of literature, painting and art in

8 *Dictionnaire de la conversation et de la lecture*, ed. by William Duckett, vol. 61 (supplement to vol. 9), (Paris: Firmin-Didot, 1847).

9 *L'Illustration*, 4 April (1908). [My emphasis. See below].

general" and defines as "the result of a combination which strikes the eyes, captivates the mind and touches the heart."[10] In other words, there exists a "narrative effect" in addition to a "lighting effect" or "visual effect."[11] The first definition of this term given by Giraud, borrowed from a lexicon of the performance arts, comes from the "father of intermediality," Arthur Pougin, who defines a special effect as an "impression produced on the audience and which, through its vivacity, excites the viewer's satisfaction, admiration or hilarity."[12] Pougin examines effects in every kind of performing art, placing himself in a sense on the side of the viewer ("impression") rather than that of how the production of the effect is obtained ("combination") as the *Littré* suggests. This difference should make us pause because, under a variety of names and formulations, it pervades discussions of illusion. What status should we grant the means of obtaining an illusion? Should we abstain from mentioning them in order to maintain the illusion, or should we speak of them, even display them?

This question becomes particularly crucial when we cross this definition with another found in the *Littré* (no. 5), which speaks of "the effect of a machine, its force, the power it transmits."[13] Indeed, in the eighteenth century, a new kind of effect arose with the introduction of the "machine," defined by Jacques Aymot as "an ensemble of mechanisms combined in such a way as to produce certain effects."[14] Here we find three relevant terms: "combination," "mechanism," and "effect."

The introduction of the machine into the performing arts was, first of all, distinct from "set design" because it brought movement. One can see it in the spectacular dispositives around royal rituals:

> There is thus this difference between Decoration and Machine: Decoration is always fixed, such as Immobile Scenes, Triumphal Arches, Pyramids, Statues, Temples, Obelisks, Paintings, Fountains, Gardens, Forests,

10 Emile Littré, *Dictionnaire de la langue française. Édition intégrale*, vol. 3 (Paris: Jean-Jacques Pauvert, 1956 [1872]), p. 486.

11 We thus find the term in the very title of essays pertaining to certain paintings, such as that on *Larousse mensuel* 65, entry: E.-L. Broquet, *L'Hiver au marais* (Salon des Artistes français, 1912): *Effet de neige*, (July 1912): pp. 458, 462.

12 Arthur Pougin, *Dictionnaire historique du théâtre et des arts qui s'y rattachent. Poétique, musique, danse, Pantomime, Décor, Costume, Machinerie, Acrobatisme, Jeux antiques, spectacles forains, Divertissements scéniques, Fêtes publiques, Réjouissances populaires, Carrousels, Courses, Tournoi, etc., etc., etc.* (Paris: Firmin Didot, 1883), p. 325.

13 Littré, *Dictionnaire de la langue française*, p. 485.

14 Jacques Amyot, *Vies Demetrius*. Quoted by the Centre National des Ressources Textuelles et Lexicales (see the entry "machine"). http://www.cnrtl.fr (accessed 9 October 2020).

> Landscapes, Perspectives and other *fixed* things. The Machine, on the contrary, aims to act, giving it an advantage over Decoration, which because it is without movement appears to be dead.[15]

The Machine, then, quickly took hold on theater stages and gave rise to a new protagonist: the *machinist*, who was also born in the seventeenth century. From the time Mazarin introduced the opera to France in 1645, the director of the establishment was assisted by a musician and someone "for the machines."[16]

The words machine and machinist (before it described the machine operator, this term referred to the inventor), when they were introduced in the theater, carried with them meanings acquired a century earlier from the military vocabulary — e.g., war machine — or from the technical vocabulary of what today we would call engineering — e.g., hydraulic machine. In France, the king's "painters and artificers," and later his actors, were hydraulic engineers, mechanics specializing in automatons, pyrotechnic specialists, architects, and fountain makers. Most came from Italy. They had a wide range of knowledge: they built theaters but also bridges and fortifications; they organized princely funerals; etc. One such Italian was Nicola Sabbattini (1575–1654), a mathematician and engineer who built the port of Pesaro, planned to channel the Foglia River, and transformed the Teatro del Sole in his hometown Pesaro, equipping it with machines described in his treatise.[17] Others included Tomaso Francini, general manager of Henri IV's waters and fountains and builder of automatons and water displays; Giacomo Torelli da Fano (1608–1678), summoned to the court by Mazarin; and Giovanni Servandoni (1695–1766), the king's decorator, an architect who obtained the Machine Room concession at the Tuileries to show his dioramas and whom Diderot, in his *Salons*, called "the great machinist."

To the extent that theatrical illusion came to rest on his work, the *machinist* took on a central role in the theater, at the same time as he was refused

15 Jacques Muguet, (Lyon, 1669), p. 144. Quoted by Hélène Visentin, "Des tableaux vivants à la machine d'architecture dans les entrées royales lyonnaises (XVIe – XVIIe siècles)," in *Dix-septième siècle* 212 (March 2001): pp. 419–428. [My emphasis].

16 When Pierre Perrin became director of the Paris Opera in 1659, he was assisted by the musician Cambert and by Alexandre de Rieux, the Marquis of Sourdeac, an aficionado of lock mechanisms, for the machines he invented. Corneille dedicated his play *Toison d'Or* (*The Golden Fleece*) to him.

17 Nicola Sabbattini, *Pratica di fabricar scene e macchine ne'teatri* (Ravenne, 1638), French trans. by M. Maria, R. Canavaggia, and L. Jouvet, *Pratique pour fabriquer scènes et machines de théâtre*. Introduction by Louis Jouvet. (Neufchâtel: Ides et Calendes, 1977). See Jouvet's introduction, which I borrow from here.

acknowledgement in that the viewer expected to be exposed to the *effects* and not to the *machinery*. The more one is unaware of who produced them, the more accomplished the effects are: in Sabbattini's treatise (1638), we find chapter headings such as this: "Fourth method for bringing men out of the floor of the stage *without anyone noticing*" (my emphasis).[18]

Nevertheless, tension appeared as early as the seventeenth century between the author of the play — the text — and the machinist. Pierre Corneille, in his comments in the introduction to *Andromeda* — a play that includes a sea monster rising up out of the water and Perseus appearing out of the sky on a winged horse — emphasized the role of Giacomo Torelli and his trick sets, describing him as having "outdone" himself:

> Here is a simple and bare description of both the Machines and the Theatres which have delighted everyone at the performance of *Andromeda*. All the glory is due to Mr. Torelli, who outdid himself in executing the designs I proposed to him, and I was often astonished at how he was able so auspiciously, without confusion, to get himself out of such a great difficulty.[19]

This contradiction is vividly described by Jean de La Fontaine in his "Epître à M. de Niert sur l'opéra" (1677), in which the poet mocks machine-made "effects":

> [this] ... astonishing spectacle
> Dazzles the bourgeois and makes them cry miracle:
> But the second time they no longer crowd around;
> They prefer El Cid, Horace, Heraclius.
> The soul is not at all moved by these objects,
> Rarely do they even content our view.
> When I hear the whistle, I never find
> The change as quick as it was promised.
> Often the counterweight resists the finest chariot;
> A god hangs from a rope, and calls to the machinist;
> A part of the forest remains in the sea,
> Or half of heaven in the midst of hell.[20]

18 Nicola Sabbattini, ibid., p. 103.

19 Pierre Corneille, *Dessein de la tragédie d'Andromède, Representée sur le Theatre Royal de Bourbon. Contenant l'ordre des Scénes, la description des Theatres & des Machines, & les paroles qui se chantent en Musique* (Rouen: 1650), p. 66.

20 Jean de La Fontaine, *Nouvelles oeuvres diverses* (Paris: Nepveu, 1820), pp.112–113.

While La Fontaine believed that the "effect" or "miracle" that amazed the bourgeois lost its force quickly, his demonstration nevertheless relies on its dysfunction and imperfections — set changes are slow to be carried out, raising a chariot into the air is a failure, an actor who is supposed to fall from the sky is left hanging in mid-air, etc. — leaving us to surmise that if these problems were overcome, the desired effects could be achieved. The role of the machinist called upon in this description ("a god hangs from a rope, and calls to the machinist" — a mischievous reversal of roles, with the actor playing a god and the machinist as a *deus ex machina*!) could henceforth only increase in importance. This is clearly what came to pass with time, because in the nineteenth century there arose the temptation to see him as "the true author of the play" and to see the story only as the means for staging trick effects: "the story consists above all in the combination of trap doors," a champion of mechanical spectacles, Théophile Gautier, wrote about *Le Sylphe d'or*.[21]

In the latter half of the nineteenth century, the name of the machinist appeared on the play's poster, alongside that of the author.

Trick Effects

Thus, nineteenth-century fairy plays began to mention the names of machinists and trick effects people in their "credits."[22] The titles of these fairy plays — such as *Le Pied de mouton* (Louis-François Ribié and Alphonse Martainville, *The Talisman or The Sheep's Foot*), a triumph as early as 1806, and *Les Pilules du diable* (Ferdinand Laloue, Anicet Bourgeois, and Laurent, *The Merry Frolics of Satan*, 1839), with their transformation scenes and surprises, flying chariots, etc. — are familiar to us because early film, and Méliès in particular, would re-use them. In his review of *Bijou*, Gautier said that all the credit was due to the set designer and machinist and that the audience, not wanting to hear the names of the authors, called out loudly for the machinist, Mr. Sacré.[23]

The term "trick" ("*truc*"), along with *truqueur* and *truqué*, respectively the person responsible for the trick effect and something that has been

21 Théophile Gautier, *Histoire de l'Art dramatique en France. Depuis vingt-cinq ans*, vol. 2 (Paris: Hetzel, 1853), pp. 67–68.

22 See Marie-Françoise Christout, "Aspects de la féérie romantique: de *la Sylphide* (1832) à *la Biche au bois* (1845), chorégraphie, décors, trucs et machines," in *Romantisme*, (12) 38 (1982): pp. 77–86.

23 Christout, "Aspects de la féérie romantique," pp.101–102. See also Paul Ginisty, who tells a similar anecdote about *Les Pilules du Diable* in *La Féérie* (Paris: Hachette, 1910), p. 128.

"tricked" or faked, appeared in the theatrical lexicon in 1847. Treatises on it were written by Boullet and Georges Moynet, an architect. The latter gave the following definition:

> We call a trick effect any modification of the appearance of an object taking place under the gaze of the viewer. The more unexpected this modification, the more successful the trick effect.

He adds:

> It is important ... that the means used to achieve the modifications be as hidden as possible. When the trick effect is crude and we see the strings involved, the viewer cries out, disappointed: "We can see the strings!"[24]

The descriptions he gives sound like those of early films, combining rapidity, acrobatics, and fake sets. In one, a large man in a hotel room drinks a full glass of strong spirits and, soon after, sees his boots walk all by themselves, climb the wall, and disappear into the ceiling; the candlestick with which he wants to light his cigar moves and takes flight; there is a diabolical head on his pillow, which goes under the bed; a black devil and a white devil pursue it, disappear and reappear; the neighbors, altered by the din, arrive along with the maid, the hotel owner, and the police. A brawl ensues, "everything is broken; an old woman is thrown out the window and the curtain falls on general mayhem."[25]

In other cases, the trick effect is optical. On stage, we see a dance class: a ballerina executes a complicated routine in front of a huge mirror, and her reflection naturally repeats the same movements and positions. But the mirror is in truth a very transparent gauze, hung in front of a backdrop, and the reflection is a second dancer with a similar build and appearance as the first. This is a feat that Max Linder and the Marx Brothers would repeat.

A list of similarities between these stage entertainments and the earliest trick films would be a long one: from the talking decapitated head to earthquakes to the *Ride of the Valkyries* projected onto a backdrop, in keeping with the techniques outlined by Henri Fourtier in 1893 in his treatise on "Tableaux de projections mouvementées," using projection lanterns and

24 Georges Moynet, *Trucs et Décors explications raisonnées de tous les moyens employés pour produire des illusions théâtrales* (Paris: À la Librairie illustrée, 1893), chapter seven, "Décors à trucs avec trappes anglaises," p. 101.
25 Ibid., pp. 107–108.

putting into practice "the methods of Messieurs Marey, Londe, Anschütz, Muybridge, etc."[26]

The Cinema Machinist

But while the effects sought and obtained were sometimes the same, can we really describe the cinema as using the same *kind* of "machinations" as those on which it drew? It quickly became apparent to those explaining these techniques that it made sense to distinguish between them when discussing the cinema.

Eugène Kress, in his *Conférences sur la cinématographie*, devoted his seventh lecture in the volume to "Tricks and Illusions" involving "applications of optics and mechanics to the Cinematograph" but also "chemistry," thereby mixing various registers.[27] Under the heading "optical means," we find the properties of mirrors, one-way mirrors, black backgrounds, and prisms, and under the heading "mechanical means" are pulleys, levers, string, etc. — techniques employed in the theater and in prestidigitation shows. "Chemical means," meanwhile, include dissolves, stop-camera techniques, cuts in the negative, slow and fast motion, and tinting and toning. In reality, however, this classification scheme quickly becomes blurred because as soon as the effect sought is recorded on the film stock, the so-called "chemical" means would constantly relay or optimize the mechanical or optical means without, however, supplanting them: the field of trick effects was one of combinations and diverse techniques and remains so to this day, as is amply demonstrated by "making of" descriptions of films employing technical feats, including those of the "digital age." It remains the case, however, that verbal accounts emphasize this optimization through specification (e.g., the inversion of a raising or lowering technique, the use of trap doors, etc.) or substitution of theatrical means by those proper to the cinematograph (the camera crank, stop-camera, dissolves, suppressing images, and "editing" ["*montage*"] — this word was not used at the time — by cutting and joining, etc.). The cinematograph is the "renderer" of this highly diverse ensemble in which mannequins co-exist alongside the most sophisticated machinery.

26 This dispositive was still being mentioned in 1922 concerning the Théâtre de Monte-Carlo in 1909 (my thanks to Stéphane Tralongo for this information).

27 Eugène Kress, *Conférences sur la cinématographie*, vol. 1 (Paris: Comptoir d'éditions de *Cinéma-Revue*, n.d. [1912]). The seventh lecture is reproduced in *Pour une histoire des trucages*, in *1895: Revue d'histoire du cinéma*, 27 (1999), pp. 7–20.

This is why it may be reductive to enlist the cinematograph in already-existing entertainment "series" — that of fairy plays, illusionism, wax museums, etc. — because it uses their procedures and techniques, for this involves seeing the cinematograph's contribution as that of a "mere recording" of something. This, as we have just seen, is far from the case and assumes, moreover, that a recording can ever be "mere."

In the theater, machine and mechanism combine with the live performance. There is thus a constant contradiction between the two and, as a result, there is a need to keep the mechanism hidden: we should not see the strings; the effects are achieved "unawares," "without anyone knowing it." In addition, as Pougin emphasizes in the entry he devotes to it, the "machinery" was something that had not changed in 200 years: pulleys, winches, trap doors. He is astounded that electricity and modern mechanisms had not become a part of it at all.[28] They would do so by the turn of the century, but this is not the essential point: what changes with the cinema is not that it introduces the use of a machine, or machines; rather it *is* a machine, through and through. As a result, the cinema itself is the "great machinist." It is not a complement but rather *produces* the very movement of the bodies, whether they are living or not — corpuscles, atoms (air, light) — and can modify their trajectory.

Moreover, we can exhibit this machinist, reveal how it works, explain its principles and procedures. And whenever we do so, we speak of the cinema *qua* machine, not of some accessory or adventitious trick. Hence the plethora of scientific and technical accounts of how cinema works, from its very beginnings up to the present day.

Fredric Jameson sees in the "special effects" of *Star Wars* the "mimesis of technology."[29] He distinguishes these effects from feats of "magic" by viewing them according to their respective modes of production and to what they answer to in a given epistemological context: we do not seek illusion in such a film so much as we seek to valorize the machinery. We can add that "making of" films dismantle the illusion for precisely this purpose or, better yet, they valorize the attainment of illusion by dismantling its artifice. But this phenomenon is neither new nor limited to today's "special effects" films.

Orson Welles, in a preface to a book by Maurice Bessy on special effects in the cinema, made a similar distinction but added a temporal aspect in

28 Jouvet, relating his initiation into the profession of stage manager where he worked with machinists in *Pratique pour fabriquer scènes et machines de théâtre*, emphasized their similarities with sailors (the use of rope, pulleys, winches, levers).

29 Fredric Jameson, *Archaeologies of the Future: The Desire Called Utopia and Other Science Fiction* (New York: Verso, 2005), pp. 70–71.

the camera's tie with the past, with an event that has already taken place in front of it:

> In the theatre, when a magician makes an egg disappear we ask ourselves *how does he make it disappear.* In the cinema, if an egg disappears, we ask ourselves *how did they make it disappear.* This difference in temporality distinguishes cinema from theatre, trick effects from magic.[30]

It would appear, therefore, that an important distinction exists between magic and trick effects, stage and film, which belongs to the machinic nature of cinema. What is admired with the latter is the technological prowess the film marshals in achieving its effect and not the skillful gesture of the magician who acts from under our very nose. The term "magic" is surely a misnomer when discussing special effects in the cinema (only by metaphor should we speak of the "magic of special effects,"[31] for the expression risks confounding two very different phenomena).

"Revealing the Secret of the Trick"

To further the above distinction, let us continue with an article by Georges Méliès published in *Phono-Cinéma-Revue* in April 1908 (Fig. 6.1).[32] In reality, the article was a letter to the editor, solicited by the journal, expressing Méliès's reaction to a two-part article published a short time before in the magazine *L'Illustration* entitled "Les Coulisses du cinématographe" ("Behind the Scenes in the World of the Cinematograph"), written by Gustave Babin.[33] After discussing the amazement to which the appearance of the cinematograph gave rise, Babin asked two "Wellesian" questions: "How can they do

30 Maurice Bessy, *Les Truquages au cinéma* (Paris: Prisma, 1951)

31 This was the title of a conference "La Magie des effets spéciaux: Cinéma-Technologie-Réception/The Magic of Special Effects: Cinema-Technology-Reception," held at the Université de Montréal, Concordia University and the Cinémathèque québécoise, in November 2013.

32 Georges Méliès, "Les coulisses de la Cinématographie. Doit-on le dire? M. Méliès, Président du Syndicat des Illusionistes de France combat ceux qui 'débinent' les Trucs," in *Phono-Cinéma-Revue*, April 1908.

33 Gustave Babin, "Les coulisses du cinématographe," *L'Illustration, Journal universel*, 28 March 1908 and 4 April 1908. This two-part article was republished in a volume of all or some of the articles devoted to the cinema appearing in this magazine from 1896 to 1939. See *Les Dossiers de* l'Illustration: *Le Cinéma* (Paris: SEFAG, 1987), pp. 22–27 and pp. 28–32 respectively. Roland Cosandey discusses these articles and a few later ones in "Cinéma 1908: films à trucs et Film d'Art. Une campagne de *l'Illustration*," *Cinémathèque*, 3 (Spring-Summer 1993), pp. 58–71.

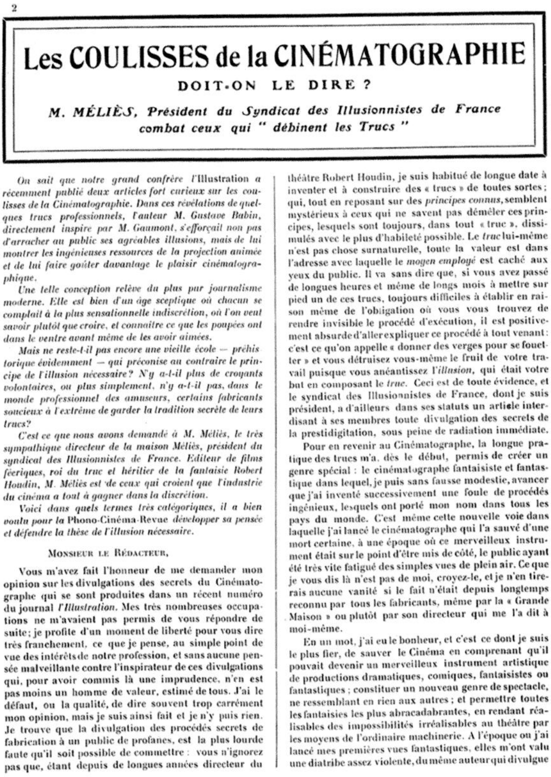

2

Les COULISSES de la CINÉMATOGRAPHIE

DOIT-ON LE DIRE ?

M. MÉLIÈS, Président du Syndicat des Illusionnistes de France combat ceux qui " débinent les Trucs "

On sait que notre grand confrère l'Illustration *a récemment publié deux articles fort curieux sur les coulisses de la Cinématographie. Dans ces révélations de quelques trucs professionnels, l'auteur M. Gustave Babin, directement inspire par M. Gaumont, s'efforçait non pas d'arracher au public ses agréables illusions, mais de lui montrer les ingénieuses ressources de la projection animée et de lui faire goûter davantage le plaisir cinématographique.*

Une telle conception relève du plus pur journalisme moderne. Elle est bien d'un âge sceptique où chacun se complait à la plus sensationnelle indiscrétion, où l'on veut savoir plutôt que croire, et connaitre ce que les poupées ont dans le ventre avant même de les avoir aimées.

Mais ne reste-t-il pas encore une vieille école — préhistorique évidemment — qui préconise au contraire le principe de l'illusion nécessaire? N'y a-t-il plus de croyants volontaires, ou plus simplement, n'y a-t-il pas, dans le monde professionnel des amuseurs, certains fabricants soucieux à l'extrême de garder la tradition secrète de leurs trucs?

C'est ce que nous avons demandé à M. Méliès, le très sympathique directeur de la maison Méliès, président du syndicat des Illusionnistes de France. Editeur de films féeriques, roi du truc et héritier de la fantaisie Robert Houdin, M. Méliès est de ceux qui croient que l'industrie du cinéma a tout à gagner dans la discrétion.

Voici dans quels termes très catégoriques, il a bien voulu pour la Phono-Cinéma-Revue *développer sa pensée et défendre la thèse de l'illusion nécessaire.*

MONSIEUR LE RÉDACTEUR,

Vous m'avez fait l'honneur de me demander mon opinion sur les divulgations des secrets du Cinématographe qui se sont produites dans un récent numéro du journal *l'Illustration*. Mes très nombreuses occupations ne m'avaient pas permis de vous répondre de suite; je profite d'un moment de liberté pour vous dire très franchement, ce que je pense, au simple point de vue des intérêts de notre profession, et sans aucune pensée malveillante contre l'inspirateur de ces divulgations qui, pour avoir commis là une imprudence, n'en est pas moins un homme de valeur, estimé de tous. J'ai le défaut, ou la qualité, de dire souvent trop carrément mon opinion, mais je suis ainsi fait et je n'y puis rien. Je trouve que la divulgation des procédés secrets de fabrication à un public de profanes, est la plus lourde faute qu'il soit possible de commettre : vous n'ignorez pas que, étant depuis de longues années directeur du théâtre Robert Houdin, je suis habitué de longue date à inventer et à construire des « trucs » de toutes sortes; qui, tout en reposant sur des *principes connus*, semblent mystérieux à ceux qui ne savent pas démêler ces principes, lesquels sont toujours, dans tout « truc », dissimulés avec le plus d'habileté possible. Le *truc* lui-même n'est pas chose surnaturelle, toute la valeur est dans l'adresse avec laquelle le *moyen employé* est caché aux yeux du public. Il va sans dire que, si vous avez passé de longues heures et même de longs mois à mettre sur pied un de ces trucs, toujours difficiles à établir en raison même de l'obligation où vous vous trouvez de rendre invisible le procédé d'exécution, il est positivement absurde d'aller expliquer ce procédé à tout venant : c'est ce qu'on appelle « donner des verges pour se fouetter » et vous détruisez vous-même le fruit de votre travail puisque vous anéantissez l'*illusion*, qui était votre but en composant le *truc*. Ceci est de toute évidence, et le syndicat des Illusionnistes de France, dont je suis président, a d'ailleurs dans ses statuts un article interdisant à ses membres toute divulgation des secrets de la prestidigitation, sous peine de radiation immédiate.

Pour en revenir au Cinématographe, la longue pratique des trucs m'a, dès le début, permis de créer un genre spécial : le cinématographe fantaisiste et fantastique dans lequel, je puis sans fausse modestie, avancer que j'ai inventé successivement une foule de procédés ingénieux, lesquels ont porté mon nom dans tous les pays du monde. C'est même cette nouvelle voie dans laquelle j'ai lancé le cinématographe qui l'a sauvé d'une mort certaine, à une époque où ce merveilleux instrument était sur le point d'être mis de côté, le public ayant été très vite fatigué des simples vues de plein air. Ce que je vous dis là n'est pas de moi, croyez-le, et je n'en tirerais aucune vanité si le fait n'était depuis longtemps reconnu par tous les fabricants, même par la « Grande Maison » ou plutôt par son directeur qui me l'a dit à moi-même.

En un mot, j'ai eu le bonheur, et c'est ce dont je suis le plus fier, de sauver le Cinéma en comprenant qu'il pouvait devenir un merveilleux instrument artistique de productions dramatiques, comiques, fantaisistes ou fantastiques ; constituer un nouveau genre de spectacle, ne ressemblant en rien aux autres ; et permettre toutes les fantaisies les plus abracadabrantes, en rendant réalisables des impossibilités irréalisables au théâtre par les moyens de l'ordinaire machinerie. A l'époque ou j'ai lancé mes premières vues fantastiques, elles m'ont valu une diatribe assez violente, du même auteur qui divulgue

Fig. 6.1 Article by Georges Méliès, "Doit-on le dire?" in *Phono-Cinéma-Revue* in April 1908.

that?" and "How do they do that?" He begins by describing astonishing scenes ("effects"): a car maims a sleeping drunkard, whose legs are cut off; "otherwise febrile" clowns, "as stunning as the Hanlon-Lees," climb ladders backwards, leap like balls from the ground onto the tops of walls, then onto kitchen shelves; a horse-drawn omnibus travels backwards full-tilt, the cab seeming to move ahead of the horses; fairy-tale pumpkins travel at top speed up a hilly street, leaping onto the mezzanine of an ordinary house; a skier flies through space like an eagle, toppling over a factory chimney, knocking it over, in his tumultuous descent; a swarm of genies, imps, and ballerinas is born more instantaneously than by a fairy wand, without any visible trap door having been opened [...] Then, thanks to Léon Gaumont, who opened the doors to his studios and behind-the-scenes facilities, like "Robert-Houdin gaily revealing the secrets behind his tricks," Babin endeavors to explain how these effects were achieved so that "any attentive spectator is capable of explaining and describing even the most mysterious ones and to reconstruct what the professionals call the 'shooting of the scene' [*la prise de vue*]."[34] He emphasized in particular "the use of resources exclusive to the device itself, the Cinematograph, [which] make it possible to achieve the most marvelous effects, which had been completely unknown and impossible until now":[35] stopping the camera, making substitution possible; having something appear gradually through a dissolve; cutting and joining shots ("a movement recorded in a hundred images at the time of the shot and reproduced by only twenty or twenty-five images at the time of projection will take a quarter or a fifth of the time and seem hurried and epileptic, like in a nightmare");[36] and reversal, achieved during shooting or projection, so that everything will unfold backwards.

We can see in these two long articles the extent to which the discursive efforts at technical explanation, the wish to strip bare the cinema-machine, become important with regards to cinema. And it would continue to be so with the arrival of sound, or films that introduced new trick effects, such as *King Kong* (Cooper and Schoedsack, 1933), or with the arrival of 3D, etc.[37]

The previous year, in his article "Kinematographic Views," Méliès himself had explained how cinema and all of its "trick effects" function. When asked to react to Babin's articles in *Phono-Cinéma-Revue*, however, he was hostile to such revelations. He was speaking in his capacity as the "President of the Illusionists' Guild of France," and only the introductory paragraph identified him as the producer (the term was "*éditeur,*" meaning publisher) of fairy films, the king of the trick effect, and heir to Robert-Houdin's fantasy world. His entire argument dealt with the illusionist's *profession* — speed, dexterity, experience — and on the time needed to acquire these qualities and thus the need to protect them: "Revealing secret production techniques to the uninitiated public is the gravest mistake one can make."

The procedure, the trick effect, he wrote, aims at illusion. "All its value resides in the fact that the public does not know how it is done."[38] He adds: "As soon as the public knows how [a trick] is achieved [...] they will say: it is not that hard."[39] It is like opening the bellies of puppets before having loved them, he lamented.

34 Babin, op. cit.,1987, p. 22.

35 Ibid.

36 Ibid., p. 30. I note in passing the reference to "editing," not named as such, and the reference to epilepsy.

37 All of these phenomena were the subject of detailed articles in *L'Illustration*.

38 Méliès, "Les coulisses de la Cinématographie. Doit-on le dire?" op. cit., p. 2.

39 Ibid., p. 3.

Conversely, Babin "stripped the procedure bare" and in so doing exhibits a different kind of relation to effects and the illusion than Méliès had. Moreover, he had anticipated and dismissed the latter's objections with the remark that:

> The cinematograph, however, will lose nothing of its attraction for them [viewers "in the know"], quite on the contrary. The marionettes that were the joy of our childhood did not cease to entertain us when one day someone with sharp eyes exclaimed from the parterre: "You can see the strings!"[40]

Without entering into a discussion of the competition or rivalry that may have existed between Méliès and Babin,[41] we can confirm that the latter's point of view was widely shared at the time: one is not interested in the cinema as a producer of illusions (real or fairy-like) without, correlatively, being also interested in the artifices and mechanisms that make this illusion possible.

Moreover, on 15 June 1908, François Valleiry responded to Méliès in *Ciné-Phone-Gazette* by distinguishing between the professional level on which Méliès works and that of the viewer.[42] The viewer will undoubtedly not use this knowledge to make trick films himself in such a way as to compete with those being made, he said, and thus there is no danger in this regard that tricks of the trade will be stolen (such theft took place, of course, but mostly through the hiring away of employees from one producer by another).

Méliès did not always demonstrate the same reserve with respect to explaining "behind-the-scene" activities in cinema and its trick effects, as I have noted. On the contrary, in his text "Kinematographic Views,"[43] he detailed the camera's mechanical functions and then discussed the effects

40 Babin, op. cit., p. 22.

41 In fact, Méliès alluded to the hostility towards him that Babin had expressed "at the time of [his] earliest fantastic pictures": "he maintained that it was shameful to use a *scientific* instrument for *buffoonery*... I had the pleasure of noting that he later changed his mind and has followed me on the path I opened up," "Les coulisses de la Cinématographie. Doit-on le dire?" op. cit., p. 3.

42 François Valleiry, "Doit-on le dire? Réponse à M. Méliès," in *Phono-Ciné-Gazette* 78 (15 June 1908).

43 Georges Méliès, "Les vues cinématographiques," in *Annuaire général et international de la photographie* (Paris: Librairie Plon, 1907), pp. 362–392. This text has been republished several times with material that had been cut or rearranged and identified by Jacques Malthête. Malthête's most recent version has been published in English translation as "Kinematographic Views," in André Gaudreault, *Film and Attraction: From Kinematography to Cinema* (Urbana: University of Illinois, 2011), pp. 132–152.

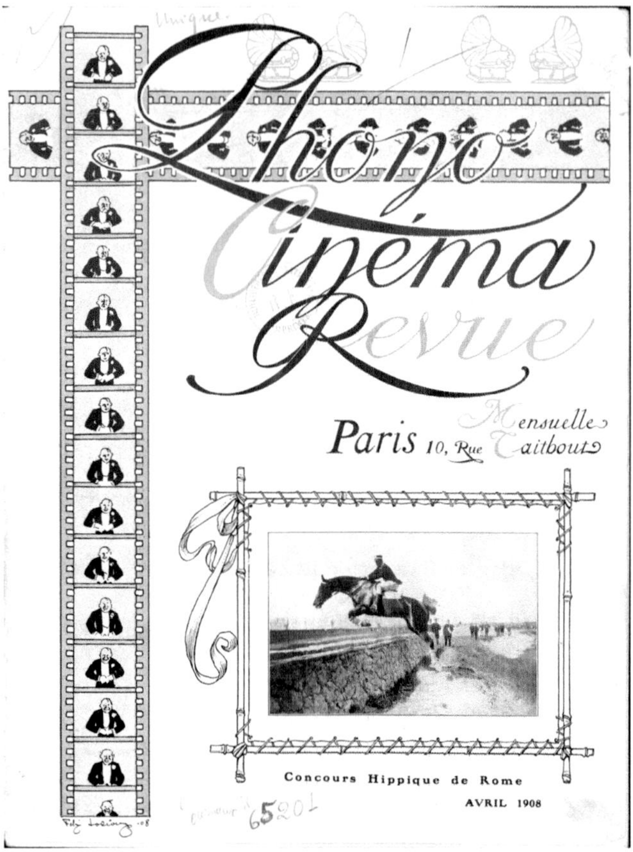

Fig. 6.2 Cover of *Phono-Cinéma-Revue*, April 1908 issue.

that could be achieved with it. His concern was certainly to maintain the "illusion" of a fantastic or fairy-play scene, and he feared technical imperfections that would "reveal the trick effect," as he said — and thus ruin the effect by revealing it. In his descriptive and explanatory approach, however, he did not "hide" any of the film techniques used to generate effects: slow motion, fast motion — and variations of speed in general — superimpositions, sleight of hand by stopping the camera. And he incorporated into the catalogue of trick effects theatrical machinery and its procedures and illusionist acts alike (we have seen that Kress took up a similar approach).

Had the situation changed in one year? Had the proliferation of explanations for what goes on behind the scenes in cinema led Méliès to protest and disapprove of such accounts? It could be noted that in the very journal where he had published his response to Babin, he was quite isolated on this point. In fact, this issue of *Phono-Cinéma-Revue* was run through with

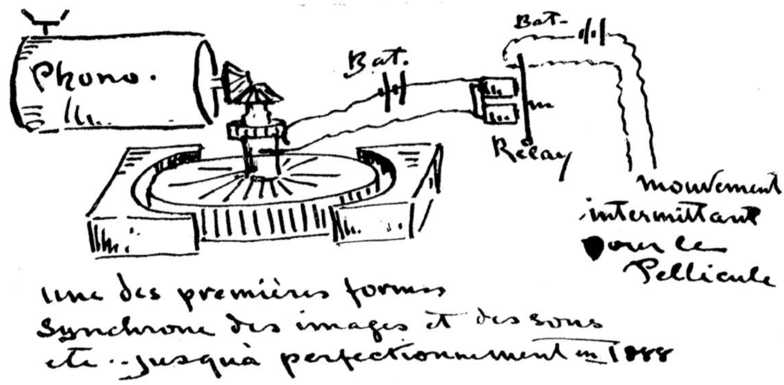

Fig. 6.3 Drawing of a phonograph and film stock from an article by Laurie Dickinson in April 1908 issue of *Phono-Cinéma-Revue.*

explanatory discussions of "cinema's mysteries": not only its "tricks" and "effects" (as Babin had done) but also its more mundane or *ordinary* artifices and techniques. Right from the journal's cover, showing two strips of film forming a right angle and duplicating the edges of the publication, the artificiality of cinema was front and center (Fig. 6.2). These strips contained some fifteen film frames (at the time they were called "small photographs" in French, and the preference was to give their size — 18 x 24mm — and their number per film — 10,400 per 200 meters). Each was a medium close-up of a figure whose image scarcely varied but which showed repetition and seriality, thereby suggesting cinemas' ability to decompose or analyze movement frame-by-frame.

On the cover, then, the material and technological reality of the cinematograph's mechanism is thus revealed metonymically from the outset. The reality of the cinematic image was also marked by a snapshot of someone jumping on a horse, nicely framed (along with the calligraphy of the title), by means of a *freeze frame* or isolated photogram (even though this image is not presented as a photogram because it is isolated) whereas the two pieces of film stock emphasized the *decomposition of movement.*

Inside the issue, just after Méliès's letter, an article by Laurie Dickinson, Edison's collaborator now working in London, revisited "Les Origines de la cinématographie" ("The Origins of Cinematography") to discuss the author's contributions to the development of moving images, demonstrating how machines for reproducing movement were modelled on the phonograph, with cylinders, etc. (Fig. 6.3). Here, we truly are "backstage," with the cinematograph as machine revealing its innermost mysteries as they lie in its very principle of operation.

Reading further on in the magazine, we find an article by Gédéo devoted to "Le Cinématographe chez soi — Les souvenirs animés" ("The Cinematograph in the Home — Moving Memories"). The author deplores the fact that the cinematograph, which has succeeded so well on so many levels — it is the "newspaper of tomorrow" — is still mostly "public," made for collective reception. It needs to be able to become, like the phonograph and photography, an instrument for the delight of the family in the private surroundings of the home. Instead, the models provided to amateurs, he deplores, are New Year's toys "created to satisfy the curiosity of future mechanics."[44]

Méliès's phrase "opening the bellies of puppets before having loved them" brings to mind Baudelaire. P. Adams Sitney has given a fine account of how Baudelaire's condemnation of photography, analyzed to death (moreover, it was the daguerreotype he condemned, not photography, and the two are quite different), contrasted with his interest in optical toys. Whereas photography produces a "truth" effect by virtue of its indexical nature, the phenakistoscope, used by a child *fully aware of what it was*, did not give rise to this illusion of reality: the child produces the images knowingly, can see what they are doing, and can act on the machine, stopping or accelerating it. The issue is one of dispositives. The amateurs described in *Phono-Cinéma-Revue* occupy the position of the child in Baudelaire's "La Morale du Joujou" ("The Moral of the Toy"), the author of the article regretting that the cinematograph goes no further than the status of a toy. Are film development, drying, and striking a negative, an obstacle? No: "most amateurs today have a darkroom, which is very easy to put together."[45]

Finally, Gédéo concludes, we can hope that recordings on film will soon be replaced by another, more practical and less dangerous means. Using paper, for example, or square glass plates, which would simplify things: the dangerously inflammable and fragile celluloid must be replaced. At that point, the tourist who does not take a train without a Kodak camera or a stereo camera will be armed with a precision portable cinema.

Further on, an unsigned article in the same journal ("Without Darkness") promises the arrival of a projector that will operate without darkness, making it possible, finally, to introduce the cinematograph to cafés, gardens, and schools in broad daylight.

In another article, Octave Uzanne sees in the cinema an adjunct to the theater for everything involving the setting and décor of a play: "it

44 Gédéo, "Le Cinématographe chez soi — les souvenirs animés," in *Phono-Cinéma-Revue*, April 1908, p. 8.
45 Ibid.

would enable us to see on stage, finally, all the unreality that until now has hampered the audience's full understanding" — choppy seas, clouds, backgrounds, dream rides in the Valkyries, marching forests. "No machinist's trick effect has been able to provide it with this."[46]

We then visit the Eastman Kodak plant with this introduction:

> For most viewers, the Cinematograph is a marvel amongst marvels. How is the eye fooled? By which mysterious means are the gestures and life itself of its characters reproduced? This is the question we all ask ourselves, without finding a solution.
> The Cinematograph's entire success lies, precisely, in this illusion. We believe that it should be respected, and yet we are not among those who refuse to provide Cinematography with a scientific explanation, and this is why we dedicate the following lines to the history of film [i.e., the author is discussing the history and manufacturing of celluloid], the very basis of the industry of animated projections.[47]

In other words, apart from Méliès, who deplored the fact that Babin "revealed the trick" (the expression was used by Babin himself), everyone in this journal pled, on the contrary, for an explanation of cinema's "mysteries": the authors untiringly seeking to explain *how it has worked* since 1895 and before, and "*How did they do it?*," to use an expression that recurs in Babin and that Orson Welles would take up 40 years later.[48]

Babin's article and the other ones from *Phono-Cinéma-Revue* were by no means exceptions. For example, in the June 1908 issue of *Lectures pour Tous*, we also find ten pages on "Les trucs du Cinématographe" ("Cinematographic Tricks"), motivated by the idea that everyone wants to know how this machine works and what techniques are used to create extraordinary effects: a diver exiting the water quicker than he entered it and, like an arrow, returning to a diving board ten meters in the air, or else someone walking down the street suddenly beginning to run backwards at a dizzying speed.

In this light, it appears inadequate to speak of a "novelty period" or of amazement and interest in the mechanics of the "cinematic illusion" as being limited to the earliest years of the cinema and as something limited to the

46 Octave Uzanne, "Un des 'premiers amis du cinéma' Monsieur Octave Uzanne," in *Phono-Cinéma-Revue*, April 1908, p. 10.

47 Anonymous, "Les installations Eastman Kodak," in *Phono-Cinéma-Revue*, April 1908, p. 28.

48 The controversy between Méliès and the other authors is discussed in detail by Frédéric Tabet in his doctoral thesis, *Circulations techniques entre l'art magique et le cinématographe avant 1906*, Université Paris-Est (2011).

discovery of the new medium. The phenomenon cuts far deeper.[49] Interest and focus on the moving picture camera, on the film stock and its hundreds, its thousands of photographs, on trick effects, etc. are all part of a discourse around the cinematograph according to which this machine leads to an upheaval in people's mode of perception and in the reconstitution of natural movement by means of its mechanical properties (stoppages, fast motion, backward motion, etc.). Better yet, this discourse also took place *on the screen* by seizing the viewers' everyday world — places, gestures, activities — and transforming them. This is obviously what filmmaker-theorists such as Jean Epstein, Marcel L'Herbier, and Dziga Vertov explicitly proclaimed in the early 1920s — when cinema was institutionalized and claimed a place as the "seventh art" — but it was also a feature of less specialized and more commonplace discourses on the cinema. As Léon Moussinac repeatedly said, "technology commands!": "We must never forget that cinema is not a finished art, that it is constantly being discovered, because technology has a primordial role in it by virtue of its complexity, that new revolutions will modify yet further its forms of creation and revelation."[50]

Trick effects for Moussinac, beyond their narrative or dramatic function, demonstrate moreover the power of technology.

Work Made Easy

Babin concluded his investigation with a film that, in his view, was "still mysterious," one that "[his] resources did not enable [him] to solve."[51] The film in question shows "a woodworking shop operating without a woodworker," a trick effect that went unexplained, along with a few other tricks ("a table laden with food which are consumed without it being clear how, by an invisible person"; "a bottle that pours its wine into a glass by itself"; "a knife that throws itself at a loaf of bread and a sausage and cuts them up"; etc.) (Fig. 6.4).[52] The Gaumont operator who revealed all the trick effects

49 See my article: François Albera, "First Discourses on Film and the Construction of a Cinematic Episteme," in Nicolas Dulac, André Gaudreault, and Santiago Hidalgo, eds., *A Companion to Early Cinema* (Oxford: Blackwell, 2012), pp. 121–140.

50 Léon Moussinac, "Diversité des films," in *Ce Soir*, (3 April 1937). Reprinted in Valérie Vignaux and François Albera, eds., *Léon Moussinac critique et théoricien des arts: Anthologie critique* (Paris: AFRHC, 2014), p. 355.

51 Babin, op. cit., p. 32.

52 Ibid. Emile Cohl uses the same examples in his 1925 article "Dessins animés," attributing them to Méliès, p. 862. See note 58 below.

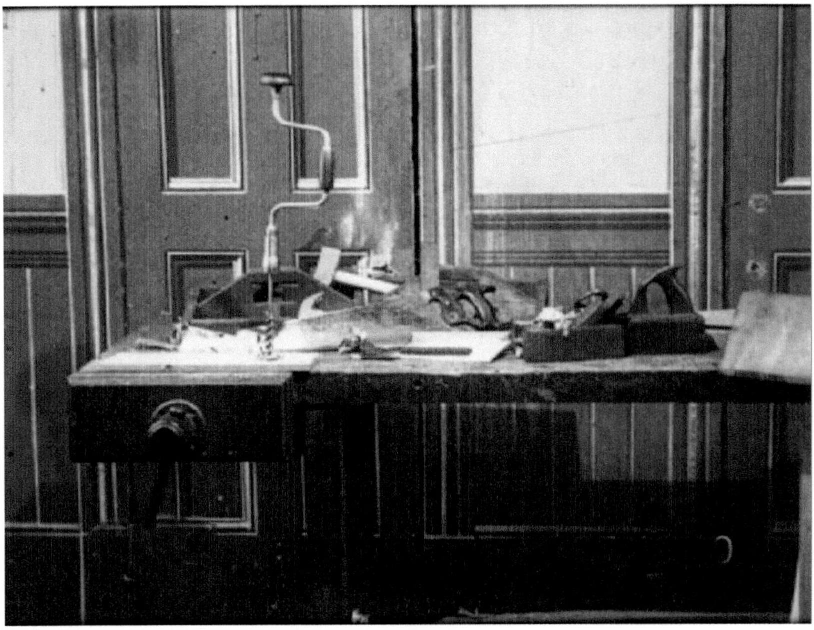

Fig. 6.4 Still from *Work Made Easy*: the workshop (Vitagraph, 1907).

that Babin explained to the reader told him: "I have been looking for months and discovered a week ago that these films which intrigue you so much are American." But it was impossible for him to "unshroud the mystery."

This film, made in a woodworking shop, is the final part of the American film *Work Made Easy*, made by James Stuart Blackton for Vitagraph and released in the United States in December 1907.[53] Its trick effects were based on what in France was called "*movement américain*" and what would later be called "*image par image*," or frame-by-frame shooting.[54]

This film, to which Kress also alluded — without naming it — in the seventh of his lectures,[55] was also the high point of the anonymous article in

53 The three photograms illustrating Babin's article are not, however, taken from this film but rather from one of its *remakes* (no doubt made by Gaumont), judging from a few elements of the decor, which differ from the original (the wallpaper in particular).

54 The expression used in French today appears in the article "Dessins animés" by J.B. de Tronquières (Emile Cohl pseudonym) for *Larousse mensuel* 222 (August 1925), pp. 861–864, where it is used to describe the making of the drawings to be filmed but not the process of filming them, which is called "turning the crank" ("tours de manivelle").

55 "Very curious results have been achieved with the Cinematograph as a result of what is called the 'mouvement américain' [...] This is how one obtains the effect of tools working on their own, etc." Eugène Kress, *Conférences sur la cinématographie*, vol. 1, (Paris: Comptoir d'éditions de *Cinéma-Revue*, n.d., 1912), p. 18.

Lectures pour Tous quoted above. There we are told how, in a woodworking shop, planes, saws, and braces can move on their own and do "their" work: "The device recorded one by one the successive positions that the objects would have taken if workers had been handling them."[56] The filmmaker thus moves the plane or saw by one or two millimeters, withdraws, and the camera records the image. Stop, new position, etc.

The Diegetic Machinist

Nevertheless, this film contains much more than this exploit of the woodworker's tools moving about on their own. It enables us to return to the *machinist*, who was hidden in the theater and is now on display in the cinema. And this time, it is on display not only through mere discursive accounts (such as Babin's and many others) but in the *space of the film itself*, whether through metaphor (as in *Work Made Easy*) or literally. For example, films exist in which the cinema-machinist is visible in the film itself, putting in motion its basic operating dispositives without seeking to create illusion (this would be a trick film) or else showing an operator-machinist operating a metonymic cinema-machine: the crank turner, the "man with a camera apparatus" (as in Vertov's film, *Chelovek s kino-apparatom* [*The Man with a Movie Camera*], 1929).

To begin, let us look at two examples from the Lumière catalogue, both from 1896, which show, almost like a mirror reflection of the motion picture camera, one or more *diegetic* "crank turners." In the first picture, *Partie de boules* (*Playing Boules*), the bourgeois playing *pétanque* in the foreground stand out from the proletarian, at first partially hidden and then becoming visible, who is turning the crank of a chaff cutter kind of agricultural machine in the background (Figs. 6.5a-c). He is carrying out the very same movement of the crank-turner in cinema during what the French aptly call the *tournage* (the film shoot) and, like the latter, has his eyes fixed on the characters playing *boules* and moving about on the "stage." He is like a machinist *setting in motion* a puppet play or a play involving mannequins.

In the second picture, *Chanteuse de rue* (*Street Singer*), the crank turner of a barrel organ (a programmed machine) acts directly on the bodies of the dancers (Fig. 6.6). The machine he sets in motion dictates their rhythm,

56 Anonymous, "Les trucs du Cinématographe," in *Lectures pour Tous* 9 (June 1908), p. 755.

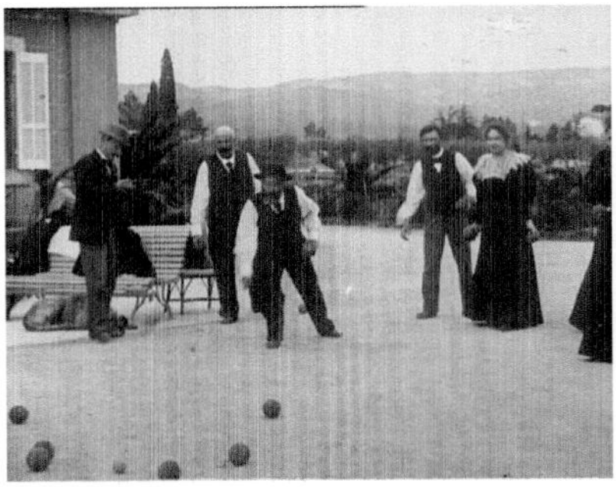

Fig. 6.5a Still from *Partie de boules* (Lumière, 1896).

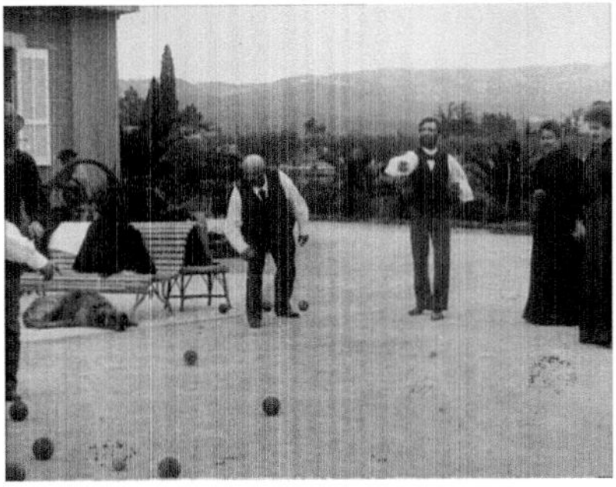

Fig. 6.5b Still from *Partie de boules* (Lumière, 1896).

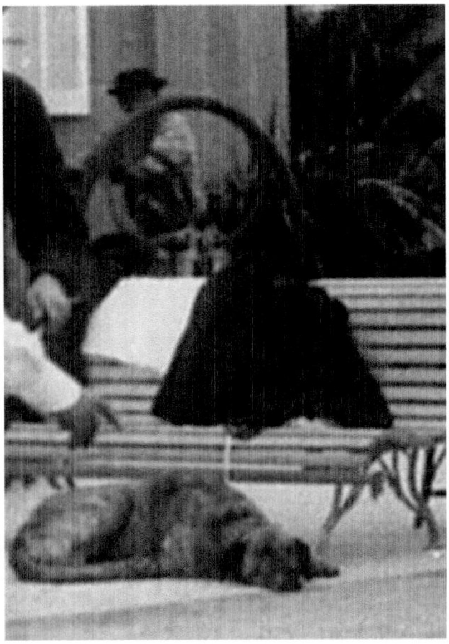

Fig. 6.5c Still from *Partie de boules* (Lumière, 1896).

speed, and movements. He is another machinist figure, and he too, his eyes fixed on the characters, is turning the crank of a device.

We find here, therefore, in simple form, the comic and fantastic mechanism that would be employed by *Work Made Easy* and other contemporaneous films. In a short Pathé film,[57] for example, an additional step is taken: a character, using a small machine with a crank that bears a strong resemblance to a moving picture camera, acts on his filmed surroundings by accelerating their action. The image thus shows us two speeds at which time unfolds, affecting the characters who appear together in the frame differently: while the crank turner moves at a normal speed, the world around him moves in fast motion: servants, passers-by, dogs, vehicles, etc.

However, this is not an example of the "frame-by-frame" method of the ending of *Work Made Easy* described above. As *Lectures pour tous* explained,

57 Anon., *Mais dépêchez-vous donc!*, Pathé, 1098. It is in the collection of the Cinémathèque française, which has done nothing more than salvage it.

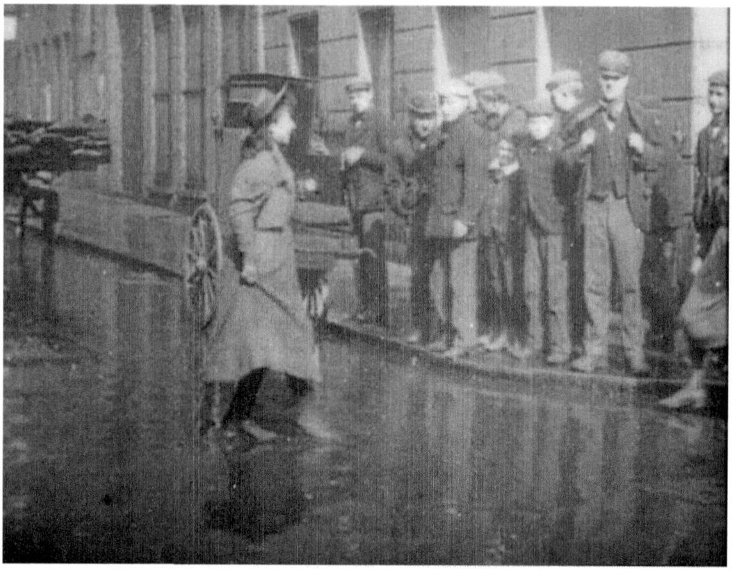

Fig. 6.6 Still from *Chanteuse de rue* (Lumière, 1896).

with respect to a film called *L'échelle* (*The Ladder*), in which a thief must run one 100 meters down a street to escape,

> By turning the crank at the normal speed of 16 frames per second, the operator would take a quantity of photographs which we could estimate at 900. Instead of taking 900 photographs, he records only 600 or 700. These 600 photographs, which still reproduce a 100-meter sprint, are projected on the screen in less time than 900. The character will no longer be running; he will be bounding forward at mad speed.[58]

In the short Pathé film, which uses the same technique, a difficulty arises because of the presence in the frame of the machine supposedly used to accelerate the action and its operator-machinist; neither is meant to be affected by the acceleration they bring about. Thus, during the shoot, the movements of the *filmed* crank turner had to be carried out extremely *slowly* in order for them to appear "*normal*" when projected, while the "*normal*" movements of the other characters will be *accelerated*. This film

58 Anonymous, "Les trucs du Cinématographe" op. cit., p. 752. If one is to believe Cohl, "the cinema device, when taking an ordinary shot, records eight photographic images per turn of the crank," in "Dessins animés," op. cit., p. 862. From this we can conclude that the operator carries out two turns of the crank per second.

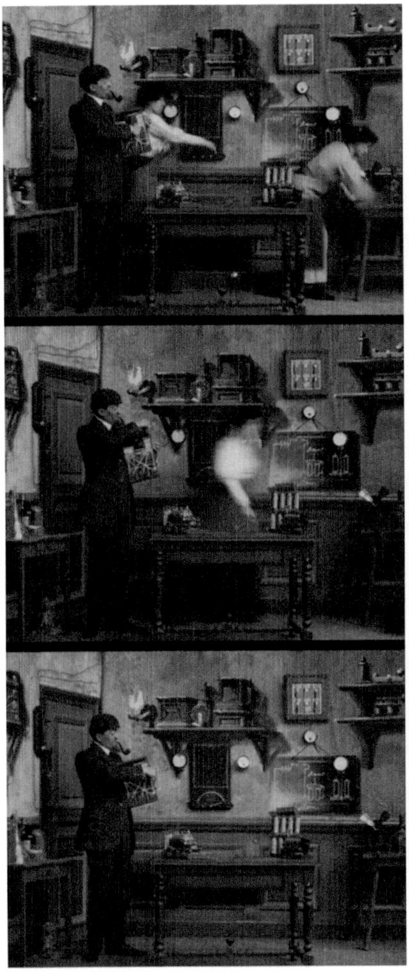

Fig. 6.7a Stills from *Mais dépêchez-vous donc!*, anon., Pathé, 1908.

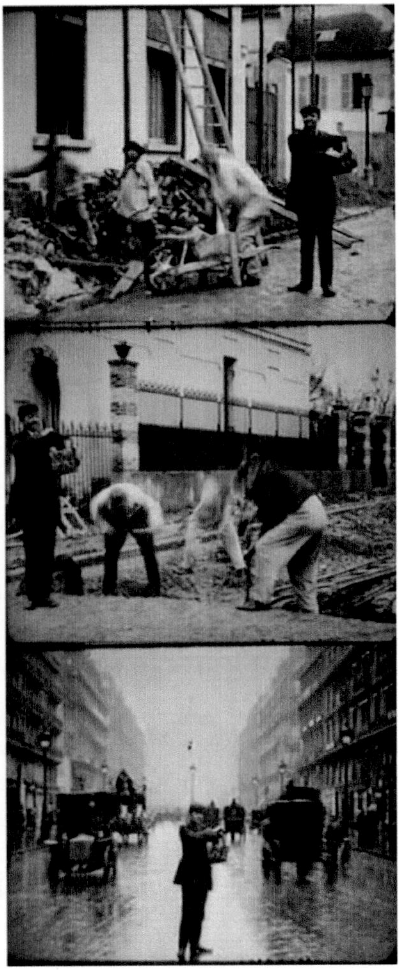

Fig. 6.7b Stills from *Mais dépêchez-vous donc!*, anon., Pathé, 1908

thus presents a paradoxical time system, not only in its image (normal/accelerated) but also in the shooting (slowed down/normal), as it inverses the usual way of achieving speed variations by manipulating the rate at which images are recorded. Here, slow motion is achieved by taking more images per second, accelerated fast motion by taking fewer. Recreated "normality" (the hand-cranking of the camera) is the result of an anomaly at the time the shot was taken (the extreme slow behavior of the diegetic machinist). As Jean Epstein would say some time later, cinema is the revealer of a restricted form of relativity that it alone makes visible and makes us feel. Here, each screen figure has its own inner clock whose differential is foregrounded through the viewer's own unchanging experience of (real) time, which provides yet a third temporal system when the film is projected.

Accelerating Work

These few films thus raise the question of *work*, of the power of the machine over the body and the role of cinema in the social dispositive employed to improve productivity. We are reminded of Paul Valéry who saw in the machine the "true government of the era" and who asked himself whether the machine's growth in "power, precision and speed" would come to influence our very being.[59] This, in the end, is the underlying theme, displaced in the Lumière pictures and made explicit in *Work Made Easy* (Figs. 6.8a-b) and *Mais dépêchez-vous donc!* (Figs. 6.9a-b).

Emile Cohl would recall these early examples in a sketch cynically entitled *Jobard n'aime pas voir travailler les femmes* (*Jobard Does Not Like to See Women Working*, 1911). Jobard, dressed as a dandy, enters a room in which his wife is doing housework (Fig. 6.10). He starts abusing her while she is grinding coffee in a manual coffee grinder so that she will work more quickly. Then he sits down in a chair, laughing, taking it easy while his wife, in fast motion, does all the housework (ironing, etc.). Here the machinist has "disappeared" in favor of the macho, who dons the magician's cast-off clothes, but we see that the first action to be accelerated is, once again, analogous with that of the crank turner, emblematic of the moving picture camera.

This sort of entertainment about accelerated work is not innocent given the context in which they appear, any less than the use of the cinematic apparatus to "accelerate" work movements. We know that Taylorism very early on developed an ingenious system for filming workers' gestures at their machine through the use of small lamps attached to their arms and body in order to devise an "optimal" arrangement and thereby increase productivity. Lenin, in one of the few texts in which he speaks of cinema, was moved by this, seeing in it "man's enslavement by the machine." He described this filming system in very precise terms.[60]

In a related field, because it also concerns performance and productivity — in this case in the realm of sports — Pierre de Coubertin also examined

59 Quoted by Jacques Guillerme in his preface to the 1972 edition of Jacques Lafitte's, *Réflexions sur la science des machines* (Paris: Vrin, 1972), p. 10.
60 Vladimir Ilych Lenin, "The Taylor System – Man's Enslavement by the Machine," *Collected Works*, vol. 20 (London: Lawrence and Wishart, 1964). Originally published in *Put Pravdy* 35 (13 March 1914).

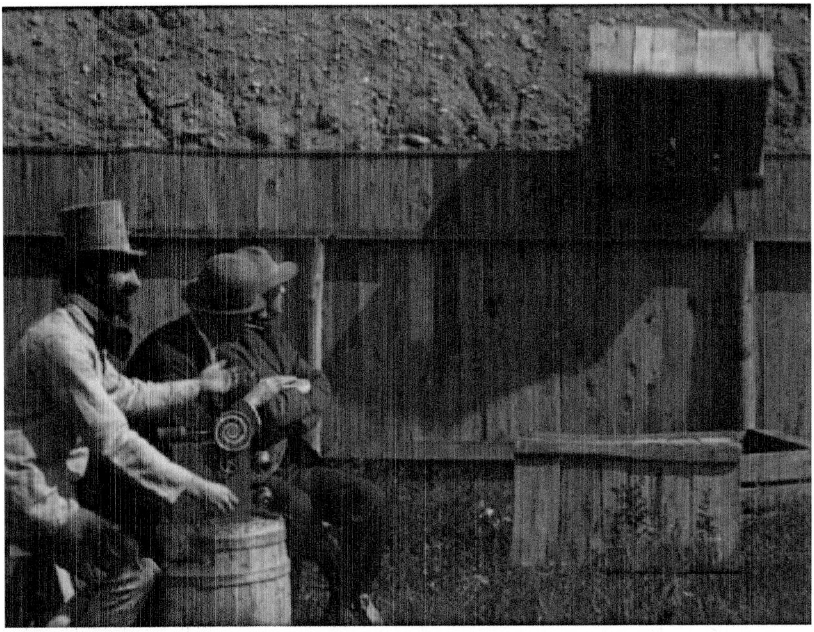

Fig. 6.8a Still from *Work Made Easy* (Vitagraph, 1907).

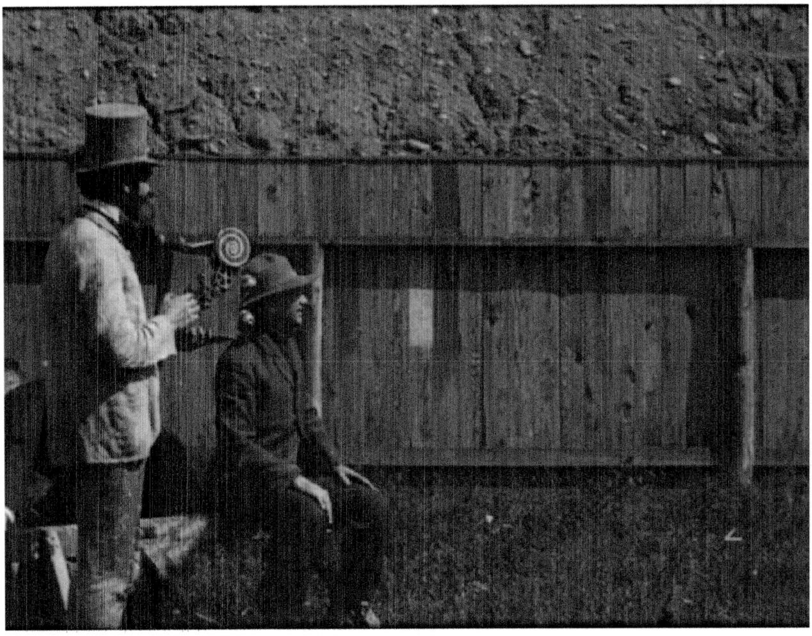

Fig. 6.8b Still from *Work Made Easy* (Vitagraph, 1907).

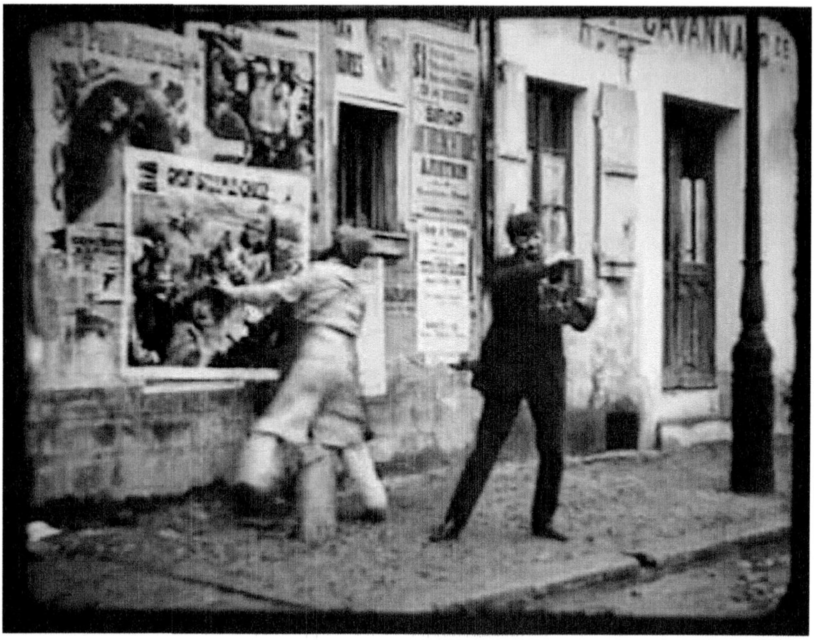

Fig. 6.9a Still from *Mais dépêchez-vous donc!* (Pathé, 1908).

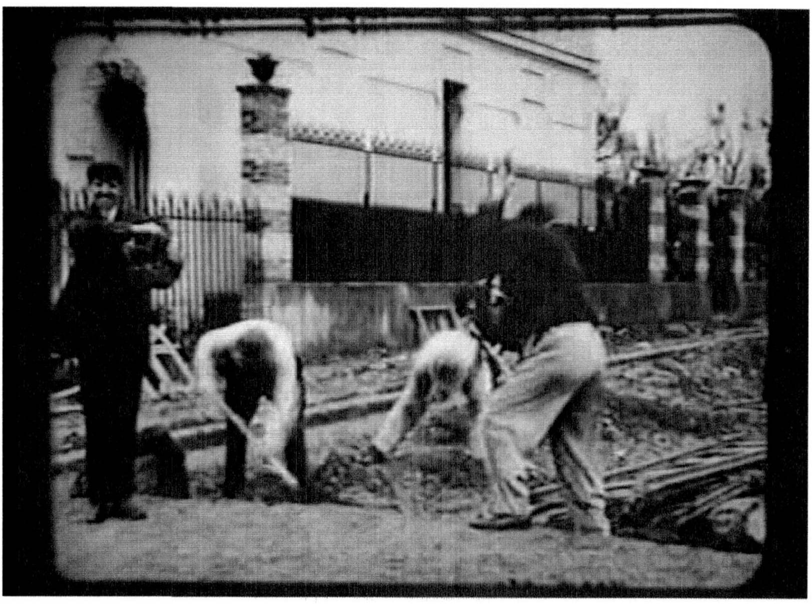

Fig. 6.9b Still from *Mais dépêchez-vous donc!* (Pathé, 1908).

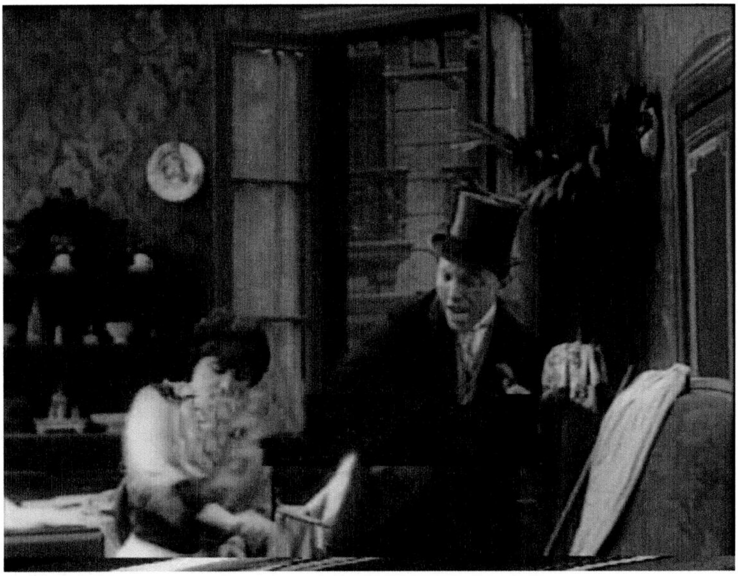

Fig. 6.10 Still from *Jobard n'aime pas voir travailler les femmes* (Émile Cohl, 1911).

the filming of athletes. He saw the cinema as a new means of perception, comparable to X-rays, showing athletes their gestures and their imperfections through the reproduction of movement. He had read Georges Demenÿ and was quite familiar with Taylorism. He was convinced of the need for this "corrective cinema" and wrote that every gymnasium, household, armory, pool, etc. should be equipped with moving picture cameras, adding that the athlete should be filmed on site and not in a studio, "caught in action" and if possible unawares: "The more the camera is 'indiscreet,' the more it will be 'corrective.'"[61] This is a cynical version of Vertov's "life caught unawares" and of the "mechanical eye" being more powerful than the human eye.

As Harun Farocki showed in his installation *Deep Play* at Documenta 12, on the World Cup of football, this kind of data for analyzing athletic gestures and for simulating better models is widely prevalent and finely tuned.

Another Genealogy

This leads us to sketch another genealogy than solely technical research into the decomposition of movement, particularly human (walking, for

61 Pierre de Coubertin, "Le sportif et le cinéma," *Bulletin du Bureau international de pédagogie sportive* 8, 1932, p. 8.

example) and its recreation through devices that make it possible to observe it and derive from it principles of controlling and "educating" movements, or correction and normality. Both the military and gymnastics were suitable experimental institutions for perfecting "techniques of the body" (Mauss) and procedures that could be generalized. There is a social and economic dimension here that can be traced back to the eighteenth century, to Vaucanson, known for his spectacular automatons, including the famous eating, digesting, and defecating duck. Vaucanson, however, quickly sought to introduce his models to industry, and he was tasked with reorganizing the silk industry in order to standardize workers' gestures and actions, which led to mechanization and to power looms.

Aristotle had said:

> For if every instrument could accomplish its own work, obeying or anticipating the will of others, like the statues of Daedalus, or the tripods of Hephaestus, which, says the poet, "of their own accord entered the assembly of the Gods"; if, in like manner, the shuttle would weave and the plectrum touch the lyre without a hand to guide them, chief workmen would not want servants, nor masters slaves.[62]

This is the utopia of *Work Made Easy*, at the center of several of the articles discussed above: for work to take care of itself. But carrying out this utopia always begins by accelerating the tasks of workers, turning them into machines that can be acted upon at will, before being passed over altogether by machines and made unemployed. Blackton's film, like the short anonymous Pathé film and Cohl's film, presents the moving picture camera as capable of acting on the body with the goal of increasing labor productivity. The camera thus becomes an accessory of capitalist exploitation, an accessory in the exploitation of the labor force. It is present *in* the film and *with* the film, thanks to the fiction of the man turning the machine's crank, whose first action is to accelerate the work of two women engaged in housework (chopping or grinding by turning a crank), with workers climbing scaffolds at breakneck speed, navvies digging holes in record time, etc. To this physiological transformation with its heightened and jerky movements is added an exercise in social reality: to convert into action the bricklayers' laziness and the women's listlessness. The lesson is clear: work rhythms are accelerated in order to correct these "weaknesses," these "inadequacies," and the worker is thereby threatened with being "replaced" by the machine if he or she does

62 Aristotle, *Politics*, trans. Benjamin Jowett (New York: Modern Library, 1943), 1.2.1253b.§ 4.

not get moving smartly. Karl Marx formulated this perfectly in *Capital*, in which the machine becomes a "weapon" for suppressing strikes and revolts:

> But machinery not only acts as a competitor who gets the better of the workman, and is constantly on the point of making him superfluous. It is also a power inimical to him, and as such capital proclaims it from the roof tops and as such makes use of it. It is the most powerful weapon for suppressing strikes, those periodic revolts of the working class against the autocracy of capital. According to Gaskell, the steam-engine was from the very first an antagonist of human power, an antagonist that enabled the capitalists to tread under foot the growing claims of the workmen, who threatened the newly born factory system with a crisis.[63]

Marx's remark concludes with a statement that reverses the "modernist" discourse in which technological evolution and technical inventions are seen only as an expression of "progress" (an idea Walter Benjamin would critique in his essay "On the Concept of History"): "It would be possible to write quite a history of the inventions, made since 1830, for the sole purpose of supplying capital with weapons against the revolts of the working class."[64]

Contemporary resonances with these phenomena are strong enough that it is pointless to belabor them: robotization; automation; massive overproduction and, correlatively, massive underemployment; a dual society.[65]

The machine that replaces human beings, liberating them from alienated labor, is the utopia that fools no one. It is found in René Clair's *À Nous la liberté!* (*Freedom for Us!*, 1931), which Charlie Chaplin would clearly criticize in *Modern Times* (1936). Clair finesses the question of production relations

63 Karl Marx, *Capital: A Critical Analysis of Capitalist Production*, vol. 1, ed. Frederick (*sic*) Engels, trans. Samuel Moore and Edward Aveling (New York: International Publishers, 1967), pp. 435–436.

64 Ibid., pp. 563–564.

65 See the recent work of Moshe Vardi, a computer science professor and director of the Ken Kennedy Institute for Information Technology at Rice University in Texas. According to Vardi, "more than half of the world's population could see their work threatened by robotization and advances in artificial intelligence" ("Smart Robots and Their Impact on Society," annual congress of the American Society for the Advancement of Science, Washington D.C., February 2016). See also the forecast of the economists David H. Autor and David Dorn, for whom "computerisation threatens in particular routine jobs whose predictable tasks are easily made automatic, whether in the case of office jobs or of manual labour. In France, a study carried out by the firm Roland Berger in 2014 found that the replacement of human labor by machines would affect 42% of occupations and three million jobs in 2015," in *Direct Matin* 1841, (Tuesday 8 March 2016), p. 5.

by projecting onto an ideal society the liberation of alienated labor by machines, creating free time for workers.[66]

Modern Times, moreover, ties its anti-Taylorist ideas, its social discourse (unemployment, police repression), to the mechanical resources of the moving picture camera in order to depict factory social relations: the boss speeds up the production line and the film shifts into fast-motion because the boss is the machinist of the alienated workers on the production line through surveillance (internal television circuit) and the chain of command relayed by telephone line (orders given to the foreman). This homothety between the acceleration of the production line and the acceleration of the film finds its demonstrative completion when the machine *absorbs* Charlie and *spits him back out* in a perfect filmic palindrome: the film goes forward and then backwards, not only to depict the action but to demonstrate the nature of mechanical movement.

There is thus no break between the films of 1908 and *Modern Times*. Viewers are led to laugh or become distressed at the machination of the world depicted; we need think only of the number of films that take up the theme we saw at work in *The Man with a Movie Camera* of the acceleration of automobile traffic in the city: *Paris qui dort* (*The Crazy Ray*, René Clair, 1924), *The Cameraman* (Edward Sedgwick, 1928) — to accelerated cranking is added accelerated editing, something that is mentioned in discussions of trick effects beginning from the turn of the century — or *County Hospital* with Laurel and Hardy (James Parrot, 1932), in which Laurel, accidentally anaesthetized by a needle that was not intended for him, takes the wheel of a car to take Hardy home with his leg in a cast and falls asleep in traffic. Here we find a third kind of trick effect in film: the rear-projected image.

In all these films, the parameters of fatigue and human vulnerability are confronted with a mechanical world and with a possible "great machinist" capable of subjugating entire populations: the absent-minded scientist in *Paris qui dort* who, with a lever, can stop or start the city of Paris. It is a pleasant, funny version of the later bosses of *Metropolis* (Fritz Lang, 1927) and *Modern Times*. More than any other form of expression, cinema turns this relation of domination into metaphors and, more seriously, into metonyms.

66 The classical method of juxtaposing these two films limits itself to examining Chaplin's possible borrowings from Clair, perhaps even his "plagiarism," without considering the reversal he carries out by bringing out a political dimension in his film that is completely absent from Clair's film. In this respect, see the reception of the latter film in the left-wing press as discussed in François Albera, "Le machinisme et la révolution dans *À nous la liberté!* de René Clair vu par trois périodiques de gauche," in *1895: Revue d'histoire du cinéma*, 72 (Spring 2014), pp. 108–127.

Conclusion

Is it really exaggerated to infer from the cinema's machinic character its "complicity," or worse, an inseverable connection with forms of subjugation that first arose with the industrial society (machinism) and are now deployed anew with the advances of the so-called "new technologies"? The computer has now replaced the machines of a previous, "antediluvian" age. These nineteenth-century machines that now appear as nothing less than "mechanical brontosauruses" have been superseded by "today's machinery [...] leading to something like a real muscular system, controlled by a real nervous system, performing complex operating programs through its connections with something like a real sensory-motor brain."[67] The "dematerialization" of today's machine operations only results from the blindness they foster toward any externalization of their sources of energy and from the way they capture and transfer massive amounts of data. We "freely" dispose of them but only insofar as we lose sight of the minute cartographies they draw and of the ways we have of using them. Indeed, of "this disparate arsenal" which, to quote Leroi-Gouhran once more, "is supplying the parts for a composite body strangely similar to the biological one,"[68] can we not say that it now clearly belongs to the specific configuration that Michel Foucault named "bio-power"?

Bibliography

Albera, François. "Le machinisme et la révolution dans *À nous la liberté!* de René Clair vus par trois périodiques de gauche." *1895: Revue d'Histoire du Cinéma*, no. 72 (Spring 2014): 108–27. https://doi.org/10.4000/1895.4808.

———. "First Discourses on Film and the Construction of a Cinematic Episteme." In *A Companion to Early Cinema*, edited by Nicolas Dulac, André Gaudreault, and Santiago Hidalgo (Oxford: Blackwell, 2021): 121–140.

Aristotle, *Politics*, translated by Benjamin Jowett (New York: Modern Library, 1943).

Anon. "Les installations Eastman Kodak," *Phono-Cinéma-Revue* (April 1908): 28-20

Anon. "Les trucs du Cinématographe," *Lectures pour tous* 9 (June 1908): 749-758.

Amyot, Jacques. "Vies Demetrius." Centre national des ressources textuelles et lexicales (see the entry "machine"). https://www.cnrtl.fr/definition/machine.

67 Leroi-Gouran, *Speech and Gesture*, p. 248

68 Ibid.

Babin, Gustave. "Les coulisses du cinématographe," *L'Illustration* 3396 (28 March 1908); "Les coulisses du cinématographe," *L'Illustration* 3397 (4 April 1908); "Le théâtre cinématographique," *L'Illustration* 3427 (31 October 1908). Reproduced in *Les Grands Dossiers de l'Illustration: Le Cinéma* (Paris: SEFAG, 1987): 22–33.

Bessy, Maurice. *Les truquages au cinéma* (Paris: Prisma, 1951).

Christout, Marie-Françoise. "Aspects de la féérie romantique: de *La Sylphide* (1832) à *La Biche au bois* (1845); chorégraphie, décors, trucs et machines," *Romantisme* 12, no. 38 (1982): 77–96. https://doi.org/10.3406/roman.1982.4577.

Cohl, Émile. "Dessins animés," *Larousse mensuel* 222 (August 1925): 861-864.

Coissac, G.-Michel. *Les coulisses du cinéma* (Paris: Paul Duval, 1929).

Cosandey, Roland. "Cinéma 1908: films à trucs et Film d'Art. Une campagne de *l'Illustration*." *Cinémathèque* 3 (Spring-Summer 1993): 58–71.

Corneille, Pierre. *Dessein de la tragédie d'Andromède, représentée sur le théâtre royal de Bourbon. Contenant l'ordre des scènes, la description des théâtres et des machines, et les paroles qui chantent en musique* (Rouen, 1650).

Coubertin, Pierre de. "Le sportif et le cinéma," *Bulletin du Bureau International de Pédagogie Sportive* 8 (1932) : 7-9.

*Direct Matin, no.*1841 (8 March 2016).

Duckett, William, ed. *Dictionnaire de la conversation et de la lecture*, vol. 61, supplement to vol. 9 (Paris: Firmin-Didot, 1847).

Ducom, Jacques. *Le Cinématographe scientifique et industriel: Traité pratique de cinématographie.* (Paris: Geisler, 1911).

———. *Cinéma muet, sonore et parlant* (Paris: Albin Michel, 1931).

Gaudreault, André. *Film and Attraction: From Kinetography to Cinema* (Urbana: University of Illinois, 2011).

Gautier, Théophile. *Histoire de l'art dramatique en France depuis vingt-cinq ans*, vol. 2 (Paris: Hetzel, 1853).

Gédéo, "Le Cinématographe chez soi – les souvenirs animés," *Phono-Cinéma-Revue* (April 1908) : 7-8.

Ginisty, Paul. *La Féérie* (Paris: Hachette, 1910).

Giraud, Jean. *Le lexique français du cinéma: des origines à 1930* (Paris: Centre national de la recherche scientifique, 1958).

Jameson, Frederic. *Archaeologies of the Future: The Desire Called Utopia and Other Science Fiction* (New York: Verso, 2005).

Jullier, Laurent. "L'ère numérique: vers l'évanescence du trucage." In *1895. Revue d'Histoire du Cinéma no. 27 – Pour une histoire des trucages*, edited by Thierry Lefebvre (Paris: Association Française de Recherche sur l'Histoire du Cinéma, 1999): 113–120.

Kress, Eugène. *Conférences sur la cinématographie*, vol. 1 (Paris: Comptoir d'éditions de Cinéma-Revue, 1912).

Lafitte, Jacques. *Réflexions sur la science des machines* (Paris: Vrin, 1972).
La Fontaine, Jean de. *Nouvelles œuvres diverses* (Paris: Nepveu, 1820).
Larousse mensuel 65. (see entry "Léon Broquet, *L'Hiver au marais*", 1912)
Leroi-Gouran, André. *Speech and Gesture*, translated by Anna Bostock Berger (Cambridge, MA: MIT Press, 1993).
Lefebvre, Thierry, "Avant-propos." *1895: Revue d'Histoire du Cinéma* no. 27 (September 1999): 7–20.
Lenin, Vladimir I. "The Taylor System – Man's Enslavement by the Machine." In *Collected Works*, vol. 20 (London: Lawrence & Wishart, 1964): 152-154. Originally published in *Put Pravdv*, no. 35 (13 March 1914).
Littré, Émile. *Dictionnaire de la langue française. Édition intégrale*, vol. 3 (Paris: Jean-Jacques Pauvert, 1956).
Löbel, Léopold. *La technique cinématographique: projection, fabrication des films* (Paris: Dunod et Pinat, 1912).
Malthête, Jacques. "Kinematographic Views." In André Gaudreault, *Film and Attraction: From Kinematography to Cinema* (Urbana: University of Illinois, 2011): 132–152.
Marx, Karl. *Capital: A Critical Analysis of Capitalist Production*, vol. 1, edited by Frederick Engels, translated by S. Moore and E. Aveling (New York: International Publishers, 1967).
Méliès, George. "Les vues cinématographiques." In *Annuaire général et international de la photographie* (Paris: Librairie Plon, 1907) : 362–392.
———. "Les coulisses de la cinématographie. Doit-on le dire? M. Méliès, président du Syndicat des illustionistes de France combat ceux qui 'débinent' les trucs," *Photo-Ciné-Revue* (April 1908).
Moussinac, Léon. "Diversité des films," *Ce soir* (3 April 1937). Reprinted in Valérie Vignaux and François Albera, eds. *Léon Moussinac; Critique et théoricien des arts; Anthologie critique* (Paris: Association Française de Recherche sur l'Histoire du Cinéma, 2014): 354-355.
Moynet, Georges. *Trucs et décors: Explications raisonnées de tous les moyens employés pour produire des illusions théâtrales* (Paris: À la librairie illustrée, 1893).
Pougin, Arthur. *Dictionnaire historique du théâtre et des arts qui s'y rattachent. Poétique, musique, danse, pantomime, décor, costume, machinerie, acrobatisme; jeux antiques, spectacles forains, divertissements scéniques, fêtes publiques, réjouissances populaires, carrousels, courses, tournois, etc., etc., etc.* (Paris: Firmin-Didot, 1885).
Sabbattini, Nicola. *Pratica di fabricar scene e macchine ne' teatri.* Ravenne, 1638. Translated by Maria and René Canavaggia and Louis Jouvet. *Pratique pour fabriquer scènes et machines de théâtre* (Neufchâtel: Ides et Calendes, 1977).
Tabet, Frédéric. "*Circulation technique entre l'art magique et le cinématographe avant 1906*," Doctoral dissertation (Université Paris-Est, 2011).

Uzanne, Octave. "Un des 'premiers amis du cinéma' Monsieur Octave Uzanne," *Phono-Cinéma-Revue* (April 1908): 9-10

Valleiry, François. "Doit-on le dire? Réponse à M. Méliès," *Phono-Ciné-Gazette*, no. 78 (15 June 1908).

Vignaux, Valérie, and François Albera, eds., *Léon Moussinac; Critique et théoricien des arts; Anthologie critique* (Paris: Association Française de Recherche sur l'Histoire du Cinéma, 2014).

Visentin, Hélène. "Des tableaux vivants à la machine d'architecture dans les entrées royales lyonnaises (XVIe – XVIIe siècles)," *Dix-septième siècle* 212 (March 2001): 419–428.

About the Author

François Albera is Professor Emeritus at the Université de Lausanne (Switzerland). He is Editor-in-Chief of *1895 revue d'histoire du cinéma*. His most recent books include *Cinema Beyond Film* (co-edited, 2010), *Cine-dispositives. Essays in Epistemology across Media* (co-edited, 2015); *Le cinéma au défi des arts* (2019); and *Fernand Léger et le cinéma* (2021).

7. Black Magic: The "Space Between the Frames" in Cinematic Special Effects

Donald Crafton

Abstract

"Special effects" is not simply an addition to the cinema system; it is integral to cinema. The fundamental concept involves the mechanical nature of cinematography, the phenomenology of the experience of film, and the implications for theories of film and theories of animation. The history of explosions as an effect is considered. The essay explores the heuristics of the "naught," the "space between frames," and the "interval" in the magic performances of Georges Méliès, the editing of Edwin S. Porter, and the writings of Henri Bergson, Sergei Eisenstein, Dziga Vertov, and Norman McLaren.

Keywords: space between frames, cinematography, explosion, nothingness, magic, animation, projection, Lumière cinématographe, time, space, duration, Dark Matter

Peindre, non la chose, mais l'effet qu'elle produit.
(To paint, not the thing, but the effect that it makes.)
—Mallarmé[1]

This is an essay about nothing. It does, however, address a very specific nothing, the nothingness that is the mechanical basis of the film experience and that exists between the flashes of photographs animated by a cinematic

1 Stéphane Mallarmé, quoted in Patricia Terry, "Mallarmé and Basho," in Robert Greer Cohn et al., eds., *Mallarmé in the Twentieth Century* (Madison, NJ: Fairleigh Dickinson University Press, 1998), p. 269.

Lefebvre, M. & M. Furstenau (eds.), *Special Effects on the Screen: Faking the View from Méliès to Motion Capture*. Taylor & Francis Group, 2022
DOI 10.5117/9789462980730_CH07

viewing device. We cannot deny that this Dark Matter of cinema exists. It is demonstrably there, yet invisible and intangible. Yet this is not your ordinary nothing; it is a special nothing without which there would be no practice called 'special effects' in cinema.[2] Its specialness lies in the way that the cinematic apparatuses of recording and re-presenting turn the system's darkness into a performance, something we know that exists, but which, although we cannot see it, we can readily naturalize.

There are actually two kinds of nothing "between the frames." There are the mechanical, measurable, constitutive lacunae that the recording and projection processes generate — which is not as simple as one might think; and there is the persistent topos of the *performative space* between the frames that filmmakers and commentators have hypothesized and theorized for many years. Let us call this performative emptiness the *naught*. Considering the cinema's history of "special effects" provides the opportunity to come to grips with this nothing that, counter-intuitively, is something: the time-space interframes of motion pictures.

If this nothing is both a space and an interval of time, then where and when does it exist? It is not physically on strips of film. Whether the frame lines of the adjacent individual photographs are touching, as in some formats (for example, Lumière and CinemaScope), or separated by black "window box" bars to preserve a certain aspect ratio (the early-sound SMPE[3] standard), the blackness derives from somewhere else. We have to explore the mechanics of cinematography and the phenomenology of projection to understand this ontology.

"Special effects" is not simply an addition to the cinema system, it is integral to cinema. This is because its technological basis is a mechanism for structurally omitting information twice: first in the taking, recording phase; secondly in the playback, viewing, projecting phase. Cinema through most of its twentieth-century history depended on the workings of complicated clockwork machines whose purpose was to generate this nothingness. Clocklike in two ways: the cameras and projectors resembled the precision engineering of the most up-to-date chronometers as they sliced, diced, and reconstituted time, translating it into structured alternations of lightness and blackness to create motion. Secondly, this technical absolute then became the basis for filmmakers and commentators' manipulation

2 I consider "special effects," when referring to a practice, to be a compound noun and treat it as singular (per US usage). When "special" refers to particular effects, like explosions, I treat "effects" as plural.

3 The Society of Motion Picture Engineers, which came into being in 1916.

of this fundamental mechanism, which is simply physics. In so doing, they created meaning and, for some, limned the boundaries of cinema's invisible propelling force. Both efforts energize cinema's black moment between images by turning it into an opportunity for a special kind of performance.

Cinema Has Always Been Digital

Sometimes taking a close look at the elements of what we take to be common knowledge can be productive. We all know how cinema works in its two distinct modes, recording and projecting. In both, the aim is to expose one still photograph at a time on moving film stock. The cameras are manufactured to both record and, crucially, to *not record,* at a regular fixed rate, so the captured light passing through the lens alternates with its blockages, and recorded images alternate with the not-photographed. The operator's steady hand or a motor should guarantee smooth motion. The presumption is that the processed image strip will then be viewed at the same rate, thereby reconstituting the original motion in its original time and optical space. Screen image movement results from the projector's dyadic alternation of white light and blackness, image and no image. This was well known long before the advent of film, when shuttered devices such as the phenakistoscope visualized motion by the principle of intermittent occlusion and revelation. The Lumière Cinématographe patent of 1895 acknowledged the fundamental play of stillness and mobility in the camera's transport system when it specified:

> The essential character of the mechanism of this device is to operate intermittently in a manner that imprints successive displacements upon a regularly perforated strip, such prints being separated by a resting time during which either the exposure or the projection of the frames occurs.[4]

Thus, the inventors described their device as one that regulated light in time through mechanical action. The possibility of trickery — an "effect" — is introduced during the taking phase, that is, the no-light time of displacement when frames are not being exposed.[5]

4 *"Le mécanisme de cet appareil a pour caractère essentiel d'agir par intermittence sur un ruban régulièrement perforé de manière à lui imprimer des déplacements successifs séparés par des temps de repos pendant lesquels s'opère soit l'impression, soit la vision des épreuves"* [author's translation]. Lumière patent, www.institute-lumiere.org (accessed 13 October 2013).

5 "As our recent research into the body of Lumière and Edison films demonstrates, an impressive number of views shot before 1900 are made up of successive frames that constitute more

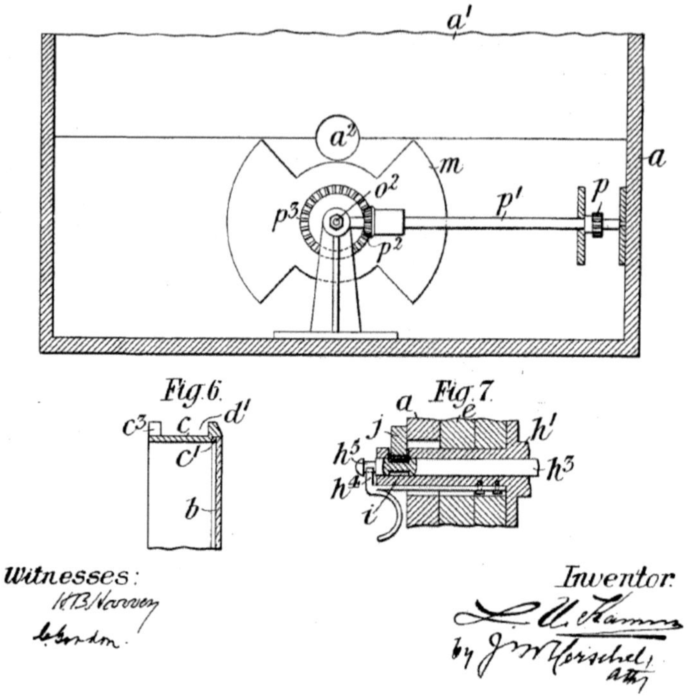

Fig. 7.1 Leonard Ulrich Kamm. Kinetoscope. U.S. Patent 643,718, filed August 22, 1898, and issued February 20, 1900. Detail of Kamm's Figure 5, showing double-blade shutter.

The role of the motion picture projector has been neglected in considering the phenomenology of film watching. In fact, their design is aesthetically constitutive as well as a mechanical prerequisite. For the recorded motion to be reproduced, the projector must create its own displacement and resting times. This procedure parallels the action during recording but is not identical. This was the basis of the first experiments in projected photographic motion by LePrince, Brouly, Edison, the Skladanowsky brothers, and others.

We know, of course, that the screen must be blacked out for the moment when the film is being pulled from one frame to the next. In projection, the transport system moves the filmstrip between a lens and a light source, then halts it momentarily while an opening in a rotating shutter allows the focused image to flash onto a screen. The mechanism then moves the film forward into darkness again, while the next adjacent frame is exposed,

than a single moment in time and thus contradict the canonical model." (André Gaudreault with the assistance of Jean-Marc Lamotte, "Fragmentation and Segmentation in the Lumière 'Animated Views'," in *The Moving Image* [3]1 [Spring 2003], p. 111).

and so on.[6] The projector mechanism effectively creates an analog to the not-photographed duration of the camera's work by arbitrarily substituting a time of not-projected duration. But this pull-down interval (displacement time) is not the only period of non-illumination.

A typical patent for an early projecting mechanism filed by Leonard Kamm describes the essentials for synchronizing the two requisite movements: the intermittent apparatus that moves and stops the film, and the continuously rotating shutter (Fig. 7.1).[7] This type of shutter, which the operators called "two bladed" or "two-winged," is an opaque disk with two sections cut out. It makes one revolution per frame, thus flashing each still image onto the screen twice. At least two bursts of light are desirable because a single one per frame causes the unpleasant sensation of intense flicker, as was the case when the cinématographe occasionally would be converted into a projector.[8]

Although film projection using this standard technology is an analog mechanical system, its *raison d'être* is digital.[9] Its flashing beam is a vehicle for content, which in fact is a binary event: one or zero, corresponding to light or no-light. Functionally, it is a vastly slowed-down version of the way digital information is relayed through light pulses in a fiber optic cable. Let us say that *F* stands for the duration of one frame of projected film. The symbol $\square$ (a white box) stands for the presence of light on the screen (resting time) and $\blacksquare$ (a black box) is no-light. The symbol $\blacksquare_s$ (black box with sub-s, for 'stationary') is the no-light period when the film frame remains locked in the gate between flashes and $\blacksquare_t$ (black box with sub-t, for 'transit') is the no- light period during the displacement time, when the frames are being

6 Gladys Bollman and Henry Bollman, *Motion Pictures for Community Needs: A Practical Manual of Information and Suggestion for Educational, Religious and Social Work* (New York: Holt, 1922), p. 208.

7 "By turning the handle, the plate will be intermittently rotated by the spiral cam and the shutter continuously rotated, the solid portions of the shutter obscuring the light when the plate is moving, while the cut-away portions allow the light to pass when the plate is stationary, one of the series of pictures being always in line with the lenses when the plate is stationary." Leonard Ulrich Kamm, *Kinetoscope*. US Patent 643,718, filed 22 August 1898 and issued 20 February 1900.

8 The Institut Lumière has produced an excellent 3D animation demonstrating the cinématographe, clearly showing how its "one-bladed" shutter operated: http://www.institut-lumiere.org/musee/les-freres-lumiere-et-leurs-inventions/cinematographe.html (accessed 29 March 2016).

9 I want to emphasize that I am not implying a teleological development from analog to digital, an error of Kittler, for example, when he emphasizes "that in the eyes of technicians film was already from the start, long before its heyday, seen as only provisional. Mechanical-chemical image recording, mechanical storage, and mechanical playback are out of place in a century that is defined primarily through the conversion of traditional media to electricity" (Friedrich Kittler, *Optical Media: Berlin Lectures* [Cambridge: Polity Press, 2010], p. 24).

transported into their next position. The projecting operation (with the two-blade shutter), then, may be expressed graphically as:

$$F(\square, \blacksquare_{s}, \square, \blacksquare_{t}), F(\square, \blacksquare_{s}, \square, \blacksquare_{t}), F(\square, \blacksquare_{s}, \square, \blacksquare_{t}), \ldots.$$

In terms of duration, the sum of $(\blacksquare_{s} + \blacksquare_{t})$ is > $(\square \times 2)$ because the amount of screen darkness is greater than that of illumination. Looking at the illustration accompanying Kamm's patent demonstrates why this is true.

The area of the opening of the cut-out sections of the shutter disk is smaller than the opaque sections; indeed, measuring the apertures with a protractor reveals that each is close to the later "Simplex standard" of 84 degrees.[10] At a projection speed of 16 frames per second, viewers are exposed to 32 flashes of image per second. Because the shutter aperture is not quite 90 degrees, however, each of these flashes lasts a little less than 1/32nd of a second. This means that viewers watching a film projected at this rate with this type of shutter experience nothing on the retina *considerably more* than they are experiencing illumination. Specifically, in a one-hour film the projector shows an image for 28 minutes and nothing for 32 minutes.[11] Although we do not see it, the screen blinks — a lot.

10 For a discussion of shutters, see this Film-Tech Cinema System Forum: http://www.film-tech.com/ubb/f1/t007944.html (accessed 25 March 2016).

11 To arrive at this result, consider: the shutter makes one complete revolution (360°) to project each frame for 1/16 second. The 84-degree open apertures of the two-bladed shutter are projecting light on the screen for 46.7% of the time (84x2=168/360=0.4666). By the corresponding calculation, the 96-degree opaque part of the disk is blocking light 53.3% of the time. Therefore, a 60-minute film would be projecting images for 27.996 minutes and showing nothing for 31.999 minutes. Because a two-bladed shutter still induces some flicker at speeds below 18fps and the flicker intensifies at higher levels of illumination, some 35mm projectors began using three-blade shutters as early as 1903 (Deac Rossell, "Projectors," in Richard Abel [ed.], *Encyclopedia of Early Cinema* [Abingdon: Taylor & Francis, 2005], p. 539). These shutters "flash" each frame three times, that is, 48 flashes per second. Lescarboura described the problem and its solution,

> For years pictures flickered, when suddenly projector designers gave mathematics a cruel blow. By multiplying the flickers they produced a zero effect. That is to say, they replaced the single-blade shutter by a two- or three-blade shutter [...] In the case of the two- and three-bladed shutters, the light was cut off while the film moved, and once or twice while the image was at rest. The result of increasing the flickers has been to cut up the light interval so as to make a less marked contrast between the dark period, when the light is shut off entirely, and the light period when the image is being projected (Austin C. Lescarboura, *Behind the Motion Picture Screen*. [New York: Scientific American Publishing Company/ Munn & Company, 1919] p. 248.)

During projection, the three-bladed, or 60-degree, shutter "cuts up" the beam into equal light and no-light times. When projection speed was standardized at 24fps after the coming of sound,

Can this nothing be directly viewed? Let us turn to that purveyor of wise and foolish knowledge, *Wikipedia*, which tells us:

> It is possible to view the black space between frames and the passing of the shutter by the following technique;
> Close your eyelids, then periodically rapidly blink open and closed. If done fast enough you will be able to randomly 'trap' the image between frames, or during shutter motion.[12]

We leave it up to the reader to verify this claim experimentally while observing that, even if one were able to maintain a steady blink rate of 24fps, keeping one's twinkling in phase with a rotating projector shutter would be quite an accomplishment. Nevertheless, the awareness of this Dark Matter in film projection has spawned much conjecture. In the early days of cinema, when shooting and projecting speeds varied from 12 to 24 frames per second subject to the operator's control when turning the machines' cranks, the viewers' awareness of the imageless intervals would have been more apparent. So, if one imagines Henri Bergson, let us say, watching a Lumière film being projected at a flickery 12-16 fps, the nature of the moving pictures as a parade of still frames could easily have been grasped.[13]

Flicker has always been the enemy of audiences and projectionists. Cinematic technology has been engineered to minimize or eliminate it. Flicker produces more than physical discomfort. The distracting scintillations foreground the off-on *0*s and *1*s underlying the cinema experience, thereby working against the goal of creating an optic of smooth, flowing motion and an eyes-wide-open approximation of vision.

Flicker, though, was the friend of the curious, creative, and predatory filmmakers who stepped in to exploit these periods of camouflaged nothingness, the zero-light time of the screen, to make their trick, scientific, and animated films, and their "special effects." They transmuted cinema's nothingness, which technically is a necessary evil, into space and time that sometimes concealed the cinema's labor and sometimes its mystery.

projector manufacturers reverted to the two-bladed 84-degree shutter, since its 48 flashes per second were sufficient to produce flickerless motion for most viewers.

12 https://en.wikipedia.org/wiki/Movie_projector (accessed 28 March 2016).

13 Bergson did not indicate specific films, but his example of a "passing regiment" ("le défilé d'un régiment") in *Creative Evolution* (New York: Henry Holt, 1911, p. 351; *L'évolution créatrice* [Paris: F. Alcan, 1908, p. 329]), is usually taken as a reference to one of Lumière's "cavalry" films, such as *(7ème) Cuirassiers: défilé (par escadron)* (1896) or *Défilé d'un régiment d'infanterie* (1898).

The Black Magic of "Special Effects"

Motion picture cameras must snap each emulsion frame only once or the image will be blurred. Therefore, their shutters must have single apertures that admit only one burst of light per frame. Many first-generation filmmakers, however, famously reverse-engineered their OEM (original equipment manufacturer) to defeat the intended operation of metering the resting and displacement times between exposures. They knew their audiences expected to see "natural," continuously moving content on the movie screens, and they learned how to subvert that expectation. Georges Méliès seized this opportunity to import cinema into his existing conjuring practice. As we see in films like *Excelsior!* (1901), the magician capitalized upon his audiences' assumptions that if something on the screen looked like continuous motion, then it was. It is this proclivity, combined with the technical loophole in the filmmaking process, that make possible "special effects." To modernize his show and seek an advantage over his competitors, Méliès had to innovate cinematic equivalents to replace the apparatuses of stage magic.[14] Perhaps this film's title, the comparative of the Latin *excelsus* (high or sublime), memorializes Méliès's desire to elevate stage magic to a new level of technical complexity and wonderment. On stage, the cape, handkerchief, puff of smoke, or lighting blackout would obstruct and misdirect the audience's view while the substitutions were made. In one sequence of this film, a magician flicks a linen sheet over an adult woman to change her into two girls. This is what André Gaudreault calls the *resumption* technique, that is, stopping the camera, scooting away the woman, bringing in the children, resuming the take, and then fine-tuning the flow of images by editing the two filmstrips together, thereby masking the cut and creating the onscreen transformation (Figures 2a and 2b).[15] These choreographed displacements in the *mise en scène* were executed with the knowledge that, later, the gaps in the recordings and the splice that covered them up, would be masked by the

14 Chéroux has made the case for the influence of the parallel cultural series, recreational trick photography, on Méliès (Clément Chéroux, "The Great Trade of Tricks: On Some Relations Between Conjuring Tricks, Photography, and Cinematography," in Laurent Guido and Olivier Lugon, eds., *Between Still and Moving Images* [New Barnet, UK: John Libbey Publishing, 2012], pp. 81–96). The scenes in *Excelsior!* discussed here, however, appear to be cinematized equivalents of theatrical stage practices.

15 For Méliès's editing and the splice as "magical" misdirection, see André Gaudreault, "Méliès the Magician: The Magical Magic of the Magical Image," *Early Popular Visual Culture* (5)2 (July 2007), pp. 167–174; Tom Gunning, "'Primitive' Cinema: A Frame-up? Or the Trick's on Us," *Cinema Journal* (28)2 (Winter, 1989), pp. 3–12; Tom Gunning, "'Now You See It, Now You Don't': The Temporality of the Cinema of Attractions," *The Velvet Light Trap*, (32) (Fall 1993), pp. 3–12.

continuous motions of the film projector and its regulated turning shutter. By editing out the mechanics of the substitutions, such as actors walking off and onto the stage, Méliès also knew that his viewers would accept the scene as a straight recording of a transformation marvelously caught on film, and not as two events with a fissure in between. Such an intervention subverts the conceptual contract between filmmakers and spectators who assume that the times of nothingness (if they know of them at all) will be insignificant, that no information will be lost (or added). More sophisticated viewers might also expect that the brief periods of black screen time would precisely correspond to the camera's moments of non-recording time. Hah! say some filmmakers ... Gotcha.

For his tricks to work, magician Méliès took great care to design his special effects. For one thing, the camera had to be mounted and maintained rock-steady so that the frame lines between the adjoining shots would match perfectly. His sets were designed to create a theatrical space consistent with the stages upon which he expected his films to be projected. They also were cleverly designed to aid the invisibility of the editing (successfully enough to fool several generations of film historians). In *Excelsior!*, for instance, the decor disguises the joints between the successive takes (Figs. 7.2a-b). The splice in the upper edge of the frame is barely visible because its edge coincides with the edge of the painted decoration on the back wall (the arrow in Fig. 7.2c). Furthermore, Méliès habitually stages the action near the bottom of the scene so as to direct the viewer's gaze — in genuine prestidigitation style — away from the "trick," that is, the splice at the top of the frame. We should also assume that the original film had hand coloring and musical accompaniment that would have aided in the distractions. (We wonder if there would have been cymbal clashes to coincide with the moments of transformation?).

The resumption effect enables multiple performances to occur. The actors (one of whom is Méliès) are skillfully trained to "freeze" into *tableaux vivants* during the dark time of shooting while the changes in the *mise en scène* are effected; there is the performance of Méliès working outside of the visible action as the film director; and there is the manipulation of the time and space between shots that, although not recorded or edited out, will be realized as a performance (a transformation) when the film is projected. Méliès often populated his films with magicians and devils, always self-portrayals, as if to imply that "black magic" was at work. Of course, it was not really black magic — everyone would have been shocked if Satan had actually materialized during the film séance. Rather, it was a *performance* of demonic realization, not unlike that in Kenneth Anger's *Invocation of My Demon Brother* (1969). The filmmakers invoke the "natural

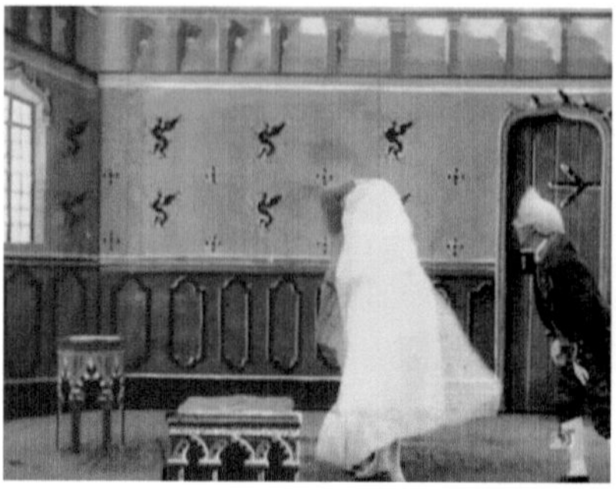

Fig. 7.2a Georges Méliès, *Excelsior!* (1901). Adult female actor is covered by a white sheet.

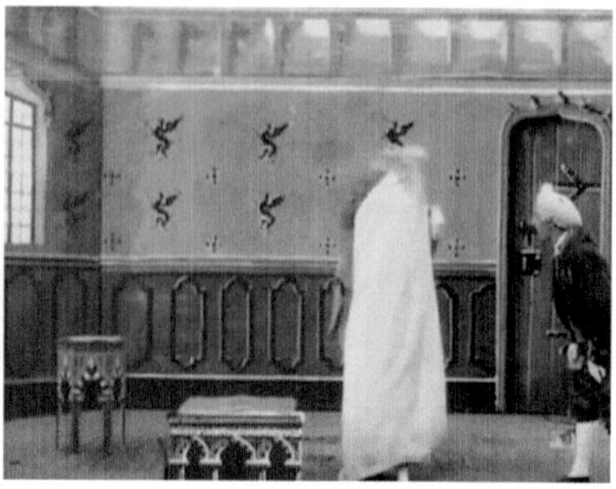

Fig. 7.2b Georges Méliès, *Excelsior!* (1901). Two female children are uncovered by the sheet.

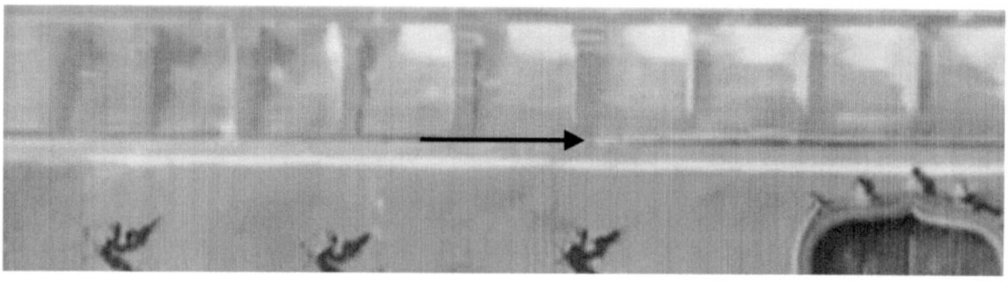

Fig. 7.2c Detail of upper right section of the frame, showing the splice line (arrow).

magic" of cinema, the forces of darkness between the frames, where the black arts and the black screen merge in a grand metaphor of cinema as a medium of the supernatural. For Méliès, the space between the frames was where magic dwelled.

The Black Between Frames Is Not an Image

After listing all the cinematic effects that he thought Georges Méliès had invented, Scottish-Canadian animator Norman McLaren singled out the fade to black: "To prevent the film being fogged at the end of shots, the custom [in the era of Méliès] was to slowly close the diaphragm of the camera. When the film was developed, this technically necessary fade into blackness was normally discarded as being useless artistically for the audience. Méliès saw it as an essential part of the film."[16] It should not surprise us that the

16 Norman McLaren, "Homage to George Méliès," Speech delivered at the Cinémathèque française, 1954; transcript published in David Shepard, ed., *Georges Méliès: First Wizard of Cinema (1896–1913)* (Booklet accompanying a DVD) (Los Angeles: Flicker Alley, 2008), p. 5. McLaren also stressed the technical centrality of Méliès's innovations: "Like the discovery of the wheel, such devices seem so obvious and simple only after they have been invented, but in fact they are really momentous, for, like the wheel, they are capable of an infinite number of applications" (p. 5).

animator obsessed with the interframe would applaud a kindred spirit's exploitation of the black side of cinema. Before attending to McLaren, for now let us point out that the special effect of fading to black is not really black. Nor is a long stretch of black frames (that is, having no pictures) really black. These moments are photographic *images* of blackness, but they are not the *naught* of the time and space between the frames. That is not an image; it is nothing.

Even in these faded-to-black circumstances, the projector still is alternating light and no-light. The darkest scenes are still flickering, despite our not being able to see them very well. Only if a strip of completely opaque black leader were to be run through the projector would the screen truly be black because all the light would be blocked. This never happens in normal movie situations. Considered abstractly though, we would have a situation where there are no frames, no binary alternation of light and dark, and therefore, no *naught*. But would this even be a film?

Much more common is the special effect of the freeze frame, when one frame is reprinted to present us with one slice of the "frozen" motion. Keep in mind, though, that although the movement is frozen, neither the film nor the projector stops moving. Indeed, a freeze frame image alternates light and dark just the same as its unfrozen brethren. This explains why one easily can distinguish between a still photograph projected on the movie screen as a slide and a movie freeze frame. The slide image does not flicker; the freeze frame does. It also probably has moving artifacts such as lint, fingerprints, or other surface debris, as well as the grain (which is just as noticeable as it is in the unfrozen frames).

Exploding the "Space Between the Frames"

The stop-action resumption trick in early cinema was often accompanied by a screen-filling burst of smoke. This was one of the tools in magicians' kit that early filmmakers found useful for masking cuts. In G.A. Smith's *Mary Jane's Mishap* (1903), the unfortunate protagonist is blown up and disappears in a cloud of smoke when she lights the stove with paraffin. Smith's tragicomedy calls attention to the uncanniness of Mary Jane's disappearance by having her body return at the end, its integrity surprisingly restored, albeit as a ghost.

Of course, everyone knows Edwin S. Porter's stop-action trick when a dummy takes the place of the human train engineer being thrown overboard in *The Great Train Robbery* (1903). But this is not the only resumption effect

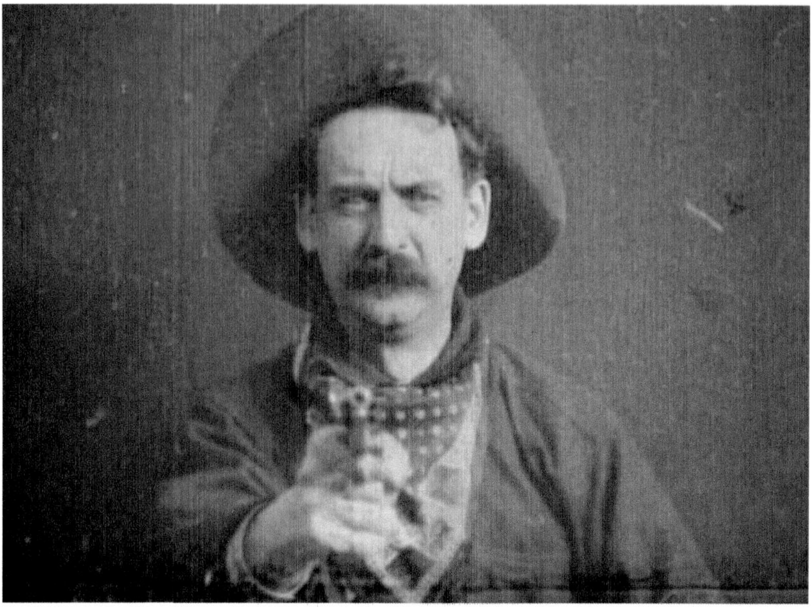

Fig. 7.3a Edwin S. Porter, *The Great Train Robbery* (1903). The bandit aims his pistol at the audience… (note the splice at the bottom of the frame).

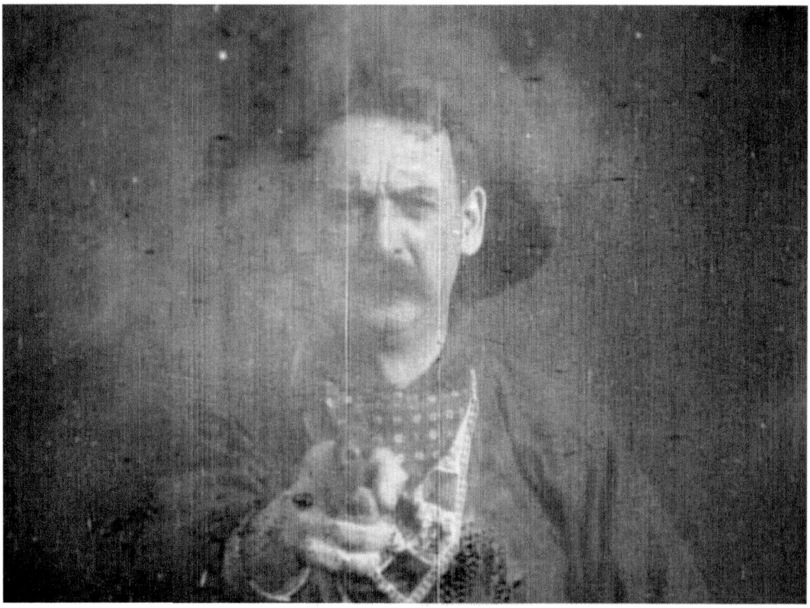

Fig. 7.3b Edwin S. Porter, *The Great Train Robbery* (1903). A puff of smoke in the adjacent frame signals that the pistol has fired.

in the film. The shot that is emblematic of the cinema of attractions, when the bandit fires his pistol into the face of the moviegoer, is also the invisible joining of two shots, one without smoke (Fig. 7.3a), and one with it (Fig. 7.3b). Note the splice line along the bottom of the first frame. Evidently the smoke puff was injected into the scene between the stopping and restarting of the camera to simulate what might issue from a (very smoky) six-shooter. The filmmaker visually creates an explosion that never existed in pro-filmic time or space.

Porter used the smoke cloud, as Méliès sometimes did, to mask the splice but also to trigger audible and olfactory sensations through synesthesia. Reviewers wrote of being able to hear the gunshots and smell the smoke. (And probably they really did hear gunshots, thanks to the "sound effects" that would have accompanied the show.) But there is a difference. Since Méliès's film is of a magician's performance, magical effects are to be expected. Each stopping and resumption adds to our knowledge that action has been omitted at the basic level, although we willingly enjoy the effect as a trick. As Tom Gunning described the illusionism of the nineteenth century, "the magic theater labored to make visual that which it was impossible to believe. Its visual power consisted of a *trompe l'oeil* play of give-and-take, an obsessive desire to test the limits of an intellectual disavowal — I know, but yet I see."[17] While Méliès's cuts may be invisible, their "effect" is not. We know these transformations are explicitly tricks, but we prefer to see them as "real," acted-out cinema performances. Porter's editing is in some ways the inverse. We see the six-shooter fire and we suspect no cinematic foul play. The editing is so well executed that it becomes apparent only when individual frames are studied, something that Porter never intended. Nor would that be something that was even possible outside the Edison studio, the lab, or the projection booth unless one could somehow examine single frames. The splice exploits the camera's temporal gap to construct a performative act that did not occur — the firing of a pistol.

Méliès and Porter probably had different intentions in these films. The former's "effect" was the equivalent of a particular stage performance. Thus, each resumption and edit must allude to what is missing, specifically the actions known to be necessary to create the illusion. Porter, however, did not wish to draw attention to the handiwork of creating the smoke but only to make us think that the weapon has fired. Though the aims were different, these nascent special-effects artists exploited a critical difference

17 Tom Gunning, "An Aesthetic of Astonishment: Early Film and the (In)credulous Spectator," *Art and Text*, 34 (Spring 1989), pp. 31–45.

between film recording and projection. In the taking phase, although it was not engineered that way, the interval between the frame exposures can be infinitely variable. Filming can pause as long as required to do whatever is needed to be done to prepare the profilmic event (acting, moving objects, adding smoke) before the camera rolls again. In projection, though, the framerate must stay the same (or reasonably so, when the machine is cranked by hand). The filmmakers take as given the stability of projection. Their differences aside, though, both filmmakers and all the ones to follow rely on everyone's understanding of the projection system as generating an uninterrupted flow of motion from individual snapshots. The "specialness" of the filmmakers' special effects derives from their ability to intervene during the recording phase and work at the interframe level either ostentatiously or quietly. So special effects rely *all the more* on the expectation that normalcy will be transformed in some way and on projected continuity.[18]

Just as magicians used to take an oath never to divulge their secrets, the crafters of special effects conceal their techniques behind the curtain — or more *à propos*, behind a puff of smoke. Now, some special effects have their self-effacement more in common with Porter's approach. Modern-day films of all genres, even those more modest non-blockbuster ones, use abundant but unobtrusive computer-generated special effects, for example in the hundreds of altered shots in *Bridge of Spies* (Steven Spielberg, 2015) that recreate 1950s Berlin. As Stephen Prince notes, though, these types of effects have not been called "special" since the 1970s.[19] Now the official Oscar-worthy term is "Visual Effects." But there are still special effects that call attention to themselves as Mélièsian magical happenings. For instance, aliens may blow up a recognizable monument or locale as they do to Manhattan in *Independence Day: Resurgence* (Roland Emmerich, 2016). The spectacular explosion occupies a privileged position in special effects because it is so violently transformative and so joyously abundant.

There seem to be three species of movie explosions: those staged as live events and filmed (often with multiple cameras to produce footage that

18 Some clarifications and disclaimers: Méliès did not always hide the transformations with smoke; rather, he seems to have timed his explosions as punctuations in a string of metamorphoses. Porter employed various non-smoking special effects in many films, including some that emulated Méliès quite closely. At some point after cameras became outfitted with spring or electric motors, some manufacturers offered a single-frame setting in their cameras' designs. Later, there were purpose-built cameras for animation and special effects (e.g., by the famous Acme Tool and Manufacturing Company).

19 Stephen Prince, *Digital Visual Effects in Cinema: The Seduction of Reality* (New Brunswick, NJ: Rutgers University Press, 2012), p. 3.

may be used to extend the temporality of the event and to change viewing angles and distance);[20] explosions manufactured by cinematic techniques, usually some elaboration on stop-action substitution, editing, or montage; and implied events not shown on screen but whose effects are indicated by actors' reactions, flying debris, and of course, percussive sound effects. Contemporary explosions, say for example the several dozen in the films of producer-director Michael Bay (e.g., *Transformers: The Last Knight*, 2017), routinely orchestrate all three categories.[21] These big bangs are amalgams of numerous ingenious mechanical and pyrotechnical innovations. Springboards and wires, later to be digitally erased, make vehicles and bodies fly through the air, preferably into the virtual space of the audience using wide-angle lenses for filming and 3D projection. Compressed air may be used to catapult things sky-high, creating the look of a blast but in reality moving things safely. Explosive powders, flammable foam, and more compressed air create brilliantly luminous balls of fire, but since these bursts are of very short duration, they cannot really burn anything if properly controlled. They produce little explosive energy but a lot of bright light. Actors in fireproof suits convulse as flames devour their bodies — harmlessly, of course. And let us not forget *mise en scène*: filmmakers usually include humans — preferably protagonists or antagonists — between the explosion and the audience to emphasize the potentially catastrophic danger to the characters, not to mention to the actors. The humans also give the explosion scale; we can see that the bang truly is big and not some little studio pop. Often the backgrounds are small maquettes of buildings or digital re-creations. This increasingly is the function of CGI effects: using green screens to combine in the same frame human actors and gigantic explosions (often artificially created with fractals), which never co-existed in the profilmic scene. There are also subtle touches like rocking the camera (via an optical effect) to show the arrival a shock wave from a distance. This is especially effective in *The Hurt Locker* (Kathryn Bigelow, 2008) when we see an explosion's force lifting the soil as it rushes toward the camera. Then there is the soundtrack, of course, which adds redundant auditory shock to the violent visual destruction. Only when we see the stunning but silent explosions in Alfonso Cuarón's *Gravity* (2013) unfold without any accompanying sonic booms do we realize how much

20 This seems to have been the case in the other explosion in *The Great Train Robbery* when the bandits blow up the strongbox. I cannot detect any frameline splices, so I assume that this was a "theatrical" pyrotechnic effect executed and filmed "live."

21 There are many fan-generated compilations on the internet. One of the best is TheBatBoy12345, *Michael Bay's Best Explosions!!!*, https://www.youtube.com/watch?v=Ua9KfZmx-Kc (accessed 6 April 2016).

the cinematic explosion has become an audio-visual phenomenon. Indeed, when properly experienced in a well-equipped theater, the explosive blasts engineered by our best sound designers should be felt all over and through one's body. The physical somatic involvement that we experience by these movie explosions are reminders of the fragility and risky contingencies of our own physical bodies. Seeing humans blown up on screen plays the game of "what if": what would it be like if Schwarzenegger blew me up, or if I blew him up? And classic cartoons, of course, take explosions to an orchestrated level of bodily violence. The "hunting trilogy" directed by Chuck Jones and featuring Bugs Bunny and the hapless Daffy Duck is a superb example that cultivates permutations of verbal jokes revolving around whether it is duck or rabbit hunting season, then always resulting in Daffy being shot up.[22] The explosion follows the same pattern: Daffy's head is replaced by a cloud of white smoke that reveals his rearranged features when it clears. Thus, the drawn explosion mimics those in the Mélièsian transformations.

Movie explosions are so effective and so ubiquitous because they play so many chords on the cinematic instrument. Often, they punctuate the film by beginning or ending a crucial narrative sequence. They are scopic bursts — remnants of the cinema of attractions — but they also carry some of the film story's affective burdens. Kristen Whissel aptly refers to these as "effects emblems." She defines an emblem as "a cinematic effect that operates as a site of intense signification and gives stunning (and sometimes) allegorical expressions to a film's key themes, anxieties, and conceptual obsessions — *even as it provokes feelings of astonishment and wonder.*"[23]

The explosion that destroys the Heart Machine in the city of *Metropolis* (Fritz Lang, 1925/1926) does all of this, plus it places all of civilization, symbolized by the workers' children, in danger. The destruction of world monuments in *The Day the Earth Stood Still* (Robert Wise, 1951) is pitiful because the space invaders threaten the global inhabitants' symbolic cultural monuments. One thinks of other ideological-laden movie-ending and world-ending explosions: the mushroom cloud finale of *Dr. Strangelove* (Stanley Kubrick, 1964) and the presumed fiery nuclear conflagration that concludes *Kiss Me Deadly* (Robert Aldrich, 1955). In Japanese anime, exploding objects, bodies, and worlds are among the frequently recurring traits of the genre. But now we are not talking about simply exploding objects, buildings, and people or

22 The trilogy is *Rabbit Fire* (Jones, 1951), *Rabbit Seasoning* (Jones, 1952), and *Duck! Rabbit, Duck!* (Jones, 1953).

23 Kristen Whissel, *Spectacular Digital Effects: CGI and Contemporary Cinema* (Durham, NC: Duke University Press, 2014), p. 6. Emphasis in the original.

simulating same on film. We are attributing meanings to these blow-ups. I suggest that audience and filmmakers' fascination with exploding things in movies is fundamental to cinema's magical appeal. I will return to this thought later, but for now, I shall observe that explosions are more than spectacle and narrative; they also are discourses about interruption and continuity, the ghostly and the uncanny, and the body's fragile wholeness that run through the history of film, film theory, and popular discussion. Hidden within the tripartite cinema apparatus — production, reception, and culture — are mechanisms that create or invoke secret spaces, that confound viewers who understand movies as vehicles for real motion, real events, emotion, and comprehensible stories but who also receive, often with a combination of dread and pleasure, the demonic, the uncanny, the ideological, the ineffable, the extravagant, the artistic. And so we come to animation...

Méliès did not use frame-by-frame animation *per se*, but his approach anticipates how animation cinema would foreground itself in the twentieth century. In the years following 1898, Arthur Melbourne-Cooper, James Stuart Blackton, and others also began interfering with the camera's carefully engineered displacement time, simply by exposing one, two, or short bursts of frames instead of turning the crank continuously. Dziga Vertov properly called this "frame-shooting," opposed to standard "shot-shooting."[24] Most cameras exposed one frame of raw stock with each rotation of the shutter, and so applying "one turn one frame" to produce animation was not a particularly sophisticated technique. The experimenters knew that slightly offsetting the objects or drawings in front of the lens between the exposures would produce moving things when the film was projected. Soon, they were innovating the film technique that underlies what we now recognize as animation cinema — before computer-generated composition replaced it with new technology.

While we have always understood that arranging the raw material for the frames of an animated film before the camera rolls is where the creative work takes place, what has not been emphasized enough is that, regardless of how it was made or its content, the film is not *animated* until it is projected. Edgar Morin recognized this, noting that "projection only puts the finishing touches to the work of movement."[25] As with

24 Mihaela Mihailova and John MacKay, "Frame Shot: Vertov's Ideologies of Animation," in Beckman, ed., *Animating Film Theory*, (Durham and London: Duke University Press, 2014), p. 149.

25 Edgar Morin, *The Cinema, or the Imaginary Man*, trans. Lorraine Mortimer (Minneapolis: University of Minnesota Press, 2005), p. 5; quoted in Maria Tortajada, "Photography/Cinema:

Méliès and the tricksters, animators assume that the magic happens in projection when the discontinuous photographs on the filmstrip spring into action. This is one reason why elsewhere I have argued that the work of animation artists and technicians is a *conditional performance*.[26] The screen movements that we see in animation cinema have never previously occurred; they will become a performance only if and when viewers witness the screening.

Many filmmakers have commented on the expressive potential of cinematic black magic, but none perhaps as powerfully as Norman McLaren. In an often-quoted remark, he declared,

> Animation is not the art of DRAWINGS-that-move but the art of MOVEMENTS-that-are-drawn. What happens between each frame is much more important than what exists on each frame. Animation is therefore the art of manipulating the invisible ~~that~~ interstices that lie between frames.[27]

The animator grasps the challenge of exploiting the Dark Matter that contains the possibility of movement. McLaren conceives of animation as a project of making each drawn or photographed frame a record of an interframe absence: the *during* of a motion that two adjacent frames represent only as its *before* and *after*. McLaren's adage recognizes that screen movement is a function of the interval and the unseen actions therein. Occasionally during filming, the work of manipulation is accidentally captured. In many early animated films, the hand of the animator sometimes would be caught for an instant, as in Blackton's *Humorous Phases of Funny Faces* (1906).[28] Such moments are a fleeting glimpse into the construction zone between the frames, a place where animators labor but where spectators are not allowed.

Complementary Paradigms in the Early Twentieth Century," in Laurent Guido and Olivier Lugon, eds., *Between Still and Moving Images* (New Barnet: John Libbey Publishing, 2012), p. 35.

26 Donald Crafton, *Shadow of a Mouse: Performance, Belief, and World-Making in Animation* (Berkeley: University of California Press, 2013).

27 Norman McLaren, handwritten note reproduced in *Cinéma 57*, 14 (January 1957), p. 12; and in Hervé Joubert-Laurencin, *La lettre volante: quatre essais sur le cinéma d'animation* (Paris: Presses Sorbonne Nouvelle, 1997), p. 52. I have retained McLaren's original emphases and markings. I felt honored, humble, and a little nervous when I originally made this presentation in the *Salle Norman-McLaren* of the Cinémathèque québécoise in Montréal.

28 One of this paper's anonymous readers observed that, while the intrusion of the animators' hands in several of these early films was accidental, the intentionally revealed "hand of the artist" would become a recurring figure in animated films for the next several decades.

McLaren seems to claim that to make animation requires doing something contradictory or even impossible. He is suggesting that the graphics that become the frames of an animated sequence are the signifiers of movements that have never before taken place but which have the potential to move and perform, if properly drawn and projected. The animator's goal, then, is to concretize something abstract or imaginary — movement — by illustrating its incremental phases. The drawings are merely a means to an end. So it is with the producers of special effects. Their craft consists of creating believable markers that will convince audiences that they are watching something that is happening or has already happened. In fact, they are watching a performance that is original with each screening; the White House is destroyed by alien invaders; Forrest Gump is almost blown up in Vietnam. But audiences accept these spectacles, not because they believe them to be true but, on the contrary, because they accept them as screen performances. As it is with animation, motion picture special effects are a belief system.

The definitions of cinema and of animation are as porous, disingenuous, and transient as a politician's promises. Defining special effects in terms of one, the other, or both, then, will draw us into a semantic morass.[29] Based on observation, common sense, and our phenomenological experience of "the movies" as an everyday activity — or as Gaudreault and others have identified it, a cultural series — we can see that for most of cinema's history, animation cinema has developed with aims that differ from cinematic special effects on the surface but fundamentally are nearly identical.

The chief difference between animation and special effects is that the former historically has tended to have been identified with and has self-identified as something apart from general, so-called mainstream or live action cinema. It has sometimes been bracketed as a "short subject" mode of production (a cartoon), avant-garde filmmaking, and sometimes as a genre. As it evolved within the classical paradigm, it usually has been framed in critical and popular discourse as a distinction, a departure, a sub-species, or even an aberration, but always at least a noticeable separation from

29 For the purpose of this essay, I am addressing animation cinema as it is commonly understood as a specialized mode of theatrical or televisual film production. This kind of animation is always cinema, but cinema is not always (indeed, seldom is) this kind of animation. The best spokesperson for the view that "not only is animation a form of film but all film, film 'as such', is a form of animation" is Alan Cholodenko ("'First Principles' of Animation," in Beckman, ed., *Animating Film Theory*, p. 98). For my semantic analysis of this proposition, see Donald Crafton, "The Veiled Genealogies of Animation and Cinema," *Animation: An International Journal*, (6)2 (July 2011), pp. 93–110.

ordinary movies. This aligns animators with Méliès, whose films for a long time were considered novel attractions but not mainstream. Analogous to McLaren's animating invisible movement, Méliès used the black magic between frames to animate metamorphosis, as in *Excelsior!* when the adult woman optically and symbolically gives birth to twin offspring. Other forms of special effects depend on the ability of filmmakers to insinuate their techniques into the filmwork as an integral component of spectacle and narration. The example from *The Great Train Robbery* highlights a paradox inherent in all dramatic performance: the final sequence is not a film of a non-smoking gun being replaced by a smoking gun but rather one of a gun firing. This aggressive effect would fail if our reaction was, "Oh, it's a trick shot." At the same time, did 1903 audiences wonder if the bandit had shot E.S. Porter (or the film's titular producer, Thomas Edison) dead on the spot? Probably not.

Now, animation may be a special effect and vice versa, as when an encapsulated sequence of animation is inserted into a live action film. Then, frame-by-frame cinematography may alter a scene of normalcy, as when Nosferatu's coffin uncannily moves of its own accord, or when a Transformer goes on a rampage. Or the effect may remain unnoticed, as when skies with stormy clouds appear on cue, thanks to CG animation.

Granted, there are no clear boundaries here, especially in today's era of media convergence. In the classical cinema, nevertheless, we might say that the aim of animators usually was to bring the dark spaces of cinema to light (as McLaren seemed to be doing), while the special effects craftspersons usually tried to keep the Dark Matter in the dark. Animators revel in the alterity of their enterprise, their "termite terrace," while special-effects technicians happily remain in the secret service of the classical cinema system. Practitioners of special effects are due recognition for their labor and ingenuity, but the mechanics of their labor usually have been designated invisible.

These animators engendered an original performance from inanimate objects and drawings, and what Disney, other animators, and a few critics would tout as the illusion of life. Thus, from the nothingness of the displacement time of non-filming and out of the darkness between projected frames emerged a powerful mythology of parthenogenesis: the animator is a god, a magician, or a father capable of creating life. He (or sometimes she) endows static objects with an *anima*, enabling these imaginary beings to perform on screen just as humans do. Even in these extreme cases of exhibitionism, the specialness of animation and special effects converge in on the effacement of their creative labor.

Interframe Aesthetics

We do not know which films Henri Bergson saw between 1900 and 1903. Did he notice how the smoke scenes in many of them contributed to the duplicity of the film? It appears to be a unitary flow, but like the magician's distracting movements, it hides a break that one may know about intuitively or learn about from reading articles on cinematography, or from examining a film strip, or from observing what happens when the projector is started or stopped with the beam on. Bergson lectured on this contradiction of cinema — that stillness is necessary to perform motion — in a course on movement he taught in 1903–04 and incorporated into *Creative Evolution* in 1907. He saw in Marey's chronophotography and then in the cinematograph an interrogation of Zeno's arrow paradox.[30] A launched arrow would never reach its target because when the arc of its trajectory is divided into smaller and smaller instants, it would have no movement because it is always frozen in each little division of time and cannot be in two divisions (present and future) at once. Because its photographic frames seized instants, cinema was posited as a similar series of micro-temporalities. This strip of film, according to Gunning, "in itself has no movement, only a succession of immobile states. The movement, as Bergson says, comes from elsewhere, from the apparatus, i.e., the cinema projector that must be cranked into motion in order to impart a mechanical rapid rhythm of replacement to the images, thereby creating an appearance of movement."[31] Bergson wrote neither of animation nor of explosions, but the cinematograph's inability to show what happens between the frames while duplicitously maintaining the flow of time and the continuity of space was precisely the origin of what we call animation and special effects.

Bergson's equivocal role of film frames as bearers of motion is similar to and different from McLaren's: the paradox for Bergson was twofold. First, while cinema could simulate movement, it could not capture true duration because of the "gap" between the frames. There cannot be a full spectrum of continuing motion — no breaks, no stasis. And second, the "cinematographic method" of thinking and behavior, while mimicking the flow of consciousness, is a falsehood, since motion, like being, is continuous,

30 Maria Tortajada, "'The Cinematographic Snapshot': Rereading Etienne-Jules Marey," in François Albera and Maria Tortajada, eds., *Cinema Beyond Film: Media Epistemology in the Modern Era* (Amsterdam: Amsterdam University Press, 2010), p. 80; Tom Gunning, "The 'Arrested' Instant: Between Stillness and Motion," in Guido and Lugon, *Between Still and Moving Images*, *op. cit.*, p. 26; Tortajada, "Photography/Cinema," op. cit., p. 38.

31 Gunning, "The 'Arrested' Instant," op. cit., p. 27.

complete, utopian, even salvational.[32] Analogous to human development, while one might speak of the different "ages" (adolescence, adulthood, etc.), real lifespans are whole, unfragmented, and indivisible into stages.[33]

According to Maria Tortajada, Bergson's still/moving conception of being and his opportunistic seizing of the cinematograph to illustrate it haunts film theory. "It can be said," she writes, "that this idea extends across the history of the cinema, and that Bergson's philosophy provides its subtext."[34] Although Bergson's postulation of the film frames as fatally incapable of yielding true motion does anticipate McLaren's thinking, the animator's conclusion is the opposite of the philosopher's. He found that using the projector to animate his drawings revealed the utopian world hiding between frames. It is not likely that he would have agreed that creating motion via film was a hopeless cause.

Norman McLaren's thinking was formed during his collegiate days in Glasgow when his film club screened and discussed the latest Soviet imports. In a late-life interview, he stated that it had been repeated viewings of *Oktyabr': Desyat' dney kotorye potryasli mir* (*October*, Sergei M. Eisenstein, 1928) that revealed to him the rhetorical strategies of editing as a way to communicate abstract ideas. Sergei Eisenstein, in his cinema and in his writing, found that filmmakers could produce emotions and control — or at least influence — meaning by the way two shots were joined. As he (now famously) wrote, "the degree of difference between the positions A and B determines the tension of the movement [...] This dynamization of the subject, not in the field of space but of psychology, i.e., emotion, produces: emotional dynamization."[35] François Albera shows that, as early as the 1923

32 Gunning argues that Bergson does not condemn the cinematograph, as is often claimed, but rather "rhetorically used the mechanics of the cinematograph to illustrate a misconception of movement.... Bergson does not attack mechanical movement, but rather a mechanical model for our understanding of movement" (ibid., pp. 27, 28).

33 See Elena Fell, *Duration, Temporality, Self: Prospects for the Future of Bergsonism* (Bern: Peter Lang, 2012), p. 63. Bergson would "pan" *Boyhood* (Richard Linklater, 2014) because we see only a few stations in the development of the boy Mason, not an uninterrupted twelve years of the life of Ellar Coltrane, the actor who played him. Conversely, Bergson might give a qualified thumbs-up to *Blow Job* (Andy Warhol, 1963). On one hand, the film captures the action as a one-to-one ratio of duration and depiction with no temporal effacement. On the other, the eponymous action (purportedly) takes place out-of-frame, and therefore, since that part of the "self" depicted has no spatial dimension, it does not exist. There may be other reasons he might not like this film.

34 Tortajada, "'The Cinematographic Snapshot'," p. 81.

35 Sergei Eisenstein, "A Dialectic Approach to Film Form," *Film Form: Essays in Film Theory* (New York: Houghton Mifflin Harcourt, 2014), p. 57. Elsewhere in the essay, Eisenstein posits that the "material concreteness of the image within the frame presents — as an element — the greatest difficulty in manipulation." He proposes that montage, the artful editing of the frames,

"Montage of Attractions" essay, Eisenstein was intrigued by the juxtaposition of images, which he called the "technical (optical) basis" of film. He reprised this theme in 1929, exploring "the phenomenon of movement."[36] Activating — or as he prefers, dynamizing — the space between frames ("montage-pieces") was in some aspects magical or shamanistic; it made the dialectics of cinema possible by energizing inert elements that by themselves had no meaning. His emphasis on optical surprises and the unexpected results of juxtapositions are akin to the shocks of miscomprehension masquerading as recognition, curiosity, and revulsion that characterized the theater of attractions and later the cinema of attractions. In his argument with Pudovkin, when Eisenstein rejects his colleague's description of cinema as shots flowing as connected linkages, he also is refuting Bergson's concept of cinema as a band of immobile images that faked natural motion. Recall that the cinematographic method, as Bergson called it, led us to falsely understand movement and therefore life as a series of discrete events and not as a holistic lived reality. Albera has pointed out that Eisenstein understood the cinematic cell to be the frame, but unlike Bergson's view, he argued that the viewers experience these photograms as superimpositions, one upon the other vertically, as it were, rather than as a linear succession. Perhaps the Russian director was recalling the experience of watching Mutoscope-like moving photographs? The energy of montage and the brilliant dialectic of suggestion and reception are activated within this stack of frames.[37] Like the animators, Eisenstein harnesses the power of the interstices between the frames when he edits his stone lions and gives them life in *Bronenosets Potyomkin* (*Battleship Potemkin,* 1925). The beasts leap to life to perform their trick. But the sequence, heralding the assault on Odessa, also is supposed to symbolize the proletariat rising against the Czar and to illustrate the explosive transformation of totalitarianism and capitalism into socialism.[38]

is one way to manipulate them because it acts like "the system of language, which is forced to use the same mechanics in inventing words and word-complexes," p. 60.

36 François Albera, "Eisenstein and the Theory of the Photogram," in Ian Christie, Richard Taylor, and Richard Taylor, eds., *Eisenstein Rediscovered: Soviet Cinema of the 20's and 30's* (Oxford: Routledge, 1993), p. 194.

37 Ibid., pp. 201–202.

38 Yuri Lotman's comments on the transformation of Russian into Soviet culture are apropos: "Marxist theorists wrote that the journey from capitalism to socialism would inevitably have an explosive character. This was based on the fact that all other formations were conceived within the framework of previous stages, whilst socialism initiates a completely new period, whose origins are only made possible by the ruins, rather than the bosom, of historical precedent" (Yuri Lotman, *Culture and Explosion*, Marina Grishakova, ed., Wilma Clark, transl. [New York: Mouton de Gruyter, 2009], p. 173).

Eisenstein's fellow filmmaker and rival, Dziga Vertov, developed his own theory of the significance of the blackness between frames. The cinema, analogous to the ways in which science was producing a new understanding of infinitesimally small as well as distant objects, could likewise yield views of the otherwise unseeable structures of society. He also was a believer in "intershot intervals" that "involve movement and meaning at once."[39] He wrote that it is not the moments themselves in film or any art that count but instead how the moments are connected, the "negative of time." Now, Vertov's written treatment of the interval as "a visual correlation of shots" sometimes closely resembles Eisenstein's definition of montage. At other times, though, in language anticipating McLaren, Vertov referred to "the movement between shots"[40] and claimed that "intervals (the transitions from one movement to another) are the material, the elements of the art of movement, and by no means the movements themselves."[41] The instrumentality of the cinema, then, what he called the Kino-Eye, was to be found in its ability to render visible "that which the eye does not see" by exposing the ideology latent in the interval.[42] This contrasts to Eisenstein's belief in the dialectic that results when the visible contents of one shot clashes with its neighbor to create meaning.

Vertov seems to have visualized his epistemology of the interval in the sequence in *Chelovek s kinoapparatom* (*Man with a Movie Camera*, 1929) when we see the young woman rising and dressing for the day. Constructed as a series of metaphors and analogies between the life of a city, a human life, and the voyeuristic, performative camera and cameraman, the sequence is a powerful demonstration of Vertov's rhetoric. As she looks out the window, the blinds open and close in apparent synchronism between her blinking eyes and the movie camera. Perhaps Vertov here is trying to make the invisible interval visible, much as our *Wikipedia* experiment aspires to do by similar means. His palimpsest might be unraveled thusly: the eyes of the woman in the scene look out upon the world but see nothing until the shutters of the window shade are opened to let in the light. The blinking eyes and the oscillating window shutters must act in unison (eyes open when shutters are open) if she is to see the world. We, the film

39 Mihailova and MacKay, "Frame Shot," op. cit., p. 147.

40 Dziga Vertov, "Kino-eye to Radio-eye" (1929), in *Kino-eye: The Writings of Dziga Vertov*, Annette Michelson, ed., (Berkeley: University of California Press, 1984), pp. 90–91.

41 Dziga Vertov, "We: Variant of a Manifesto" (1922), quoted in Simon Cook, "'Our Eyes, Spinning like Propellers": Wheel of Life, Curve of Velocities, and Dziga Vertov's 'Theory of the Interval,'" in *October* 121 (Summer 2007), pp. 85–87.

42 "Definitions of kino-eye such as 'kino-eye = cinema-analysis, kino-eye = the theory of intervals, kino-eye = the theory of relativity on the screen,' etc., emerge. ...Kino-eye is defined as 'that which the eye does not see....'" (Vertov, "Kinopravda" [1934], in *Kino-eye*, op. cit., p. 131).

viewers, correspondingly, must synchronize our vision with hers to see what she sees, and that can only be facilitated during our identification with the camera and the projector's spinning shutters during the performance of *Man with a Movie Camera*. The phenomenal existence of truth (a social truth) can only be revealed this way: we can see seeing when we film seeing seeing, film the act of filming seeing seeing, and project the film of the film of the act of seeing seeing. Vertov's thinking is idealist: The interval for him is neither the Eisensteinian frame that fakes motion nor the Bergsonian cinema's false duration but rather the ideological truth that the filmmaker extracts from the rising awareness of the interframe. This meaning is abstract and as such aspires to calculus.[43] We should point out that Vertov does not reject the interval's magical potential. In the scenes showing the Chinese magician performing for the children, the tricks and sleight-of-hand represent atavistic prestidigitation, a backward-looking fascination with magic that Soviet truth is driving underground, specifically by the special effects of cinema, of montage, and the Kino-Eye.

Conclusion

But finally, what is this space between frames that intrigues us? It seems to derive from an asynchronism between the mechanics that produce the film image and the phenomenology of film watching, between the recording and projection of the frames and the physical and psychological engagement of viewers. At the very least, the commonly held opposition between *still* and *moving* that informs so much recent scholarship on photography and cinema should be reconsidered as a triad to account for the role of the *naught*. Clearly, as a topos of cinema, for those who believe in it, the naught is more than hypothetical. The Dark Matter of cinema can be an inspiration, a justification for a certain understanding of how film operates, or a rhetoric aimed to persuade us to accept particular views of film's value or purpose. Bergson might have been the first but not the last to find the interframe a source of awe and trouble. There is Laura Mulvey, for whom the fundamental paradox of cinema is movement produced from a stasis that must remind us of its photographic origins:

> Stillness may evoke a "before" for the moving image as filmstrip, as a reference back to photography or to its own original moment of registration. Although the projector reconciles the opposition and the still frames

43 Cook, "'Our Eyes,'" op. cit., p. 90.

> come to life, this underlying stillness provides cinema with a secret, with a hidden past that might or might not find its way to the surface. The inanimate frames come to life, the unglamourous mechanics are covered over and the entrancing illusion fills the screen.[44]

But is cinema's repressed secret only its origin in still photography? The discovery of this Dark Matter in the cinematic apparatus that manifests itself in trick films, animation, special effects, and didactic cinema is more difficult to identify, and yet this naught has spawned generations of creative applications and interpretations. For these filmmakers, writers, and theorists, it has evolved into a metaphor for transcendence according to his or her pragmatic definition. We may think of "special effects" in cinema merely as an attraction or utilitarian device, but at the same time we cannot fail to see in the blackness of the interframe a Romantic expression of the sublime and the ineffable. It is a way to reinsert the human into the machine.

Which brings us back to explosions. Certainly, this primordial and privileged special effect, which for over a century has infiltrated the entire history of cinema, new media, and now gaming without any extensive notice or commentary, must have a viable social and artistic function beyond simply operating as an eye-popping and ear-splitting spectacle. More than just a punctuation mark in a story, explosions grip us as performances of threats to somatic integrity and, metonymically, to the social order. They upend complacency and rock the comfortable support systems of dominant society and everyday life. This audio-visual assault on the positivist and empiricist underpinnings of the twentieth and early twenty-first centuries is inherently violent and weaves into the historical record the paralyzing fear of anarchism, revolution, and terrorism, the traumatic experience of wars, of catastrophic annihilation, and Armageddon. But in the cinema of special effects, when the show is over, we survive.

Bibliography

Anon. "2 blade and 3 blade shutter difference." *Film-Tech Cinema System Forum*. Accessed 13 November 2021. http://www.film-tech.com/ubb/f1/t007944.html.

Anon. "Movie Projector." *Wikipedia*. Accessed 13 November 2021. https://en.wikipedia.org/wiki/Movie_projector.

44 Laura Mulvey, *Death 24x a Second: Stillness and the Moving Image* (London: Reaktion Books, 2006), p. 67.

Albera, François. "Eisenstein and the Theory of the Photogram." In *Eisenstein Rediscovered: Soviet Cinema of the 20's and 30's*, edited by Ian Christie and Richard Taylor (Oxford: Routledge, 1993): 194–203.

Bergson, Henri. *Creative Evolution* (New York: Henry Holt, 1911).

Bollman, Gladys, and Henry Bollman. *Motion Pictures for Community Needs: A Practical Manual of Information and Suggestion for Educational, Religious and Social Work* (New York: Holt, 1922).

Chéroux, Clément. "'The Great Trade of Tricks': On Some Relations Between Conjuring Tricks, Photography, and Cinematography." In *Between Still and Moving Images*, edited by Laurent Guido and Olivier Lugon (New Barnet, UK: John Libbey Publishing, 2012): 81–96.

Cholodenko, Alan. "'First Principles' of Animation." In *Animating Film Theory*, edited by Karen Redrobe Beckman (Durham, NC and London: Duke University Press, 2014): 98–110.

Cook, Simon. "'Our Eyes, Spinning like Propellers': Wheel of Life, Curve of Velocities, and Dziga Vertov's 'Theory of the Interval,'" *October* 121 (Summer 2007): 79–91.

Crafton, Donald. "The Veiled Genealogies of Animation and Cinema," *Animation: An International Journal* 6, no. 2 (July 2011): 93–110.

———. *Shadow of a Mouse: Performance, Belief, and World-Making in Animation* (Berkeley, CA: University of California Press, 2013).

Eisenstein, Sergei M. *Film Form: Essays in Film Theory*, edited and translated by Jay Layda (New York: Houghton Mifflin Harcourt, 2014).

Gaudreault, André, with the assistance of Jean-Marc Lamotte. "Fragmentation and Segmentation in the Lumière 'Animated Views'," *The Moving Image* 3, no. 1 (Spring 2003): 110–131.

Greer Cohn, Robert et al., eds. *Mallarmé in the Twentieth Century* (Madison: Farileigh Dickinson University Press, 1998).

Gunning, Tom. "An Aesthetic of Astonishment: Early Film and the (In)credulous Spectator," *Art and Text* 34 (Spring 1989): 31–45.

———. "The 'Arrested' Instant: Between Stillness and Motion." In *Between Still and Moving Images*, edited by Laurent Guido and Olivier Lugon (New Barnet, UK: John Libbey Publishing, 2012): 21–31.

Joubert-Laurencin, Hervé. *La lettre volante: quatre essais sur le cinéma d'animation* (Paris: Presses Sorbonne Nouvelle, 1997).

Kittler, Friedrich. *Optical Media: Berlin Lectures* (Cambridge, UK: Polity Press, 2010).

Lescarboura, Austin C. *Behind the Motion Picture Screen* (New York: Scientific American Publishing Company/Munn & Company, 1919).

Lotman, Juri. *Culture and Explosion*, edited by Marina Grishakova, translated by Wilma Clark (New York: Mouton de Gruyter, 2009).

McLaren, Norman. Handwritten note reproduced in *Cinema* 57, no. 14 (January 1957): 12.

———. "Homage to George Méliès." In *Georges Méliès: First Wizard of Cinema (1896-1913)*, booklet accompanying a DVD, edited by David Shepard (Los Angeles: Flicker Alley, 2008).

Mihailova, Mihaela, and John MacKay. "Frame Shot: Vertov's Ideologies of Animation." In *Animating Film Theory*, edited by Karen Redrobe Beckman (Durham, NC and London: Duke University Press, 2014): 145–166.

Morin, Edgar. *The Cinema, or the Imaginary Man*, translated by Lorraine Mortimer (Minneapolis, MN: University of Minnesota Press, 2005).

Mulvey, Laura. *Death 24x a Second: Stillness and the Moving Image* (London: Reaktion Books, 2006).

Rossell, Deac. "Projectors." In *Encyclopedia of Early Cinema*, edited by Richard Abel (Abingdon: Taylor & Francis, 2005): 539.

Tortajada, Maria. "'The Cinematographic Snapshot: Rereading Etienne-Jules Marey." In *Cinema Beyond Film: Media Epistemology in the Modern Era*, edited by François Albera and Maria Tortajada (Amsterdam: Amsterdam University Press, 2010): 79–96.

Vertov, Dziga. *The Writings of Dziga Vertov,* edited by Annette Michelson (Berkeley, CA: University of California Press, 1984).

About the Author

Donald Crafton has received the Distinguished Career Award from the Society for Cinema and Media Studies and the Jean Mitry prize presented at Pordenone, Italy. He has been recognized by the Jean Vigo Institute (France), the International Animation Festival in Zagreb, and the Academy of Motion Picture Arts and Sciences. Crafton held the first endowed professorship in film studies at the University of Notre Dame.

8. Photography and the Composite Image, or A Portrait of Méliès as Bergsonian Filmmaker

Benoît Turquety

Abstract

The prominence of Georges Méliès in the history of cinema is indelibly linked to his use of the 'substitution trick', often described as a forebear of editing. Yet Méliès also exploited a different sort of trick, one stemming from 'recreational photography': multiple exposures with a black background. While the tricks achieved with this technique have received less attention, they are no less crucial for two reasons. First, the technique pertains to what, for many theorists, has been conceived as an essential feature of photographic and cinematographic images, namely their *synchronicity*, or the fact that, as Henri Van Lier put it, "whatever the exposure time and the moment of impact of each particular photon, the arrival of all, in the end, is dated by the arrival of the last one." Secondly, Méliès's use of the black background enables us today to reassess a distinction that digital cinema has undermined, namely that between *editing* and *compositing*.

Keywords: Georges Méliès, film editing, compositing, digital cinema, Henri Bergson

What Did Méliès Invent?

The second volume of Georges Sadoul's *Histoire générale du cinéma*, entitled *Les Pionniers du cinéma* ("The Film Pioneers"), was published in 1947, one year after the first tome on the "Invention of cinema." That second book is divided into two parts: the first deals with the "Méliès era," that of craftsmanship, from 1897 to *Le voyage dans la lune* (*A Trip to the Moon*, 1903); the second

Lefebvre, M. & M. Furstenau (eds.), *Special Effects on the Screen: Faking the View from Méliès to Motion Capture*. Taylor & Francis Group, 2022
DOI 10.5117/9789462980730_CH08

covers the "Pathé era," the time of the growing industrialization of film production, from 1903 to 1908. In the third chapter of the first part, "*Les vues fantastiques de Georges Méliès,*" Sadoul introduces, in parentheses, an important subtitle: "*Le truquage, premier aspect de la technique cinématographique*" ("Tricks, the first aspect of film technique").[1] So it appears that, for Sadoul, something called "film technique" — or perhaps "cinematic technology," as these French terms remain ambiguous in English[2] — was born with the first works of Georges Méliès. Such technique begins, moreover, as "truquage," derived from "trucs" ("tricks"), the term that would come to designate "special effects" in France. That same year, Sadoul published a contribution in the first issue of the *Revue internationale de filmologie* under the title, "*Georges Méliès et la première élaboration du langage cinématographique*" ("Georges Méliès and the first elaboration of film language"). In Sadoul's terms, this is probably a stronger claim, if slightly weakened when phrased as a "first elaboration." Still, Sadoul's move from Méliès as an inventor of film technique to Méliès as pioneer of film language is a crucial one for a historian and theoretician such as Sadoul.

In fact, the founding role attributed within traditional historiography to Méliès with regards to film technique, to the language or the art of film, is related first and foremost to a particular trick: the "*truc à arrêt ou truc par substitution*" ("stop trick or substitution trick"), as he named it in the famous 1907 article "*Les vues cinématographiques*" ("The Kinematographic Views").[3] The fact that this particular trick, among all the techniques developed by Méliès, has been systematically emphasized as the most important can be explained by several factors. The first is certainly the mythology of its invention, promulgated originally by Méliès himself. Supposedly discovered by accident when his camera became jammed as he was "photographing as usual in the Place de l'Opéra,"[4] it opens the way to Méliès's success and to his subsequent explorations of the possibilities of the film medium.

1 Georges Sadoul, *Histoire générale du cinéma.* vol. 1. *L'invention du cinéma.* (Paris: Denoël, 1946). *Histoire générale du cinéma.* vol. 2. *Les pionniers du cinéma* (Paris: Denoël, 1947).

2 On this question, see: Rick Altman, "Toward a Theory of the History of Representational Technologies," *Iris* (2) 2, (1984), pp. 111–125; and Benoît Turquety, "On Viewfinders, Tape Slicers and Video Assist Systems: Questioning the History of Techniques and Technology in Cinema." In *Technology and Film Scholarship: Experience, Study, History,* André Gaudreault and Santiago Hidalgo, eds. (Amsterdam: Amsterdam University Press, 2015).

3 Georges Méliès, "Les Vues cinématographiques," in *Annuaire général de la photographie.* Reprinted as "Kinematographic Views: A Discussion," in André Gaudreault, *Film and Attraction. From Kinematography to Cinema.* (Urbana, Chicago, Springfield: University of Illinois Press, 2011), as Appendix B, pp. 133–152.

4 Méliès, "Kinematographic Views," in Gaudreault, op. cit., p. 148.

A second factor is that this trick is immediately connected to the question of editing, one of the major problems shared by both the traditional historiography and the new film history.[5] Sadoul's move from "technique" to "film language" relies on that articulation: the "birth" of film as art must be understood, according to him, as synonymous — or contemporaneous — with its birth as a language, and both then must be associated with the beginnings of editing or montage. Editing appears here as the major, specifically cinematic technique, and the elaboration of editing techniques as the moment when a specifically cinematic domain would emerge.

As Jean Mitry puts it, for example, in the first volume of his 1967 *Histoire du cinéma: Art et industrie*:

> [Méliès's] tricks pertain to photography. An *animated* photography that remains, we agree, the initial cinematic fact but that is not in any way a *filmic* fact.[6] The only filmic fact discovered by Méliès (it is quite important...) is the *crank stop*. It is this stop which, combined with superimpositions (of a photographic nature), will allow him to realize his wonders.[7]

Discovering cinema — as either an art or a "language" — is, in such an account, equivalent to discovering editing as a technique and montage as a conceptual operation. From a historiographical perspective based on intermediality, film splicing techniques derive from the theater, but that does not change the fact that one can find in the act of editing an essential feature specific to the cinema. Jacques Maltête still insisted on that point in 1999, writing that there is a "specifically Mélièsian technique, whose main characteristic remains the act of joining together pieces of film [in French: *coller*]. For, let us repeat it, to achieve his tricks, Méliès has edited his films, a lot and with minute precision."[8]

5 André Gaudreault, for instance, has been working on editing as early as his contribution to the 1978 Brighton conference on crosscutting (1979) and until his later works on the presence of editing in the Lumière films (2002). André Gaudreault, "Les détours du récit filmique (Sur la naissance du montage parallèle)," in *Cahiers de la cinémathèque* 29 (1979), pp. 88–107; and "Fragmentation et segmentation dans les 'vues animées': le corpus Lumière," in *Stop Motion, Fragmentation of Time*, François Albera, Marta Braun and André Gaudreault, eds., (Lausanne: Payot, 2002), pp. 225–245.

6 Jean Mitry (*Histoire du cinéma: art et industrie*, [Paris: Éditions universitaires, 1967]) refers here to the distinction established by Gilbert Cohen-Séat (*Essai sur les principes d'une philosophie du cinéma I. Introduction générale: Notions fondamentales et vocabulaire de filmologie*, [Paris: Presses Universitaires de France, 1946]) between cinematic and filmic facts, though his use of the terms differs considerably.

7 Mitry, op. cit., p. 370.

8 Jacques Malthête, "Quand Méliès n'en faisait qu'à sa tête," in *1895: Revue d'histoire du cinéma*, (27) (1999), p. 26.

But as Clément Chéroux reminded us in 2010, after Georges Sadoul had already formulated it in 1947, the "stop trick" is not the only basis of the Mélièsian technique. Other methods, other genealogies, must be taken into account. Sadoul in fact divides Mélièsian procedures into two sets: those that he associates with theatrical machinery, "substitution"; and those that he relates with photographic practices, "composite photography" and "photomontage."[9] Chéroux specifies that this photographic filiation had to be distinguished from prestidigitation, because it belonged to another cultural constellation of practices, that of "photographic recreations."

Thus, if the "stop trick" has been emphasized in most accounts, it is always in a logic of specificity — specifically cinematic, specifically Mélièsian. Other techniques that derived from photographic manipulation do not possess such specificity, for neither the cinema nor Méliès invented them. Yet Méliès did not even invent the "stop trick," nor was he its sole user. It is already visible in the 1895 Edison film, *The Execution of Mary, Queen of Scots* (Alfred Clark), as is well known.[10] Specificity issues notwithstanding, these photographic techniques are nevertheless rich with theoretical implications, and perhaps more so today, in the context of the transition of cinema to digital technology.

Méliès and the Black Backgrounds

In Sadoul's history, the substitution trick comes first in the chronology of Méliès's work. Only subsequently do other types of tricks appear:

> It is in 1898 that Méliès uses for the first time three new tricks: superimposition, multiple exposure and mattes. These three methods are all borrowed from photographic technique and appear, in quick succession, in the Star Film catalogue.[11]

To Sadoul, these two consecutive beginnings "prove that Méliès [had] just bought a treatise on photography and [applied] the tricks described within."[12]

9 Clément Chéroux, "The Great Trade of Tricks: On Some Relations Between Conjuring Tricks, Photography, and Cinematography," in *Between Still and Moving Images*, Laurent Guido and Olivier Lugon, eds., (New Barnet: John Libbey, 2012), p. 52.

10 On this film, see Charles Musser, *Edison Motion Pictures, 1890-1900: An Annotated Filmography*, (Washington/Pordenone: Smithsonian Institution Press/Le Giornate del Cinema muto, 1997), pp. 189–190.

11 Sadoul, *Histoire générale du cinéma*. vol. 2 , op. cit., p. 52.

12 Ibid.

These three tricks have a common point; their basic procedure consists in having the strip of film pass several times in the camera. For Sadoul, multiple exposures or superimposition can be connected to the tradition of "spirit photography," dating back in France to the Second Empire.[13] The use by Méliès of mattes, or "blackboard screens placed in front of the lens [...] masking part of the camera's field,"[14] are an extension of the principle of multiple exposure that Sadoul also names "composite photography"[15] and which consists of a superimposition or a series of superimpositions in front of a black background.

The main application of this trick, already known in the tradition of "photographic recreations," was the multiplication of all or part of a subject. Whether in *Récréations photographiques*, written by A. Bergeret and F. Drouin in 1893,[16] or in *Photographie récréative et fantaisiste* by C. Chaplot in 1904, both published by Charles Mendel, the "black background" tricks appear in a chapter titled "*Images multiples*" ("Multiple images"). Of course, multiplication is not mandatory when using the black background and re-exposing the film.[17]

Black backgrounds make use of a property of photochemical emulsions that is as obvious as it is finally quite extraordinary. Chaplot explains the principle of the trick in the following manner:

> Absolute blackness being, as is well known, inactive on the sensitive surfaces used in photography, if a more or less lit object is placed in front of a black background constituted either by a cloth absorbing as completely

13 These photographs purported to show the ghosts that haunted the late nineteenth-century houses and were associated with the spiritualist and occult practices that were an important component of that period's culture. See Mireille Berton, "The Magism of Cinema and Imaginary Spiritism in France at the Beginning of the Twentieth Century," in *Early Popular Visual Culture* (13)2 (2015).

14 Sadoul, *Histoire générale du cinéma*. vol. 1, op. cit., p. 60.

15 Ibid., p. 53. (Note that the mention "Composite photograph" is found in a caption that accompanies the reproduction of a photo engraving showing a man juggling with his head originally published in *La nature* in 1886. The 1973 re-edition of the book jettisons this photograph and, with it, the mention "composite photograph.")

16 Albert Bergeret and Félix Drouin, *Les Récréations photographiques* (Paris: Charles Mendel, 1893).

17 Sadoul notes that "one of the themes [of multiple exposure spirit photography] was characters juggling with several copies of their own heads" (*Histoire générale du cinéma*. vol. 2, op. cit., p. 56). An early example of this is the series of seven slides preserved at the Cinémathèque française along with John Rudge and William Friese-Greene's magic lantern and dated around 1881. They show a man, in front of an entirely black background, playing with his head (the sole "original" one — there is no multiplication in this case) detached from his body.

> as possible the light rays or better by an opening toward a dark room, the positive print shall present the image of that object precisely outlined.[18]

Having photographed or filmed something black, that is, is to have done exactly nothing: the film is not less "blank" (e.g., unaltered, pristine) after shooting than it was before the shot was taken. It is not only black, it is blank. That is a very important distinction. Much later, experimental cinema — particularly perhaps that which belongs to the tradition named "structural" by P. Adams Sitney — made use in its own ways of that ambivalence of the cinematic dispositive: on a white screen, nothing (nothingness itself) is white (e.g., only "light" appears in a purely translucent piece of film); whereas on the celluloid or in the darkened hall of a movie theater, nothing is black (the absence of light). These "nothings" bring along with them different connotations, but they really are perceived as nothing by the spectators and thus play important roles within the filmic system. Christian Metz wrote in "*Trucage* and the Film" that in the fade to black, "the black rectangle is viewed far less as such than as a brief instant of filmic void" (Metz uses the term *néant filmique* or filmic nothingness).[19]

So filming a lit object in front of a black background is to only have recorded the object itself — not its black surroundings — leaving the rest of the photo-sensitive emulsion untouched. The photographic act then is not based on the framing of a field limited by the edges of the camera gate but on recording only the image of that object which is thus outlined with precision against photographic or filmic nothingness. The edges of the frame are not pertinent anymore, the field is only defined by the foregrounded object's luminance. That constitutes a specific and, in fact, quite singular property of photographic emulsion.

In some essays as early as the 1910s, the use of black backgrounds was already perceived in a lineage that strangely weds Etienne-Jules Marey's fixed-plate chronophotographs with much less scientific and more 'obscure' practices. E. Kress wrote around 1912: "We have seen that the properties of black backgrounds have been put to use by Marey for his photographic studies of movement. The black background constitutes the essential element of Black Magic."[20] In Marey already, the black background allows

18 Charles Chaplot, *La Photographie récréative et fantaisiste. Recueil de divertissements, trucs, passe-temps photographiques,* (Paris: Charles Mendel, 1904), p. 89.

19 Christian Metz, "*Trucage* and the Film," trans. Françoise Metzler. *Critical Inquiry*, (3)4 (1977), p. 661.

20 Eugène Kress, *Conférences sur la cinématographie*, tome 1, (Paris: Comptoir d'édition de *Cinéma-Revue*, 1999 [c. 1912]), "Trucs et illusions. Applications de l'optique et de la mécanique au cinématographe." 7th conférence. In *1895: Revue d'histoire du cinéma*, 27 (1999), p. 12.

to indefinitely multiply the exposures of the plate. Uncovering the lens, triggering the shutter, one exposes the whole plate, but only the non-black parts are actinic: the black areas surrounding whatever is delineated by light are still unexposed and therefore remain perfectly exposable.

The tradition of composite photography proposed a notable variation on this principle. Instead of isolating a subject from a wholly black background, it makes holes in profilmic space with "reserved" dark areas, destined to be exposed during a second run in the camera. Méliès — like others at the time, but he was probably the first in cinema — conceived of the image not as a homogeneous whole but as built by zones, to be filled in succession. That is what the French term "*réserve*," which was already in use at the time, refers to. As E. Kress explained:

> The black background is of essential, fundamental use in filmmaking, especially in fairy filmmaking. [...] The black background allows the photograph of bright objects while sparing around them a reserve of sensitive surface which in its turn can be exposed to other objects, the space cut by the first ones having been carefully marked to avoid any unforeseen confusion of movements or of objects themselves.[21]

"*Réserver*" ("To keep in reserve") was defined by the French dictionary *Littré* in the 1860s as "Retaining something from a whole, or an object among several," or "Keeping for another time, for another use, for another occasion, for another treatment." The term suggests both the withdrawing of a part, and a temporal delay. In photography, a section of the overall light-sensitive emulsion is here reserved for later use. If the visual "field," or profilmic space, is still determined by the edges of the frame, by the angle covered by the lens, it is no longer homogeneous, rather it is structured in discrete areas, exposed separately, composed (or composited) at different moments.

Photography as a Synchronous Medium

Black background practices in fact interestingly seem to contradict what stands today as the dominant conception of photography.[22] Indeed, several contributions to the theory of photography have formulated the implications

21 Kress, "Trucs et illusions," op. cit., p. 12.

22 For reasons that will become clearer, I would argue that this conception must be historicized, though dating it with any precision would necessitate a very large discourse analysis on its own.

of a fundamental technical aspect of its apparatus that has more or less explicitly led today to the construction of an epistemological framework concerning it. According to that framework, photography — unlike painting, which is produced sequentially in time — is *synchronous*. As Philippe Dubois explains:

> Remaining as close as possible to the technical procedure and the chemical reaction, it can be said [...] that the photographic emulsion [...] reacts at once in its entirety to the light information that literally hits it. [...] It is in this sense that the photochemical information can be described as synchronous. All the silver halide crystals that compose the pointillist weft of the sensitive surface are affected simultaneously [...][23]

Thus, for Dubois, "photography really is a slice of the space-time continuum."[24] Henri Van Lier, a few months before in 1983, had defined photography as a "synchronous imprint" that is "dated to the billionth of a second." "Whatever the exposure time and the moment of impact of each particular photon, the arrival of all, in the end, is dated by the arrival of the last one."[25]

If the exposure time is long, the successive arrival of photons inscribes in the image, besides the forms of objects, the motion blur that characterizes the duration block of which the image is the imprint. But whether snapshot or long exposure, the photograph as imprint appears as fundamentally simultaneous in all of its parts, and thus as essentially synchronous.

This is a fundamental trait of photography. It is what constituted for Muybridge or Marey the very scientific usefulness of photography. It is what allowed a photograph to function as evidence that the four legs of a given horse have been in this position at the same instant, an instant that can be precisely determined. It is what allowed, in Marey's work, for the representation of the relative simultaneous positions of all the muscles of the human body in each phase of walking. Marey may have accumulated several instants on the same plate, but each instant, each exposure, must preserve its complete homogeneousness, such preservation being at the very core of the process. For Christian Metz, "photographic representation [is] itself, theoretically strictly faithful since it was obtained, so to speak, in one stroke."[26]

23 Philippe Dubois, *L'Acte photographique*, (Paris/Bruxelles: Nathan/Labor, 1983), pp. 158–159.

24 Ibid., p. 104.

25 Henri Van Lier, *Philosophie de la photographie* (Paris: *Les Cahiers de la photographie*. 1983), p. 18.

26 Metz, "*Trucage* and the Film," op. cit., p. 671.

A Bergsonian Example: Cutting Out Figures

Synchronicity, though specific to the photographic medium, may not have been as central to the history of photography theory as other aspects of the dispositive, such as the problem of the trace for example, and it has not pervaded film studies to any significant level. For instance, it does not play any appreciable role in André Bazin's "Ontology of the Photographic Image." Interestingly though, this quality of simultaneousness makes its way in Henri Bergson's famous description of the "cinematographical mechanism of thought," in Chapter IV of *Creative Evolution*.[27] Yet if the brevity of exposure time constitutes the explicit theoretical definition of the series of instantaneous photographs that make up the film strip, this brevity in fact plays no role whatsoever in the example that Bergson uses as a starting point to develop his concepts of "becoming" and "duration." The elementary photographs (e.g., the film frames) are instead characterized by the overall simultaneousness of each one.

As a matter of fact, Bergson supports his argument with an example that can only strike us as strange and paradoxical. He writes: "Suppose we wish to portray on a screen a living picture [*une scène animée*], such as the marching past of a regiment."[28] In choosing a regiment of soldiers, Bergson opts for the Mareysian motif of the human walk, though it is now multiplied, disseminated in a sorted set of subjects composing a global synchronous motion. Bergson then opposes two possible methods for achieving a moving image. One would be to isolate each of these subjects, "to cut out jointed figures representing the soldiers, to give to each of them the movement of marching, a movement varying from individual to individual although common to the human species, and to throw the whole on the screen."[29] But that implies "an enormous amount of work," practically impossible, and the results he says would be "poor": "how could it, at its best, reproduce the suppleness and variety of life?"[30]

The second method, writes Bergson, is "more easy and at the same time more effective,"[31] and it corresponds to what the cinematograph does. This comes as no surprise, of course, since the example itself was obviously constructed with the cinema in mind! Bergson describes the

27 Henri Bergson, *Creative Evolution*. Authorized transl. by A. Mitchell. (New York: Random House, 1944 [1907]).

28 Ibid., p. 331.

29 Ibid.

30 Ibid.

31 Ibid.

cinematographic method in this way: "With photographs, each of which represents the regiment in a fixed attitude, it reconstitutes the mobility of the regiment marching."[32] The opposition is thus articulated here between an impossible method, which would consist in isolating each individual flow of motion, and an efficient one which takes synchronous slices of the whole and which "extract[s] from all the movements peculiar to all the figures an impersonal movement, abstract and simple, movement in general, so to speak."[33] What a strange description! The difference between the two methods stems from the fact that photography is essentially synchronous, that the emulsion records the attitudes of the regiment in its entirety at once, thus transferring over the instantaneous nature of photography to the temporal nature of cinema! In short, it is the photographic property of simultaneity that proves central to the coherence of the example, even though it does not play any role in the theoretical development that follows. According to Maria Tortajada, "Bergson's thinking as a whole rests on the opposition between the continuity of a movement associated with duration and the immobility of [individual] sections."[34] Yet careful consideration of Bergson's example shows it to be almost a separate or autonomous device in his overall argument, one that produces problems that well exceed the philosopher's conceptual line of reasoning. There is no doubt that this section of *Creative Evolution* has come to be of key importance in the history of film theory where it has usually been connected with the dialectics of continuity and discontinuity that is crucial both for Bergson and for the cinematic dispositive.[35] Recently, though, Elie During has argued that, for Bergson, the cinematograph allowed for a "fine-grained differentiation between two senses of time, in conformity to a duality of tendencies running at the heart of the scientific view of the universe: global time (homogeneous, absolute, generic) and local time (differentiated, relational, individual)."[36] It follows, therefore, that as a conceptual device, Bergson's example of the walking regiment grounds the photographic in a globally synchronic temporality, based on homogeneousness and coherence. Thus, whether "instantaneous"

32 Ibid.

33 Ibid., p. 332.

34 Maria Tortajada, "Photography/Cinema: Complementary Paradigms in the Early Twentieth Century," in *Between Still and Moving Images*, Laurent Guido and Olivier Lugon, eds., op. cit., p. 37.

35 On this question, see Guy Fihman, "Le cinéma date du jour où," in *Du cinéma selon Vincennes* (Paris: L'Herminier, 1979), pp. 181–188.

36 Elie During, "Notes on the Bergsonian Cinematograph," in *Cine-dispositives*, F. Albera and M. Tortajada, eds., (Amsterdam: Amsterdam University Press, 2015), p. 128.

or not, photography (including cine-photography) possesses as an essential attribute the remarkable synchronicity of all its points or discrete elements, its "all at once" quality that makes it precisely dated/dateable as a singularity in all of its entirety.

Cutting Images: Editing and Compositing

That coherence of the photographic image — its essential temporal cohesion — is at the heart of a technical distinction: the difference between editing and compositing. Editing implies cutting the celluloid strip between the images while leaving the images themselves untouched. Compositing, to follow Martin Lefebvre and Marc Furstenau, "is usually defined as a 'special effect' consisting in the layering of two or more image-elements to produce the impression of a single visual field."[37] In the contemporary division of labor within filmmaking, both editing and compositing are a part of postproduction. However, the differences between the two are significant enough that they are still thought of as two different professions. Editors are not — or not yet? — special effects supervisors. Nevertheless, both processes involve some form of relationship between images: "While composites don't rely on the linear or temporal juxtaposition of shots — the domain of editing, per se — they nonetheless require the juxtaposition of elements taken from at least two separate shots."[38] Editing and compositing are two forms of juxtaposition, interweaving multiple images into a single, more or less continuous flow. The only difference may be that editing deals with whole images, whereas compositing breaks them into image elements.

Montage is traditionally connected, in film theory, with editing, but in fact, photomontage — a technique that belongs to the realm of photographic recreations — technically belongs to compositing. It can still be, at least in certain instances, described in terms of montage, as François Albera has proposed regarding the photographer Gustave Le Gray. From 1856 on, Le Gray's photographs of the sea usually consisted of two juxtaposed image-elements: the sea itself came from one original photograph, while the sky came from another. That allowed him to use different exposure times for the two elements, and thus have them both sharp and contrasted, but also to associate different places or reuse a particularly beautiful sky with different

37 Martin Lefebvre and Marc Furstenau, "Digital Editing and Montage: The Vanishing Celluloid and Beyond," *CiNéMAS*, (13)1–2 (2002), p. 82.
38 Ibid., p. 82.

seas, etc. Le Gray "makes some montage," as Albera comments: "The time of picture taking — here there are two, dissociated — becomes distinct from the time of montage and of the reading of the image by the spectator — here, there is but one."[39] Le Gray's photographs are clearly composed: the horizon line joins together two separate (but coherent) moments. The fundamental simultaneity of photography is altered, in a way that may or may not be perceived by the spectator, who may notice the "trick" on the horizon line itself or else detect that two different photographs share the exact same sky. In any case, the images rely on the sharp juxtaposition of two instants, each remaining precisely framed by the photographic apparatus and within the final image.

Returning to the use of black backgrounds, we can see that it further complicates photographic simultaneity, disrupting the concrete homogeneity of the time and space of the shooting, and playing with the spectator's expectations.

Here, several consecutive images of, say, Méliès against a black background are a possibility — in fact, this is always what happens in the cinema as one frame follows another. This is what is disrupted by a special effect that would show us what now appears to be several images/views of Méliès at once, simultaneously. In this case, the cinema adds something new to composite photography: in film, on a black background, subjects are not only multiplied but they can all be moving at the same time, each performing a different and synchronized action, for a certain duration. The apparatus produces not only a non-homogeneous instant but heterogeneous blocks of duration flowing at the same time and appearing in the same shot (though they were not shot at the same time). The black background trick was well known among photographers, but what Méliès invented, or understood, was the process whereby film can composite in a single image and a single duration several distinct temporal flows.

Marey had used the black background for the multiple exposures of his subject in his fixed plate chronophotographs, but for his machines that were based on moving strips of celluloid, each image was exposed only once. In Méliès's use of the black background, however, we are not given to see only one isolated bright object but a space that is divided in distinct areas and whose final synchronization into a single visual field is an illusion that requires patient, difficult construction. Whereas the stop/substitution trick was "*facile*" for Méliès, in both meanings of the French term — not

39 François Albera, "Pour une épistémographie du montage: préalables," in *CiNéMAS*, (13) 2 (2002), p. 24.

difficult and cheap — choreographing and compositing multiple exposure shots was not. As he wrote in 1907:

> I will conclude by saying that the simplest tricks, much to my chagrin, make the greatest impact, while those achieved through superimpositions [*superpositions*], which are much more difficult, are hardly appreciated, except by those who understand the problems involved. Views [*vues*] performed by a single actor, in which the film [*pellicule*] is exposed up to ten times in the camera, are so difficult that they become a veritable Chinese puzzle. The actor, playing different scenes ten times, must remember precisely to the second, while the film is running, what he was doing at the same time in previous passages and his exact location on stage. [...] If, at some point in the procedure, the actor makes an untoward gesture, if his arm moves in front of a character already photographed, this will be superimposed and out of focus, which reveals the trick.[40]

For Méliès, the visual field perceived by the viewer as a single shot can be fragmented into several duration-space blocks, the continuity of which must be strictly maintained. That is the difficulty, the challenge that makes the beauty of the successful trick. The level of complexity evoked here by Méliès can be observed when carefully watching the films. *Un homme de tête* (*The Four Troublesome Heads;* 1898) is probably the earliest use of a black background in what has been preserved of Méliès's *oeuvre.*

On one hand, the film nicely illustrates the great care that has gone into compositing into a single visual field the various fragments, into hiding the cuts as well as choreographing the performance. But it also highlights the technical challenges that go into creating such an effect: on the left side of the frame, the table-top wobbles as if it were eager to live an autonomous life, sometimes appearing to detach itself from its legs. Over to the right, parts of the severed head — the chin and the neck — appear superimposed over the white of the tabletop. The instability of the various parts of the composited image thus reveals the difficulty, given the contemporary state of technology in Méliès's time, of keeping the camera absolutely steady. Yet these internal quivers bestow to Méliès's images a distinct liveliness or even "life," diverse and teeming: a singular form of animation. Hand-painted colors, so wonderfully important in Méliès's films, also contribute, with their own quivers, to this specific internal life of the animated photographs of the period.

40 Méliès, "Kinematographic Views,' in Gaudreault, op. cit., pp. 148–149.

The frame is not, for Méliès, a fundamental and indivisible unit of the film; nor is the shot. He plans his "fantastic views" as arrangements of distinct durations, distributed with precision in the visual field. Cuts and splices play an important role in these compositions, but no more so than the distribution of the black backgrounds within the set or the mattes in front of the camera's lens. Hence these beautiful and fantastic Mélièsian sets, such as the one designed for *L'homme à la tête de caoutchouc* (*The Man With the Rubber Head*, 1901) where an archway with large barn doors can be seen in what is otherwise a science laboratory. The doors open onto a mysterious dark room, in front of which the scientist appropriately places the table on which the experiment (and the special effects) will transpire.

It could be argued, in fact, that Méliès combines the two methods that Bergson presents as opposite procedures. Méliès, as it were, would not "cut out jointed figures" representing the marching soldiers but would film them one by one in front of a black background — probably playing each soldier himself! In other words, Méliès turns the cinematograph (almost) on its head, making possible the otherwise "impossible" non-cinematographic method mentioned by Bergson. The black background allows him to isolate the moving objects, yet without resorting to the "cutting out of figures," the unworkable and indeed rather eccentric solution proposed by Bergson. It is still the case that the isolation and multiplication of objects involve precisely the great complexity of a "Chinese puzzle," as opposed to the efficient simplicity of the "normal" cinematograph.

The "Normal" and the "Special"

We may ask what, precisely, does the qualification of "normal" mean here? One could argue that it means first and foremost "efficient simplicity," the simplest use of the cinematic machinery, or else — though this is not exactly the same thing, at least theoretically — using it in the way it seems was "meant" to be used. This is what Christian Metz appears to be suggesting in his essay "*Trucage* and the film." A photograph, he says, is either "normal" or tricked.[41] However, this is not an obvious distinction to make, and perhaps should not be one, given that "normal" is not an obvious aesthetic and theoretical criterion. To give an example from a different artistic field, most would probably agree that John Cage's Sonatas for prepared piano do not represent a "normal" use of the piano, but there would hardly be any such

41 Metz, "*Trucage* and the Film," op. cit., p. 671.

consensus that Cecil Taylor's use of the instrument in *Unit Structures* — "unprepared" though the thing may be — represents a "normal" use of a piano. Indeed, some listeners might well assume that this is not what a piano is "made" for. Yet any sense of "normalcy" here would refer to a culturally and historically defined context which, while it includes technological aspects — e.g., the relation between how a piano sounds with how it was fabricated at a given moment in history — is mainly grounded in an aesthetic history of music and listening practices.

Should there be a "normal" use of the photographic or cinematic apparatus, it might appear at first as more clearly oriented by technological criteria. When facing a machine, it seems logical to consider as "normal" a use based on its switching on and off. This is what it was made for. Any other use has to be considered not normal — perhaps we should say "pathological," following the famous Canguilhemian opposition, which could account for the depreciation undergone by trickery in twentieth-century photography. If not "pathological," it is in any case what defines the "special" of "special effects." In the passage cited from Metz above, "normal" appears as simply defined by its opposition to "tricked." Here, the "normal" is both the simple and the sane; special effects are characterized by a complication of the process, exceeding the "efficient simplicity" of the basic apparatus. But this technological basis for the definition of the "normal" is in fact no less historically and culturally grounded than in the case of the piano. Yet, a few pages later in the same essay, Metz further nuances this opposition when it comes to cinema as he considers how our perception of editing has changed. Indeed, he claims that editing was initially perceived as a special effect or *trucage* before being integrated into the "normal" practices of the medium. "Today," he writes, "we are so accustomed to montage that it would occur to no one to add to the list of *trucages* (or "special" effects) such an ordinary and general manipulation."[42] For photography, such a techno-cultural distinction implies, for instance, that, as explicitly formulated in Dubois, the "photographic act" would consist entirely in the gesture of switching the camera on and off (e.g., activating the shutter mechanism). All the other actual or possible gestures involved in the photographic process would therefore be secondary or accessory, whether done by the photographer herself or by technicians in the laboratory during "postproduction."

We should realize, however, that the enormous privilege that is given to this act is attached to an ideological conception of photography. It is also dependent on a certain historical state of the medium within culture, as a

42 Ibid., p. 672

network of technologies and practices, a state that is not intrinsically linked with photography itself but appeared much later. If one wishes to date it by reference to some turning point, the famous slogan "You press the button, we do the rest" forged by Eastman for his Kodak in 1888 — capturing the popular imagination and laying the ground for a business empire — could be considered a major event in the history of photography, more important indeed than, for instance, the emergence of "instantaneous photography" as such. It led to that strongest of distinctions: the art of photography is pressing the button (at the right time, at the right place, etc.); the rest is something else (not "normal" photography, and not really art either). Before the Kodak, a photographer would have had to dabble in chemistry, and photographic gestures involved a complex set of procedures occurring at different moments, as the earlier quote from Albera indicated. "Pressing the button" was but one of them and there was no essential difference between that instant and all the other possible and actual manipulations constituting "photographic recreations." It follows that the "normal" use of the photographic apparatus became the simplest one in great measure because photography itself came to be defined by simplicity, which was not the case at the beginning and is not an intrinsic or essential quality of the photographic machinery in itself.

Given the above, it is interesting to see that the institutionalization of "special effects" within the Hollywood industry has occurred precisely on this basis, as has been demonstrated by Julie Turnock.[43] Special effects were "special" inasmuch as they required "specialists" for their realization. As Turnock points out, within the Hollywood technical literature, "the term 'intricate' (or 'intricacies') appears often as a transition term between 'trick work' and 'special effects photography' in the 1920s."[44] Special effects are thus defined by their very complexity, a complexity whose criteria pertain first and foremost to the organization of labor in the movie production industry. In the classical institutionalized division of labor, the cinematographer was an expert of the "normal" use of the film apparatus; he — they were men[45] — was not supposed to master those other, "intricate" uses for which independent workers would be enlisted.

In parallel, Turnock also reminds us that, from the start, "special effects" have had to do with the composite image. The first technicians who have

43 Julie Turnock, "Patient Research on the Slapstick Lots: From Trick Men to Special Effects Artists in Silent Hollywood," *Early Popular Visual Culture*, (13) 2 (2015), pp. 152–173.
44 Ibid., p. 154.
45 The first female cinematographer to join the US cinematographers' guild, the American Society of Cinematographers (ASC), was Brianne Murphey in 1980.

been recognized as special effects pioneers in the United States, Carroll Dunning and Frank Williams, for instance, had invented and promoted compositing techniques. Their processes were variants of the black background principle, allowing for the integration within a single shot of images that came from different times and places. The images produced were not "normal" in part because they were complicated to achieve and required specialized labor other than that of the cinematographer, and also because they broke the fundamental temporal and spatial homogeneity of the photographic image. It follows that the definition of "special effects," which is largely based on the status of the composite image within photography, is as much a cultural one as it is a technical one. In other words, it must be understood as a historical phenomenon, but not one that is strictly determined by technological shifts. But if this is the case, then the role of the "digital revolution" in the field of special or visual effects may not be as central as it has been claimed to be by film scholars, or if it is, it may be so for cultural more than strictly technological reasons.

Digital Photography and the Composite Image

It is difficult to describe the Mélièsian techniques of the black background without noticing surprising similarities with the specificities of digital photography and particularly of digital cinema. Silver grains and pixels have been compared many times in recent "film" theory,[46] while Philippe Dubois has already insisted on the essential discontinuity of the (analog) photographic image, "the silver halide crystals [composing] the pointillist weft of the light-sensitive surface."[47] But in truth, this rather "conceptual" discontinuity has not trickled down to photographic practice. As we have seen, the traditional photographic image could in principle only be produced as a whole, in its entirety, "without the operator being able to change anything in the course of the game (during the sole exposure time)."[48] Dubois insists on this aspect, presenting it as a medium-specific feature:

> [...] that trait of synchronism radically distinguishes photography from painting. [...] For the photographer, there is only one choice to make, a

46 See, for instance, the title of Giovanna Fossati's book, *From Grain to Pixel: The Archival Life of Film in Transition*, (Amsterdam: Amsterdam University Press, 2009).

47 Dubois, *L'Acte photographique*, op. cit., p. 159.

48 Ibid.

> global, single choice, which is irreversible. For once the shot (the section) [*sur le coup* (*la coupe*)] is made, all is said and done, inscribed, fixed. That means one cannot interfere anymore with the image while it is being made. If manipulations are possible — see the pictorialists —, it is afterwards [*après-coup*(*e*)], and precisely by treating the photo as a painting.[49]

Dubois's position, we have seen earlier, addresses a specific photographic paradigm founded on synchronism because it considers simultaneity to be an essential characteristic of the photographic image as opposed to other technical modes of image production. Given this perspective, anything that exceeds the basic simultaneity of the photograph is considered non-photographic, as relating rather to painting. It should be added, however, that Dubois's stance on the issue originates in a theoretical context in which medium specificity was a major aesthetic problem. As a result, adulterated or doctored photographs almost fall out of the photographic realm altogether and are perceived as having more in common with paintings or drawings than with undoctored photographs.

The novelty of digital technology has consisted, in great measure, in breaking down this a priori homogeneity of the photographic image in cinema — if we admit that these images can still be called photographic, a claim that some seem inclined to question. Consider color grading. In photochemical postproduction, color grading had to find a balance of color rendition that each time took into consideration the entire image. Digital grading, by contrast, can isolate mobile areas in the visual field and apply to each a different color correction. Compression software now routinely breaks down the picture fields that make up the images in areas that correspond neither to the scale of the "frame" nor to that of the shot. In *The Virtual Life of Film*, D.N. Rodowick quotes a letter written by John Belton:

> In looking at a movie that has been digitized, temporal and spatial data have been omitted via the sampling that is part of both quantification and compression. Part of the image that remains constant over several frames is therefore given to us in frame one, then replaced in successive frames by a numerical code that refers us back to frame one. For that particular part of the image, we are seeing one brief moment of time and space again and again.[50]

49 Ibid.

50 D.N. Rodowick, *The Virtual Life of Film*, (Cambridge, MA/London: Harvard University Press, 2007), p. 137.

Digital compression — in its most widespread forms — replaces the classical structure based on film frames and shots by a more specific topology: it locates neighborhoods of spatial and sometimes temporal similarities and associates them under a single value in order to reduce the transport and the storage of "useless" data.

In broad terms, if 30 neighboring pixels show the same blue (or approximately), its coordinates in the red-green-blue system will be stored once and attributed to these 30 points. If the shot remains static and the hues stable (or approximately) for 30 frames, that single blue will be attributed to a time-space block of 900 points. For that given area, the perceived blue at the end of the interval dates back 30 frames before the rest of the image that we see. Compressed digital images are thus, by definition, immediately composite, conceived of and worked on as juxtapositions of jointed areas, originating from several different moments. A compressed digital cinema shot has several distinct durations flowing alongside each other, all parts of the image yet unfolding at different speeds. Even in so-called uncompressed digital images, the colors of each pixel are reconstructed from data obtained most of the time through a Bayer matrix, a system of pixel-size primary color filters placed in front of the sensor's photosites. Individual colors are then calculated by a complex procedure involving weighted adjacent zones. In that perspective, it is only artificially that a single pixel can be isolated within a digital cinema image: a pixel is in fact never independent from its neighboring pixels The digital image is partitioned in mobile areas and not strictly in pixels.

Of course, fundamental differences remain between digital images and the black backgrounds used by Méliès. Firstly, it was Méliès himself who organized the distribution of the reserves (the darkened areas) in his image, whereas it is the compression software that arranges the areas of coherence within the pictorial field according to the filmed object. In postproduction, nevertheless, it is the digital colorist who models the topology of the images according to her needs – and/or the cinematographer's, the director's, etc.

Now if digital cinema is in fact always made of composite images, does this imply that it is always tricked, that it is always a special or visual effect? The question leads us astray, however, since it would be difficult to attribute to a compression algorithm the sort of intentionality that characterizes Méliès's efforts to achieve trick shots. Yet does this mean that there is now no difference between "photographic" (e.g., live-action) cinema and animation, as suggested today by many scholars — in the manner of Philippe Dubois who intimated that doctored photos were in fact (almost) paintings?

(Tricked) Photography and Scientific Truth

As I have already mentioned, determining which practices do or do not fall into the category of tricks or special effects should not be based on technological criteria alone. The cultural conditions of the reception of the images and of the dispositive as a whole must be taken into account. In the Mélièsian practice of multiple exposures, the shot is not a "synchronous imprint" anymore. The use of black backgrounds turns on its head that conception of photography — or at least, deals with it on its own terms. The resulting shot shall be perceived a priori as synchronous by the spectators because it appears photographic in nature. And *de facto* the shot is indeed photographic in all of its parts. Yet the whole of what is seen departs from undoctored photographic imagery if one takes the simultaneous production of the entire image as an essential or defining aspect of photography and its dispositive. But what if we were to consider instead that the trait of global synchronicity attributed to photography as a fundamental technological attribute of its machinery is predominantly a cultural construction, a paradigm, which could then be historicized and whose evolution could be dated and traced? This is the paradigm described implicitly by Christian Metz when he claims that "the photographer only has the choice between a 'normal' shot [*prise de vue*] and, if he really wants to fool his audience, the characteristic lie, the fraudulent practice [...] a brazen falsehood."[51] This inescapable either/or alternative relies on an implicit conception of photography in which the "truth" of the imprint is the sole mode of justification and existence of photography. For D.N. Rodowick, the emergence of digital media would today reaffirm this truth value of the photographic process, in quite similar terms:

> In terms of market differentiation, computer-generated imagery codes itself as contemporary, spectacular, and future-oriented [...] At the same time, the photographic basis of cinema is coded as "real," the locus of a truthful and the authentic aesthetic experience of cinema.[52]

Georges Sadoul had begun commenting on Méliès's *oeuvre* in the first volume of his *Histoire générale du cinema*. The first aspect he had considered was Méliès's reconstructed topical events, which remain to this day by far the lesser-known part of his work, for reasons probably connected with the

51 Metz, "*Trucage* and the Film," op. cit., p. 671.
52 Rodowick, *The Virtual Life of Film*, op. cit., p. 5.

already mentioned dominant conceptions of Méliès's place in the history of film language and techniques. Sadoul meant to reconsider the apparent paradoxical character that the expression "*actualités reconstituées*" ("recreated news") may have assumed for his reader in 1948, in the context of the practices of the press at the end of the nineteenth century. The historian wrote that "Méliès's aim was not to counterfeit, but to reconstruct."[53] Sadoul reminded us that photography held but a trifling place in the newspapers of the time, reserved solely for luxury publications. The press was barely illustrated, and when it was, it was mostly with drawings:

> When, exceptionally, photography was employed to represent historical events [...] extras and photomontage were used without scruples to reconstruct scenes [...] that no photographer happened to record in reality. These reconstructions, accepted as such, did not shock anyone in the least, and were even reproduced in historical books without taking the trouble to emphasize that they were not authentic. Since such a state of mind was at the time universal, since it was not yet admitted that only candid photographic documents were valid, the cinema, in its turn, would also very naturally reconstruct topical events.[54]

This was a state of culture in which the differences between a drawing and a photograph, between an "authentic" photograph and a "tricked" one, were not essential ontological differences. Sadoul's comment on the lack of validity of these reconstructions shows how foreign this conception was to his time. They constitute a historical use and conception of photography that Sadoul, as a historian, recognizes as important to the understanding of that era but does not accept on a broader theoretical level. Interestingly, Sadoul's argument here connects photography with history. If such indistinction between authentic and reconstructed photographs seems undesirable for Sadoul, it is not so much because he holds dear to a conception of photography or cinema as entertainment or art based on its ability to serve as a trace or recorded imprint but rather because such a use of photography is incompatible with his conception of historical work. Sadoul's is a conception of photography that involves a conception of history and whose epistemological coherence risks contamination in a culture that does not clearly distinguish between authentic and reconstructed photographs. As for cinema, it is made of photographs, and photographs can always end up in history books...

53 Sadoul, *Histoire générale du cinéma*. vol. 1, op. cit., p. 330.

54 Ibid., p. 328.

But photographs can also end up in other kinds of scientific books. Charles Darwin was among the first to use photographic illustrations in a scientific volume, *The Expression of Emotions in Man and Animal*, in 1872.[55] In fact, these illustrations played a major role in the book, constituting its very subject. Darwin described, analyzed, compared, and commented at length on these portraits of various individuals of all ages, genders, and kinds, human or non-human. Darwin's project was partly based on Guillaume Duchenne de Boulogne's photographic and physiognomic research, published as *Mécanisme de la physionomie humaine ou analyse électro-physiologique de l'expression des passions* in 1862.[56] Duchenne's method consisted in electrically stimulating isolated muscles on human faces to produce facial configurations that could be read as manifestations of emotions. Duchenne's book was accompanied by an atlas presenting a series of photographs of his experiments. His tool for obtaining the expressions could, of course, be considered as some kind of trick, were it erased from the photographs; but in fact, the method was the very subject of the study. Darwin's project in this lesser-known book of his remains today as uncanny as it is fascinating — if only for the juxtaposition of man and animal in its title. Beyond Duchenne's mechanized bodies, Darwin was interested in expression itself. The book's illustrations had to show explicit and clear examples of these manifestations that the scientist would analyze. Contingent upon any given purpose and working with the limited materials at his disposal, Darwin's study used both photographs and drawings. To produce this new sort of scientific illustrations which were to be so central to the book as a whole, Darwin hired a renowned photographer, Oscar Rejlander. Rejlander was not a scientist nor was he known for having developed rigorous experimental protocols such as required for proper scientific research. On the contrary — but is this really contrariwise? — Rejlander was famous as an art photographer, one with a particular talent for composite photography and photomontage. One of his greatest successes was the allegory *The Two Ways of Life*, a combination of 32 image-elements he made in 1857. In fact, nearly all the photographs illustrating *The Expression* were staged or tricked. If the book was of considerable historical importance, later scholars have not failed to emphasize the rather questionable scientific status of that constructedness. But, according to Phillip Prodger, such a "deconstructionist" perspective,

55 Charles Darwin, *The Expression of Emotions in Man and Animal*. (London: John Murray, 1872).

56 Guillaume Duchenne De Boulogne, *Mécanisme de la physionomie humaine ou analyse électro-physiologique de l'expression des passions*. (Paris: J.-B. Baillière et fils, 1862).

laudable and productive though it may be, cannot entirely account for the historical complexity of the photographs Rejlander made for Darwin's book. Moreover, Prodger claims: "The deconstructionist view also grants special status to photographic representations that would not have been acknowledged by either Darwin or his audience."[57]

To put it simply, neither Darwin nor his reader would have understood that it ought to be required, for a photograph to be considered as scientifically valid, that it be entirely devoid of all retouching, of all "trick." The problem was for them one of accuracy, not one of "veracity": the expressions depicted by the photographs had to be accurate, and photography allowed for this accuracy, though not necessarily because it was a direct "imprint." Before being conceived as a trace, photography was perceived as a most precise kind of drawing possible, one that offered minute details that were otherwise practically unattainable through 'classical' (hand) drawing. That accuracy, combined with synchronicity — whether it affected the entire image at once or in discontinuous fashion as a result of multiple exposures — were the real advantages of photography over drawing. In short, the idea of photography at the time could accommodate image manipulations so that, to Darwin and his contemporaries, there was no essential difference between a photograph and a composite photograph. That distinction would only appear later, in a different cultural and epistemological context. As François Albera has pointed out about another genre, that of mid-nineteenth-century landscape photography, where images were routinely and widely doctored: "Far from fetishizing the moment of picture taking — the fixed instant, or even the imprint — Le Gray [granted] a decisive importance to his interventions in the studio, where he [made] use of chemistry and [proceeded] to combinations [composites]."[58]

It follows then that photography defined as a "synchronous imprint," as "a slice of the space-time continuum," is not the result of the technical constitution of the photographic dispositive: it is but a moment of its cultural and epistemological history. This means that we must distinguish our dominant conception of photography today from an earlier one that could encompass, without any contradiction, "photographic recreations" and instantaneous photography. This is not to say that the dominant conceptions of either photography or cinema have succeeded and superseded one another in a linear fashion. Various conceptions have usually co-existed with each

57 Phillip Prodger, *Darwin's Camera: Art and Photography in the Theory of Evolution*. (Oxford: Oxford University Press, 2009), pp. 223–224.

58 Albera, "Pour une épistémographie du montage: préalables," op. cit., p. 25.

other in different guises, co-present in almost any given state of culture in the modern era. Yet, if one looks at the dominant threads, it appears clearly that Méliès's work and his conception of the cinema derive from that earlier moment while developing its potentialities in specific directions. And we need to recover this moment whose alterity can elude us. Thus, undoing the temporal and spatial homogeneity of the shot, Méliès conceived of his films as blocks of "space-duration" that corresponded neither to a film frame nor to a shot, and these blocks were joined in ways that differed from "normal" film editing.

The emergence of digital cinema should lead us reevaluate this aspect of Méliès's work, whose importance in the history of ideas concerning the cinema digital protocols of image calculation and special effects may actually help us understand. It would certainly please Méliès to know that we can now begin to fully appreciate the true value of his contributions — not just his use of the substitution trick but those most difficult ones that required the use of the black background.

Bibliography

Albera, François. "Pour une épistémographie du montage: préalables," *Cinémas* 13, no. 2 (2002) : 11–32.

Altman, Rick. "Toward a Theory of the History of Representational Technologies," *Iris* 2, no. 2 (1984): 111–125.

Bergeret, Albert, and Felix Drouin. *Les Récréations photographiques* (Paris: Charles Mendel, 1893).

Bergson, Henri. *Creative Evolution*, translated by A. Mitchell (New York: Random House, 1944).

Berton, Mireille. "The 'Magism' of Cinema and Imaginary Spiritism in France at the Beginning of the Twentieth Century," *Early Popular Visual Culture* 13, no. 2 (2015): 113–133.

Chaplot, Charles. *La Photographie récréative et fantaisiste. Recueil de divertissements, trucs, passe-temps photographiques* (Paris: Charles Mendel, 1904).

Chéroux, Clément. "The Great Trade of Tricks: On Some Relations Between Conjuring Tricks, Photography, and Cinematography." In *Between Still and Moving Images*, edited by Laurent Guido and Olivier Lugon (New Barnet: John Libbey, 2012): 81–96.

Cohen-Séat, Gilbert. *Essai sur les principes d'une philosophie du cinéma I. Introduction générale: Notions fondamentales et vocabulaire de filmologie* (Paris: Presses Universitaire de France, 1946).

Darwin, Charles. *The Expression of Emotions in Man and Animal* (London: John Murray, 1872).
Dubois, Philippe. *L'Acte photographique* (Paris/Brussels: Nathan/Labor, 1983).
Duchenne De Boulogne, Guillaume. *Mécanisme de la physionomie humaine ou analyse électro-physiologique de l'expression des passions* (Paris: J.-B. Baillière et fils, 1862).
During, Elie. "Notes on the Bergsonian Cinematograph." In *Cine-dispositives*, edited by François Albera and Maria Tortajada (Amsterdam: Amsterdam University Press, 2015).
Fihman, Guy. "Le cinéma date du jour où," In *Du cinéma selon Vincennes*, edited by Département d'études cinématographiques et audio-visuelles de l'Université Paris 8 (Paris: L'Herminier, 1979): 181–188.
Fossati, Giovanna. *From Grain to Pixel: The Archival Life of Film in Transition* (Amsterdam: Amsterdam University Press, 2009).
Gaudreault, André. "Les détours du récit filmique (Sur la naissance du montage parallèle)," *Cahiers de la cinémathèque* 29 (1979): 88–107.
———. "Fragmentation et segmentation dans les 'vues animées': le corpus Lumière." In *Stop Motion, Fragmentation of Time*, edited by François Albera, Marta Braun, and André Gaudreault (Lausanne: Payot, 2002): 225–245.
Kress, Eugène. "Trucs et illusions. Applications de l'optique et de la mécanique au cinématographe." 7e conférence. *1895 Revue d'histoire du cinéma* 27 (1999): 7-32.
———. *Conférences sur la cinématographie*. vol. 1 (Paris: Comptoir d'édition de *Cinéma-Revue*, 1912).
Lefebvre, Martin, and Marc Furstenau. "Digital Editing and Montage: The Vanishing Celluloid and Beyond," *Cinémas* 13, no. 1-2 (2002): 69–107.
Malthête, Jacques. "Quand Méliès n'en faisait qu'à sa tête," *1895. Revue d'histoire du cinéma* 27 (1999): 26.
Méliès, Georges. "Les vues cinématographiques." In *Annuaire général et international de la photographie* (Paris: Librairie Plon, 1907), pp. 362–392. Reprinted as "Kinematographic Views: A Discussion." In Gaudreault, André, *Film and Attraction: From Kinematography to Cinema* (Urbana: University of Illinois Press, 2011): 133–152.
Metz, Christian. "*Trucages* and the Film," translated by François Metzler, *Critical Inquiry* 3, no. 4 (1977): 657–675.
Mitry, Jean. *Histoire du cinéma: art et industrie* (Paris: Éditions Universitaires, 1967).
Musser, Charles. *Edison Motion Pictures, 1890–1900: An Annotated Filmography.* (Washington/Porderone: Smithsonian Institution Press / Le Giornate del Cinema muto, 1997).
Sadoul, Georges. *Histoire générale du cinéma. Vol. 1: L'invention du cinéma* (Paris: Denoël, 1946).

Tortajada, Maria. "Photography/Cinema: Complementary Paradigms in the Early Twentieth Century." In *Between Still and Moving Images*, edited by Laurent Guido and Olivier Lugon (New Barnet: John Libbey, 2012): 81–96.

Turnock, Julie. "Patient Research on the Slapstick Lots: From Trick Men to Special Effects Artists in Silent Hollywood," *Early Popular Visual Culture* 13, no. 2 (May 2015): 152–173.

Turquety, Benoît. "On Viewfinders, Tape Slicers and Video Assist Systems: Questioning the History of Techniques and Technology in Cinema." In *Technology and Film Scholarship: Experience, Study, History*, edited by André Gaudreault and Santiago Hidalgo (Amsterdam: Amsterdam University Press, 2015): 239–259.

Van Lier, Henri. *Philosophie de la photographie* (Paris: Les Cahiers de la photographie, 1983).

About the Author

Benoît Turquety is Associate Professor of cinema history and aesthetics at the Université de Lausanne. He is Director of the SNF research project on Bolex and amateur cinema and of the EPIMETE/Digital Media Epistemology research axis. He is a founding member of the Material Archival Studies Network and a member of the Dispositives research group, of the Network for Experimental Media Archaeology, as well as the Technology and the Humanities project. His recent publications include *Inventing Cinema: Machines, Gestures and Media History* (Amsterdam University Press), *Medium, Format, Configuration: The Displacements of Film* (Meson Press), and *Des avant-dernières machines. Cinéma, techniques, histoire* (co-edited with Selim Krichane, L'Âge d'Homme).

9. From Trick to Special Effect: Standardization and the Rise of Imperceptible Cinematic Illusions

Katharina Loew

Abstract
Although special effects are typically thought of as tools that facilitate astonishing spectacles, for over a century their principal purpose has been to economize and enhance while blending in unnoticeably with a given filmic universe. Special effects established themselves as a routine facet of filmmaking between 1910 and 1930. Standardized compositing methods such as matte paintings, travelling mattes, or rear projection held out the prospect of an effective and cost-effective way to inconspicuously manipulate the filmic image. Taking a transnational and comparative approach, this essay examines developments in the French, German, British, and American film industries. A modified version of Christian Metz's taxonomy of special effects renders it possible to discriminate between different types of effects. By focusing on what Metz calls "imperceptible" tricks, this essay traces historical changes in applications of and discourses on special effects, as well as their impact on film style.

Keywords: Christian Metz, silent cinema, set extensions, imperceptible special effects

Special effects are usually associated with attention-grabbing spectacle.[1] However, grand sensations constitute but a small fraction of all the effects

1 I am grateful to Ariel Rogers, Sarah Keller, Tom Gunning, and Martin Lefebvre for their feedback and generous advice.

Lefebvre, M. & M. Furstenau (eds.), *Special Effects on the Screen: Faking the View from Méliès to Motion Capture*. Taylor & Francis Group, 2022
DOI 10.5117/9789462980730_CH09

employed in mainstream filmmaking for the past century. The vast majority was (and is) supposed to remain unnoticed. This was not always the case: in early cinema, tricks were primarily used to astonish audiences. Between the 1910s and the early 1930s, however, their dominant function changed. Imperceptible effects became ubiquitous, and techniques such as matte paintings, travelling mattes, and rear projection emerged as key tools for discreetly shaping the filmic universe.

To date, special-effect scholarship has predominantly focused on spectacular effects and corresponding genres.[2] Researchers have generally limited themselves to Hollywood[3] and, with the exception of mainly fan-driven explorations of sensational all-time favorites like *King Kong* (Merian C. Cooper and Ernest B. Schoedsack, 1933) or artists like Ray Harryhausen, little attention in special effects scholarship has been paid to the period before 1970.[4] Aspects of early special effects have usually been investigated in the context of the work of figures such as Georges Méliès and topics such as slapstick, stunts, action serials, and animation.[5] The technical development,

2 For major works in special-effects scholarship that are mainly concerned with spectacular effects and corresponding genres, see Kristen Whissel, *Spectacular Digital Effects: CGI and Contemporary Cinema* (Durham, NC: Duke University Press, 2014); Scott Bukatman, *Matters of Gravity: Special Effects and Supermen in the 20th Century* (Durham, NC: Duke University Press, 2003); Michele Pierson, *Special Effects: Still in Search of Wonder* (New York: Columbia University Press, 2002); Julie Turnock, *Plastic Reality: Special Effects, Technology, and the Emergence of 1970s Blockbuster Aesthetics* (New York: Columbia University Press, 2015); Lisa Purse, *Digital Imagining in Popular Cinema* (Edinburgh: Edinburgh University Press, 2019). For important work on imperceptible effects, see Stephen Prince, *Digital Visual Effects in Cinema: The Seduction of Reality* (New Brunswick, NJ: Rutgers University Press, 2012); Mark Cotta Vaz and Craig Barron, *The Invisible Art: The Legends of Movie Matte Painting* (San Francisco: Chronicle Books, 2002), pp. 31-32; Dominique Païni, "The Wandering Gaze: Hitchcock's Use of Transparencies," in *Hitchcock and Art: Fatal Coincidences*, eds. Domique Païni and Guy Cogeval, trans. Guy Conolly et al. (Montréal, QC: The Montreal Museum of Fine Arts, 2000), pp. 51-78; Laura Mulvey, "A Clumsy Sublime," *Film Quarterly*, (60) 3 (2007), p. 3; Julie Turnock, "The Screen on the Set: The Problem of Classical-Studio Rear Projection," *Cinema Journal*, (51) 2 (Winter 2012), pp. 157–162; Ariel Rogers, *On the Screen: Displaying the Moving Image, 1926–1942* (New York: Columbia University Press, 2019).

3 Exceptions include Rolf Giesen, *Special effects: Vom Spiegeleffect bis zur Computeranimation* (Ebersberg: Edition 8½, 1994); Réjane Hamus Vallée and Caroline Renouard, *Les effets spéciaux au cinema: 120 ans de créations en France et dans le monde* (Vanves: Armand Colin, 2018); Laura Lee, *Japanese Cinema Between Frames* (London: Palgrave Macmillan, 2017), pp. 17–50; Katharina Loew, *Special Effects and German Silent Film: Techno-Romantic Cinema* (Amsterdam: Amsterdam University Press, 2021).

4 See, for instance, Orville Goldner and George Turner, *The Making of King Kong: The Story Behind a Film Classic* (New York: A.S. Barnes, 1975).

5 Donald Crafton, *Before Mickey: The Animated Film 1898-1928* (Chicago: University of Chicago Press, 1993); Elizabeth Ezra, *Georges Méliès: The Birth of the Author* (Manchester: Manchester

uses, aesthetic qualities, industrial organization, and reception of early special effects remain largely unexplored.

This essay seeks to contribute to our understanding of special effects by examining their modes of production and reception in the era of the silent feature film. In particular, it tracks the establishment of standardized composite techniques, the concomitant proliferation of imperceptible effects in mainstream filmmaking, and the specialization of effects labor. This investigation is guided by a modified version of Christian Metz's well-known taxonomy of special effects. Because it allows for a differentiation of types of effects, the proposed categorization makes it possible to reconstruct in greater detail historical changes in the applications of and discourses about special effects as well as their impact on film style. Wherever possible, this essay adopts a transnational perspective, exploring conditions and trends comparatively on both sides of the Atlantic.

The establishment of special effects as a routine part of filmmaking was made possible by two interdependent developments: starting in the late 1910s, numerous special-effect techniques were patented that lent themselves to imperceptible effects, for instance, set extensions. Simultaneously, in contrast to earlier trick effects, such patented processes were not implemented by regular cinematographers or art directors but by specialized technicians. As we will see, the process of standardizing special effects was burdened with numerous fraught issues like trade secrets; patent rights; the status of the effects artist; and ideals of creativity, honesty, and realism. Nonetheless, given that standardized techniques reduced costs and specialized personnel could easily be integrated into the studio apparatus, imperceptible special effects quickly became a regular feature of mainstream filmmaking.

Imperceptible special effects call attention to the fact that cinema is a powerful means of manipulating human perception. Their study underscores to what extent virtually every mainstream film is composed of countless separate pictorial elements that are brought together by means of editing and inconspicuous image compositing to simulate a unified whole. The filmic image, which we often mistake for a mirror of nature, reveals itself as an artifice.

University Press, 2000); Jennifer Bean, "Technologies of Early Stardom and the Extraordinary Body," *Camera Obscura*, 48 (2001), pp. 8–57; Rob King, *The Fun Factory: The Keystone Film Company and the Emergence of Mass Culture* (Berkeley, CA: University of California Press, 2009); Jacob Smith, *The Thrill Makers: Celebrity, Masculinity, and Stunt Performance* (Berkeley, CA: University of California Press, 2012).

Revising Metz's Taxonomy of Special Effects

In his 1971 essay "*Trucage* and the Film," Christian Metz develops a typology of special effects that provides a useful framework for thinking about cinematic illusions.[6] In the context of narrative cinema, Metz distinguishes three types of "*trucages*" — i.e., optical effects resulting from a manipulation of filmic images — which he categorizes as "invisible," "visible," or "imperceptible." As this terminology indicates, Metz's classification of effects is based on audience perception: on whether viewers can detect and diagnose the manipulation of the filmic image. Metz is interested in the relationship between effect and viewer. The decisive question is how the spectator "reads" a given effect.[7]

In the case of the first category — "invisible *trucage*" — viewers recognize that the filmic image has been modified, but the method remains enigmatic to them. Metz points out:

> The spectator could not explain how it was produced nor at exactly which point in the filmic text it intervenes. It is invisible because we do not know where it is, because we do not see it (whereas we see a blurred focus or a superimposition). But it is perceptible, because we perceive its presence, because we 'sense' it and this feeling may even be indispensable, according to the codes, to an accurate appreciation of the film.[8]

The label "invisible" is confusing because these are usually attention-grabbing effects. Metz gives the example of *The Invisible Man* (James Whale, 1933). Even the most inexperienced viewers will assume that some type of specialized process was involved to make a man appear invisible, but they are oblivious as to how it was done.

The second type — "visible *trucage* " — remains vague, given that Metz does not define it in any detail. Unlike "invisible *trucage*," this kind does not represent a diegetic reality and is "meant to be discernible (accelerated motion, slow motion, etc.)."[9] Metz gives the examples of *Ballada o soldate* (*Ballad of a Soldier*, Grigori Chukhrai, 1959) and *Staroye i novoye* (*The General Line*, also known as *Old and New*, Sergei M. Eisenstein, 1929).

6 Christian Metz, "*Trucage* and the Film," trans. Françoise Meltzer, *Critical Inquiry*, (3)4 (Summer 1977), pp. 657–675.

7 Ibid., p. 664.

8 Ibid.

9 Ibid., p. 663.

In the former, a young soldier travels on a train and mentally evokes (by means of superimposition) a girl he has encountered shortly before. In the latter, a peasant girl and a worker literally speed up (by means of time-lapse photography) inert bureaucrats.

"Imperceptible *trucage*," the third category, is meant to be completely indiscernible. If successful, the viewers will not notice the trick at all. Metz points out that this type corresponds perfectly with "what is called the 'realistic film'."[10] His prime example is the use of stand-ins. He also invokes films co-starring Ingrid Bergman and Charles Boyer, whose difference in height was allegedly harmonized by having Boyer wear platform shoes.

Metz's is undoubtedly the best taxonomy of special effects available. However, certain aspects limit its utility as a general framework for special-effect scholarship. First, Metz describes his proposed categories in rather loose terms, using examples in lieu of precise definitions. This is particularly true in the case of "visible *trucages*," which he addresses only implicitly. Second, his association of individual processes (like superimposition or slow motion) with categories of reception is not plausible. The effects of trick techniques are always contextual. For example, depending on the situation in which it appears, a superimposed figure could represent a diegetically true ghost ("invisible *trucage*") or a character's hallucination ("visible *trucage*"). Similarly, slow motion can be evident ("visible *trucage*") or, when for instance used in conjunction with miniatures, indiscernible ("imperceptible *trucage*"). Third, Metz's categories pertain specifically to spectatorship. They classify how in fictional works, viewers perceive and apprehend manipulations of the filmic image. Metz is less concerned with the characteristics of the effects themselves or what tasks they fulfill in the context of a film or genre, which diminishes the usefulness of his taxonomy for special effects scholarship.

Finally, Metz's labels are confusing. To be sure, they characterize quite accurately his object of investigation: the ways in which viewers perceive manipulations of the filmic image in a narrative context. The *manipulation* can be "invisible," even if the resulting effect attracts attention. However, when used to describe the effects themselves, Metz's terminology becomes obfuscating. For example, both "visible" and "invisible" effects are highly visible. There is little difference in common use between the meaning of "imperceptible" and "invisible," particularly when referring to visual phenomena. Because Metz's labels are not easily manageable, they seem inexpedient as a general basis for examinations of special effects.

10 Ibid., pp. 663–664.

To make Metz's categories more useful for special-effect scholarship, they need to be expanded, refocused, and specified. Metz's emphasis of spectatorship is important, but it neglects another key dimension, namely the narrative, generic, and aesthetic features, and functions of the effects themselves. In what follows, I propose a revision of Metz's taxonomy that pays closer attention to these aspects. I seek to clarify the individual categories of special effects by developing them, providing definitions, and establishing subcategories. I will also suggest modified, less ambiguous labels. The proposed types are neither absolute nor universal, and how to classify a specific trick is ultimately always a question of interpretation. However, these new categories allow for more specificity when considering "special effects." In this essay, they will facilitate a better understanding of how uses of and attitudes toward special effects changed during the first decades of the twentieth century.

The most conspicuous attribute of the first group of effects is not that they originate in invisible manipulations of the filmic image but that they present striking visual displays. I will therefore rename this category "spectacular effects." These tricks portray characters, objects, and events that are understood to be objectively present within the fictional universe. They look believable, but it would have been physically impossible or unsafe to achieve them by conventional means. Spectators are invited to simultaneously enjoy the spectacle facilitated by these tricks and marvel at the technical feat that deceives them. We can distinguish two subcategories: first, the fantastic is staged as spectacle. Spectacular-fantastic effects have become constitutive for the horror and fantasy genres. Second, tricks are employed to represent extreme versions of reality. Spectacular-naturalistic effects facilitate horrendous car crashes, falls from high places, explosions, or encounters with wild animals. This subcategory is particularly prevalent in action, adventure, and disaster films.

The second type of effects does not simulate a believable reality. It comprises phenomena that exist outside of the objectively perceivable fictional universe. Like fantastic or naturalistic spectacles, these effects call attention to themselves. The label "visible" therefore does not sufficiently differentiate between the first and second categories. Because the latter group concerns the use of tricks to represent thoughts, intentions, attitudes, sensations, or emotions, I will label it "expressive effects." In contrast to the spectacular type, which represents fictional facts, expressive effects compel the viewer to provide interpretations such as "this effect illustrates the protagonist's hallucination" or "this effect indicates the passage of time." Again, we can differentiate two subcategories. First, there are subjective

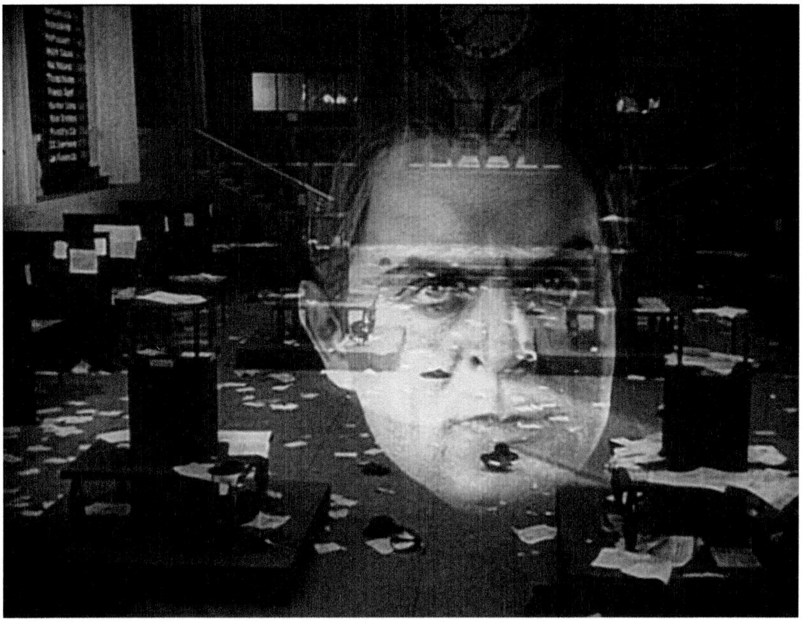

Fig. 9.1 Superimposition in *Dr. Mabuse the Gambler* (Fritz Lang, 1921/1922).

effects, which visualize what a character remembers, envisions, perceives, feels, thinks about, or dreams. Second, we have extradiegetic effects, which embody an intervention from the narrating instance. Such effects may be used to emphasize dramatically significant aspects, represent the non-representable (like sound in a silent film), or comment on characters or events, for instance by means of symbols or visual gags. For example, in *Dr. Mabuse, der Spieler* (*Dr. Mabuse the Gambler*, Fritz Lang, 1922), the eponymous villain manipulates the stock market disguised as a trader. Dramatic stock market turbulences ensue. After hours, a long shot shows the trading floor left abandoned, littered with leaflets, papers, and hats. A close-up of the mysterious trader is superimposed on the trading room and quickly dissolves to a close-up of Mabuse (Fig. 9.1). The effect announces the mastermind behind the events and simultaneously serves as an emblem of Mabuse's dominance.

I will maintain Metz's terminology for the third category of effects, which seeks to efface itself. Imperceptible effects replicate physical reality or blend in unnoticeably with a stylized fictional universe. The audience is not meant to become aware of the effect. Typically, similar effects could be achieved by conventional means, but it is more practical and/or economical to employ a specialized process. For example, in the history of cinema, countless driving scenes have been facilitated by means of rear projection and countless ceilings have been added by means of glass and matte paintings. Although the primary function of imperceptible effects is to eliminate practical obstacles, they also have much bearing on film aesthetics. In contrast to spectacular and expressive effects, which are employed in the service of narration, they shape the look of the shot. As we will see, imperceptible effects became omnipresent during the 1920s. Ever since, most tricks used in narrative filmmaking belong in this category.

Tricks in Early Cinema

The various types of trick effects can already be found in early cinema. Initially, most tricks were of the spectacular-fantastic type. For instance, in *Upside Down, or The Human Flies* (Walter R. Booth, 1899), a magician levitates a hat, disappears into thin air, and leaves his audience dancing on the ceiling. An inversed camera and backdrop made the actors appear to stand upside-down on the ceiling. Indeed, the *raison d'être* of early trick films and *féeries* was above all the display of fanciful effects.

Subjective effects are likewise prevalent in early cinema. In the first years, dreams, visions, and memories were typically represented by means of superimpositions or vignettes within the frame. In *Histoire d'un crime* (*History of a Crime*, Ferdinand Zecca, 1901), the filmmakers resorted to a theatrical device. They placed a smaller stage behind a hole in the backdrop of the main set. The prisoner's memories are represented on the insert stage above his head. In *The Little Match Seller* (James Williamson, 1902), a multiple exposure against a black background was employed to depict the freezing girl's visions. An early attempt to visualize individual perception can be found in *Dreams of a Rarebit Fiend* (Edwin S. Porter, 1906). Here, whip pans of city footage in opposite directions are superimposed on a drunk clinging to a lamppost on a rocking base. Unlike subjective effects, extradiegetic ones are rare before the 1910s. In *College Chums* (Edwin S. Porter and J. Searle Dawley, 1907), a multiple exposure composite is used to represent both sides of a phone conversation (Fig. 9.2). Here, a painted aerial view of a city on the lower part of the frame is combined with two vignettes against a black background showing characters speaking on the phone with their dialog written out in stop-motion animated letters between them.

While filmmakers clearly intended audiences to recognize fantastic and expressive effects as tricks, the situation is more ambiguous in the case of spectacular-naturalistic trick effects used in re-enactments of non-fiction subjects. In Edison's *Execution of Mary, Queen of Scots* (Alfred Clark,

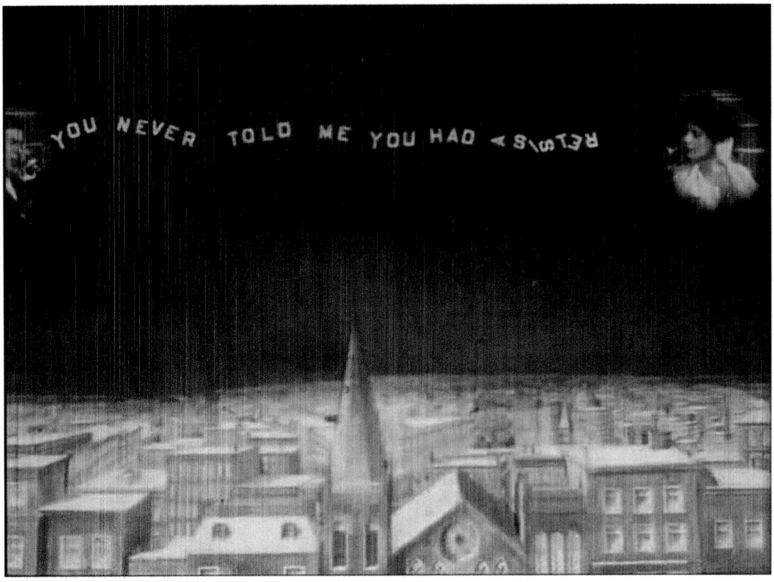

Fig. 9.2 Multiple exposure composite to represent both sides of a phone conversation in *College Chums* (Edwin S. Porter and J. Searle Dawley, 1907).

1894), for example, Mary's on-screen beheading was achieved by means of stop-action, and *A Railway Collision* (Robert R. Booth, 1900) makes use of a miniature. Did contemporary viewers recognize such effects as tricks? Regarding *A Railway Collision*, Frederick Talbot claims that "[m]any people who saw this film marveled at Paul's good fortune in being on the scene to photograph such a disaster."[11] Talbot's account obviously bears resemblance to well-worn cinema myths according to which certain audiences lacked the sophistication to distinguish between cinematic illusion and reality. Examples include the assumed panic in view of *L'arrivée d'un train en gare de La Ciotat* (*The Arrival of a Train at La Ciotat*, Lumière, 1896) or the situation portrayed in metacinematic rube films such as *The Countryman and the Cinematograph* (Robert W. Paul, 1901).[12] Nonetheless, it cannot be ruled out that some viewers mistook *A Railway Collision* for an actuality film. As Frank Kessler has shown, re-enactments of factual scenes, a common

11 Frederick A. Talbot, *Moving Pictures: How They Are Made and Worked* (Philadelphia: J.B. Lippincott Company, 1912), p. 205.

12 For a discussion of cinema's "myth of origin," see Tom Gunning, "An Aesthetic of Astonishment: Early Film and the (In)Credulous Spectator," in *Viewing Positions: Ways of Seeing Film*, ed. Linda Williams (New Brunswick, NJ: Rutgers University Press, 1995), pp. 114–133. For a discussion of rube films in American cinema, see Miriam Hansen, *Babel & Babylon: Spectatorship in American Silent Film* (Cambridge, MA: Harvard University Press, 1997), pp. 25–30.

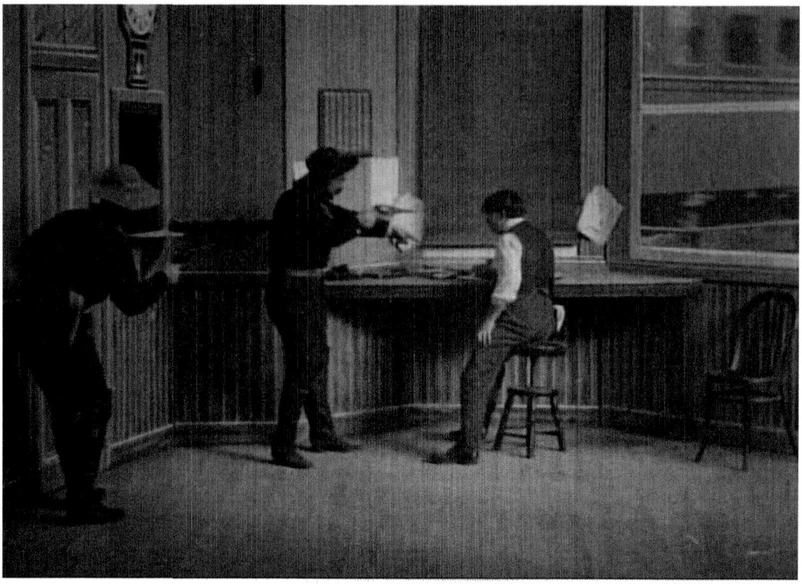

Fig. 9.3 Imperceptible effect in *The Great Train Robbery* (Edwin S. Porter,1903).

practice up until around 1907/1908, were not always distinguished from strictly documentary material.[13] Such ambiguity, which is characteristic for early cinema, can make it difficult to differentiate spectacular-naturalistic and imperceptible tricks. In the case of *A Railway Collision*, however, the situation seems clear. Contemporary newspapers treated the film discretely from both factual views and re-enactments and instead described it in the context of trick pictures: "A good class of humorous subjects was also dealt with by the machine, and many trick pictures were shown, *A Railway Collision* being especially acceptable to the audience."[14]

Imperceptible tricks constitute only a small fraction of all tricks employed in early cinema. Devising trick techniques is time-consuming, troublesome, and expensive. Early filmmakers therefore limited their use of illusions to depicting what was impossible to show otherwise. It was not until standardized techniques, which could be churned out relatively quickly and cheaply, became available that non-essential tricks — for instance, those that aimed at adding production value — became economically worthwhile. Despite constituting a considerable investment with limited proceeds, imperceptible effects can occasionally be found in early cinema, for instance in *The Great*

13 Frank Kessler, "Actualités," in *Encyclopedia of Early Cinema*, ed. Richard Abel (London: Routledge, 2005), pp. 6–7.

14 "Anglo-European Panorama," in *The Daily News* (Perth), (19 December 1903), p. 8.

Train Robbery (Edwin S. Porter, 1903), where portions of the frame representing windows or doors, such as the stationmaster's office window, were masked during the first exposure and the view outside added in a second exposure (Fig. 9.3). These effects aim at making studio interiors appear more authentic, but they do not call attention to themselves.

Trade Secrets and Ideologies

The proposed categorization of special effects casts new light on audience and industry attitudes towards special effects. As we will see, ideological considerations, which affected the different categories of effects rather unequally, became a key factor in the use or renunciation of special effects. Whether they are employed on the theatrical stage, in the darkroom, or in commercial conjuring, illusionary techniques and devices rely on concealing their workings. By the same token, most cinematic trick techniques have (at least initially) been treated as trade secrets. Technicians, producers, and exhibitors had different reasons for trying to keep methods of trick production a mystery. For technicians, specialized knowledge about trick techniques constituted a bankable advantage and improved their employability. Producers and exhibitors, on the other hand, were fearful that the disclosure of production methods would diminish audience interest. Keystone Studios, as Rob King has shown, made a virtue out of necessity and aggressively publicized the studio's secretiveness.[15] Others tried to fend off persistent disclosure requests from the public by claiming that most viewers really did not want to know. In 1912, for example, *Motion Picture Story Magazine*, an industry-controlled fan publication, explained:

> And we cannot explain some pictured tricks, not only because many of these are trade secrets, but because the explanation would spoil the effect for many others. You may belong to those who want to get behind the scenes and see how it's done, but it's a lot nicer to sit out front and see the effect without realizing that the fairy flies because a man in a dirty undershirt is pulling on the other end of a rope she's tied to. The majority want to believe that the fairy flies, even tho [sic] she doesn't and the will of the majority rules.[16]

15 King, op. cit., pp. 182–190.

16 "A Little Letter to the Inquisitive by the Inquiry Editor," in *The Motion Picture Story Magazine* 3 (April 1912), p. 164.

Because the industry was determined to keep trick techniques top-secret, few details ever reached the public. Only occasionally did individual producers break the code of silence for publicity purposes. In 1908, Léon Gaumont invited Gustave Babin to his studio. The resulting two-part article "Les Coulisses du Cinématographe," which appeared in the respected journal *L'Illustration,* explained spectacular effects in several Gaumont productions, richly illustrated with production photographs and diagrams.[17] One year later, J. Stuart Blackton and Albert E. Smith employed the same approach to promote Vitagraph's *Princess Nicotine; or, The Smoke Fairy.* Six weeks before the film's release, they provided *Scientific American* with a detailed account of the mechanics of that film's spectacular effects, again with numerous illustrations.[18] Both articles caused a sensation. They were reprinted nationally and internationally and became the main source for countless subsequent publications on trick films. Illustrations from Babin's article even came to adorn a series of French and German collection cards advertising Liebig's Meat Extract; the explanations of the tricks were given on the reverse side. The unusually long and detailed section on tricks in British science writer Frederick A. Talbot's popular *Moving Pictures: How They Are Made and Worked,*[19] which subsequently became a reference work for curious spectators, relied heavily on Babin and the *Princess Nicotine* article in addition to the author's personal connection to Robert William Paul. However, the film industry on both sides of the Atlantic did not take kindly to the revealing of trade secrets. As Donald Crafton has shown, Babin's revelations were met with sharp criticism from within the French film industry.[20] In the same vein, an American exhibitor complained in the trade press about

> producers [who] invite people into their places to show them how pictures are made. It takes all the mystery out of the thing. And time and again in my theaters I have heard young folks pooh-pooh scenes where trick photography was employed. They had seen how it was done so that it no longer interested them.[21]

17 Gustave Babin, "Les coulisses du cinématographe, " *L'Illustration* 3396 (28 March 1908), pp. 211–215 and 3397 (4 April 1908), pp. 238–242. Reprinted in *Les Dossiers de* l'Illustration: *Le Cinéma* (Paris: SEFAG 1987), pp. 22–32.

18 "Some Tricks of the Moving Picture Maker," *Scientific American* (26 June 1909), pp. 476-477, 487.

19 Talbot, *Moving Pictures*, op. cit., 1912.

20 Crafton, *Before Mickey*, op. cit., pp. 29–30.

21 "Split Reel Notes' for Theater Men," in *Motography* 5 (4 August 1917), p. 262.

Initially, trick effects were approached in a similar fashion on both sides of the Atlantic. However, by the 1910s, the discourse began to diverge noticeably. In the United States, trick photography was increasingly framed as incompatible with a new American "aesthetic of authenticity." As Richard Abel and Richard Slotkin have shown, this prescription for American cinema was part of an emergent and highly influential form of "virile" American culture with xenophobic and supremacist undercurrents.[22] Red-blooded American "realism," with its emphasis on wholesome storytelling, was construed as an antithesis to the visual spectacle and artifice of French magic films. Tricks, an essential element of the most recognizable French genres, were publicly discredited and portrayed as inconsistent with American tastes: "The public wants realism in pictures, not papier mache [sic] effects and trick photography."[23] The discussion about cinematic illusions also had strong moralistic undertones. Tricks were often portrayed as ultimately fraudulent and in discordance with praiseworthy honesty and moral integrity. Consequently, trick photography was, supposedly already in 1914, a thing of the past. *Motion Picture Magazine* pointed out: "Trick photography was also indulged in in the olden days. Today, scenarios calling for trick photography are barred by nearly all manufacturers. The tendency has been to portray life, not the fanciful affairs called 'magic pictures' a decade ago."[24]

However, trick photography continued to be employed consistently in American cinema. Nonetheless, because of the American fixation on "realism," fantastic films and therefore spectacular-fantastic effects became indeed rare. Tricks were mainly confined to two, less prestigious genres to which the "realism" requirement did not apply: slapstick comedy and spectacular melodrama. They were typically of the spectacular-naturalistic type. In slapstick comedies, extradiegetic effects were also frequently employed, specifically for visual gags. Many comedy and serial producers used trick effects as a selling point and openly advertised the "splendid trick photography"[25] of their films or the fact that their "authors and directors [...] have studied these possibilities for years and have gone further into trick photography than has been dreamed of before."[26] Simultaneously,

22 Richard Abel, *The Red Rooster Scare: Making Cinema American, 1900–1910* (Berkeley, CA: University of California Press, 1999), pp. 122–126; Richard Slotkin, *Gun Fighter Nation: The Myth of the Frontier in Twentieth-century America*, (New York: Atheneum, 1992), pp. 156–193.

23 "Pauline Periled in Auto Race," in *Motography* 16 (17 October 1914), p. 525.

24 Irving Crump, "Interesting Picture Figures," in *Motion Picture Magazine* (August 1914), p. 113.

25 Advertisement for the serial *Our Mutual Girl* (1914) in *Reel Life* 9 (14 November 1914).

26 Advertisement for the Black Diamond comedy *Nearly a Deserter* (United States Motion Pictures Corporation/Paramount Pictures, 1916) in *Moving Picture World* (16 September 1916).

it also became common among comedy and serial producers to publicly disavow the use of trick effects: "George Larkin wants his admirers to know that the thrills he creates for the screen are strictly on the level — there is no trick photography or quick changes there."[27] While producers were acutely aware that spectacular effects continued to have great audience appeal, the dominant trade discourse discounted tricks as old-fashioned, childish, and dishonest.

The bifurcation in American cinema between "straight" respectable films on the one hand and trivial trick genres on the other had no parallel in Europe. There was no insistence on cinematic "realism," nor were tricks discredited. Instead, they were valued as tools for creative experiments with the cinematic apparatus. Marina Dahlquist has pointed out that even at a time when the popularity of trick films and *féeries* in France subsided after 1909, their reception remained positive.[28] In fact, tricks occupied an important position in the critical discourse about film art. As I have shown for the German context, trick technology served as key evidence that the medium of film could provide new scope for the imagination and for *creating* images rather than simply reproducing them.[29]

Tricks in the 1910s

The role of tricks in silent feature films has received little attention in film scholarship, and the 1910s are virtually uncharted territory. Tricks — ranging from spectacular-fantastic and spectacular-naturalistic to imperceptible, subjective, and extradiegetic — featured prominently in several genres that became major assets of European film industries of the 1910s. The action-driven short comedies of performers like Max Linder (Max), André Deed (Boireau/Cretinetti), Raymond Frau (Kri Kri), Ernest Bourbon (Onésime), Marcel Fabre (Robinet), and Ferdinand Guillaume (Tontolini/Polidor), which had a strong bearing on the development of American slapstick, relied on spectacular and expressive effects as a standard feature. Following Éclair's effect-filled and hugely successful *Zigomar* (Victorin Jasset, 1911), crime thrillers such as Gaumont's *Fantômas* series (Louis Feuillade, 1913–1914),

27 "Who's Who and Where," *Film Fun* 335 (February 1917), n.p.

28 Marina Dahlquist, *The Invisible Seen in French Cinema before 1917* (Stockholm: Aura förlag, 2001), p. 23.

29 Katharina Loew, "The Spirit of Technology: Early German Thinking about Film," in *New German Critique*, 122 (Summer 2014), pp. 129–148.

Itala's *Tigris* (Vincenzo Denizot, 1913), Éclair's *Protéa* (Victorin Jasset, 1913), or Nordisk's *Dr. Gar-el-Hama* series (Eduard Schnedler-Sørensen and Robert Dinesen, 1911–1918) frequently employed particularly extradiegetic trick effects. The creators of *Der Student von Prag* (*The Student of Prague*, 1913), director Stellan Rye, author Hanns Heinz Ewers, cinematographer Guido Seeber, and actor Paul Wegener, fostered the development of a tradition of fantastic film in Germany and produced at least eight more feature-length fantastic films in 1913/1914.[30] On his own following the death of Rye and the absence of Ewers and Seeber due to the war, Wegener continued to cultivate the fantastic genre during the war with influential trick-heavy films like *Rübezahls Hochzeit* (*Rübezahl's Wedding*, Paul Wegener, 1916) and *Der Yoghi* (*The Yogi*, Paul Wegener, 1916; considered lost). Partly because it allowed for playful experiments with the cinematic apparatus, fantastic subject matter remained popular in the German cinema throughout the silent era.

The most astounding array of special effects during the 1910s, however, was displayed in Italian historical epics like *L'Inferno* (*Dante's Inferno*, Francesco Bertolini and Adolfo Padovan, 1911), *Gli ultimi giorni di Pompeii* (*The Last Days of Pompeii*, Mario Caserini, 1913), and most prominently *Cabiria* (Giovanni Pastrone, 1914), which left audiences and critics "speechless with awe."[31] Yet the Italian epics are not only noteworthy on account of their extraordinary spectacle, they also anticipated future developments in special-effect production. They increasingly employed techniques that allowed filmmakers to extend sets and actual landscapes with models and paintings, both for spectacular-naturalistic effects (such as the eruption of the Etna in *Cabiria*) and imperceptible effects (for instance to represent the walls of the city of Dis across the Styx in *L'Inferno*).

Italian studios were also ahead of their time in terms of professional specialization. In 1912, Itala offered Segundo de Chomón a permanent position as a trick specialist: "For the development of our studio we need an operator specializing in trick films, who can also provide trick films for scenes by other directors."[32] More research is needed to reconstruct the

30 *Ein Sommernachtstraum in unserer Zeit* (*A Midsummer Night's Dream*, Stellan Rye, 1913); *Kadra Sâfa* (Stellan Rye, 1913); *Die Eisbraut* (Stellan Rye, 1913); *Die Augen des Ole Brandis* (*Old Brandis's Eyes*, Stellan Rye, 1913); *Erlkönigs Töchter* (*Erlking's Daughters*, Stellan Rye, 1914); *Der Verführte* aka *Geheimnisse des Blutes* (*The Seduced*, Max Obal, 1913); *Das Haus ohne Tür* (*The House Without a Door* , Stellan Rye, 1914) and *Der Golem* (*The Golem*, Henrik Galeen and Paul Wegener, 1914). All are considered lost.

31 "World's Greatest Film is Screened," in *Motography* 2 (11 July 1914), p. 38.

32 See Itala's offer letter to Chomón, quoted in Joan M. Minguet Batllori, *Segundo de Chomón: The Cinema of Fascination* (Barcelona: Biblioteca de Catalunya, 2010), p. 131.

emergence of a distinct position of studio trick technician, yet Itala's hiring of Chomón, who worked on several of the studio's effects-heavy feature productions including *Padre* (Dante Testa and Gino Zaccaria, 1912) *Tigris*, *Cabiria*, and *Il fuoco* (*The Fire*, Giovanni Pastrone, 1916), may well have been one of the earliest initiatives in this regard.

From what we know, the creators of special effects during the 1910s were rarely studio-employed experts during the 1910s. Early trick technicians like Guido Seeber and Fred W. Jackman, chief cameraman at Mack Sennett, were cinematographers whose trick technological expertise attested to their exceptional mastery of the camera. In the United States, the low esteem of special-effects work also found expression in renowned cinematographers like Billy Bitzer delegating occasional trick work to an assistant.[33] During the 1910s, trick expertise was typically seen as an additional skill rather than a standalone specialization. Even Chomón, whom Itala recruited as a trick specialist, was ultimately hired as a "director and operator."[34]

Towards Standardized Trick Techniques

During the 1920s, the number of tricks employed in mainstream filmmaking increased exponentially. This was largely the result of a growing prevalence of imperceptible effects. Newly developed methods allowed filmmakers to seamlessly integrate live action with models, photographs, paintings, and eventually also filmed footage. These techniques were typically patented, which made them publicly available and encouraged technicians to specialize. As we will see, the most influential special-effects techniques of this era did not originate in studio workshops but were developed by autonomous artists. However, studios soon hired specialized technicians and began to establish dedicated special-effects production units. Standardized techniques were identified as an effective and economical way to inconspicuously manipulate the filmic image. As a result, imperceptible effects, which until then had not been a regular feature in mainstream filmmaking, became ubiquitous on both sides of the Atlantic.

A limited number of techniques was employed on a routine basis: initially miniatures and glass shots, subsequently also matte paintings, the Schüfftan process, and travelling matte techniques, and finally, starting in the early 1930s, rear projection and optical printing. The foremost pioneers of

33 Karl Brown, *Adventures with D.W. Griffith*, (New York: Farrar, Straus and Giroux, 1973).
34 See Chomón's contract with Itala, quoted in Batllori, op. cit., p. 132.

composite methods that came to be primarily used for set extensions, i.e., for imperceptible effects, were Norman Dawn, Ferdinand Pinney Earle, and Walter L. Hall in Hollywood and Walter Percy Day and Eugen Schüfftan in Europe. At the same time, numerous others — both independent inventors and studio employees like Hollywood art directors Ben Carré and Irvin J. Martin, British set designer Edward Charles Rogers, and German-Hungarian artist Stefan Lhotka — developed set extension methods. However, in the absence of patented processes, their contributions are often difficult to ascertain: an overwhelming majority of films is lost, only the most salient effects were ever mentioned in trade and fan publications, and set extensions are by nature inconspicuous. The travelling matte processes by Frank D. Williams and C. Dodge Dunning and rear projection replaced (rather than extended) the on-set decorations and were mainly used for imperceptible effects. Again, a range of technicians worked on (and sometimes patented) analogous techniques,[35] yet the early history of travelling matte processes and rear projection is still poorly understood and warrants further investigation.[36]

Pioneers of Standardized Techniques in the United States

Today, the best-known pioneer of techniques primarily used for imperceptible effects is Norman Dawn (1886–1975). Dawn's role has been widely recognized for two reasons: he kept a meticulous record of the effects he created and found a vocal supporter in special-effects historian Raymond Fielding. As a result, and although claims about original uses in special effects are usually not tenable, Dawn has often been cited as the "inventor" of glass shots and matte paintings. Both techniques employ glass paintings to modify the visual appearance of a live-action scene. Glass shots are recorded in one exposure, matte paintings in two. Dawn, who according to his own account had been inspired by Georges Méliès and his own work as a still photographer,[37] first used a glass painting in his scenic short *Missions of California* (1907). Many of Dawn's films feature picturesque and exotic

35 See, for instance, Max Handschiegl, 1932, *Process of Making Double Exposures*, US Patent 1,840,669, issued 12 January 1932; and Max Handschiegl, 1932, *Trick Method of Producing Composite Negatives*, US Patent 1,840,670, issued 12 January 1932. Handschiegl applied for these patents in 1923 and 1926.

36 For important work on early rear projection, see Rogers, *On the Screen*, op. cit.

37 Cotta Vaz and Barron, op. cit., pp. 31–32; George E. Turner, "The Evolution of Special Effects," in *The ASC Treasury of Visual Effects*, ed. George E. Turner (Hollywood, CA: ASC Holding Co., 1983), p. 26.

locales, captured during his frequent travels from Alaska or the Great Barrier Reef to Sumatra or Bolivia and augmented by numerous composite shots. Dawn devised and employed imperceptible effects primarily for aesthetic reasons. For instance, for a film commissioned by the amusement park attraction *Hale's Tours of the World* (1907), Dawn used a glass shot to add a Mayan temple to a scene of bathing Mexican women to render the scene more picturesque.[38]

In 1911, Dawn began working in Hollywood, yet he remained an outsider in the American film industry. Characteristically for early special-effect artists, Dawn developed trick techniques for use in his own films and subsequently began to offer his services to other filmmakers. Between 1917 and 1921, he was under contract with Universal where he produced effects for about 20 films. Most of them Dawn directed himself, but he occasionally also created effect shots for films of other directors, most notably for *Blind Husbands* (Erich von Stroheim, 1918), *The Kaiser, the Beast of Berlin* (Rupert Julian, 1918), and *Girl in the Dark* (Stuart Paton, 1918). As a special-effects artist, Dawn seems not to have been in high demand at the time, which suggests that only a minority of producers recognized the potential of set extension techniques.

In June 1918, Dawn received a patent for a multiple exposure matte painting technique.[39] However, he ultimately failed to defend his patent in a 1922 lawsuit that antagonized the entire film industry.[40] It was generally agreed that Dawn had "merely succeeded in gaining a patent on an idea which even the real cinema pioneers [...] considered long before 1918 as being common property of the motion picture industry."[41] Such claims were extremely common in patent litigation lawsuits at the time. Even though few inventions are truly unprecedented, filmmakers who had already used a technique before it was patented also risked being sued for patent infringement.

A patent grants the inventor exclusive rights in exchange for the public disclosure of his or her invention. These rights prevent others from making, using, selling, or distributing the patented invention without permission. Between the 1910s and the 1930s, a patenting mania seized the special-effects

38 "Hale's Tours, Effect No. 9 (1907)," in *Norman O. Dawn Collection of Special Effects Cinematography*, Harry Ransom Center, The University of Texas at Austin, Austin, TX.

39 Norman O. Dawn, *Cinematographic-Picture Composition*, US Patent 1,269,061, issued 11 June 1918.

40 "Producers Line-Up: Director's Suit Calls Up Strong Opposition," in *Los Angeles Times* (17 August 1922): II11.

41 "Attempts to Control Double Exposure Method," *American Cinematographer* 6 (September 1922), p. 4. For a comprehensive account of Dawn's lawsuit, see Cotta Vaz and Barron, pp. 54–59.

trade, followed by a deluge of patent litigation lawsuits. The situation was particularly precarious in the United States, where lengthy legal battles over special-effect patents even came to endanger the existence of major studios like Warner Bros. and Paramount.[42] In 1936, after years of arbitration, the MPPDA ended these patent wars with a compromise: studios agreed to pool their special-effects patents and make them available to the entire industry.[43]

While special-effects patent litigations became a menace for producers, patenting first facilitated the establishment of standardized special-effects techniques and consequently the increasing pertinence of imperceptible effects in commercial filmmaking. It allowed technicians to profit from innovations beyond improving their personal employability. Patenting turned special effects into a commodity. Dawn ultimately failed to enforce his patent. It took until May 1924 for the case to be dismissed. In 1925, Dawn's patent was purchased by the Motion Picture Producers Association, which made it available to its members.[44]

The defendant in Dawn's patent litigation was his arch-rival, the painter and poet Ferdinand Pinney Earle (1878–1951). In contrast to Dawn, who mostly worked on shorts and smaller-scale productions, Earle created matte shots for more prominent films during the late 1910s and early 1920s, including *The Blue Bird* (Maurice Tourneur, 1918), *Daddy-Long-Legs* (Marshall Neilan, 1919), *The Miracle Man* (George Loane Tucker, 1919), *Robin Hood* (Allan Dwan, 1922), and *Ben Hur: A Tale of Christ* (Fred Niblo, 1925). Having received his training in Europe, Earle joined the film industry in 1915 as a title artist. Simultaneously, he began experimenting with techniques that combined paintings and live action, which he termed "motion painting."[45] Despite multiple press reports about motion painting (that testify to Earle's talent for self-promotion), the process's principles remain obscure. Media stories indicate the use of either multiple exposure[46] or multiple printing,[47] but it is unclear whether Earle always employed mattes. The most detailed

42 Birk Weiberg, "Roy J. Pomeroy, Dunning Process Co., Inc., and Paramount Publix Corporation vs. Warner Bros Pictures, Inc., Vitaphone Corporation, and Frederick Jackman: How the Movie Industry Turned to Rear Projection," unpublished conference paper, Society for Cinema and Media Studies Annual Conference, Seattle, WA, 2014.

43 William Stull, "Producers Pool Composite Process Patents," in *American Cinematographer*, 11 (November 1936), p. 461.

44 "300 Important Cameramen Becoming Recognized as Photographic Marvels," in *Variety* (4 January 1928), pp. 6, 13.

45 See, for instance, "Metro Uses Motion Painting," in *Motion Picture News* (6 April 1918), p. 2065.

46 Jeanne Redman, "Art Effects in Cinema," in *Los Angeles Times* (11 August 1920), p. III4.

47 Gene Copeland, "In a Persian Garden," in *Picture Play Magazine*, 4 (June 1920), pp. 34–35; 92.

account by Earle himself describes a single exposure composite technique akin to the glass shot, which however makes use of canvas instead of glass.[48]

Given that Norman Dawn sued Earle for infringement of a patent that describes a multiple exposure matting technique, it can be assumed that Earle worked with a variety of composite techniques. The term "motion painting" most likely does not refer to a single process but rather to different related set-extension techniques. In 1919, Earle began preparations for his magnum opus *The Rubaiyat of Omar Khayyam* (completed in October 1921), an eight-reeler with allegedly 500 motion paintings.[49] This number cannot be verified since *Omar Khayyam* is considered lost. However, even a fraction would imply that *Omar Khayyam* was indeed a milestone in the history of special effects: no other production of the silent era featured a comparable number of composite shots. It was likely the first time that a feature-length film consisted predominantly of composite shots. It was not until the 1930s that standardized techniques, particularly rear projection, made such extensive use of composite shots economically viable.

Earle employed motion paintings simultaneously for economic and aesthetic reasons: "Earle paints on a piece of academy board about 14 by 20 inches a castle, say, that would cost $50,000 to build as a set. But the painting in composition and lighting reaches planes of artistry impossible with an actual set."[50] For Earle, imperceptible effects not only reduced production costs by substituting for straight photography, they amounted to an aesthetic improvement, allowing him to achieve images that would have been impossible without the use of trick photography.

With funding from Pittsburgh industrialist Theodore Ahrens, Earle formed the production company The Rubaiyat, Inc. and an art studio that employed about a dozen matte painters — an unparalleled number throughout the silent era.[51] Notably, it was not one of the established studios but an independent producer who first utilized standardized trick techniques on an industrial scale and employed droves of specialized matte artists. Earle harbored grandiose plans for future projects that would include motion-painting-based film adaptations of Wagner's *Der Ring des Nibelungen* (*Ring of the Nibelung*), Goethe's *Faust*, Homer's *Iliad*, Milton's *Paradise Lost*, and

48 Ferdinand Earle, "Screen Renaissance Through Motion Painting," in *The Blue Book of the Screen*, ed. Ruth Wing, (Hollywood: Pacifict Gravure Company, 1924), pp. 345–348.

49 "Omar Khayyam Bows to Screen," in *The Lewiston Daily Sun* (16 August 1922), p. 4.

50 Earle, "Screen Renaissance," op. cit., p. 345.

51 "Earle to Make Artistic Production of *The Rubaiyat of Omar Khayyam*," in *Exhibitors Herald* (2 July 1921), p. 106.

Dante's *Divina Commedia* (*Divine Comedy*).[52] None came to fruition. Nor did *Omar Khayyam* ever appear before the public. Starting in late 1921, Earle found himself entangled in a legal battle with Ahrens and subsequently in the patent dispute with Dawn.[53] Meanwhile, his brother William P.S. Earle utilized motion paintings in his lost epic *The Dancer of the Nile* (1923). The *Los Angeles Times* reported that the film made use of "three different processes and four kinds of paintings,"[54] which supports the assumption that a number of techniques were grouped together under the label "motion painting."

It was not until 1925 that Ahrens, bypassing Earle, released *Omar Khayyam* in a truncated version under the title *A Lover's Oath*. The *Los Angeles Times* critic Anthony Anderson called the film a "desecration" and a "horrid butchery" of Earle's original work.[55] In fact, Anderson had seen footage of *Omar Khayyam* in 1921 and recalled: "The film, uncut, unassembled, untitled, held us entranced for hours, a thing of amazing beauty, truly a moving picture, or rather an endless gallery of moving pictures, each more lovely than its predecessor."[56] In an industry that considered "realism" as an intrinsic value, Earle employed standardized special-effect techniques to aestheticize the filmic image.

To what extent Earle's approach to special effects left traces in Hollywood remains to be determined. However, although his career as a producer-director ended in failure, he at least indirectly influenced the future development of matte and glass painting practices in Hollywood. In contrast to Dawn, who worked solitarily, Earle served as mentor to influential special-effects artists like Paul Detlefsen, in charge of glass and matte paintings at RKO and later Warner Bros., and Robert Sterne, illustrator in Disney's story department. Earle did not contribute to the establishment of standardized special-effects techniques by patenting an influential process. Instead, he demonstrated that independent producers could employ standardized special-effects techniques successfully on a grand scale to simultaneously economize and increase production value. However, like Dawn, Earle had difficulties subjecting himself to industry hierarchies and left Hollywood permanently in the mid-1920s.

52 "Norse Saga in Pictures," in *Los Angeles Times* (22 August 1920), p. 23.

53 For the legal controversies that ultimately prevented the release of *Omar Khayyam*, see David S. Shields, *Still: American Silent Motion Picture Photography* (Chicago: University of Chicago Press, 2013), pp. 224–226.

54 "Painted Sets Take Place of Costly Fabric," in *Los Angeles Times* (23 December 1923), p. 25.

55 Anthony Anderson, "*Omar Khayyam* A Picture Gallery," in *Los Angeles Times* (20 December 1925), p. III 44.

56 *Ibid.*

Very few facts have been established about the third Hollywood-based pioneer of set-extension techniques, Walter Limond Hamilton Hall.[57] Even though he worked exclusively in the United States, Hall was better known in Europe, where his technique for creating set extensions had significant impact on the leading film industries. Together with the Schüfftan process, the Hall process was instrumental in establishing imperceptible effects as a standard feature of European filmmaking. A British citizen born in India, Hall attended school in England and subsequently immigrated to the United States, presumably in the mid-1890s.[58] Having worked for scenic companies that provided scenery and stage equipment for theaters and schools, he joined the film industry as the art director on *Intolerance* (D.W. Griffith, 1916). Karl Brown, the camera assistant on the film, offers the only known description of Hall. In his memoirs, he depicts "Spec Hall" as an eccentric man of extraordinary talents and "two all-absorbing interests: the science of perspective and an alert fearfulness for the safety of his teenage daughter."[59] The spectacular Babylon set of *Intolerance* was constructed full-scale, but Hall simultaneously began developing a matting process that he patented in 1921.[60] The patent details how to match opaque and glass paintings, photographs and three-dimensional hanging miniatures to a live-action scene. It is unclear how the Hall process was received in Hollywood and to what extent it ended up being employed in the United States. Following the publication of his patent, Hall's traces become sparse. He appears to have remained active in the film industry as a special-effects artist during the 1920s. Hall reportedly worked for Cecil B. DeMille[61] and designed glass paintings and miniatures at Paramount.[62]

Pioneers of Standardized Techniques in Europe

Unlike most American special-effect inventors at the time, Hall patented his process not only in the United States but also in five leading European film

57 For Hall's probable date of death, see "Obituary," in *Los Angeles Times* (6 December 1932), p. 16.

58 *Wellington College Register: 1859–July 1905* (Thomas Hunt, Wellington College, Berkshire, 1906), p. 139.

59 Brown, op. cit., p. 151.

60 Walter L. Hall, 1921, *Method of Making Pictures*. US Patent 1,372,811, issued 29 March 1921. Hall subsequently patented his invention also in France, Germany, Austria, and Denmark.

61 Earl Theisen, "The Evolution of the Motion Picture Story,Parts I and II" in *The International Photographer* (April 1936), pp. 16–17; 22–23 and (May 1936), pp. 12–13, 27.

62 "Rear Projection Big Advantage," in *International Photographer* (April 1938), pp. 30–33.

nations. There, the publication of his patent was met with a large response. In countries like France, Britain, and Germany, set-extension techniques were previously not widely known. Under the name "Hall process," they quickly became commonplace. Nonetheless, many filmmakers also criticized glass and matte painting techniques for being evocative of "painted canvas" and unfit for conveying a sense of depth.[63] In contrast to their American colleagues, European filmmakers preferred three-dimensional set-extension techniques such as foreground models.

The only European country where painted set extensions thrived during the 1920s was France. This flourishing was owed to the work of Walter Percy Day (1878–1965), the *doyen* of European matte art. During this period, standardized special effects were often tied to one independent artist or workshop in a given national film industry; they had not yet become a common industry standard. In France, glass and matte painting techniques were pioneered by Day's workshop. By his own account, Day first employed the Hall process at Maurice Elvey's and Fred Paul's Ideal Company in the UK[64] before moving to France in 1922, where he worked for a firm that commercialized the Hall process.[65] He created his initial glass paintings for *Les opprimés* (*Flanders Under Philip II,* Henry Roussel, 1923) and subsequently worked on approximately 35 French productions, including *Nana* (Jean Renoir, 1926), *Napoléon* (Abel Gance, 1927), and *Au bonheur des dames* (*The Ladies' Delight,* Julien Duvivier, 1930).[66] Seeking to refine the Hall process, Day independently developed his own matte painting method that became known in France as *procédé Day* (the Day process). Upon his return to the UK in 1932, Day established a reputation as "the magician of British films."[67] Day's glass and matte paintings facilitated some of the most visually arresting British productions of the 1930s and 1940s like *Things to Come* (William Cameron Menzies, 1936), *The Thief of Bagdad* (Ludwig Berger, Michael Powell,

63 See, for instance, Robert Chessex, "F.W. Murnau dreht Faust. Filmstadt Berlin um 1925," in *Film Bulletin*, 153 (April/May 1987), p. 59.

64 Walter Percy Day, "The Origin and Development of the Matte Shot," in *The Photographic Journal* (October 1948), pp. 209–211.

65 Réjane Hamus-Vallée mentions that the file on Jean Renoir's *Nana* at the Cinémathèque française contains an exchange with a Paris-based company named "Hall's Process," which employed Walter Percy Day at the time. See Réjane Hamus-Vallée, "Peindre comme un cinéaste. Pour une esthétique du Matte Painting." Thèse d'HDR (Université Paris III Sorbonne Nouvelle, 2013), p. 59.

66 Susan Day, "Walter Percy Day: Biography," http://www.walterpercyday.org (accessed 9 October 2020).

67 Egon Larsen, "Here is the Inside Story of the Magician of British Films," in *Cavalcade*, 21 (31 May 1949), p. 9.

and Tim Whelan, 1940), *A Matter of Life and Death* (Michael Powell and Emeric Pressburger, 1946), and *Black Narcissus* (Michael Powell and Emeric Pressburger, 1947). Like Earle in the United States, Day ran an independent matte painting workshop where he trained numerous special-effects artists, including his sons Thomas and Arthur Day, George Samuels, Wally Veevers, as well as his stepson and future Disney matte artist Peter Ellenshaw.

Day, a special-effects artist rather than an entrepreneur, created made-to-order glass and matte paintings in his private workshop. Another European pioneer, in contrast, set new standards for the commercialization of special-effects processes. The Schüfftan process, developed by painter and future cinematographer Eugen Schüfftan (1886–1977), became the most commonly used compositing technique in European cinema during the 1920s and 1930s.[68] It relies on partially transparent mirrors for combining live-action footage with a variety of visual components, including photographs, models, and moving images. Schüfftan originally intended his process as a tool to "visualize the imagination,"[69] to represent spectacular-fantastic effects like magical transformations and extreme size differences as between Gulliver and the Lilliputians. However, the technique's success ultimately resulted from its suitability for set extensions.

Schüfftan was arguably the first special-effects technician to commercialize his invention globally and on a large scale. Between 1922 and 1930, he applied for over 40 patents in at least eight countries for variations of his process. Universum Film AG (Ufa), Universal Pictures, and British National Pictures acquired licenses to commercialize the patents in Germany, North America, and the rest of the world respectively. Specialized companies in Germany and in the UK offered made-to-order Schüfftan shots to filmmakers. The Schüfftan process was used in many high-profile European films, including *Metropolis* (Fritz Lang, 1927), *Napoléon*, *Blackmail* (Alfred Hitchcock, 1929), and *Münchhausen* (Josef von Báky, 1943), as well as numerous smaller-scale productions. While European critics enthusiastically praised the seemingly unlimited potentials of the new technique, the Schüfftan process was met with considerable resistance in the United States. During the 1920s, the technique was utilized only in a handful American productions, including *Love Me and the World is Mine* (E.A. Dupont, 1927)

68 See also Katharina Loew, "Magic Mirrors: The Schüfftan Process," in *Special Effects: New Histories, Theories, Contexts*, eds. Dan North, Bob Rehak and Michael S. Duffy, pp. 62–77 (London: BFI, 2015).

69 Gertrud Isolani, "Gespräch mit Eugen Shuftan," in *Basler Nachrichten* (19 October 1965), p. 9.

and Into Her Kingdom (Svend Gade, 1926). As they had done with Dawn's matte painting technique, American commentators called into question the novelty of the Schüfftan process. Carl Louis Gregory, for instance, claimed that what "lately has been heralded as a wonderful German invention under the name of the Schuefftan process [...] is antedated by several American users."[70] Gregory and others may well have used mirrors for compositing previously. However, attempts to challenge the Schüfftan patents in court were invariably unsuccessful.

Despite its prevalence in Europe, the Schüfftan process also encountered problems there. Due to mismanagement, Spiegeltechnik GmbH & Co., the German licensor of the Schüfftan patents, ran into financial difficulties. Although Ufa had invested significant sums in the technique, the studio underutilized it and eventually divested itself of Spiegeltechnik. Rumors circulated that Ufa employees had sabotaged the Schüfftan process, possibly for fear of standardization and outsourcing.[71] The introduction of sound, on the other hand, sparked new interest in the Schüfftan process. As a staff member of Spiegeltechnik pointed out, the "'Talkies' [...] increased the demand for the process tremendously, because it was possible to portray large scenes in a small compass and defeat the bugbear of echo in the microphone."[72] Eventually, however, the larger European studios shifted to high-tech composite systems (above all rear-projection) like their American competitors. The Schüfftan process consequently only played a minor role in post-war special effects but continued to be employed where resources were scarce: in low-budget film production, television, and amateur filmmaking. Even in the digital age, it has been occasionally employed.[73]

Travelling Mattes and Rear Projection

Glass and matte paintings, foreground models, and the Schüfftan process allow for an imperceptible integration of live action with models, photographs, paintings, and even moving images. However, they do not permit actors to enter the "virtual" part of the frame. Starting in the

70 Carl Louis Gregory, "Trick' Photography Methods Summarized," in *American Cinematographer*, 3 (June 1926), pp. 9–10, 16–17, 20–22.

71 Bericht über Zwischenrevision der Geschäftsbücher und der Bilanz per 28 February 1927. *Files of the Deutsche Spiegeltechnik GmbH & Co. K.-G. Berlin* (R109/I2455). Bundesarchiv Berlin.

72 Hans Nieter, "The Schüfftan Process of Model Photography," in *The Photographic Journal* (January 1930), pp. 16–18.

73 For instance, in *Lord of the Rings: The Return of the King* (Peter Jackson, 2003).

early 1920s, new processes became available that facilitated not only the combination of two moving images side by side but also the "embedding" of one moving image into another one. With the help of travelling mattes, separately filmed fore- and background footage can be printed onto the same filmstrip without appearing superimposed. In the case of rear projection, previously recorded background footage is projected behind the live action during filming. The "embedding" of one moving image into another had long been an aspiration of filmmakers; among other things it reduced the necessity of costly location shooting and made exterior settings more controllable.

Numerous travelling matte processes were devised during the 1910s and 1920s. Two basic types can be distinguished. The earlier form, referred to as black (or white) backing travelling matte systems, involve black (or white) backdrops and the creation of a silhouette duplicate of the foreground scene. The latter, known as color separation or "self-matting" processes, relies on color complementary light for combining two independently recorded moving images. While technicians on both sides of the Atlantic devised related techniques,[74] those by Frank D. Williams and C. Dodge Dunning became most widely used.[75] Neither the Williams nor the Dunning process was patent protected outside of the United States, and both received attention in Europe.[76] However, what role these processes came to play in the different European film industries warrants further investigation.

By his own account, Frank D. Williams (1893–1961) started exploring travelling mattes in 1910.[77] Like many of the leading American trick specialists of the 1920s and beyond, including Hans F. Koenekamp, Fred W.

74 For instance, German cinematographer Guido Seeber patented a photochemical method to create travelling mattes in 1924. Guido Seeber, 1924, *Verfahren zum Erzeugen von Kombinationsbildern*, Deutsches Reichspatent 407593, issued 19 December 1924.

75 See also Julie Turnock, "Patient Research on the Slapstick Lots: From Trick Men to Special Effects Artist in Silent Hollywood," in *Early Popular Visual Culture*, (13) 2 (May 2015), pp. 152–173.

76 See, for instance, Guido Seeber, *Der praktische Kameramann*. Vol. 2, *Der Trickfilm in seinen grundsätzlichen Möglichkeiten* (1927) (Frankfurt am Main: Deutsches Filmmuseum, 1979), pp. 143–146; Günther Herkt, "Dunning-Verfahren," Film-Kurier 38 (19 September, 1931); and (rather belatedly) André Bazin, "The Life and Death of Superimposition," in *Bazin at Work: Major Essays and Reviews from the Forties and Fifties*, trans. Alain Piette and Bert Cardullo, ed. Bert Cardullo, (New York: Routledge, 1997), pp. 73–77.

77 Earl Theissen, "In the Realm of Tricks and Illusions," in *The International Photographer* (June 1934), pp. 8–11. Other inventors also worked on travelling mattes around the same time. French film historian E. Kress had described a basic travelling matte technique in a 1912 booklet on trick techniques. Reprinted as E. Kress, "Trucs et illusions. Applications de l'optique et de la mécanique au cinématographe (c. 1912)," in *1895: Revue d'histoire du cinéma*, 27 (September 1999), pp. 7–20.

Jackman, and Phil Whitman, Williams initially worked as a cinematographer in comedy production. In 1918, he patented his black backing travelling matte system, which became commercially available four years later.[78] The Williams process was mainly utilized for imperceptible but also expressive and spectacular effects. It first attracted attention in Hollywood for its imperceptible use in *Beyond the Rocks* (Sam Wood, 1922) and *Manslaughter* (Cecil B. DeMille, 1922). It was prominently employed for expressive effects in *He Who Gets Slapped* (Victor Sjöström, 1924) and *Sunrise* (F.W. Murnau, 1927). Probably the most spectacular application of the technique can be seen in *The Invisible Man* (James Whale, 1933). Through most of the 1920s, Frank Williams created Williams shots as an independent contractor.

Williams was a pugnacious and unscrupulous character, regularly engaged in lawsuits an also sought to capitalize on the special-effects patents of others. He surreptitiously obtained control over nineteen patents for compositing techniques, including Walter L. Hall's, and assigned them to the newly formed company Patents Process Inc., presumably to reduce alimony payments to his estranged wife.[79] Following the bankruptcy of Patents Process Inc. and himself, Williams set out in 1930 to capitalize on his own travelling matte process by establishing a specialized company in the fashion of Schüfftan's Spiegeltechnik. In contrast to Spiegeltechnik, however, Composite Laboratories Co. not only offered made-to-order Williams shots but also general laboratory services, which became the company's main source of income. As a result, Composite Laboratories Co. was able to survive bankruptcy in 1932 and continued to exist until 1956, long after the Williams process had become obsolete.

The second widely employed travelling matte process of the late 1920s and early 1930s was the Dunning process. Owner of the patent was C. Dodge Dunning (1907–1959), the teenage son of Carroll H. Dunning (1881–1975), former vice-president of the company that marketed Prizmacolor.[80] The Dunning process was first employed in low-budged FPO productions like

78 Frank D. Williams, 1918, *Method of Taking Motion Pictures*, US Patent 1,273,435, issued 23 July 1918.

79 "How Williams got hold of my patent," in J. Searle Dawley Notebooks, J. Searle Dawley Collection, folder 20, book 3, 314-324, Margaret Herrick Library, Beverly Hills, CA; See also "United States Circuit Court of Appeal For the Ninth Circuit in the Matter of Patents Process, Inc., a corporation, bankrupt, In the Matter of Frank D. Williams, bankrupt, Patents Process Inc., a corporation and Frank D. Williams, Appellants, vs. Walter C. Durst, Trustee in Bankruptcy of the Estates of Patents Process, Inc., a corporation, bankrupt, and Frank D. Williams, bankrupt, Appellee." Transcript of Record, filed 14 December 1933, p. 27.

80 Caroll D. Dunning, 1927, *Method of Producing Composite Photographs*, US Patent 1,613,163, issued 4 January 1927.

Silver Comes Through (Lloyd Ingraham, 1927). In 1927, the Dunnings founded the Dunning Process Co., Inc., whose production facilities were promoted as "the first private studio ever built for special process work exclusively."[81]

Meanwhile, Roy Pomeroy (1892–1947), head of the experimental department at Famous Players Lasky (and later of special effects at Paramount), had also sought a patent for a color separation system. Pomeroy's process was virtually identical with Dunning's. Even though Pomeroy had applied for patent protection five months before his competitor, his patent was issued eighteen months after Dunning's. Compelled to come to an arrangement with Pomeroy, the Dunnings acquired an exclusive license to the Pomeroy process in 1930 and subsequently commercialized both processes jointly. Like all standardized techniques, the Dunning-Pomeroy process was particularly suitable for imperceptible effects, as seen in *Anna Christie* (Clarence Brown, 1930), but like the Williams process it was also employed for spectacular-fantastic ones as in *Just Imagine* (David Butler, 1930) and spectacular-naturalistic ones as in *Mutiny on the Bounty* (Frank Lloyd, 1935). The Dunning process established itself at a time when sound made expressive effects appear expendable in American cinema. Unlike the older Williams process, it was therefore rarely utilized for that purpose.

In Hollywood around 1930, the Dunning-Pomeroy process was the dominant method for combining separately filmed images. However, interest soon began to wane. Hollywood's new special-effects method of choice was rear projection, which was more flexible and more quickly executed. The sudden success of rear projection also reflects studios' aspiration to reduce their dependence on techniques that had to be licensed from and produced by independent patent holders. As Birk Weiberg has shown, at Warner Bros., the decision to move from color separation travelling mattes to rear projection was triggered by patent litigation lawsuits filed in April 1931 by the Dunning Process Co., Roy Pomeroy, and Paramount Pictures.[82] In contrast to the processes offered by independent effects houses, studio-developed methods like rear projection held out the prospect of full legal, financial, and technical control. The Dunning process's momentarily all-too-powerful position may well have contributed to its precipitous demise. At this point, the Dunnings developed a three-color process and, in 1939, transformed the

81 "Expansion with a Big E," in *International Photographer* (February 1930), pp. 34–35.

82 Birk Weiberg, "Roy J. Pomeroy, Dunning Process Co., Inc., and Paramount Publix Corporation vs. Warner Bros Pictures, Inc., Vitaphone Corporation, and Frederick Jackman: How the Movie Industry Turned to Rear Projection," unpublished conference paper, Society for Cinema and Media Studies Annual Conference, Seattle, WA, 2014.

Dunning Process Co. into the Dunningcolor Corporation, which remained in business until 1955.

In contrast to most set extension and travelling matte techniques, which originated in independent workshops, rear projection was largely a product of studio research and investment in new and expensive equipment. In the early 1930s, rear projection established itself as Hollywood's preferred method and for many decades dominated special-effects work all over the world. Like travelling matte processes, rear projection facilitates the combination of foreground performances with pre-filmed backgrounds. Known variously as projected background process, projection process, or transparency process at the time,[83] the technique transformed special-effects production in Hollywood virtually overnight. Farciot Edouart, since 1932 head of Paramount's Transparency Department, recalled that Fox spearheaded experiments with rear projection in 1929 and only three years later the technique had become the industry standard: "[S]carcely a picture is released by any such major producer, which does not include at least two or three such scenes — and some few productions have used this process to make 75 or 80% of their total footage."[84]

Rear projection initially faced major technical obstacles, above all a lack of emulsion sensitivity, sufficiently large and highly transmitting projection screen material, electronic syncing systems for camera and projector, and high-intensity lighting. As a result, the technique was not suitable for a broad range of applications until 1930. Still, rear projection was nothing new. Already in 1912, German photographer Hugo Sontag had received a patent for capturing foreground object and rear-projection in one exposure.[85] Norman Dawn reportedly employed (still) rear projection for two shots in *Ghost of Thunder Mountain* (Louis Kleine, 1912), and German film pioneer Jules Greenbaum concurrently experimented with the filming of rear-projected moving images.[86] Prominent early applications in *Faust* (F.W. Murnau, 1926) and Lang's *Metropolis* employed the technique for spectacular effects. As an industry standard, however, rear projection was mostly used imperceptibly, above all for the representation of distant background motion, for instance in driving scenes.

83 Travelling matte techniques were also referred to as transparency process. Similarly, both travelling mattes and rear projection were called "process shots."

84 Farciot Edouart, "The Transparency Projection Process," in *American Cinematographer*, (July 1932), pp. 15, 39.

85 Hugo Sontag, 1912, *Verfahren zur Herstellung photographischer Kombinationsbilder durch Aufnahme des Objekts in Verbindung mit einem von hinten auf einen durchscheinenden Schirm projizierten Hintergrunde*, Deutsches Reichspatent 246,940, issued 14 May 1912.

86 Guido Seeber, "Gefilmte Projektion," in *Filmtechnik*, 4 (5 August 1925), pp. 76–78.

Tricks Become Special Effects

The increasing prevalence of imperceptible effects in mainstream filmmaking since the late 1910s traces back to producers' aspiration to economize while simultaneously enhancing the appearance of their pictures. At the same time, tricks continued to be viewed with suspicion in the American film industry. For instance, sheets for glass paintings were known as "cheaters" in Hollywood studios, which suggests that even filmmakers themselves perceived set extensions as somewhat dishonest.[87] Fearful that the public might feel betrayed by imperceptible effects, studios made every effort to conceal their use. Norman Dawn recalled:

> Even when I was at Universal they didn't believe in telling anybody about effects [...] They considered anything that was a drawing or a glass shot a fake. So they didn't want to let the exhibitors know that this was a cheap picture full of fakes. They kept that all quiet [...] no matter if it was nothing more than an ordinary double exposure.[88]

Little information on special effects transpired. Evidently, the industry gag order not only concerned studio employees but also the press. Occasionally, press articles revealed (not always accurately) production methods of one-of-a-kind spectacular effects in lavish productions with wide audience appeal like the parting of the Red Sea in *The Ten Commandments* (Cecil B. DeMille, 1923) or the flying carpet in *The Thief of Bagdad* (Raoul Walsh, 1924). Usually, however, the trade and fan press remained rather silent about special effects, particularly imperceptible ones.[89]

Around 1926, discussions about standardized special-effects techniques became more common in American technical trade journals like the *Journal of the Society of Motion Picture Engineers* (SMPE, published since 1916) and *American Cinematographer* (published since 1921). Nonetheless, even at a time when imperceptible effects became routine in film production, uneasiness persisted. Producers and technicians felt compelled to assert that special

87 Carl Louis Gregory, "Trick Photography," in *Journal of the Society of Motion Picture Engineers* (SMPE) (September 1926), pp. 99–107.

88 Norman O. Dawn, *Innovations in Cinematography* (Los Angeles: American Film Institute, 1977), pp. 65–66.

89 Rare early discussions of techniques that lend themselves to imperceptible uses in the fan and trade press include G. Harrison Wiley, "In Lieu of Genii: Modern Cameraman Surpasses Magician of Ancient Days," in *Photodramatist*, 7 (December 1922), pp. 11–12, 42; and "Does the Camera Lie?" in *Photoplay Magazine*, 4 (September 1923), pp. 30–33, 115–119.

effects were both inevitable and by no means geared towards deceiving the audience. Cinematographer William Stull, for instance, emphasized:

> [T]he true aim of the various special effects technicians involved is rarely to foist a fake or trick upon the public. Save in fantastic films like *King Kong* or *One Million B.C.* [Hal Roach and Hal Roach Jr., 1940], modern camera trickery is intended solely to put a better picture on the screen for the audience, and to do it easily, economically, and above all, without endangering human life.[90]

As Stull's statement suggests, the public image of special effects remained contested, and film technicians sought to rebrand special effects — traditionally deemed old-fashioned, un-American, inauthentic, and inferior — as highly efficient, economical, and indeed as an essential aspect of industrial film production. Fred Jackman, supervising director of the Trick Department at Warner Bros., proudly contrasted the contemporary special-effects engineer with the sorcerer of yesteryear:

> The modern special-effects engineer performs his work on the basis of thoroughly known principles, and he is able in advance to predict both the result itself and the cost of obtaining it. In the earlier days of the industry, the few cinematographers who practiced what was then called "trick camera work" were regarded almost as so many magicians. They did not always know how successfully their magic was going to turn out, but they tried, and learned — and in learning laid the foundation of today's knowledge and practice.[91]

Jackman's special-effects engineer is a specialist. The increasing specialization of effects labor was a direct consequence of a standardization of techniques. Experimental one-of-a-kind effects, which are usually of the spectacular or expressive type, are time-intensive and costly and do not require specialization. Standardized techniques, in contrast, which are commonly used for imperceptible effects, aim at economizing and can be churned out more effectively by specialists.

90 William Stull, "Process Cinematography," in *The Complete Photographer* volume 8, Willard D. Morgan, ed., (New York: New York Educational Alliance, 1943), pp. 2994–3005.

91 Fred W. Jackman, "The Evolution of Special-Effects Cinematography from an Engineering Viewpoint," in *Journal of the Society of Motion Picture Engineers* (SMPE) (September 1937), pp. 293–302.

Between the 1910s and the 1930s, special effects were created in three different labor configurations (of special effects): as a secondary occupation, by independent contractors, and by long-term studio employees. Initially, effects were devised by filmmakers for use in productions that they were working on in another, more prominent capacity, for instance as cinematographer, director, or art director. A growing need for special-effects experts prompted artists like Percy Day or former Keystone cinematographer Phil Whitman to freelance as trick specialists. Simultaneously, owners of marketable patents like Schüfftan, Williams, and Dunning established independent companies that produced made-to-order effects shots and employed specialized staff. When studios set up special-effects departments, some trick experts became long-term employees. Figures like Hans F. Koenekamp (Warner Bros.), Vernon Walker (RKO), or Farciot Edouart (Paramount) created effects at the same studios for decades.

In Europe, the trend towards specialization in the special-effects trade was initially less robust than in the United States. Schüfftan and Day demonstrate that similar dynamics were at work, but throughout the 1920s, the leading special-effects artists like Günther Rittau, Carl Hoffmann, and Erich Kettelhut were concurrently cinematographers and art directors. In European cinema, permanent special-effects units were probably rare until the 1930s. Ufa, Europe's largest studio, established a Schüfftan department in 1925 but quickly outsourced it. Even Guido Seeber, the *doyen* of German trick technology, did not commit exclusively to special effects until 1935, when he took charge of Ufa's newly established special-effects department.

The fact that the various labor relations represent different historical phases may suggest a straightforward teleology towards the full integration of special-effects labor into the studio apparatus. However, such a view would be overly simplistic. In the period under investigation, the three types of effects labor existed side-by-side on both sides of the Atlantic, and most technicians worked in different professional configurations. Even after major studios had established permanent special-effects units, alternative models remained common. Julie Turnock has rightly emphasized the prevalence of independent special-effects houses and contractors, who were hired on a fixed-term or by-production basis throughout the studio era.[92] Simultaneously, special effects also continued to be created as a secondary occupation, particularly at less affluent studios.

The establishment and operation of special-effects studio departments in Hollywood are still poorly understood. As we have seen, studios had no

92 Turnock, "Patient Research on the Slapstick Lots," op. cit., pp. 152–173.

interest in alerting the public of the existence of such units, and complex and unstable studio structures make them difficult to track down. Also, their names, organization, and function varied from studio to studio and changed periodically. Warner Bros., for example, had a "Miniature and Trick Department" in 1926, after the merger with First National a joint "Scientific Research Department" in 1928, a "Trick Department" in 1930, and a "Camera Effects Department" in 1932. Finally, not all special-effects techniques were necessarily housed in the same department. For instance, miniatures, optical printing, matte painting, and rear projection were often organized as separate divisions. Trick departments operated like second units separately from the main unit and covered all shots that could be cheaper and more efficiently produced with the help of special effects.

The first Hollywood studio to institute a permanent unit concerned with special effects was probably Universal. In 1920, art director Elmer Sheely established an "Experimental Department" at Universal, "for the purpose of making research into the realm of the unusual in photography."[93] Universal was evidently at the vanguard with respect to special effects. The studio also employed Dawn at the time, who however does not seem to have been involved in Sheely's endeavor. The next studios to institutionalize special effects were Lasky and Fox, which formed "Experimental Departments" around 1923, and MGM, which installed a "Trick Department" in 1925. By the mid-1920s, all major Hollywood studios had established special-effects units, yet their sizes differed greatly. When newly established RKO set up its "Special Effects Department" in 1929, it consisted of merely one artist and one cameraman. In contrast, Hollywood's largest trick unit, Warner's "Scientific Research Department," already had eleven permanent staff members in 1926.[94] With the establishment of specific departments, studios valorized special effects as an essential part of filmmaking. In-house units furthered the integration of special effects into an increasingly industrial mode of film production and held out the prospect of far-reaching control over techniques and personnel.

This essay has argued that the introduction of standardized methods that lent themselves to the discrete manipulation of filmic images fundamentally transformed the role of special effects in mainstream filmmaking. Early cinema was dominated by one-of-a-kind effects devised by generalists with

93 Philip H. Whitman, "Pans and Tilts," *American Cinematographer,* (2) 20 (1 November 1921), p. 8.

94 "Biggest Stage on Earth Devoted Entirely to Special Process Work," *American Cinematographer* (April 1929), pp. 20–21, 35.

the intention of exciting awe and wonder. By the 1930s, however, most special effects were created by specialized personnel according to established methods and was not supposed to be noticed. As I have shown, the move towards standardization and specialization proceeded in analogous ways in leading Western film industries, albeit not always simultaneously. The taxonomy derived from Christian Metz has facilitated a more nuanced understanding of the metamorphoses in the special-effects trade. In particular, it has helped specify what types of effects played more prominent roles at particular historical junctures and why.

For Metz, the notion of cinema as an uninterrupted, "total" image is fundamentally flawed. He describes the medium as characterized by interstices — gaps at different levels that permit interventions.[95] Interstices between shots are affected by montage, those between frames can result in effects like slow motion, composite shots exploit interstices at the level of individual images, and, in the digital age, even individual pixels have become susceptible to manipulation. Metz's notion of the cinema as interstitial reverberates in Stephen Prince's observation that "[c]inema is a composited medium, whether analog or digital, and this singular condition has been undervalued in our existing theories."[96] The verisimilitude of cinematic images ostensibly substantiates theoretical claims about the medium's inherent indexicality and realism. In many cases, however, it primarily attests to the mastery of filmmakers in concealing the constructedness of their work. Throughout film history, imperceptible effects, mostly undetected by audiences and critics, have shaped film aesthetics and consequently our understanding of the nature of the cinematic medium. Increased awareness of their workings and prevalence prompts us to reassess our conception of the filmic image. Imperceptible effects encourage us to construe filmic images as inherently malleable — less as representations of reality and more as artifice, craft, and technique.

Bibliography

Anon. "300 Important Cameramen Becoming Recognized as Photographic Marvels," *Variety*, (4 January 1928): 6, 13.

Anon. "A Little Letter to the Inquisitive by the Inquiry Editor," *The Motion Picture Story Magazine* (April 1912): 162

95 Metz, "*Trucage* and the Film," op. cit., pp. 671–672.

96 Prince, *Digital Visual Effects in Cinema*, op. cit., p. 53.

Anon. "Anglo-European Panorama," *The Daily News* (Perth), 19 December 1903.

Anon. "Attempts to Control Double Exposure Method," *American Cinematographer* 6 (September 1922): 4.

Anon. "Biggest Stage on Earth Devoted Entirely to Special Process Work," *American Cinematographer* (April 1929): 20–21, 35.

Anon. "Earle to Make Artistic Production of *The Rubaiyat of Omar Khayyam*," *Exhibitors Herald*, (2 July 1921): 106.

Anon. "Expansion with a Big E," *International Photographer* (February 1930): 34-35.

Anon. "Metro Uses Motion Painting," *Motion Picture News* (6 April 1918): 2065

Anon. "Norse Saga in Pictures," *Los Angeles Times* (22 August 1920): III23

Anon. "Obituary," *Los Angeles Times* (6 December 1932): 16

Anon. "Omar Khayyam Bows to Screen," *The Lewiston Daily Sun* (16 August 1922): 4

Anon. "Painted Sets Take Place of Costly Fabric," *Los Angeles Times* (23 December 1923): 25

Anon. "Pauline Periled in Auto Race," *Motography* 16 (17 October 1914): 525

Anon. "Producers Line-Up: Director's Suit Calls Up Strong Opposition," *Los Angeles Times* (17 August 1922): II11.

Anon. "Rear Projection Big Advantage," *International Photographer* (April 1938): 30–33.

Anon. "Some Tricks of the Moving Picture Maker," *Scientific American* vol. 100, no 26, (26 June 1909): 476-477, 487.

Anon. "Split Reel Notes for Theater Men," *Motography* 5 (4 August 1917): 262.

Anon. "Who's Who and Where," *Film Fun* 335 (February 1917): n.p.

Anon. "World's Greatest Film is Screened," *Motography* 2 (11 July 1914): 38

Abel, Richard. *The Red Rooster Scare: Making Cinema American, 1900–1910* (Berkeley, CA: University of California Press, 1999).

Anderson, Anthony. "*Omar Khayyam* A Picture Gallery," *Los Angeles Times* (20 December 1925):III 44..

Babin, Gustave. "Les coulisses du cinématographe," *L'Illustration* 3396 (28 March 1908); "Les coulisses du cinématographe," *L'Illustration* 3397 (4 April 1908); "Le théâtre cinématographique," *L'Illustration* 3427 (31 October 1908). In *Les Grands Dossiers de l'Illustration: Le Cinéma* (Paris: SEFAG, 1987): 22–33.

Bazin, André. "The Life and Death of Superimposition." In *Bazin at Work: Major Essays and Reviews from the Forties and Fifties*, translated by Alain Piette and Bert Cardullo, edited by Bert Cardullo (New York: Routledge, 1997): 73–77.

Bean, Jennifer. *The Fun Factory: The Keystone Film Company and the Emergence of Mass Culture* (Berkeley, CA: University of California Press, 2009).

Brown, Karl. *Adventures with D.W. Griffith* (New York: Farrar, Straus & Giroux, 1973).

Bukatman, Scott. *Matters of Gravity: Special Effects and Supermen in the 20th Century* (Durham, NC: Duke University Press, 2002).

Chessex, Robert. "F.W. Murnau dreht Faust. Filmstadt Berlin um 1925," *Film Bulletin* 153 (April/May 1987): 53–63.

Copeland, Gene. "In a Persian Garden," *Picture Play Magazine* 4 (June 1920): 34–35, 92.

Cotta Vaz, Mark, and Craig Barron. *The Invisible Art: The Legends of Movie Matte Painting* (San Francisco: Chronicle Books, 2002).

Crafton, Donald. *Before Mickey: The Animated Film 1898–1928* (Chicago: University of Chicago Press, 1993).

Crump, Irving. "Interesting Picture Figures," *Motion Picture Magazine* (August 1914): 113-114.

Dahlquist, Marina. *The Invisible Seen in French Cinema Before 1917* (Stockholm: Aura förlag, 2001).

Dawn, Norman O. *Innovations in Cinematography* (Los Angeles: American Film Institute, 1977).

Day, Susan. "Walter Percy Day: Biography." Last modified December 2017. http://www.walterpercyday.org.

Earle, Ferdinand. "Screen Renaissance Through Motion Painting." In *The Blue Book of the Screen*, edited by Ruth Wing (Hollywood: Pacific Gravure Company, 1924): 345–348.

Edouart, Farciot. "The Transparency Projection Process," *American Cinematographer* (July 1932): 15, 39.

Ezra, Elizabeth. *Georges Méliès: The Birth of the Author* (Manchester: Manchester University Press, 2000).

Giesen, Rolf. *Special Effects: Vom Spiegeleffect bis zur Computeranimation* (Ebersberg: 8½, 1994).

Goldner, Orville, and George Turner. *The Making of King Kong: The Story Behind a Film Classic* (New York: A.S. Barnes, 1975).

Gregory, Carl Louis. "Trick Photography," *Journal of the Society of Motion Picture Engineers* (SMPE) (September 1926): 99–107.

———. "Trick' Photography Methods Summarized," *American Cinematographer* 3 (June 1926): 9–10, 16–17, 20–22.

Gunning, Tom. "An Aesthetic of Astonishment: Early Film and the (In)Credulous Spectator." In *Viewing Positions: Ways of Seeing Film*, edited by Linda Williams (New Brunswick, NJ: Rutgers University Press, 1995): 114–133.

Hamus-Vallée, Réjane. *Peindre comme un cinéaste. Pour une esthétique du Matte Painting*. Thèse d'HDR, Université Paris III Sorbonne Nouvelle, 2013.

Hamus Vallée, Réjane, and Caroline Renouard, *Les effets spéciaux au cinéma: 120 ans de créations en France et dans le monde* (Vanves: Armand Colin, 2018).

Hansen, Miriam. *Babel & Babylon: Spectatorship in American Silent Film* (Cambridge, MA: Harvard University Press, 1997).

Herkt, Günther. "Dunning-Verfahren,"*Film-Kurier* 38 (19 September 1931): 1.

Hunt, Thomas, ed. *Wellington College Register: 1859–July 1905* (Wellington College, Berkshire, 1906).

Isolani, Gertrud. "Gespräch mit Eugen Shufta," *Basler Nachrichten* (19 October 1965): 9.

Jackman, Fred W. "The Evolution of Special-Effects Cinematography from an Engineering Viewpoint,"*Journal of the Society of Motion Picture Engineers* (SMPE) (September 1937): 293–302.

Kessler, Frank. "Actualités." In *Encyclopedia of Early Cinema*, edited by Richard Abel (London: Routledge, 2005): 6–7.

Kress, Eugène. "Trucs et illusions. Applications de l'optique et de la mécanique au cinématographe (c. 1912)," *1895. Revue d'histoire du cinéma* 27 (September 1999): 7–20.

Larsen, Egon. "Here is the Inside Story of the Magician of British Films," *Cavalcade* 21 (31 May 1949): 9.

Lee, Laura. *Japanese Cinema Between Frames* (London: Palgrave Macmillan, 2017).

Loew, Katharina. "The Spirit of Technology: Early German Thinking about Film," *New German Critique* 122 (Summer 2014): 129–148.

———. "Magic Mirrors: The Schüfftan Process." In *Special Effects: New Histories, Theories, Contexts*, edited by Dan North et al. (London: British Film Institute, 2015): 62–77.

———. *Special Effects and German Silent Film: Techno-Romantic Cinema* (Amsterdam: Amsterdam University Press, 2021).

Metz, Christian. "*Trucage* and the Film," *Critical Inquiry* 3, no. 4 (1977): 657–675.

Minguet Batllori, Joan M. *Segundo de Chomón: The Cinema of Fascination* (Barcelona: Biblioteca de Catalunya, 2010).

Mulvey, Laura. "A Clumsy Sublime," *Film Quarterly* vol. 60, no. 3 (2007): 3.

Nieter, Hans. "The Schüfftan Process of Model Photography," *The Photographic Journal* (January 1930): 16–18.

Païni, Dominique. "The Wandering Gaze: Hitchcock's Use of Transparencies." In *Hitchcock and Art: Fatal Coincidences*, edited by Dominique Païni and Guy Cogeval, translated by Guy Conolly et al. (Montréal: The Montreal Museum of Fine Arts, 2000): 51–78.

Percy Day, Walter. "The Origin and Development of the Matte Shot," *The Photographic Journal* (October 1948): 209–211.

Pierson, Michele. *Special Effects: Still in Search of Wonder* (New York: Columbia University Press, 2002).

Prince, Stephen. *Digital Visual Effects in Cinema: The Seduction of Reality* (New Brunswick, NJ: Rutgers University Press, 2012).

Purse, Lisa. *Digital Imagining in Popular Cinema* (Edinburgh: Edinburgh University Press, 2019).

Redman, Jeanne. "Art Effects in Cinema," *Los Angeles Times* (11 August 1920): III4.

Rogers, Ariel. *On the Screen: Displaying the Moving Image, 1926–1942* (New York: Columbia University Press, 2019).

Seeber, Guido. "Gefilmte Projektion," *Filmtechnik* 4 (5 August 1925): 76-78.

———. *Der praktische Kameramann. Vol. 2: Der Trickfilm in seinen grundsätzlichen Möglichkeiten*. (Frankfurt am Main: Deutsches Filmmuseum, 1979).

Shields, David S. *Still: American Silent Motion Picture Photography* (Chicago: University of Chicago Press, 2013).

Slotkin, Richard. *Gun Fighter Nation: The Myth of the Frontier in Twentieth-Century America* (New York: Atheneum, 1992).

Smith, Jacob. *The Thrill Makers: Celebrity, Masculinity, and Stunt Performance* (Berkeley, CA: University of California Press, 2012).

Stull, William. "Producers Pool Composite Process Patents," *American Cinematographer* 11 (November 1936): 2994–3005.

———. "Process Cinematography." In *The Complete Photographer*, vol. 8, edited by Willard D. Morgan (New York: New York Educational Alliance, 1943): 2994–3005.

Talbot, Frederick A. *Moving Pictures: How They Are Made and Worked* (Philadelphia: J.B. Lippincott Company, 1912).

Theisen, Earl. "In the Realm of Tricks and Illusions," *The International Photographer* (June 1934): 8–11.

———. "The Evolution of the Motion Picture Story, Parts I and II" *The International Photographer* (April and May 1936): 16–17; 22–23; 12–13, 27.

Turner, George E. ed. *The ASC Treasury of Visual Effects* (Hollywood: ASC Holding Co., 1983).

Turnock, Julie. "The Screen on the Set: The Problem of Classical-Studio Rear Projection," *Cinema Journal* 51, no. 2 (Winter 2012): 157–162.

———. "Patient Research on the Slapstick Lots: From Trick Men to Special Effects Artist in Silent Hollywood," *Early Popular Visual Culture* 13, no. 2 (May 2015): 152–173.

———. *Plastic Reality: Special Effects, Technology, and the Emergence of 1970s Blockbuster Aesthetics* (New York: Columbia University Press, 2015).

Weiberg, Birk. "Roy J. Pomeroy, Dunning Process Co., Inc., and Paramount Publix Corporation vs. Warner Bros Pictures, Inc., Vitaphone Corporation, and Frederick Jackman: How the Movie Industry Turned to Rear Projection." Unpublished conference paper (Seattle, WA: Society for Cinema and Media Studies Annual Conference, 2014).

Whissel, Kristen. *Spectacular Digital Effects: CGI and Contemporary Cinema* (Durham, NC: Duke University Press, 2003).

Whitman, Philip H. "Pans and Tilts," *American Cinematographer* 2, no. 20 (1 November 1921): 8.

Wiley, G. Harrison. "In Lieu of Genii: Modern Cameraman Surpasses Magician of Ancient Days," *Photodramatist* 7 (December 1922): 11–12, 42.

———. "Does the Camera Lie?," *Photoplay Magazine* 4 (September 1923): 30–33, 115–119.

About the Author

Katharina Loew is Associate Professor of German and Cinema Studies at the University of Massachusetts, Boston. She is the author of *Special Effects and German Silent Film* (Amsterdam University Press, 2021). Her work on silent cinema and film technology has also been published in *New German Critique*, *Film Criticism*, *Early Popular Visual Culture*, and several edited collections.

10. Special Effects and Spaces of Communication: A Semio-Pragmatic Approach

Roger Odin

Abstract

The central idea of this essay is the following: one cannot hope to understand how special effects function without first taking into consideration the communicative spaces inside which they operate. After arguing how this idea was already implicit in Christian Metz's famous article "*Trucage* and the Film," the author turns to consider a new phenomenon, namely the massive presence of special effects in the commonplace space of our everyday lives. The discussion centers on the development of special effects that are specific to this space through the use of smartphones and on the need to constantly historicize our analysis of them as their "specialness" tends to wane with time. In his conclusion, the author insists on the heuristic value of the notion of 'communicative space,' since taking the commonplace space of the everyday as a field of inquiry requires one to break with the habit of considering special effects solely in the communicative spaces pertaining to cinema, and thus to investigate them from a different perspective altogether.

Keywords: semio-pragmatism, communicative spaces, everyday life, cinema, smartphone

In "*Trucage* and the Film," Christian Metz remarks that "the meaning" of the term *special effects* "is rather vague."[1] We must thus begin by trying to see

1 Christian Metz, "*Trucage* and the Film," trans. Françoise Melzer, *Critical Inquiry* (3)4 (Summer 1977), p. 665.

Lefebvre, M. & M. Furstenau (eds.), *Special Effects on the Screen: Faking the View from Méliès to Motion Capture*. Taylor & Francis Group, 2022
DOI 10.5117/9789462980730_CH10

the matter a little more clearly. The presupposed (in the linguistic sense of the term) of the expression "special effects" is that there are effects that are not "special." The problem thus consists in knowing by what means, and in relation to what, an effect may be called special. It appears to me that the concept "communicative space" as I have defined it in my semio-pragmatic approach, may be useful in offering a response to these questions.[2]

The reader will recall that by communicative space, I understand a tool constructed by the theorist whose function is to make possible describing how communication functions (or does not function) within a given context. I define it as a space within which a range of constraints impels the *actants* (*sender* and *receiver*) to share the same communicative experience. This range of constraints governs the construction of *actants* as well as the *relations* between them; the choice of the mode or *modes* of producing meaning and affect; and the construction of the communicative *operator* (out of which meaning is produced).[3]

1. Re-reading Metz

The article "*Trucage* and the Film" occupies a singular place in Christian Metz's work: whereas his theoretical ideas on the whole adopt quite deliberately an immanent (or "object-oriented") perspective, his approach to some of the problems raised by special effects has a strong pragmatic dimension. Here I wish to carry out a re-reading of this essay, keeping in mind the concept of the communicative space, which, it appears to me, will make it possible to bring out this pragmatic aspect quite clearly.

From the outset, Metz asserts that we ought to distinguish between two ways of constructing the concept of special effects: that of the technician, whose communicative space frames the activity making films, and that of semiotician, whose communicative space pertains to elaborating of theories about the cinema. The concept of *trucage* as presented here must

2 Roger Odin, *Spaces of Communication. Elements of Semio-Pragmatics* (Amsterdam: Amsterdam University Press 2021 [2011]).

3 To put it simply, the *operator* is whatever semiotic material a given communicative relation uses. However, its value as operator is not determined in advance, or *immanently*, according solely to the formal character of the sign-system it uses. Rather, its role as operator is determined *pragmatically* by the function it comes to occupy in a given communicative space. For instance, in a familial communicative space, a painted portrait of a family member often functions much like a photograph would as a mnemonic operator. The operator here is a hybrid object: a "photo-painting." See Odin, *Spaces of Communication*, op. cit., pp. 110–114.

not be confused with the "special effects" of which studio technicians speak. For technicians, special effects can be defined as "all those effects which they must create specially and which demand, in addition to the normal work of filming, a small, particular technique." "Thus loosely defined," Metz remarks, "the rubric of special effects will obviously form, for the semiologist, a heteroclitical group."[4] In the space of theory, we might thus expect that the construction of special effects as a concept responds to the criterion of coherence.

The approach Metz adopts to give coherence to his analysis is based on the nature of the cinematic sign: "What defines all process-effects is a kind of *divergence from 'photographicity'*."[5] If we define photographicity as the indexical production of an image *qua* likeness (an *analogon*), then special effects, for their part, belong to a category of non-analogous techniques. Their referent is not the world but the image itself: "they are visual but not photographic effects."[6] From this perspective, special effects are not limited to trick effects but include optical processes and syntactical signs (dissolves, irises, etc.); in short, "all process-effects which are particular and localized, which do not merge with the normal movement of the photograms."[7]

As always, Metz then embarks on a labor of taxonomy: distinctions between kinds of special effects (slow motion, fast motion, freeze frames, etc.) and syntactical signals (dissolves and various transitions); between manipulation of the camera and techniques for manipulating the film in the laboratory (postproduction work); between *taxemes* (fade to black) and *exposants* (the iris, wipe, cross fade, fast motion, slow motion, superimposition, etc.); between profilmic special effects and cinematographic special effects; and, finally, between *imperceptible* special effects (such as the use of doubles), *invisible* special effects (when we know that there is a special effect but we do not see it, for example the "invisible man"), and *visible* special effects.[8]

Metz emphasizes that in the case of these latter distinctions, the analysis changes perspective. Whereas until then, his analysis was situated in the space of the film as an immanent object, it now passes into the communicative space of reception. Metz further underscores the fact that his distinction

4 Metz, "*Trucage* and the Film," op. cit., p. 659.
5 Ibid.
6 Ibid., p. 665.
7 Ibid.
8 Ibid., p. 663.

between *imperceptible*, *invisible*, and *visible* effects is generated by the specific communicative space that is classical cinema:

> In classical cinema (diegetic cinema), a detailed and coded procedure which is part of the *cinematographic establishment* [institution] prescribes the different types of relationships which the spectator can have with *trucages*. Here we are touching on a real regulation of perception.[9]

However, when we leave this space of communication, this distinction becomes moot. This is the case, for example, when special effects operate in the communicative space of scientific research, where slow motion, fast motion, the breakdown of movements into phases, etc. now become *operators of discovery*. In this space, the cinema belongs to a different *cultural series* that includes the microscope, the telescope, the X-ray, etc.[10] Special effects then become instruments of analysis. The same could be said for the communicative space of art and the art installation, where the artwork is the *dispositive* itself, in that it creates a special effect, by which is meant a specific corporeal experience whose goal is to bring the viewer to undergo a particular aesthetic experience.

Metz also demonstrates that we should take into account the historical evolution of the cinematic space. Thus, the communicative space of early cinema is not that of cinema today; there, certain effects were seen as special, whereas today they have become completely routine: "what is experienced as a simple figure of speech today was quite frequently, for the first spectators of the cinematograph, a magic 'trick,' a small miracle both futile and astonishing."[11] Conversely, in early cinematic space, cinema as a *dispositive* was itself experienced as a special effect, engaging the viewer in a new experience (one unlike the experience of the theater, concerts, vaudeville shows, etc.).

Similarly, in his book *Limites de la fiction* (2014), Jacques Aumont demonstrates how the perception of special effects has changed from the 1950s to today: taking the example of the Western *The Far Country* (Anthony Mann, 1955), which includes a scene with an avalanche (quite obviously of polystyrene!) and a fake day-for-night scene with a moon, he notes that in

9 Ibid.

10 Consideration for these and like phenomena as "cultural series" stems from the work of André Gaudreault and Philippe Gauthier: "Les séries culturelles conférence-avec-projection et projection-avec-boniment: continuités et ruptures," in Marta Braun et al., eds., *Beyond the Screen: Institutions, Networks and Public of Early Cinema* (Eastleigh: John Libbey, 2012), pp. 233–238.

11 Ibid., p. 665.

1955 (and even ten years later), it was quite possible that viewers would see "nothing special" and be content to watch "the avalanche take place in the landscape and the seductive James Stewart on a horse at night."[12] Today, however, a different approach has mostly taken the place of this participatory attitude, "that of the knowledgeable viewer who is attentive to the logic of the representation, familiar (thanks to the many 'making of' videos and specialized websites) with the technical aspects of filmmaking, and who sees in the latter example a 'day for night' shot and in the former example a special effect through editing."[13] The cinematic space of communication today is no longer that of the 1950s: the viewer actant is constructed and acts differently.

In another section of his article, Metz insists on the constraints connected to genres:

> The most captivating fantasy films, the most amusing of burlesque films, offer us *trucages* which always remain perceived as instruments of discourse, for that is in fact of what these genres are composed. They can function as such only because they elicit a double and contradictory reaction in the public: a belief in the reality, wonderful or comical, of the events presented, and an interest in the tour de force of which cinema demonstrates it is capable.[14]

Indeed, we can view genres as communicative spaces with specific rules, leading to a specific characterization of special effects. Thus, in the classical fiction film, we speak of special effects to designate the effects that produce a particularly powerful relationship between the film and the body of the viewer, or, in any event, one that is different from that which is produced by the cinematic *dispositive* itself and by the rest of the film ("non-special"). This relationship is tied to the narrative and thereby functions in the service of the narrative *phasing* (or, *mise en phase*).[15] In documentary films, special effects function rather in the service of the discursive *mise en phase*: to make us see and understand better. As a result, the special effects are most often visible because they are operators of didactic monstration (analyzing

12 Jacques Aumont, *Limites de la fiction: considérations actuelles sur l'état du cinéma*, (Montrouge: Bayard, 2014), pp. 9–10.

13 Ibid.

14 Metz, "*Trucage* and the Film," op. cit., p. 668.

15 I call *phasing* or "*mise en phases*" the specific cinematic labor carried out by the film as it seeks to set us "vibrating" to the rhythm of the events that are the object of the narrative. See Roger Odin, *De la fiction* (Brussels: De Boeck, 2000), pp. 37–47.

and breaking down). Here the cinema is a part of a cultural series that also includes maps, plans, scientific schema, graphics, and illustrations. Finally, to give one last example, in the musical comedy, special effects are often spectacular effects: their goal is not so much to place viewers *in phase* with the story as it is to provoke in us astonishment and admiration; these are thus visible special effects.

These thoughts demonstrate that it is only in a given space of communication that we can determine what is and is not a special effect, what is special and what is not, and to understand how these effects function.

I would now like to dwell on what appears to me to be the most important question today, the massive presence of special effects in our "everyday space." Clearly, this presence is not entirely new — we need think only of the anamorphic parlor toys that appeared as early as the eighteenth century, the cultural series of optical toys (kaleidoscopes, prisms, deforming lenses), flip books, and the fashion for stereoscopic images under the Second French Empire — but today the phenomenon is more extensive and diverse than it has ever been before. I also hope to demonstrate the heuristic value of working on the everyday communicative space.

2. Special Effects in the Everyday Space of Communication

Today, special effects not only interpose themselves in our private space through video games, television, and the computer; we can no longer go out into the street, take a subway, or enter a train station or a shopping mall without being assailed by manifold special effects in the form of animated videos on advertising screens. One can also, at any time in any place, play video games and watch a film with special effects on one's tablet or smartphone. But there is more: with a mini-camera, tablet, or especially a smartphone, we carry with us, practically at all times, in our pocket, a machine for making special effects — effects that are directly accessible through the device's menu or downloadable with a simple click on the Internet.

What a difference from what existed before in the days of amateur filmmaking! Amateur filmmakers have always loved special effects, but because they could not develop and print their films, they were forced to produce these effects at the time of filming. Manufacturers, moreover, understood this demand well and came to incorporate various special-effects functionalities into their cameras (rewind, variable speed, superimpositions, dissolves), but, apart from the fact that these were not always easy to operate, the range of options on offer was of necessity very limited, and it was impossible to *manipulate the image itself.* It was impossible, for instance, to produce a sepia tone or highlight a color (for example, highlighting an element by desaturating the background color so as to show a yellow New York taxi against a black-and-white background); today, any smartphone provides a whole range of effects (called "artistic" or "creative" effects) that affect the very substance of the image: effects of grain, texture, color, brightness, etc. Some effects go even further and affect the form itself of the thing being depicted: the tilt shift function gives landscapes or urban scenes the quality of a miniature. The frame effect, meanwhile, makes it possible to place any image in an old-fashioned gilded frame, or a heart, and to border it with flowers or stars, etc.

Other special effects have been designed to function in relation with past spaces of communication that most of us still possess as mental (or memory) spaces:[16] the photographic space of the era of celluloid (sepia effect, black and white, etc.); the space of silent cinema (black and white, flickering, sped-up motion, the absence of sound, etc.); the space of the home movie or the Super-8 effect (scratches, jumps, predominant colors [blue, yellow, green], the noise of the projector, the noise of splices passing through the projector's gate, sometimes visible perforations, etc.). Reference to the space of Polaroid leads to a particularly interesting effect. What the Polaroid image made possible, as is well known, was to see the picture as soon as it had been taken. But that today is no longer a special effect at all, or even an effect of any kind; it is the norm, hence the very nature of the Polaroid effect has now changed. What it means now is simply a return to the square format with white borders and (saturated) colors, an effect that concerns the form and visual qualities of the image but not its distinct status as instantaneous. These special effects seek to recreate the experience of a certain materiality, of a kind of flesh-and-blood quality of the image that has disappeared with digital technology. Here we enter the cultural series of "retro" operators (in some menus, these effects are indeed called retro): 33 rpm records, rural objects (an old bellows transformed into a coffee table, an earthenware jug serving as a lamp base), "authentic" products ("grandma's" jam, "rough wool" sweaters, etc.).

Finally, one finds in smartphone menus a whole series of effects whose peculiar status is worthy of mention: red eyes, interior, back lighting, dusk,

16 On this concept, see my article "Espaces de communication physiques, espaces de communication mentaux," in *D'un écran à l'autre: les mutations du spectateur,* Jean Châteauvert and Gilles Delavaud, eds., (Paris: L'Harmattan, 2016), pp. 331–342.

through a window, the sea, the starry sky, fireworks, babies, food, etc. If one can speak of these visual functionalities as special effects in the communicative space of image production (they imply a specific mode of shooting), the situation is quite dissimilar in the space of reception of the image: their role is chiefly to correct the distortions introduced by photography with respect to *normal* vision (eyes should not be red; indoor lighting should not be too yellow; back lighting should not create a silhouette; what is in focus when shooting through a window should be what one sees through the window, not on the window, etc.) and, more generally, to procure a more "correct" or normatively acceptable representation of reality. In fact, these effects are the *opposite of special effects*: they are *operators of conformity with the nature of things* (or rather, we might say, with the supposed nature of things, for we are often in the presence of a widely held fantasized "natural" vision: the sea should be blue, a baby's skin should be soft, food should look appetizing, etc.).

3. Specific Special Effects in Everyday Space

Apart from the effects discussed at the end of the preceding section (effects that are not special effects in the reception space of the image), the other effects that I have mentioned are special effects in the classical sense of the term: they produce an effect that makes something stand out or be differentiated with respect to the images (fixed or moving) we usually get to see (advertisements for them speak of "stylistic effects"). The novelty of the digital resides in the fact that they can be produced in the everyday space with astonishing ease (it is truly a case of being "child's play"; this expression, moreover, should be taken literally, as children love to play with these effects!). We could not say, however, that these special effects are specific to the everyday space. And yet these effects exist.

This is the case, for example, of special effects that affect everyday space in real time. With one click, I see on the screen of my smartphone a dinosaur entering my living room; I hear its heavy steps as it moves round my furniture; another click and it is the band of seven dwarves from *Snow White* (David Hand, 1937) who have come to circle around my dining room table; another click transforms my office into an aquarium: bubbles rise up from the ground, making a curious gurgling sound, while fish swim between the bookcase and the computer. Or else consider morphing, which works on faces rather than spaces and enables the user to endlessly deform someone's features at their leisure. This special effect has become a true group game, provoking great mirth while sharing an experience that invites friendly mockery.

With the selfie, we take another step towards specificity. Unlike morphing, the selfie is not, in fact, a special effect, and yet it nonetheless functions as a special effect in the everyday space. It is an effect tied to a *dispositive* that enables me to include myself in the gaze I cast on the everyday space. Photographing oneself was a difficult task before the selfie: one had to use a timer, determine where to place the camera, etc. As for mirrors, one rarely has one to hand; in addition, they give an inverted image of ourselves. All this is now extremely easy: a lens in my smartphone is designed for this kind of shot. All I have to do is choose the context in which I wish to appear (in front of a monument, a painting or a landscape, alone or with one or more friends, etc.) and then monitor, via the screen, the image I wish to give of myself before pressing the button. And then with one click I can post my selfie on social media. As Laurence Allard has demonstrated quite well,[17] (drawing on the work of Michel Foucault), an effect such as this is a "technique of the self," by which is meant "the procedures which [are] suggested or prescribed to individuals in order to determine their identity, maintain it, or transform it."[18] The selfie is an operator in the construction of the ego. It is not surprising that it is as popular as it is.[19]

Even more specific are the effects that affect my relations with images. I mentioned at the outset of this article the fact that one can watch special-effects films on one's tablet or phone. But I still recall my wonder when I discovered that I could see films (films with or without special effects) on the screen of my phone as I journeyed about (on the bus, on the subway, in a park) and not in a movie theater or at home (on a television): for me, that was a completely new experience. But there is more: in addition to the effect tied to relocating film viewing, there was another no less astonishing effect: that of holding in the palm of my hand the screen on which the images appeared, of being able to touch it and even with the movement of my thumb and forefinger enlarge the image to see a detail I had never seen before. These are special effects I wanted to repeat for their own sake

17 Laurence Allard, "*Express Yourself 3.0*! Le mobile comme technologie pour soi et quelques autres entre double agir communicationnel et continuum disjonctif soma-technologique," in Laurence Allard, Laurent Creton, and Roger Odin, eds., *Téléphone mobile et création*, (Paris: Armand Colin, 2014), pp. 139–162.

18 Michel Foucault, "Subjectivity and Truth," in *Ethics: Subjectivity and Truth*, ed. Paul Rabinow, trans. Robert Hurley and others (New York: New Press, 1997), p. 87.

19 "Selfie" was proclaimed new word of the year for 2013 by the *Oxford English Dictionary*. The selfie has given rise to one of the most important research projects of the past few years: "Selfie Cities: Investigating the style of self-portraits (*selfies*) in five cities across the world: Bangkok, Berlin, São Paulo, Moscow, New York" (Lev Manovich).

regardless of the film I was viewing (just as early film viewers went to the cinema to see the cinematic apparatus). And I am undoubtedly not the only person to have felt this way.

Another effect is the possibility I have henceforth been given to communicate live by way of sounds and images with a space very distant from that where I find myself: the magic of seeing from Brazil, comfortably seated in a chair facing the sea, my children at home in Rennes in their living room (for example with software such as Skype). The magic also of being able to send with one click, using Bluetooth, photos to a friend with whom I am having a beer on the patio of a café.

A remark, however: these effects need to be historicized. Just as the act of going to the movies became routine and was no longer an effect in itself (except in special cases such as an Omnimax cinema or certain films in 3D), today their special nature has for the most part dimmed. We might even say that they are no longer effects: I watch a film on my phone, I communicate by Skype and Bluetooth. They are part of our everyday routine.

On the other hand, it seems to me that the Snapchat effect still functions as a special effect: I take down a photograph that someone has just sent me and then it disappears forever from my smartphone after a few seconds (unless one uses computer forensics tools!). That still astounds me. This experience of an ephemeral image runs contrary to such a degree to the entire logic of data storage that has been drilled into us that I am still stupefied. But for how much longer? Special effects tend to become worn out more and more quickly, no doubt because of the way we use and abuse them.

In his book *La vie esthétique*, Laurent Jenny raises the question of another special effect:

> As is often the case, my eye is drawn to the picturesque displays in one of those New York grocery stores open day and night and run by Pakistanis who offer up a jumble of merchandise ranging from ballpoint pens to bouquets of flowers… I instinctively get out my cell phone … and, as always to see a little more, find myself gripped by the obsession with the digital and extractive zoom on the transparent effects of ice cubes and pieces of pineapple. The result, which I see immediately, fills me with wonder. The subject has become completely unrecognizable and given way to an unquestionable Cubist composition from that marvelous period in the years 1908–1912 when Braque and Picasso sought to outdo one another, on the brink of abstraction.[20]

20 Laurent Jenny, *La vie esthétique: Stases et flux* (Lagrasse: Verdier, 2013), pp. 89–90.

For Jenny, looking at the world through the zoom is a special effect that modifies his view of it and leads him to make reference to a pictorial mental space. He notes, moreover, that this experience is not new, quoting a passage from Proust's *In Search of Lost Time* in which Marcel, on the train taking him to Balbec, suddenly sees the sky through the window of the train car: "I glimpsed in the window-pane, above a little black copse, serrated clouds of downy softness in a shade of immutable pink, dead and as seemingly indelible now as the pink inseparable from feathers in a wing or pastel dyed by the fancy of the painter".[21] "Every element of the landscape," Jenny remarks "appears to respond to a 'call of the frame' and to resonate there." And he adds: "Naturally there is nothing here that can astonish us, we who look at the world no longer through windows but by passing over it the digital screens of our cameras and smart phones."[22]

In an interview with the newspaper *Le Monde*, the same author remarked: "Have you noticed that people use their smart phones not for taking pictures or for archiving but to look at something as soon as they have shot it? In a sense, they want to see 'framed,' they want to see themselves or what they observe placed in a frame."[23] To see framed is, most likely, the most specific and most frequent special effect in the everyday communicative space today. Natural vision is not framed: seeing framed changes our relationship with the world. I have often been fascinated by the way children use a small camera or a smartphone that has been given to them: when visiting a museum, they turn around and around the artworks to choose which one they will place in their personal archive; when walking in nature, they set off in search of insects or flowers to photograph. The relationship with the frame stimulates a seeking gaze. The search for framing leads to a keen, inquisitive, precise gaze. Never has the world been scrutinized with so much attention.

In certain circumstances, this gaze is doubled up by a will to communicate: we have entered into the logic of the witness. This is the case, for example, with public protests or demonstrations: it is clear that their participants are concerned to frame so as to show what the news media often do not (such as police brutality), to locate in order to authenticate (they film street signs with street names), to make people react by filming close-ups with a strong emotional charge (the swollen face of a beat-up protestor). It appears to me

21 Laurent Jenny, citing Marcel Proust, ibid., pp. 67–68. (Translation used from *In the Shadow of Young Girls in Flower*, trans. James Grieve, [London: Penguin Classics], 2002, p. 233).

22 Ibid., pp. 67–69.

23 Laurent Jenny, "Libre comme l'art," *Le Monde*, 15 March 2013. In *La vie esthétique*, a section is entitled *Visions cadrées* ("Framed Visions"), op. cit., pp. 67ff.

that most often, however, this desire to see things framed illustrates a wish to transform the world into an aesthetic space. The frame modifies our perception of reality in that it introduces points of reference (the edges of the frame) that lead us to construct relations that do not exist in reality. If we were to observe ourselves looking at the world through the screen of our phone, we would see ourselves shifting this frame until the play of relations appears satisfactory to us and appears "beautiful"; the frame is an operator of beauty. In doing so, we eject from our field of vision everything that is outside the frame. The frame cuts and eliminates, indicating our desire to elude a total vision. But there is more: the frame functions as a *deictic, index*, or *shifter* ("look how beautiful *this* is"), but instead of fixing presence in the moment, it seeks to construct "a state that will be experienced as beautiful in the future."[24] Tomorrow, it will have been beautiful. Fix the world in beauty: now there is a very special effect. We can see in this an opening up to astonishment and wonder but at the same time a way of removing oneself or at least of protecting oneself from the world and its ups and downs. Producing distance, the frame can also lead to cutting oneself off from the world: here the "frame effect" is transformed into a "screen effect."

Conclusion

Just as working on the home movie enables us to raise new questions for cinema as a whole, taking everyday communication space as a field of study requires that we break with our habits when analyzing special effects in the space of cinema and leads us to approach this concept differently. Not only does this re-examination confirm what Metz intuited, that one can only speak of special effects by placing oneself inside a given space of communication, it also demonstrates that in a given space, it is possible to identify degrees of specificity of special effects and extremely diverse modes of functioning (including effects that function in an opposite manner to special effects). This analysis has also brought out forms of special effects that had hitherto not been identified: effects that interpose themselves directly in and on our everyday space, effects tied to our relations with images and their circulation, and effects that change the way we see the world (the zoom, the frame).

24 Robert Musil, "Hier ist es schön," in *Prosa und Stück* (Reinbek, 1978), p. 523. Quoted by Karl Sierek, "C'est beau ici: se regarder voir dans le film de famille," in Roger Odin, ed., *Le Film de famille, usage privé, usage public* (Paris: Méridiens Klincksieck, 1999), pp. 75–76.

More generally, this brief analysis has shown the heuristic value of the concept of communicative space: like changing disciplines, changing the space of communication enables us to "turn the object around."[25]

Bibliography

Allard, Laurence. "*Express Yourself 3.0!* Le mobile comme technologie pour soi et quelques autres entre double agir communicationnel et continuum disjonctif soma-technologique." In *Téléphone mobile et création*, edited by Laurence Allard, Laurent Creton, and Roger Odin (Paris: Armand Colin, 2014), pp. 139–162.

Aumont, Jacques. *Limites de la fiction: considérations actuelles sur l'état du cinéma* (Montrouge: Bayard, 2014).

Foucault, Michel. "Subjectivity and Truth." In *Ethics: Subjectivity and Truth*, edited by Paul Rabinow, translated by Robert Hurley et al. (New York: New Press, 1997).

Gaudreault, André, and Philippe Gauthier. "Les séries culturelles conférence-avec-projection et projection-avec-boniment: continuités et ruptures." In *Beyond the Screen: Institutions, Networks and Public of Early Cinema*, edited by Marta Braun et al. (Eastleigh: John Libbey, 2012): 233–238.

Jenny, Laurent. *La vie esthétique: Stases et flux* (Lagrasse: Verdier, 2013).

Metz, Christian. "*Trucage* and the Film," translated by François Melzer, *Critical Inquiry* 3, no. 4 (Summer 1977): 657–675.

Musil, Robert. "Hier ist es schön." In *Prosa und Stück. Kleine Prosa, Aphorismen. Autobiographisches. Essays und Reden. Kritik* (Reinbek bei Hamburg : Rowohlt Verlag, 1978).

Odin, Roger. "Espaces de communication physiques, espaces de communication mentaux." In *D'un écran à l'autre: les mutations du spectateur*, edited by Jean Châteauvert and Gilles Delavaud (Paris: L'Harmattan, 2016): 331–342.

———. *De la fiction* (Bruxelles: De Boeck, 2000).

———. *Spaces of Communication. Elements of Semio-Pragmatics* (Amsterdam: Amsterdam University Press 2021 [2011]).

Percheron, Daniel, and Marc Vernet. "Entretien avec Christian Metz," *ÇA cinéma* 7-8 (May 1975): 18-51.

Selfiecity. "Introduction: Investing the style of self-portraits (selfies) in five cities across the world." 2014. http://selfiecity.net/.

25 Christian Metz, commenting on the shift from semiology to psychoanalysis, in "Entretien avec Christian Metz," Daniel Percheron and Marc Vernet, *ÇA cinéma* 7–8 (May 1975): p. 34.

Sierrek, Karl. "C'est beau ici: se regarder voir dans le film de famille." In *Le film de famille, usage privé, usage public*, edited by Roger Odin (Paris: Méridiens Klincksieck, 1999): 63–77.

About the Author

Roger Odin is Professor Emeritus of Sciences de la communication at Université de Paris 3 — Sorbonne-Nouvelle, where between 1983 and 2003 he headed the Institut de recherche en cinéma et Audiovisuel (IRCAV). A proponent of semio-pragmatism, he has used this approach to investigate narrative discourse and home movies, and now uses it to investigate the production of moving images with the use of smartphones.

11. Image Capture, or The Control of Special Effects

Philippe Marion

Abstract

This essay begins by considering the tension that exists between special effects and "*trucages.*" Indeed, while the latter seeks concealment and integration, the former are far more exhibitionist in openly revealing themselves. This distinction then leads the author to reflect on a type of special effect that has now become commonplace, namely motion/performance capture imagery. Using the example of *The Adventures of Tintin* (Spielberg, 2011), he shows how this effect maintains a constant "photorealist pressurization," all the while managing to emancipate itself from the photographic nature of camera-produced images. Next, images produced by motion capture are compared with those created with the use of GoPro cameras, as both stand together in our digital age as means to exercise mastery (as "capture") over some exhaustively recordable reality.

Keywords: performance capture, "captation", GoPro, profilmic, controlling effect

How does the cinema, bright star in our digital semiophere, carry out today what contributed so much to establishing its identity: the *captation*-restoring of space-time? That is a big question. The following lines are an attempt to respond to it by means of another paradigm in the process of taking captation's place: that of *capture.*[1] The term is deceptive because it suggests only a slight slippage with respect to captation. We will see that this is not

1 This term is not commonly used in English. The Oxford English Dictionary explains that it entered the lexicon of late middle English and defines it as "a catching at, an endeavor to get, especially by address or art." In French, the verb *capter* is used to mean the picking up of a television or radio signal or the harnessing of some energy. It stems from the Latin verb *captare*

Lefebvre, M. & M. Furstenau (eds.), *Special Effects on the Screen: Faking the View from Méliès to Motion Capture*. Taylor & Francis Group, 2022
DOI 10.5117/9789462980730_CH11

necessarily the case and that, to take up Dany Laferrière's fine idea with respect to literature,[2] an insignificant detail is often a revealing detail that is simply unaware of that which it discloses. My angle of approach may at first also seem unexpected or paradoxical because I will propose the hypothesis that capturing is a perpetual special effect.

The Paradox of Special Effects

Before deciphering the imaginary of the capturing used in the new photo-realist practices of the digital age, I propose to reconsider briefly a conventional distinction, one that makes it possible to distinguish *special effects* from *trick effects*. First of all, let us consider the paradox of the routine nature of special effects as a technique — or at the least their *in-the-process-of-being-made-routine* nature — but also their impact on audiences. These effects are still called "special," even though they are standard practice in films for the general public today. The "special" nature of these now *routine* effects thus requires some thought.

Among other things, this paradox merits analysis from a socio-semiotic perspective. We could place the principle of the erasure or effacement of special effects at the far end of a continuum: they are erased, masked, made fluid, or regulated by the diegesis or by narrative economy. At the other end of this continuum is found the reverse principle: that of a kind of exhibitionism of special effects. Here they stand out as spectacular effects and are asserted as high-impact departures from the norm. Their mission is to place themselves front and center and to describe themselves as effects capable of placing themselves front and center. And here is where a socio-semiotic perspective can come into play. Recall simply that this approach rests on two complementary and connected analyses: that of the "discourse of" and that of the "discourse on." If we look at mainstream cinema today, it appears that the continuum of the integrative erasure of special effects vs. their exhibitionist standing out tends to follow the line that places in tension *discourse of film* and *discourse on film*.

meaning "to catch," to strive to obtain," or to "seek after." *Captare* is also the source of the verb "to capture." The distinction between *captation* and *capture* will be discussed below.

2 "And this is something that appears to have no importance for the moment. But which will quietly take hold, without our being aware, in our memory [...]. [The detail] has a life of its own over which our intelligence, that great decoder, has no power. The brain does not succeed in classifying this invader." Dany Laferrière, *L'art presque perdu de ne rien faire* (Paris: Grasset, 2014), p. 47.

Thus, with respect to the *discourse of film*, popular productions such as *Avatar* (James Cameron, 2009) or *Gravity* (Alfonso Cuarón, 2013) place their special effects under the sign of narrative and diegetic integration. These effects function as a cosmogonic invitation. They are, so to speak, diegetic lubricants. *Gravity* in particular mobilizes its effects to this end: to place the viewer in an immersive space flight experience, to have the sensation of being somewhere in the stratosphere in a tense state of weightlessness. In our habits of participating in the fiction, this diegetic-narrative function on the part of special effects commits our experience of them to integrative fluidity. If it creates too much unevenness, then the viewer is pulled by another force of… gravity, that of meta-language or reflexivity. This is what the critic Ted Hardy-Carnac deplores about the film *L'écume des jours* (*Mood Indigo*, 2013), adapted from Boris Vian's novel and directed by Michel Gondry:

> In film, for a gimmick to be visible, it must occupy one or several shots, it must take its place, take the time to exist, to enter into a rhythm. The film viewer can't re-read a sentence three times and then gobble up what comes after in one glance. If a shot slows up the viewer, the story suffers. While viewers are enjoying a cool effect, they leave the character behind a little.[3]

We see quite clearly here, in passing, the outlines of a now well-known opposition between *attraction* and *narration*.

Necessarily blunt, this bi-polar schema would, of course, require the addition of some nuance. After all, these are only the extreme points of a continuum, and the gulf between "erasing" special effects for the purpose of immersion and consciously celebrating them as high-impact effects is more complex, if only from the users' perspective. Indeed, viewers can let themselves be immersed in the diegesis, in the cinematic illusion, by letting themselves be carried along by the special effects while at the same time enjoying the illusionism and effectiveness of such "trick effects." This is shown clearly in this remark by a Belgian film critic commenting on *Gravity* when it was released: "Watching *Gravity*, viewers are like children astonished by a magic show and obsessed with the question: 'How do they do it?'" The critic adds: "Cuarón is constantly pushing back the boundaries of cinema, with

3 Ted Hardy-Carnac, *"L'Écume des jours" de Gondry: une romance écrasée sous un déluge d'effets visuels*, published 29 April 2013, http://leplus.nouvelobs.com/contribution/849216-l-ecume-des-jours-de-gondry-une-romance-ecrasee-sous-un-deluge-d-effets-visuels.html, most recent consultation on 20 November 2016.

long takes to depict the uncertain movements of his astronauts."[4] Special effects can thus take us into the diegesis even as they "astonish" us. But, at the same time, this fascination can go alongside the reflective question "How do they do it?," showing that the film's visual feats are quite well-concealed behind the photo-realist effect it produces. We perceive no "trick," just as we do not understand how the *prestidigitateur* (conjuror, magician) works, or rather, if the reader will allow this somewhat unpronounceable French word play, how the *presti-digital-isateur* does it.

The effacing-concealing mentioned above thus persists insofar as we do not understand and remain impressed while letting ourselves slip into the fiction. This is the interrogative fascination of a viewer who grants the mystery of this attraction-effect, a slight variation on Octave Mannoni's famous formulation, "*Je sais bien, mais quand même* [...]"[5] (I know well, but all the same [...]). I am quite aware that this weightless scene is a trick effect, but just the same I believe it, I believe in this illusion, and I take advantage of its photo-realist effectiveness.

Turning now to *discourses on film*, it remains the case that the production of hit films such as *Star Wars* (George Lucas, 1977; and all subsequent iterations of the franchise) is deliberately accompanied by marketing and media discourses that take pleasure in celebrating what goes on behind the scenes and in decoding to a degree the special effects. This, moreover, is one of the principal aims of the "making of" documents accompanying and seeking to provide added value to DVD and Blu-ray releases. Frank Kessler has proposed this colorful formulation for describing this discourse on special effects: "An explanatory meta-discourse and display of the machination."[6] The question then is to verify to what extent the adjective *special* is warranted and whether such effects justify the attention and even the viewer's decision to go out to the movie theater. For *Gravity*, viewers were even advised to go to "selected theaters" in order to fully benefit from the 3D immersion and the dynamic "womb-like" enveloping of the sound system. But here, too, the dichotomy between effacement and exhibition should be qualified. The example of *Gravity* illustrates well that meta-discourses on technological feats should not overshadow the film's goal of "holistic" immersion as its "overall" special effect, so to speak.

4 Hugues Dayez's film program on the radio news show *La Première*, RTBF radio station (French-speaking Belgium) on 21 October 2014.

5 Octave Mannoni, *Clefs pour l'imaginaire*, Paris : Seuil, p.10sq.

6 Comment made by Frank Kessler at the "Magic of Special Effects" conference (Montréal 2013).

A Special Effect Made Routine: Performance Capture

In the copious number of trick and visual effects brought about by digital technology, there is one technique that has not gone unnoticed: motion capture and its somewhat more advanced spin-off, performance capture. With reason, Jean-Baptiste Massuet grants this technique the status of the "synecdoche of virtual cinema."[7] Recall that this technique makes it possible to use digital encoding to record the positions of objects or living beings in order later to manipulate a virtual replica or avatar of them by computer. "MOCAP," "encoded recording," "virtual avatars," "animation": the terminology says a lot about the hybrid or composite nature of the technique, in keeping with present-day digital transformations. Here we find, in this very first description, the elements of a grasping of the world governed by a photo-realist intention mixed with a filmic ethos inherited from animation.[8] In fact, motion capture gives rise to a kind of "digital animation."[9] In short, it is a kind of hyper-trick effect seeking to make the captured world entirely manageable by controlling the space-time of the characters. As Peter Jackson noted with respect to the adaptation of *The Adventures of Tintin: The Secret of the Unicorn* that he co-orchestrated and produced with Steven Spielberg in 2011: "You can choose the best actors for each role without worrying about whether they really resemble Hergé's character. They are as perfectly and faithfully alive as actors in real shots, the synthetic imagery functioning like digital make-up in a sense."[10]

Thus, the Great Image Maker (or *grand imagier*) of *The Hobbit* trilogy (2012; 2013; 2014) brings out the mimetic dissociation effect brought about by motion capture. According to Jackson himself, the shooting-recording is of an "it's as if" nature, or "they said that," because the filmmaker is content to simulate real shots — in a way not unlike that described some time ago by Lev Manovich.[11] In addition, this simulation is described as "faithful" and "perfect" in the sense that the image quality is equivalent (at least in terms of facial and bodily movements) to what real shots would produce while

7 Comment made at the "Magic of Special Effects" conference (see note 6 above).

8 These elements are discussed at greater length in particular in Philippe Marion, "Les avatars du cinéma: De la caméra GoPro à la *Performance Capture*," in *Communication & langages*, (182) 4 (December 2014), p. 69ff.

9 This is the expression used by Peter Jackson himself. "Preface" by Steven Spielberg and Peter Jackson in Chris Guise, *Artbook, Les aventures de Tintin* (Wellington and Brussels: Éditions Weta/ Moulinsart, 2011), p. 12.

10 Jackson, "Preface" to *Artbook, Les aventures de Tintin*, op. cit., p. 14.

11 Lev Manovich, *The Language of New Media* (Cambridge, MA: MIT Press, 2001).

avoiding the constraints or even the botched shots that can be associated with live action shooting.

Actors are thus disassociated from their photo-realist appearance, liberated from "real world" features under the "analog" regime of captation-restoring, while the photo-realist illusion is maintained. We are faced, then, with an *imitation* of real shots. Through the power of a permanent special effect fused with the technique itself, performance capture is thus able to maintain a constant *photo-realist pressurization.*

Here we find all the qualities of the control and management of the means proper to the digital. A film can now be managed like a digitized storyboard. But what is more, by preserving the singular kinetic "anima" of each actor — his or her *kinetic fingerprint*, in Renée Bourassa's terms[12] — a new verisimilitude is achieved which we might call "anima-realism."[13] This mimetic dissociation implies that the captation of movement — "as close as possible to the actor's body" — is then given "digital makeup" through which a photo-realist effect is revealed (is recreated) in the image. In this sense, the "makeup" is no longer a simple, secondary cosmetic operation or even a special effect both remarkable and occasional: it is the decisive process that gives back a body to captured motion. It is what enables animation to be re-embodied.

We are thus dealing not with captation-restoring, as in analog cinema, but rather with captation-*reconstituting*. This gives us a realism of performance, an "animatic" realism that is no longer truly that of photo-realist captation, too haunted perhaps by the memory of photographic fixing. Yet all this is not without a kind of paradox: liberated from photo-realist *captationing*, actors have to focus on their acting style, on the way they embody a character. Thus Gad Elmaleh, part of the cast for Spielberg's *The Adventures of Tintin*, insists on the fact that "motion capture brought back to the very essence of the actor's job, which consists in making the body speak."[14] Andy Serkis, in the role of Captain Haddock, underscored this by drawing attention to the need to be parsimonious with a technique as

12 Comment made at the "Magic of Special Effects" conference (see note 6 above).

13 See Philippe Marion, "Tintin façon Spielberg: Une nouvelle manière de penser le cinéma?," presentation at the conference "The Impact of Technological Innovations on the Theory and Historiography of Cinema," forthcoming in Nicolas Dulac, ed., *Du média au postmédia: continuités, ruptures/From Media to Post-Media: Continuities and Ruptures* (Lausanne: L'Âge d'Homme).

14 Gad Elmaleh's comments were reported by Philippe Manche in an interview with Jamie Bell in the French-language Belgian daily newspaper *Le Soir*: "Jamie Bell: 'Tintin est très complexe'," in *Le Soir*, Wednesday 26 October 2011, http://archives.lesoir.be/jamie-bell-tintin-est-tres-complexe-_t-20111026-01MVR5.html

precise as motion capture in its direct "recording synergy" with the body: "the technology is very precise and responds to the slightest movement, so you can't overact. You also have to be extremely conscious of the way you use your own body."[15]

Photo-Realist Pressurization

Motion capture, as an extended and now routine special effect, thus ensures the stability of the photo-realist pressurization discussed above. This is summed up in his own fashion by Peter Jackson: "We wanted them [the actors] to resemble Hergé characters, but in a visibly realist fashion."[16] It seems significant in this respect that several articles in the press at the time of the film's release described Jamie Bell as the actor who "dubbed" Tintin. "Dubbed" and not "played"? Another revealing detail: that of refusing to let an actor play Tintin, in the spirit of a classical photo-realist recording. The idea of "dubbing" suggests, on the contrary, respecting a character already embodied in outline in the peculiar graphic-narrative system that makes up the "clear line" typically attributed to Hergé. The illusion of control is maintained, along with the pleasure tied to working with the overall aesthetic of the clear line. In a sense, Jackson and Spielberg's crew could not just settle for putting on film — and thus on the screen — a *simple profilmic reality.* Because Jamie Bell's face could never have resembled the oval mirror with a tuft of hair that is Tintin's head as drawn by Hergé. And Andy Serkis could never have expressed Captain Haddock's grimaces.

Thanks to this particular use of motion capture, and without discussing the quality or "success" of Spielberg and Jackson's work, Hergé's characters remain graphic puppets ready to seek adventure.[17] All the while maintaining through simulation the effect of a captation-restored profilmic world. At the same time, the integration of these characters into an environment

15 Andy Serkis's comments were reported by Philippe Manche in a different article also published in the same issue of *Le Soir* on Wednesday 26 October 2011: "Tintin au pays de Spielberg" ("Tintin in the Land of Spielberg"), (http://www.lesoir.be/archives?url=/culture/cinema/2011-10-26/tintin-au-pays-de-spielberg-872686.php; accessed 9 October 2013).

16 Jackson, "Preface" to *Artbook, Les aventures de Tintin*, op. cit., p. 14.

17 For a more in-depth discussion of this question, I take the liberty of referring the reader to my essay "Spielberg au pays de la ligne claire: de la graphiation à la performance capture," in Marco Grosoli and Jean-Baptiste Massuet, eds., *La capture de mouvement ou le modelage de l'invisible* (Rennes: Presse Universitaire de Rennes, 2013), pp. 203–224.

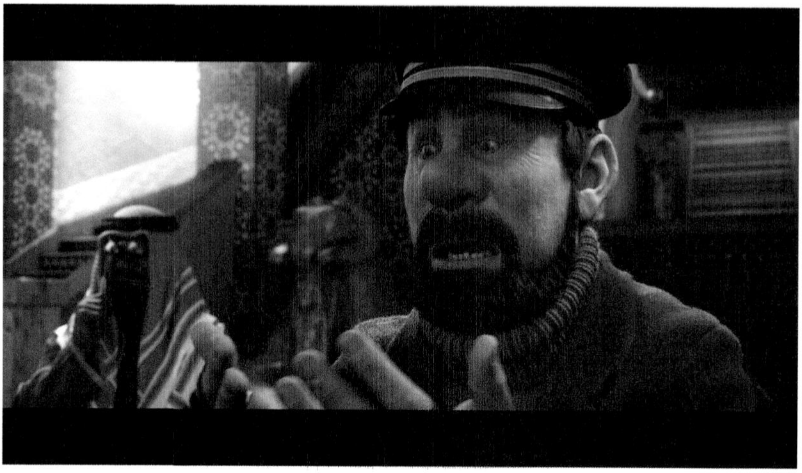

Fig. 11.1 Integration of 'real' photographic shots and CG imagery in *The Adventures of Tintin: The Secret of the Unicorn* (Steven Spielberg, 2011).

"with the credibility of a film made out of real shots"[18] (though the Unicorn sails on a sea bluer than nature) makes it possible, in tandem with the 3D effect, to respect the specific tension of Hergé's graphic poetry: that of reconciling the manifest realist power of some of the motifs depicted (the vehicles and modes of transportation, for example) with the highly simplified representation of the characters facial features (such as Captain Haddock's large nose) (Fig. 11.1).

The Memory of Animation

If we were to step back a bit and try to detect the sense of a technique such as performance capture, it becomes apparent that it crystalizes quite well the effect of control and malleability encouraged by digital encoding while at the same time making it possible to create a strong illusion — a "special avatar" — of photo-realism. But let me stress once more that such photo-realism has been simulated by means of a digital-based reconstitution. The profilmic — mimicked, simulated, and not *"captated"* — is now controlled. The inevitable surplus chaos proper to the photographic captationing of reality is brought under control once the very principle of captationing is modified. The untamed profilmic is tamed and contained in a manageable image. On the theoretical and conceptual level, moreover, the very notion of

18 Jackson, "Preface" to *Artbook, Les aventures de Tintin*, op. cit., p.14.

the profilmic (in the sense of what is in front of the camera and is captured by it) is taking on water everywhere and losing much of its relevance. As Martin Barnier remarks, this evanescence of the profilmic goes hand in hand with the widespread refusal of camera-centrism.[19] As we shall see below, digital captation is always-already digital "manipulation" through encoding. In this sense, there is no longer any "pure" profilmic; there is only a composite profilmic, mixed with the "filmographic" (the work associated with the filmic). Unless it is the reverse!

In a sense, we can also see in this new demiurge-like control offered by the digitization the traces of animation's return to power in the cinematic sphere. Thanks to motion capture, each movement and facial expression is, as we have seen, captated and transferred to the digital realm. Thanks to this special technique — this expanded and now routine special effect — we are thus witnessing a dynamic hybridization of photo-realism and animation. It is in this spirit that André Gaudreault and I have proposed the neologism "*animage*."[20] One of the effects of the digital in general and of motion capture in particular thus consists in generating a new system of expressive coherence. This system, and the demiurge-like potential for control that accompanies it, can serve as a remarkable tool for creators and filmmakers wishing to control their work and avoid the uncertainties of a photo-realist profilmic captationing. We might think of the virtual delight of someone like Hitchcock, for whom the film shoot was, by his own admission, a kind of test of his obsession with complete control. This is something he acknowledged, moreover, in one of his famous interviews recorded by Claude Chabrol and François Truffaut: "I always make my films on paper [...] And when I begin shooting the film, it is finished. Finished to such an extent that I would prefer not to have to shoot it. I have the whole thing in my head: topic, framing, dialogue, everything."[21]

Captivating Capture...

A short detour through semantics will make it possible to glimpse a few significant issues raised by the term "capture" itself, which in this case we

19 Comment made at the "Magic of Special Effects" conference (see note 6 above).

20 See André Gaudreault and Philippe Marion, *The End of Cinema? A Medium in Crisis in the Digital Age* (2013), trans. Timothy Barnard (New York: Columbia University Press, 2015).

21 André Bazin, Serge Daney et al., *La politique des auteurs: les entretiens* (Paris: Cahiers du cinéma, 2001), pp. 155–156.

can contrast with the related term "captation." Both are common terms in French, and while they share the same etymon, they are quite different when it comes to the connotative nuances that separate them and the imaginary to which they give rise. In the case of "captation," when the word is applied to the profilmic reality of analog cinema, it suggests a kind of preservation of the autonomy of this captated reality (like the airwaves one can pick up or "*capte*" with a radio). In the case of "capture," on the contrary, the word suggests a lack of autonomy, a loss of freedom, a subjugation. Captated reality is not captured reality: by forcing the matter a little, we might think of captated reality as the recording of something that otherwise exists independently of the recording, in the form of what Étienne Souriau called the "afilmic."[22]

Captured reality, for its part, no longer exists outside the prison house (or perhaps window in this case!) of the image. The taking of an impression, found in photo-realist captation of the profilmic,[23] is contrasted with the idea of appropriation, of an authoritarian control of reality which is then encoded without striking a blow. Capturing is thus located on the side of a digital "subjugation" of the profilmic. Here again we see the figure of the demiurge, of the reassuring exercise of a control over what is found in front of the camera lens. This is why, I would suggest, we can construe the concept of capture as a kind of emblem for digital filmmaking.

This reading makes it possible, in particular, to give "making of" documents a further interpretive turn. This kind of meta-communicational discourse on what goes on behind the scenes as it pertains to the "secrets" of filmmaking abounds in the current ecosystem of film consumption. A great number of DVDs, Blu-rays, YouTube videos, and websites offer viewers "making of" documents, as if they were now inseparable from the film itself, especially in the case of blockbuster films. These "making of" films could be summed up by this formulation: "give voice to the captured!" Indeed, the para-filmic ceremonial has a function that has become important in the dawn of the all-digital: to reassure us that there still exists a profilmic independent of the finished film.

Consider the following publicity image of Sandra Bullock and George Clooney used to promote Cuarón's *Gravity* (Fig. 11.2). Face to face, wearing

22 Used for the first time by Étienne Souriau in "La structure de l'univers filmique," in *Revue Internationale de filmologie* (7–8) 2e trimestre, (1951), p. 240. To the extent, as we have seen, that the concept of the profilmic remains relevant (and in force), we will see below that digital "capturing" is always-already digital "manipulation" through encoding. If anything of the profilmic remains, it is now managed only in a composite state.

23 This section summarizes and adapts an argument presented in André Gaudreault and Philippe Marion's *The End of Cinema?* op. cit.

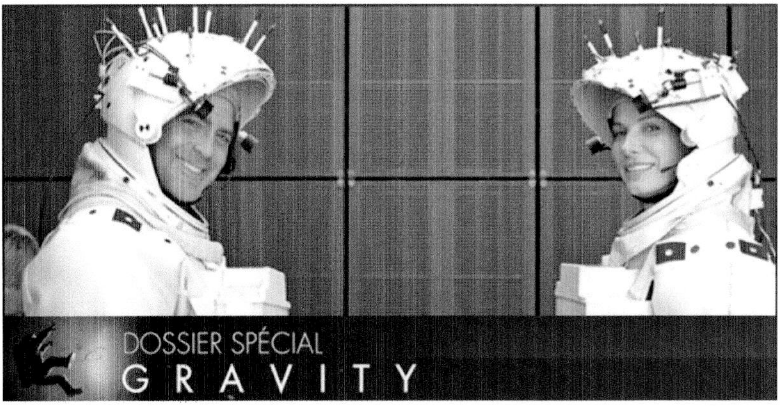

Fig. 11.2 Publicity shot of Sandra Bullock and George Clooney for *Gravity* (Alfonso Cuarón, 2013).

their astronaut helmets and sensors: "Yes, they really were on a film set!" "Yes, they had to work hard to play their character's part!" "No, technology and digital trick effects did not do everything!" This is what the "making of" tells us: look at the labor, the bloopers, the second thoughts! There really were actors who "lived in" this good old profilmic world and acted before a movie camera! Here we can see a kind of nostalgia for the lost profilmic.

Indeed, today we know that the digital affects every stage of the film shoot, which is not without consequences for the ontological status of the resulting image. In fact, digital recording is not the result of the impression of light on film stock but the product of encoding that makes possible all kinds of manipulations from the outset. We can even think of the digital captation of the image as merging so to speak with the sort of "ontological trick effect" that is encoding itself.[24] This is what makes possible such easy and unlimited transfers and treatments: encoding, by nature, enables every kind of manipulation. All these features define what D.N. Rodowick calls the "digital event," which "corresponds less to the duration and movements of the world than to the control and variation of discrete numerical elements internal to the computer's memory and logical processes."[25] Immediately encoded and hence dematerialized, data thus lend themselves at their very source to every possible kind of manipulation and rearrangement. In this sense, we can say that there is a kind of *plasticity* intrinsic to the digital image.

Thus, the traditional conception of editing is also being modified. First of all because a kind of editing we could describe as *intrinsic* exists from

24 D.N. Rodowick, *The Virtual Life of Film* (Cambridge, MA: Harvard University Press, 2007), p. 166.
25 Ibid., p. 173.

the outset in the production of every digital image: even when it is not retouched, it is always-already a "translation" through encoding and hence the result of a kind of editing in itself. Second, because these data, already manipulated from the outset by dematerialization, lend themselves to all kinds of other "modular" treatments and reconstructions. As Rodowick explains, "here montage is no longer an expression of time and duration; it is rather a manipulation of the layers of the modularized image subject to a variety of algorithmic transformations. This is what I call the digital event."[26]

Seen from this perspective, it is possible to confirm the idea that the digital renders trick effects, or even special effects, ontological. Indeed, special effects are no longer an optional supplement or one inherent to certain genres but a practice inseparably linked to the very elaboration of film images. This may mean that mass cinema has returned to a Mélièsian aesthetic, as Jacques Aumont points out: "in its social form as entertainment, cinema has made a massive return to the path set out by Méliès, that of the trick effect, of direct intervention in the image, of retouching, control, drawing."[27] Today, the widespread use of special effects is no longer the result of a magical manipulation whose sources are often hidden, as in Méliès's day; it is now incorporated into and "part and parcel," so to speak, of the film medium itself, as part of its routines.

The GoPro Camera, or the High-Performance Performer

At this point in my discussion, a fairly obvious connection can be made with the imaginary of the popular GoPro video camera with respect to my remarks on "making of" film segments. Here I will focus on the camera's users and the interpretation one can give to their way of capturing images with it.

"Wear it. Mount it. Love it." The slogan's expeditious prosody effectively embodies the formidable ease with which one can use the GoPro, the self-proclaimed most multi-functional video camera in the world. Since its invention in 2001 by the California surfer Nick Woodman, the camera has been conceived as a recording device that is protected from shocks and can be easily attached to the body of athletes in action while giving a high-quality recording. In 2016, the HD Hero camera offered the ability to record up to 60 frames per second in high-definition video. High-definition video, another form of magic and extended special effects, also invites

26 Ibid., p. 166.

27 Jacques Aumont, *Que reste-t-il du cinéma?* (Paris: Vrin, 2013), p. 62.

brief contextualizing. We know that the rate of 60 frames per second is well beyond the classic sound-era rate of 24 frames per second. This is also what makes possible the High Frame Rate (HFR) system of 48 frames per second in professional film recordings. HFR has been used in particular by Peter Jackson, beginning in 2012 with *The Hobbit: An Unexpected Journey*. With James Cameron's *Avatar 2*, HFR will attain 60 frames per second. Like the amateur GoPro camera, the HFR technique has benefited from often catchy media coverage, which provides a good definition of the "novelty effect": "Twice as many images for a heightened sensation of ultra-realism and fluidity. The Gaumont and Pathé cinemas invite you to visit their movie theatres to discover this major upheaval in the history of cinema."[28]

HFR is thus part of an ultra-realism approach that André Bazin, with his idea of total cinema, would not have rejected because this technology gives hitherto unseen precision to movements. With HFR, we thus gain in transparency by way of an effect of fluidity while at the same time heightening the realism of the image (and here we have again the magic illusion of realism).

This same special effect of extended functional ultra-realism, moreover, is vaunted in most advertisements for the GoPro camera. Apart from the quality of its HD images, this device features numerous accessories to make it easy to attach onto your gear or onto a helmet, handlebars, dashboard, wing of a remote-controlled plane, or even a drone. Unlike a smartphone, the GoPro is a camera that you "put on" or wear more than you carry, which furthers the device's immersive qualities and its exceptional image quality.

Meeting with keen success amongst athletes, and especially among "extreme sports" aficionados, the GoPro camera makes it possible to join an experience with its recording-capturing: an athletic act — or more broadly any event occurring in daily life — is seized, recorded, and even broadcast at the same time as it is experienced. In "real time." The images produced with this camera are high-resolution and often take in a singular, stunning view, while the performance is transformed into a sort of non-stop special effect made routine. The pirouette of the skier or skater is experienced from within and merges with the gaze of the "performer"; we slip into the trough of the wave in the company of the surfer. As Roger Odin points out more generally, "we now find ourselves making and experiencing the special effect."[29] In a much less amusing way, this "performer" can also be a real

28 http://www.toulouseblog.fr/actualite-19231-hobbit-dans-salles-toulousaines-48images-seconde.html (accessed 9 October 2013).

29 Comment made at the "Magic of Special Effects" conference (see note 6 above).

murderer or a terrorist carrying out an attack. Think, for example, of the cold-blooded killings that took place at the Jewish Museum in Brussels in May 2014. The murderer wore a GoPro camera that recorded his aggression as it unfolded, providing the special effect of killing.

The important originality of the GoPro system thus lies in the way it merges (and confuses) the actor and the person shooting the video, and, especially, in the user's attitude and practice. Here the special effect is not so much seen as it is experienced. Or rather, it is experienced and seen in the same "making and experiencing" identified by Odin. Making images and the spectacular are no longer seen as inseparable from the performance as experienced. With the GoPro camera, we immerse ourselves in the representation of experienced-shown-recounted performances. We can thus understand much better why these cameras have the significant first name Hero, with a prepossessing slogan aimed at one's individual identity: "Be a Hero," as if the subjective image were now a condition of recognition for heroism.

This is a major trend in our media culture, made routine for a younger generation especially through the widespread practice of selfies: life merges with its shared image. These images, acting as new vectors of media credibility, often with a high emotional quotient (for the GoPro, this involves the thrill of risk and daring-do), set themselves up as irrefutable proof that the I-filmer-actor have completely experienced this exploit. Even if this performance can sometimes be limited to the mere fact of having simply been in the right place to capture snatches of the world.

From a narratological perspective, and with reference to Paul Ricoeur's famous narrative theory of "three-fold mimesis,"[30] we can see here a form of symptomatic acceleration, or rather merging, of the three interconnected regimes that, Ricoeur maintains, underlie the essence of narrative. The lived time of preconfiguration (mimesis 1) now blends with emplotment — the visual narrative of lived performance (mimesis 2). Understood that these captated-lived performances are increasingly transmitted live on various networks or platforms, mimesis 3, the configuration that pertains to reception, is now also merged with the real time of the lived-captated-recounted performance.

There remains at least one question to consider: the perpetual state of special effects engendered by digitization, this ontological *trucage* that Metz anticipated, is it not also that which defies digitization? Special effects sought to conceal themselves by becoming routine, by becoming banal

30 Paul Ricoeur, *Time and Narrative*, 3 vols. (1983–85), trans. Kathleen McLaughlin and David Pellauer (Chicago: University of Chicago Press, 1984–1988).

through ubiquity. Indeed! But paradoxically the effect of this was to taint the entire visual system with artificiality. This, in any event, is what those nostalgic for the "silver (gelatin) age" deplore, namely the coldness of digital films and digital effects:

> At the core of the difference between silver-based film and digital is the absence of the shutter. No more flicker. No more heartbeat. The persistence of vision isn't called to the rescue to make possible the reproduction of movement using photograms. Film is made of still photographs after all. But the digital film is not. Underneath there is a grid of pixel-size slots, and it is fixed. Somehow the pixel makes what you see an icon; it is graphic and not sensorial[31]

No more flicker. No more heartbeat! It is as if the perpetual special effect generates meta-artificiality. Indeed, this may be the syndrome of the digital's lack of differentiation.

Bibliography

Aumont, Jacques. *Que reste-t-il du cinéma?* (Paris: Vrin, 2013).

Dulac, Nicolas, ed. *Du media au postmédia: continuités, ruptures / From Media to Post-Media : Continuities and Ruptures.* (Lausanne: L'Âge d'Homme, Forthcoming.).

Elmaleh, Gad. "Jamie Bell: 'Tintin est très complexe'," Interview by La Rédaction, *Le Soir* (26 October, 2011). https://www.chicagomanualofstyle.org/tools_citationguide/citation-guide-1.html#cg-interview.

Gaudreault, André, and Philippe Marion. *The End of Cinema? A Medium in Crisis in the Digital Age*, translated by. Timothy Barnard. (New York: Columbia University Press, 2015).

Guise, Chris. *Artbook, Les aventures de Tintin.* (Wellington and Brussels: Weta/ Moulinsart, 2011).

Hardy-Carnac, Ted. "'L'Écume des jours' de Gondry: une romance écrasée sous un déluge d'effets visuels," *Nouvel Obs* (29 April, 2013). https://leplus.nouvelobs.com/contribution/849216-l-ecume-des-jours-de-gondry-une-romance-ecrasee-sous-un-deluge-d-effets-visuels.html.

31 Babette Mangolte, "Analog Versus Digital, The Perennial Question of Shifting Technology and Its Implications for an Experimental Filmmaker's Odyssey," in *Camera Obscura, Camera Lucida: Essays in Honor or Annette Michelson*, eds., Richard Allen and Malcolm Turvey (Amsterdam: Amsterdam University Press, 2003), p. 264.

La politique des auteurs: les entretiens. (Paris: Cahiers du cinéma, 2001).
Laferrière, Dany. *L'art presque perdu de ne rien faire*. (Paris: Grasset, 2014).
Mangolte, Bebette. "Analog Versus Digital, The Perennial Question of Shifting Technology and Its Implications for an Experimental Filmmaker's Odyssey." In *Camera Obscura, Camera Lucida: Essays in Honor of Annette Michelson*, edited. by Richard Allen and Malcolm Turvey (Amsterdam: Amsterdam University Press: 2003): 261–274.
Mannoni, Octave. *Clefs pour l'imaginaire*. (Paris: Seuil, 1969).
Manovich, Lev. *The Language of New Media*. (Cambridge, MA: MIT Press, 2001).
Marion, Philippe. "Les avatars du cinéma: De la caméra GoPro à la *Performance Capture*," *Communication & Langages* 182, no. 4 (December 2014): 61–-76.
Marion, Philippe. "Spielberg au pays de la ligne claire: de la *graphiation* à la *performance capture*." In *La capture de mouvement ou le modelage de l'invisible*, edited by Marco Grosoli and Jean-Baptiste Massuet (Rennes: Presses Universitaires de Rennes, 2013): 203–224.
Ricoeur, Paul. *Time and Narrative*, 3 vols. Translated by Kathleen McLaughlin and David Pellauer. (Chicago: University of Chicago, 1984-88).
Rodowick, David Norman. *The Virtual Life of Film*. (Cambridge, MA: Harvard University Press, 2007).
Souriau, Étienne. "La structure de l'univers filmique," *Revue internationale de filmologie* 7-8 (2e trimestre, 1951): 231-240.

About the Author

Philippe Marion is Professor emeritus at the École de communication of the Université catholique de Louvain (UCL). Co-founder of the Observatoire du récit médiatique (ORM) and of the Groupe interdisciplinaire de recherche sur les cultures et les arts en mouvement (GIRCAM), he serves as Director of the research unit on Media Analysis at UCL and administrator to the Collectiana Foundation. A specialist in the areas of media narratology and visual culture, he has authored several volumes, including *Schuiten, filiation* (2009) and *The End of Cinema? A Medium in Crisis in the Digital Age* (2015 [2013], with André Gaudreault).

Films

12. Murnau's *Sunrise*: In-Camera Effects and Effects Specialists[1]

Janet Bergstrom

Abstract

During the late 1920s, cinematographers in the US insisted that they should supervise outside specialists. An unforgettable "special effects" shot in *Sunrise* was photographed according to the needs of outside specialist Frank Williams's patented process and his own lab. Understanding the Williams shot encourages us to see other unusual technical processes throughout *Sunrise* as they served Murnau's drama rather than innovation for its own sake. Studying the achievement and function of specific shots in *Sunrise* leads to the larger debate about creative technical-aesthetic control that is similar to modern-day arguments about who should determine the final look of a film: digital postproduction specialists or the cinematographer, production designer, and director working together.

Keywords: F.W. Murnau, *Sunrise*, Frank Williams, special effects

Introduction

Sunrise (1926–27) was made toward the end of the era in which in-camera "special effects" — achieved through the cinematographer's close collaboration with the director and the art director — were about to give way to "effects specialists," then to Visual or Special-Effects Departments, and

1 Grateful acknowledgements to Kevin Brownlow; Karl Thiede; Richard Koszarski; Werner Sudendorf; Dave Kenig; and Jonathan Erland, a founder of the Visual Effects Society, who brought the Williams shot in *Sunrise* to my attention and has been an enthusiastic, erudite commentator. Most sincere thanks to the Academy Film Archive for frame enlargements from their restored 35mm print of *Sunrise*: Michael Pogorzelski, May Haduong, and especially D.J. Ziegler for her care in taking the photographs.

Lefebvre, M. & M. Furstenau (eds.), *Special Effects on the Screen: Faking the View from Méliès to Motion Capture*. Taylor & Francis Group, 2022
DOI 10.5117/9789462980730_CH12

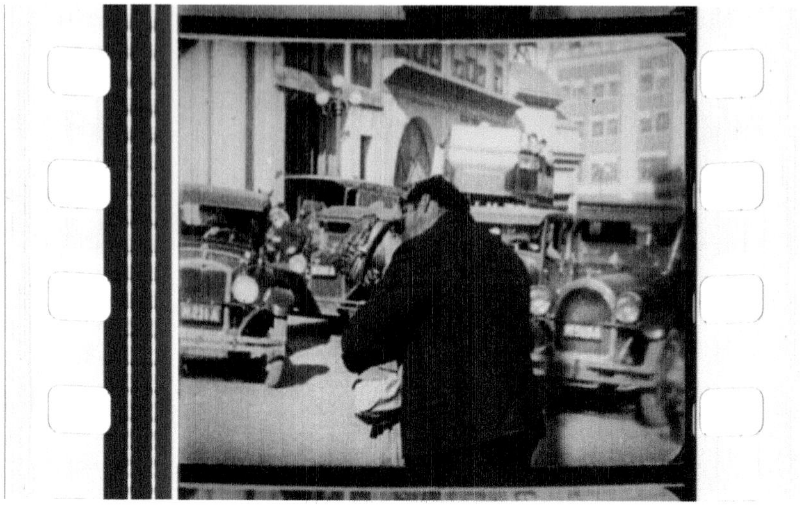

Fig. 12.1 Still from *Sunrise* (F. W. Murnau, 1927).

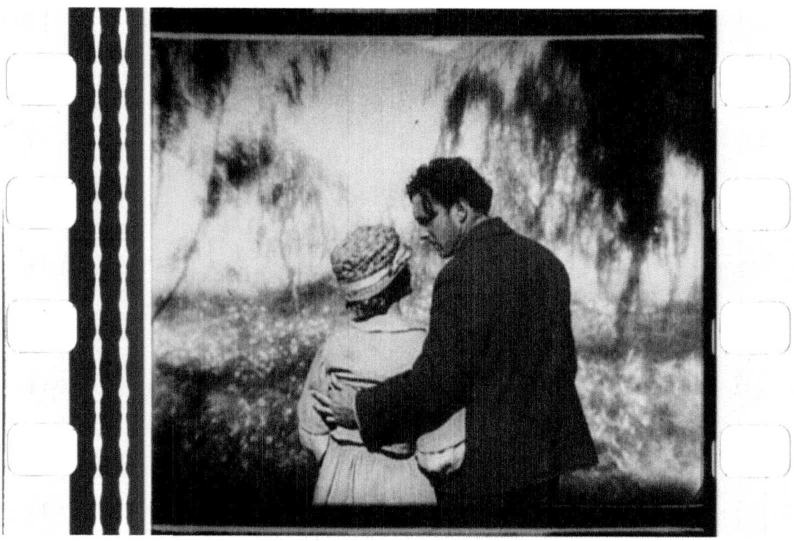

Fig. 12.2 Still from *Sunrise* (F. W. Murnau, 1927).

then to departments in which optical printing quickly became essential, particularly in view of the impossibility of accomplishing many in-camera effects because of sound recording (Figs. 12.1 and 12.2). One of the consequences of these changes was the potential displacement of supervisory control previously held by Directors of Photography.

In George E. Turner's historical overview "The Evolution of Special Visual Effects," "The Specialists" were the sign of a new phase:

> Photographic effects (like the motion picture itself) had by the mid-1920s gone far beyond the "novelty" stage of development to become an integral part of mainstream film production. In the United States, but not abroad, the studios established special departments made up of artists, cinematographers, model builders, mechanical effects technicians, and optical cameramen. The savings in production costs made possible by these departments made the "high salaried experts" very worthwhile investments.[2]

In a later section titled "The German Influence," Turner clarified one of the differences between American and German cinematography in this regard that is directly relevant to *Sunrise*. Murnau's first film for Fox was proudly advertised by William Fox himself as a German-American production, bringing German technical practices into a major American studio. In Germany, wrote Turner, "[t]he production cameraman, with the collaboration of mechanical effects experts, was responsible for all visual effects supervision as well as principal photography [...] When [Karl] Freund relocated in Hollywood in 1930, he was shocked to learn that visual effects work was handled by specialists."[3]

According to Charles G. Clarke, extending the use of techniques that he and Fred Sersen had used at Fox in 1929 compositing filmed objects with miniatures and matte painting that had proved effective and low-cost resulted in the establishment of "Photographic Effects Departments." "At one studio," wrote Clarke, "this Department is now under the supervision of its Art Department! To completely eliminate the cinematographer, they have retitled it the 'Visual Effects Department.' What a switch from those men who originally conceived it!"[4]

2 George E. Turner, "The Evolution of Special Visual Effects," in *The ASC Treasury of Visual Effects*, George E. Turner, ed., (Hollywood: American Society of Cinematographers, 1983), p. 35.

3 Ibid., pp. 37–38. In an interview conducted in 1964, Karl Freund was asked about how he worked in Berlin, if he was "free to experiment," about the facilities available to him there. He replied: "At the time we did all these ourselves, because don't forget, at that time, we didn't have any optical printers. We had to do everything in-camera. Every lap dissolve had to be done in-camera. We didn't have any kind of trick stuff that you have today, when you don't know actually, was that the cameraman or the process department doing it?" Karl Freund, interviewed by Frank Camino, 1 June 1964, UCLA Special Collections, Collection 543, Box 2, audio tape transcription by the author.

4 Charles G. Clarke, A.S.C., *Highlights and Shadows: The Memoirs of a Hollywood Cameraman* (Metuchen, NJ: Scarecrow Press, 1989) p. 88. His example was *The Sin Sister* (1929). Several years earlier, Murnau's art director, Rochus Gliese, devised complex compositing for the first moving shot in *Sunrise* (1926–27), as described later in this paper. Although both films were made for

Gordon A. Chambers, West Coast Division, Motion Picture Film Department, Eastman Kodak Company, in his essay "Process Photography" in the American Society of Cinematographers' 1931 *Cinematographic Annual*, indicated the importance that the optical printer had gained by that year within specialized effects departments. "With increased development in the application of process photography has come the creation in the various studios of a 'Special Effects Department,' which devotes its time entirely to the creation of these illusions. One of the most useful tools of the specialists who comprise these departments is the optical printer."[5]

And in his article "Optical Printing" in the same volume, Lloyd Knechtel, Head of the Special Effects Department at RKO Radio Studios, wrote: "Within the past few years the motion picture industry has almost universally adopted the system of optical printing in its many and varied uses and forms, and has found it invaluable as both an artistic and an economic aid."[6] It is worth noting, in the movement toward specialization within cinematography, that Knechtel was called "Supervising Cinematographer of Special Effects for RKO Pictures Studio" in a full-page ad in the same book.[7]

During the late 1920s, cinematographers in the US insisted that they should be supervising outside specialists. One of the most famous effects shots in *Sunrise* employs one such specialist's patented process, called a Williams shot. Interrogating that shot in *Sunrise* can help us pay attention to other kinds of special effects throughout the film — with emphasis on how they were achieved technically in view of their dramatic purpose, not technical innovation for its own sake. This essay is about the technical achievement and function of specific shots in *Sunrise* as well as the larger debate about creative technical-aesthetic control. That debate is not so different from modern-day arguments about who should determine the final look of a film — digital postproduction specialists or the cinematographer, production designer, and director. This separation of responsibilities was new and became the model for film production to follow. The issues that this division raised are reflected in today's films that require the skills of multiple effects specialists.

Fox, the personnel were different and the techniques used to achieve the special effects in *Sunrise* were kept quiet. Clarke emphasized a change in decision-making and a separation of technical work that changed the studio system, whereas the combination of forced perspective art direction and cinematography in *Sunrise* was new to Hollywood and assimilated slowly and piecemeal, if at all.

5 Gordon A. Chambers, "Process Photography," in *Cinematographic Annual*, ed., Hal Hall (Hollywood: American Society of Cinematographers, vol. 2, 1931), p. 224.

6 Lloyd Knechtel, "Optical Printing," *Cinematographic Annual* (vol. 2, 1931), p. 267.

7 Advertisement 30, in *Cinematographic Annual* (vol. 2, 1931) [n.p.].

Sunrise and Visualization

Before it won Academy Awards, and even before it was released in the fall of 1927, *Sunrise*, F.W. Murnau's first American film, was said to be required viewing for Hollywood's technical personnel.[8] At the Los Angeles premiere, the principal actors were repeatedly applauded, and there was still more applause, according to *Los Angeles Times* reviewer Edwin Schallert, "chiefly for camera effects, for the poignancy of certain scenes and for performances. It was frequent and a tribute."[9] The stunning cinematography so much praised in Murnau's German films, especially *Der letzte Mann* (*The Last Laugh*, 1924) and *Faust* (1926), as well as his American *Sunrise*, was in the service of a total cinematic experience.[10] What was not emphasized but what was just as true, was that the visual beauty that made such an impression on spectators could not have been achieved by cinematographers alone.[11] The immense number and variety of visual effects in *Sunrise* were as much a function of the physical design of spaces thought through shot by shot.

Many sets were built in forced perspective in both small interior and huge exterior spaces. Sometimes very elaborate ones were on the screen for only a single shot. Shooting in forced perspective required the camera to be placed in a precise position with respect to the set, at the distance, height, and

8 Anon., "'*Sunrise*' Paves the Way," in *Los Angeles Times*, 8 December 1927, A9: "It is safe to say that the whole motion-picture industry had awaited the coming of F.W. Murnau's *Sunrise* with much more than casual interest due not only to the fact that it is Murnau's first Hollywood-produced picture, but also represents a deliberate attempt to set a new standard." Both cinematographers for *Sunrise* — Charles Rosher and Karl Struss — won Awards of Merit at the first Academy Awards ceremony on 16 May 1929. *Sunrise* won the award for "Unique and Artistic Achievement," the only time that category was used, Janet Gaynor for Best Actress in *Sunrise*, *Seventh Heaven* (Frank Borzage, 1927), and *Street Angel* (Frank Borzage, 1937), and Rochus Gliese was one of two nominees for Art Direction.

9 Edwin Schallert, "'Sunrise' Rare Art Feature. Murnau Production Ultra in Treatment. Camera Effects Exercise Dazzling Influence. Brilliant Audience Applauds Principals," *Los Angeles Times* (1 December 1927), p. A9.

10 Kann, "'Sunrise' and Movietone": "There is an emotional appeal all through '*Sunrise*' that is terrific. Murnau has succeeded in boring his camera lens into the very brain of his players and shows you in picture form the thoughts that surge through their heads." In *Film Daily* (41) 72 (25 September 1927), p. 4. Mordaunt Hall, "Murnau's Drums and Fifes of Life": "Other directors have shown their prowess with the camera, but Mr. Murnau surpasses all. The spectator feels almost as if he were on the tram car with the Man and his Wife. He is seeing what their eyes see, like an invisible being following them." In *New York Times* (2 October 1927), p. X7.

11 Longer, thoughtful reviews did praise not only the cinematographers and Murnau but also Rochus Gliese.

angle that corresponded to the art director's design, necessitating creative solutions to visualization with light and lenses. The cinematographer's work had to match the art director's calculations exactly in order for the shot to look natural yet visually resonant and for an effective transition to the following shot or set. Moving shots were designed to survey specialized sets in a similarly precise manner, coordinated with possible movements within the shot of people, cars, streetcars, horses, or whatever else might move, arranged in perspectival harmony with backgrounds near or distant.

Sometimes visualizations were designed to be imperceptible, at least on first viewing; at other times, the effects were strong and bold, meant to fascinate or astonish. Hermann Warm, film architect (as art directors were called in Germany) for Murnau's *Phantom* (1922), later wrote a description of how two difficult special-effects sequences had been accomplished and drafted diagrams to scale to illustrate them.[12] He put the critics' emphasis on cinematography this way: "The great technical effort required for special effects like that is not noticeable to the viewer, and indeed it must not be. Moreover, the press usually credits these effects to the cinematographer."[13]

Because *Sunrise* quickly gained immense prestige, shooting with forced perspective, motivated camera movement, chiaroscuro lighting, and unexpected in-camera visual effects — or whatever it was perceived that Murnau and his creative technicians had done — became known in Hollywood as "the German technique."

To achieve his goals, Murnau brought together the best technical people of his day who were artistically inclined and who could spark each other's inventiveness. His agreement with William Fox was designed to allow him to work in the way he had been used to doing in the Berlin studios that had resulted in *The Last Laugh*, the film so highly prized by Fox. Film historian Lotte Eisner described the essential role played in the top-tier silent German film productions by planning meetings among the creative and technical personnel far in advance of shooting, during which time the group coalesced as a mutually inspiring team: "One of the secrets of the success of

12 Hermann Warm's statement and diagram, along with models created to demonstrate how this worked, can be seen in Janet Bergstrom's documentary essay *Invitation to Phantom*, published on the US DVD and BluRay of Murnau's *Phantom* (1922) (Flicker Alley, 2003; revised 2019).

13 Hermann Warm, from "Erläuterung für die rekonstruierte technische Zeichnung BAR TAUMELBILD", an undated handwritten statement signed "Architekt Hermann Warm regarding *Phantom*," Hermann Warm Collection, Deutsche Kinemathek, Berlin. Lotte H. Eisner included a different but similar description by Warm in her book *Murnau*, (Berkeley, CA: University of California Press, 1973), p. 140.

the classical German film was the perfect technical harmony achieved by long *Regiesitzungen*, discussions on the *mise en scène* of the film to be made which sometimes lasted for two months or more before the actual filming began."[14] This intense collective work on Murnau's high-budget films like *The Last Laugh* continued during production, shot by shot, scene by scene.

For *Sunrise* in Hollywood, not only did William Fox famously give Murnau free rein at his studio, he also allowed him to bring his creative team with him from Germany, effectively allowing him to make a German film in an American studio.[15] That was unheard of. Although not all of the key technical professionals named in Murnau's contracts agreed to leave Germany, the director's substitutes were approved — men he had worked with before and who were used to his standards of perfectionism. Murnau's favorite screenwriter, Carl Mayer, who was written into Murnau's contract, completed his script before the director left for the US. He remained in Germany and did not contribute to script changes during shooting, such as we can see in Murnau's handwritten notes in his shooting script.[16]

Art directors Robert Herlth and Walter Röhrig were hired away by Ufa with "very favorable long-term contracts"[17] just before Fox's contract for them came through. In their place, Murnau chose Rochus Gliese, who had been another trusted collaborator on previous films.[18] Murnau's contract

14 Lotte H. Eisner, *The Haunted Screen*, trans. Roger Greaves, (Berkeley, CA: University of California Press, 1969), pp. 36–37.

15 In a full-page ad in *Motion Picture News* on 8 May 1926, Fox announced: "Theatre Audiences Everywhere Are Waiting for This Creation! Europe's Greatest director reveals his mastery in an All-American production. WILLIAM FOX announced The First Motion Picture to be made in America by F.W. MURNAU. The flower of directorial genius adjusts his art to the needs of American photoplay audiences — backed by American equipment, unlimited resources and An All-American Cast Selected Especially for this Production by Mr. Murnau. Scenario by Dr. Karl Meyer [sic]. No Director has been so greatly praised by photoplay critics as F.W. Murnau — your patrons will be eager to see this greatest of all European directors at his best in an All-American production." Principal photography on *Sunrise* would not begin until months later, in September 1926; *Sunrise* premiered in New York City's Times Square Theater on 23 September 1927 as the first film in Fox's 1927-1928 season.

16 A facsimile of Murnau's shooting script, recto and verso of each page, was published as *Sunrise (Sonnenaufgang), Ein Drehbuch von Carl Mayer mit handschriftlichen Bemerkungen von Friedrich Wilhelm Murnau* (Wiesbaden, Deutsches Institut für Filmkunde, 1971).

17 Julius Aussenberg to Winfield Sheehan, 16 June 1926, William K. Everson Collection, New York University.

18 Gliese had worked with Murnau as his art director/film architect for *Der brennende Acker* (*The Burning Soil*, 1921/1922), *Die Finanzen des Großherzogs* (*The Finances of the Grand-Duke*, 1923), *Die Austreibung* (*The Expulsion*, 1924), and had co-scripted a film with him under the pseudonym Murglie (standing for Murnau-Gliese), *Komödie des Herzens* (*Comedy of the Heart*), directed by Gliese in 1924.

included a clause to hire cameraman Karl Freund, famous for what he could do with a moving camera in *The Last Laugh*, but Freund did not want to go to Hollywood.[19] Instead, Murnau decided on Charles Rosher, with whom he had been in close contact during the production of *Faust*. Rosher had started his career as a portrait photographer in his native England, and after an adventurous entry into American filmmaking, including filming Pancho Villa in combat in Mexico in 1913, he became Mary Pickford's exclusive cameraman. As Rosher told Kevin Brownlow,

> After Doug and Mary got married, I was with them in Europe and Doug went with me to the Ufa studios in Berlin. I did special tests of their stars to demonstrate glamour lighting. They always lit them with heavy, dramatic lighting and deep shadows. Erich Pommer was there then, and he put me under contract for a year, with Mary's consent. I acted as consultant on Murnau's *Faust*. I didn't do anything on the picture, but Murnau expected to go to America, and he kept asking, "how would they do this in Hollywood?"[20]

For his part, Rosher wanted to see how the top German cinematographers obtained the effects that had made them internationally famous.

In Hollywood, Rosher brought in Karl Struss, who had worked with him most recently as second cameraman on Pickford's *Sparrows* (William Beaudine, 1926), responsible for shooting the foreign version.[21] Each of them had a strong background in fine arts still photography; each had invented his own portrait lens. Struss had studied photographic art with Clarence White in New York, had been part of Alfred Stieglitz's Photo-Secession group, and continued to exhibit and publish his photographic work as a member of the Los Angeles Pictorialists and elsewhere after he became a cinematographer.[22]

In 1928, Struss published an essay about working on *Sunrise* with detailed examples of shot construction, and in later years, Rosher, Struss, and Gliese

19 Freund wrote to Herman Weinberg on 2 May 1947: "At the time that Murnau came to America I was not interested in anything that Hollywood could offer, so I made none of his American pictures with him." New York Public Library, Lincoln Center, Herman Weinberg Scrapbook Collection, vol. 13.

20 Kevin Brownlow, *The Parade's Gone By…* (Berkeley, CA: University of California Press, 1968), pp. 224–232.

21 Karl Struss interviewed by Richard Koszarski, August 1976, courtesy Richard Koszarski.

22 John Dorr, interviewer, "Recollections of Karl Struss: An Oral History of the Motion Picture in America," UCLA Special Collections 1969 (interviews carried out 9, 24, 30 July and 13, 27 August 1968), pp. 65–67; Richard Koszarski, "Karl Struss: The Cinematographer," in *Karl Struss: New York to Hollywood* (Fort Worth: Amon Carter Museum, 1995), pp. 167–205.

were often asked how particularly memorable and difficult shots had been accomplished.[23] We will look at descriptions of several of them and also at a particular shot that I have not, so far, seen any of them — or anyone else involved with the film — refer to, either in response to a question or spontaneously. That shot — the Williams shot — could not have been accomplished by the cinematographers and the art director alone, although they were involved in preparing it for a different kind of treatment. It required Frank Williams's specialized printing process and laboratory for his patented travelling matte compositing system. That shot introduces the role of the "effects specialist" into this story.

Technical Collaboration on *Sunrise*

Professionals from that era had their own vocabulary to describe what they did, what was difficult, and what they were most proud of. They help evoke the challenges and passion with which Murnau aspired to translate his pre-visualizations shot by shot into a cinematic world.

In 1928, Karl Struss published "Dramatic Cinematography," a lecture he had given to the most important technical association for film, the Society for Motion Picture Engineers (SMPE), emphasizing how drama and mood had to be part of the creative decision-making that motivated solutions to technical challenges faced by the cinematographers of *Sunrise*, and the necessity of working closely with both the director and the art director to achieve the integration of their goals.

According to Struss:

> Some two hundred sketches were made by Rochus Gliese, the Art Director or Visualizer, representing the two hundred scenes [shots] from the script, from which small plaster models were made of the more important sets. These gave the cinematographers [Charles Rosher and himself] first-hand information before actual shooting commenced so that all the lighting possibilities would be arranged and prepared, before the real sets were finished [...] It is difficult to express moods and thoughts — but here the camera does it. This was not a result of the camera work alone, but of the intelligence used in the preparation and lighting, and an understanding of the effects desired; again I say, complete co-operation among the three

23 Karl Struss, "Dramatic Cinematography," *Transactions of the S.M.P.E.*, (9–14 April 1928), pp. 317–319.

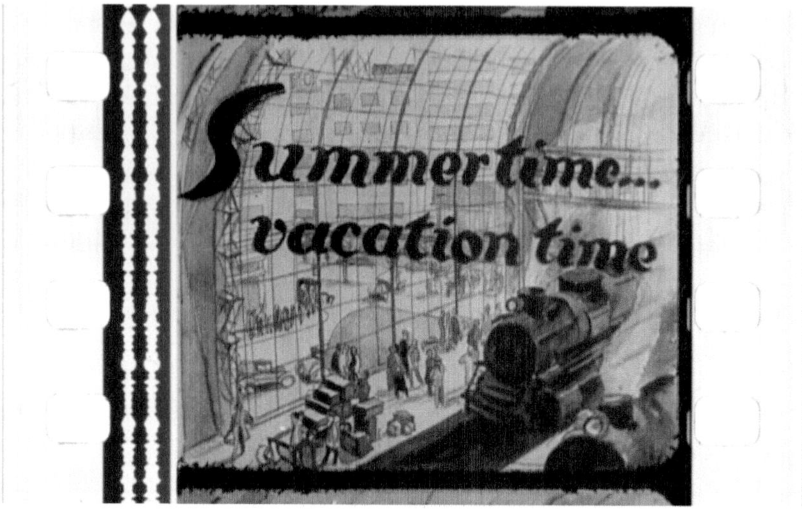

Fig. 12.3 Still from *Sunrise* (F. W. Murnau, 1927).

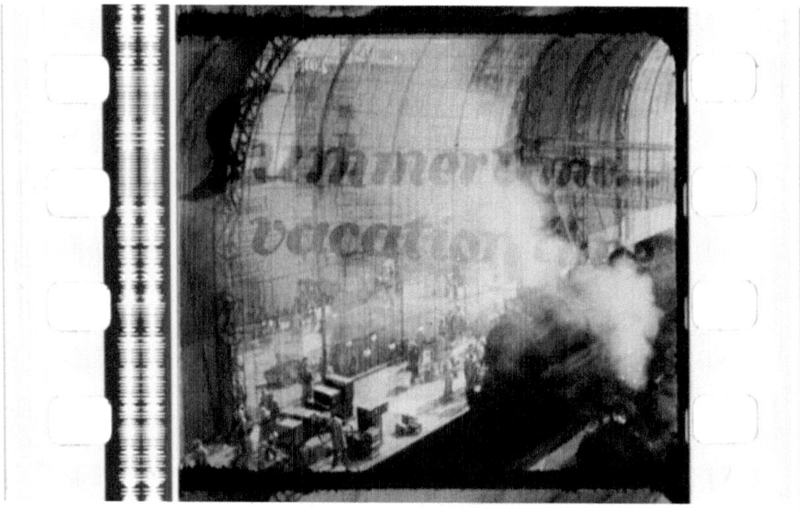

Fig. 12.4 Still from *Sunrise* (F. W. Murnau, 1927).

people most concerned in creating drama in picture form for the screen; the director, art director or visualizer, and cinematographer. In this group I did not include the author or scenarist [Carl Mayer] because his work previously had been so thoroughly done that his further co-operation was unnecessary.[24]

24 Ibid., pp. 317–318.

The first image in *Sunrise* is a drawing that looks like a contemporary magazine ad for urban dwellers to leave the city for relaxation (Fig. 12.3). Three words are written over it: "Summertime... vacation time." That static image is transformed into a cinematic image already in motion as the drawing dissolves into the first shot of the film, matching its lines perfectly, and the words disappear (Fig. 12.4).

Movements in different parts of the frame begin at different times, partly filtered by the steam from two trains, large in the foreground: first one crosses the screen, then the other begins to move. The station's huge glass wall, curving toward the ceiling, shows skyscrapers and busy street life through its slight distortions, implying levels of depth impossible to calculate. Then an elevated train speeds through at an oblique angle somewhere near the background. There is so much to look at in so many parts of the frame, the movements change so quickly and at different rates of speed, and it all seems so perfect, that it is impossible even to think about how this might have been accomplished (miniatures? mirrors? real people? models?).

Some commentators, most significantly Karl Freund, later maintained that Carl Mayer's scripts were written in such a cinematic manner that all the cameraman had to do was to follow Mayer's script: it was all there. Freund recalled that Mayer had often asked him questions about how something could be filmed while he was writing.

On the other hand, Rochus Gliese, referred to as the Visualizer by Struss, described the challenges of using Mayer's script as the basis for realizing the first moving shot in *Sunrise*:

> Carl Mayer wrote in the script's prologue: 'Summer – time for traveling. A few shots, a train starts moving out, a man and a woman.' And we had to create that! This station — it's a very brief shot, you hardly notice it. Here's how we did it. When our huge city plaza was ready, I had a platform built above it, about 10 meters high. On that platform were the lanes for the travelers and the side walls of the station. But the window panes, full-sized, only went half-way up [the glass wall]. The camera was on a parallel, a tower 25 or 30 meters high. Attached to the front of it was a model of the station ceiling, the rails, and the trains, which were children's toys. One locomotive left in a cloud of steam, and the people walked underneath, as if they were entering or leaving the trains. An insane amount of work! It was only possible with the closest, most friendly collaboration with the cameramen. We worked with passion![25]

25 Interview with Rochus Gliese by Bernard Eisenschitz held in German in West Berlin, May 1970, partly published as "Le voyage à Hollywood," ed., interviewer, trans., Bernard Eisenschitz,

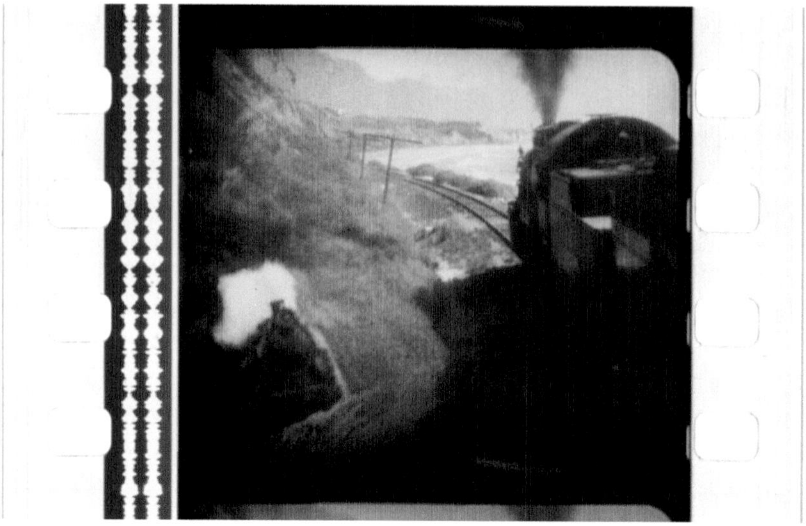

Fig. 12.5 Still from *Sunrise* (F. W. Murnau, 1927).

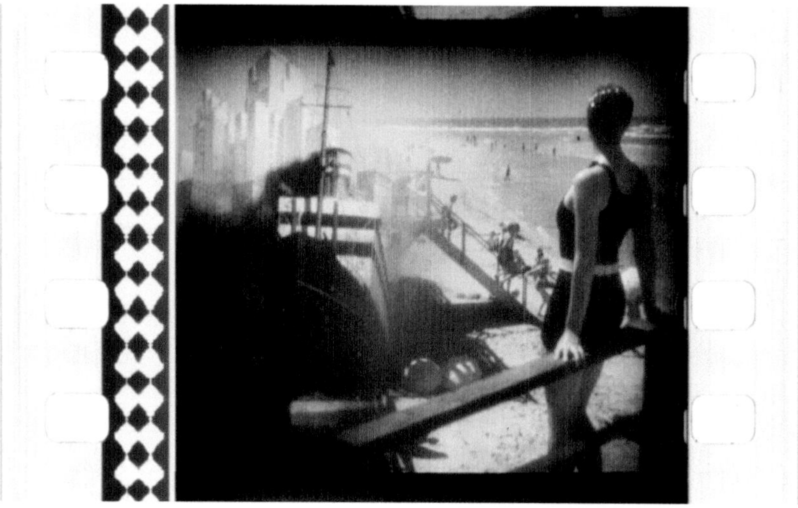

Fig. 12.6 Still from *Sunrise* (F. W. Murnau, 1927).

That set is never seen again.

In his 1928 essay, Struss wrote about the series of split-screen images that immediately follows the opening train station shot, often called the "vacation montage" (Figs. 12.5. and 12.6). "To me," wrote Struss, "the most

Cinématographe, 75 (February 1982), p. 16. My thanks to Bernard Eisenschitz for his permission and assistance with the transcript. [My translation].

impressive [examples of camerawork in *Sunrise*] are the opening atmospheric shots which give pictorially the impressions of 'vacation time' that no titles could give. *Sunrise* should be seen to realize some of the problems involved in picturizing stories."[26]

Later, he explained in interviews how he worked with Charles Rosher to execute those split-screen shots:

> We had done some miniature work before we went to Lake Arrowhead, split screen and so forth. [Rosher] did that with his Mitchell camera and I also did it with my Bell & Howell camera. And we kept track where there was split screen. We might have two entirely different exterior scenes made into two verticals. The frame could be split half and half. We had something with miniature ships, and we had live trains down near [San Juan] Capistrano, and other things of that sort, combined. It might be a week or ten days after we shot one half, and then we shot the other half. We had to keep track of it, to see that it matched and that the exposures balanced too. Nowadays they leave that to the special effects department. Then, the director of photography always did those things himself.[27]

In *Sunrise*, Janet Gaynor runs away from her husband (George O'Brien) in terror up a hill on the other side of the lake after he has almost drowned her and gets on a streetcar that just then emerges through the trees, stopping for her. As it is moving away, O'Brien jumps on board, and they are gradually taken into the city plaza. The first part was shot on Lake Arrowhead in the San Bernardino Mountains, not an easy location to transport and power heavy equipment; the second part was shot at Fox Hills in Los Angeles. Each shot required a special effort, designed so that the streetcar, with windows all around, showed off the surroundings and appeared to move continuously from the countryside into the city as the couple's emotions intensified. For the forest scene, according to Charles Rosher and as reported in the press, a mile-long track was built at Lake Arrowhead, streetcar rails were laid, and a real streetcar was brought in. Unlike the streetcar in the city set, which was powered electrically like those on the streets outside the studio (an amazing feat, as reported by the press and in the words of actors and professionals involved), the Arrowhead streetcar was mounted on an automobile chassis.[28]

26 Struss, "Dramatic Cinematography," op. cit., pp. 18–19.

27 Dorr, "Recollections of Karl Struss," op. cit., pp. 48–49.

28 Brownlow, *The Parade's Gone By...*, op. cit., p. 232. Sometimes one reads that Rosher was not involved in the Arrowhead shooting at all because of a statement Scott Eyman published from

Film historian Bernard Eisenschitz asked Rochus Gliese how he worked with the cinematographers. Gliese replied, "Today cinematographers create the image. It wasn't like that then. The painter, the architect" — and Gliese referred to himself as both:

> collaborated with the cinematographer to create the image. The cameraman — that is, the first cameraman — looked, the [matte] painter verified the framing precisely, the director controlled and requested other things. For the streetcar in the country to go into the center of the city [after the first part shot at Lake Arrowhead, where the lake is visible through the windows], we had a small part of a hill on Fox's back lot. We couldn't go beyond that area, because Tom Mix was shooting a western next to it. So I designed a track that would zigzag and use all the space we had to the maximum. With each successive camera position, we stopped, I looked through the viewfinder and I painted the background on glass. Then we built it. Everything was determined by the camera angles, we could not deviate. It would have been impossible otherwise: they would not have been able to position the camera other than the way I had constructed the set because it had been built precisely for the camera.[29]

Murnau had become world-famous for moving camera shots that seemed unique, even impossible, sweeping the audience's emotions with them: he had "unchained the camera" in *The Last Laugh* and again in *Faust*. Karl Struss had high praise for the way Murnau thought out camera movements as inseparable from their dramatic purpose in *Sunrise*. "After *Sunrise*," said Struss, "everyone was moving the camera, but mechanically, not for dramatic effect. Murnau really knew how to do that."[30] In another interview, he added, "He was the only director who, right in the beginning, knew how to use the moving camera. He didn't use it as so many of them do, who don't know what they're doing with it. He moved that camera so that it came

his interview with Struss: "In the beginning of the picture, Charlie [Rosher] was ill, up at Lake Arrowhead and he didn't do any of that part of the picture." Scott Eyman, *Five Cinematographers: Interviews with Karl Struss, Joseph Ruttenberg, James Wong Howe, Linwood Dunn and William H. Clothier* (Metuchen, NJ: Scarecrow Press, 1987), p. 8. In other interviews, Struss says that Rosher missed the first week of shooting at Arrowhead because of illness. Off-camera photographs and multiple accounts, including Rosher's on-site descriptions, testify to his work during the Arrowhead shoot.

29 Gliese, "Le voyage à Hollywood," op. cit., pp. 15–16.

30 Dorr, "Recollections of Karl Struss," op. cit., p. 70.

to a climax at the end of a shot. And he stayed with that, he didn't cut the moment it was over with."[31]

The shot when the husband walks into the marsh to meet the city woman in the moonlight was one of the most important for establishing the mood of the film and one of the most difficult to accomplish of all the moving shots in *Sunrise* (Figs. 12.7, 12.8 and 12.9). It begins by following the husband, then moves alongside and then ahead of him so that he walks more or less toward the camera, and then the camera moves away from him to the city woman waiting: not waiting with him, as if sharing his emotions, but waiting for him. He is no longer on-screen, and we lose a sense of where he is in the space off-screen. Exactly when and where he will re-enter the field of vision is unpredictable, to her and to us.

The realization of this shot was described many times by the cinematographers. Karl Struss's motorized Bell & Howell made it possible for him to operate precise, complicated camera movements under cramped conditions that would have been impossible with Rosher's hand-cranked Mitchell.

Rosher described how cinematographer Carl Hoffmann, when he was shooting *Faust* in Berlin, had mounted rails in the ceiling of the studio so that a camera could be suspended from it and track from above. He used that idea, with modifications, building an overhead track at Fox's Western Avenue studio. But he faced problems lighting Murnau's complex moving shots. "For some scenes, such as the swamp sequence, the camera went in a complete circle. This created enormous lighting problems. We built a railway line in the roof, suspended a little platform from it, which could be raised or lowered by motors. My friend and associate, Karl Struss, operated the camera on this scene. It was a big undertaking; practically every shot was on the move."[32]

Struss remembered it vividly and described it in detail many times.

> The scene opens on a night landscape and the camera photographs the moon in the distance. The camera was suspended from an overhead curving track, and I used my Bell & Howell camera, which was motorized. I had to look straight down into the viewing part of the camera, which is about four and a half inches to the right of the photographing lens, a two-inch focal length lens. I had put a magnifier on the ground glass on the viewing side, and of course the shutter was flickering through there, and the image was upside down and reversed right and left. I had to hold

31 Susan and John Harvith, "Karl Struss Remembers," in *Karl Struss: Man with a Camera*, eds., Susan and John Harvith (Cranbook, MI: Cranbrook Academy of Art/Museum, 1976), p. 17.

32 Charles Rosher, in Brownlow, *The Parade's Gone By…*, op. cit., p. 232.

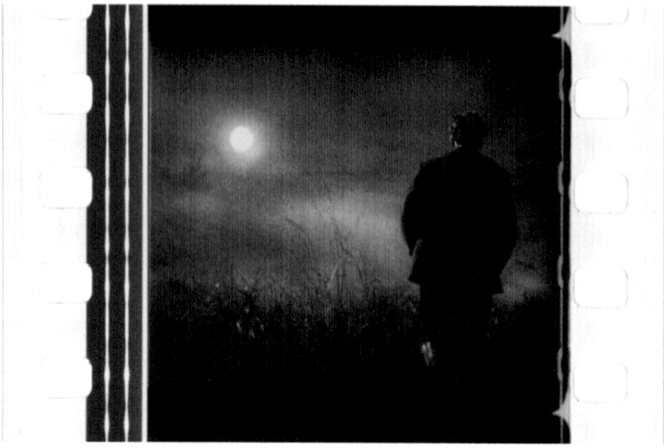

Fig. 12.7 Still from *Sunrise* (F. W. Murnau, 1927).

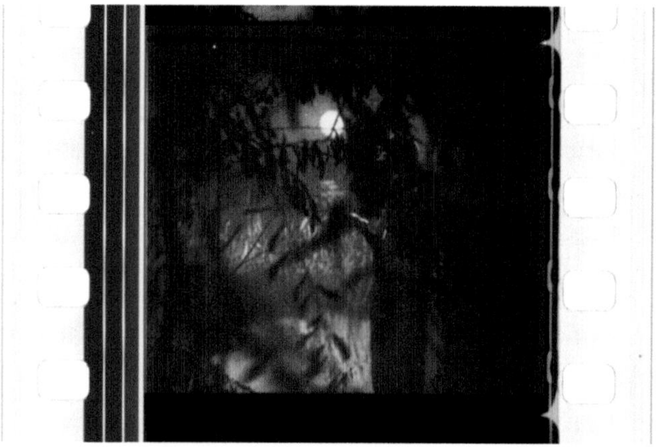

Fig. 12.8 Still from *Sunrise* (F. W. Murnau, 1927).

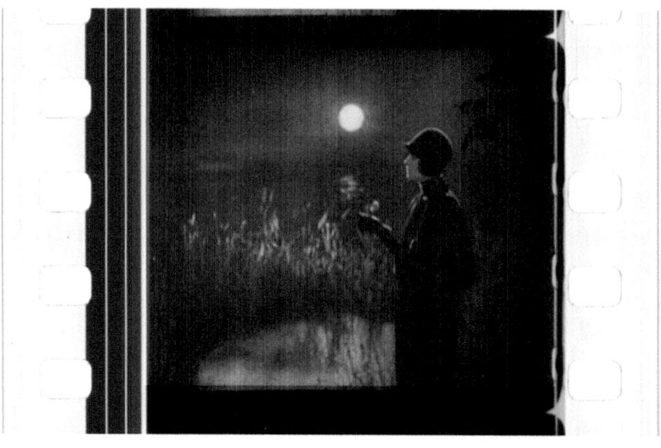

Fig. 12.9 Still from *Sunrise* (F. W. Murnau, 1927).

> a handle which operated the pan between my knees and bend over and look straight down, and the camera started to move. It swung over to the right a little bit to go straight through some very small trees that we had to separate as we went through them with a sort of plow in front of the camera. We follow George O'Brien from the beginning as he goes along the path. Then he goes out of the shot to the right. Then we plow through the trees and come to a stop. And here we see Margaret Livingston powdering her nose. Now one of the problems was the fact that we couldn't work any handles to tilt up or down. I had to rig up a wooden upright so that the two tubes that held the matte box in front of the lens would come out to a certain point as we went through the trees and stop, so I wouldn't be shooting into the top stage lights in the distance that were giving an effect of moonlight on the water and the marsh. We then come to a stop and find her in the right position. Then George O'Brien walks into the scene. It's one of those things which are awfully difficult mechanically — to keep that all smooth [...] That was one of the most difficult shots I ever contended with. We had to do it a number of times for safety.[33]

In his 1928 article, Struss emphasized to his S.M.P.E. audience that he was not describing such a difficult shot as an example of technical achievement for its own sake, but that it was necessary for the scene's emotional and subjective purpose.

> This was one of the most difficult scenes to photograph and to light, and probably one of the most dramatic as well as pictorial. The impression of that scene must last a long time, for numerous people have asked me about it before anything else. Such a scene could not have been made effectively with a stationary camera in the usual 'long shot' — but here we move with the man and his thoughts to his objective, and come upon it in a dramatic manner. The effect is breath-taking, so great and impressive is the pictorial beauty revealed; the hazy moon — the misty marsh atmosphere — the lazy movement of the water — and in the silhouetted foreground — the girl! No words are needed here to tell what is happening — we seem to be surreptitiously watching the love scenes — and so it goes throughout the picture.[34]

When Karl Struss was asked about the "German influence" on Hollywood filmmaking, the suggestion was that lighting would have been the most

33 Dorr, "Recollections of Karl Struss," op. cit., pp. 29–30.

34 Struss, "Dramatic Cinematography," op. cit., p. 318.

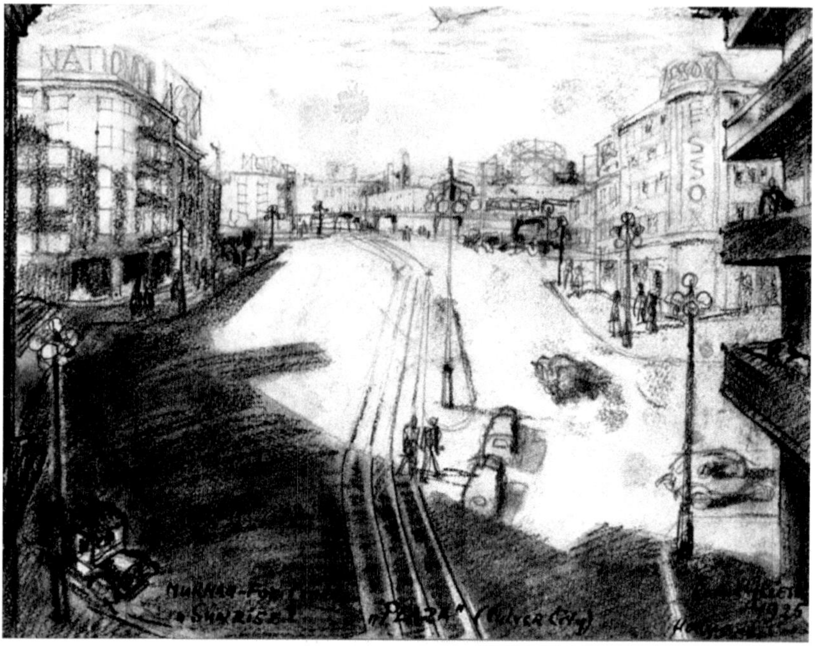

Fig. 12.10 Drawing of city plaza by Rochus Gliese for *Sunrise* (city segment).

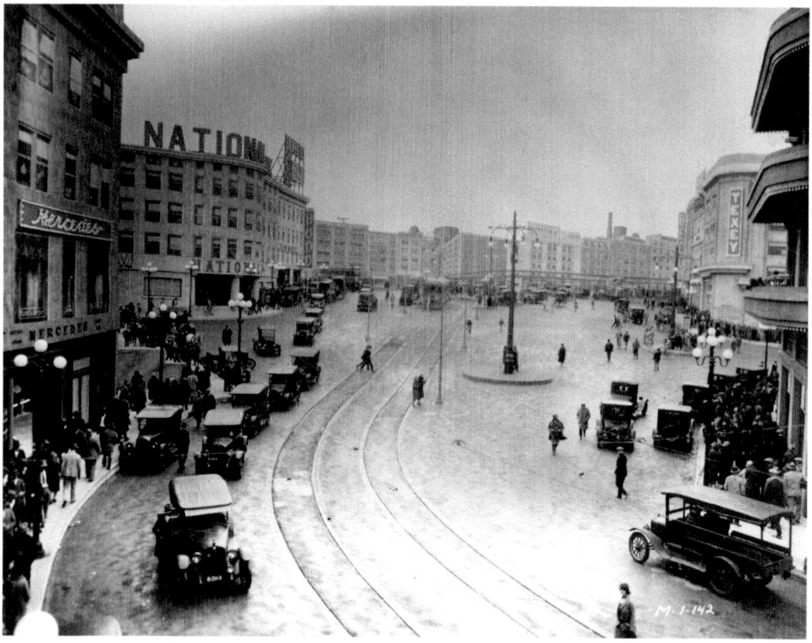

Fig. 12.11 Production still of the city plaza set in *Sunrise* (city segment).

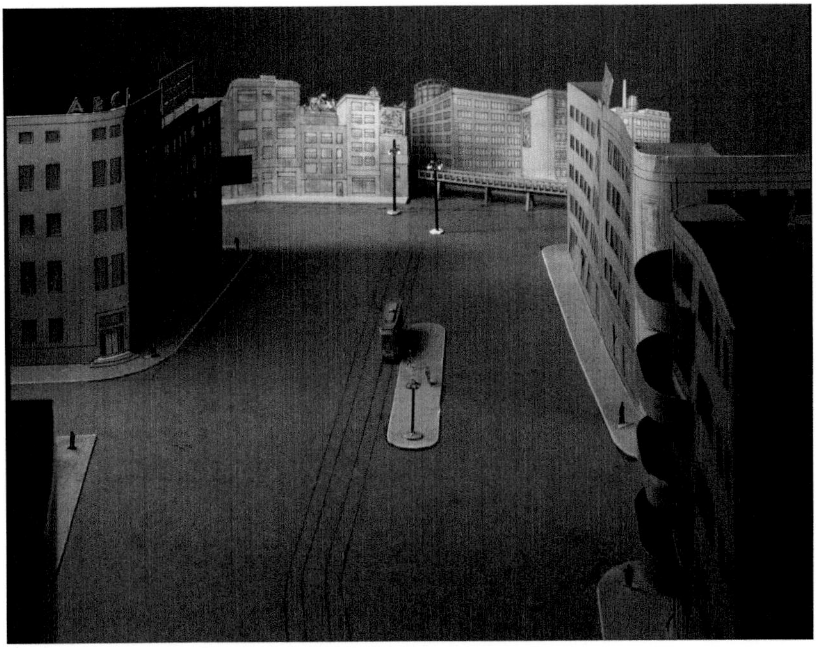

Fig. 12.12 Model showing the entire city plaza for the set of *Sunrise* (city segment).

important new element. That was not surprising, because lighting was and still is the most frequently cited stylistic influence of German silent cinema on American cinema. But Struss replied that he did not learn much about lighting from working with the German team: "On *Sunrise* the greatest difference that we ran into was the fact that the sets were built in perspective."[35] Rochus Gliese recalled his happiness about Fox's agreement to the creation of a "fantastic, but more European city."[36] He must have been thinking not only that the city would not imitate an American city but how the city would be constructed and shot, namely, in the way they had become experts in doing in Berlin (Figs. 12.10, 12.11 and 12.12). In the words of Karl Struss, "The whole set for the city was at the Fox [Hills] studio, that tremendous set, acres of it. The buildings in the background were only 25 feet high; they were miniatures, but they were so far away and all uphill with the traffic moving down. That was to help perspective."[37]

35 Dorr, "Recollections of Karl Struss," op. cit., p. 57.

36 "Zusammenarbeit mit Murnau: Auszug aus einem Interview von Gerhard Lamprecht mit dem Bühnenbildner und Regisseur Rochus Gliese" (11 August 1956), Werner Sudendorf, ed., SDK-Tonträgersammlung, Deutsche Kinemathek, Berlin. (My translation).

37 Harvith, "Karl Struss Remembers," op. cit., p. 15.

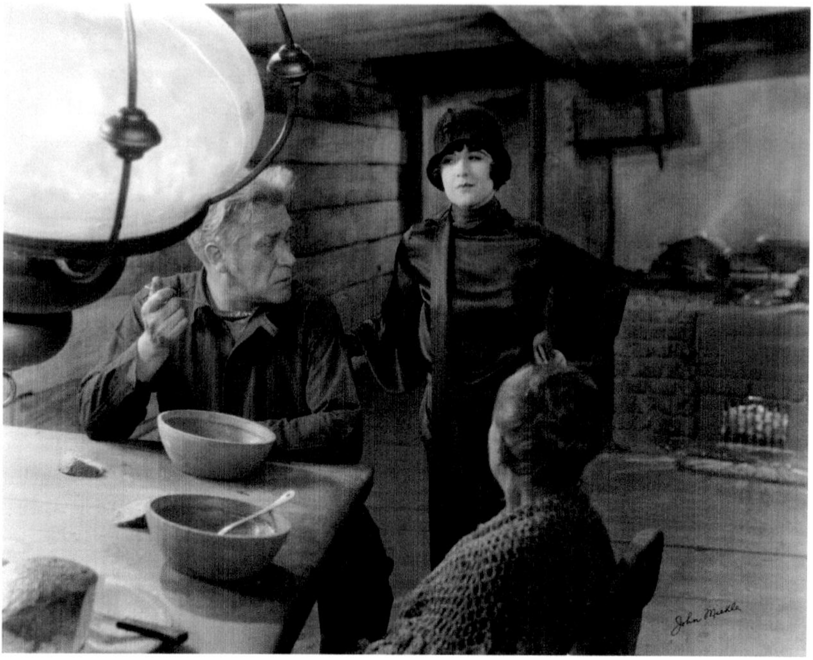

Fig. 12.13 "Cézanne shot." Production still of *Sunrise* showing forced perspective.

Perspective shooting was also used for interior sets, for instance, the modest home where the city woman rented a vacation bedroom (Fig. 12.13). She enters the main room from a door at the back of the frame, dressed in fashionable satiny black as if for a night on the town, and walks directly toward the foreground (along the same camera axis) where a country couple, her hosts, are seated at a table eating dinner. Their table is obviously not level: it is tilted strongly, unrealistically downward toward mid-frame, possibly offset by an exaggeratedly huge lit globe ceiling lamp in the upper left quadrant of the frame. The forced perspective in the foreground of this daring "Cézanne shot," as I think of it, masks the steep downward slope in the floor from the back of the room to the foreground as well as the left side wall that angles outward as it nears the table, subtly making the space seem larger. The ceiling also tilts down from right to left. The action in the shot, however, does not let us dwell on this set of unlikely, beautiful perspective structures: the woman walks quickly to the couple at the table, interrupting their dinner and gestures toward her high heels that need to be shined. The off-kilter architecture and pictorial composition on the left side of the frame is partly balanced on the right by drawing the eye toward the light flickering in the fireplace and steam rising from a pot on the stove, and that is where the action moves to accommodate the city woman's request.

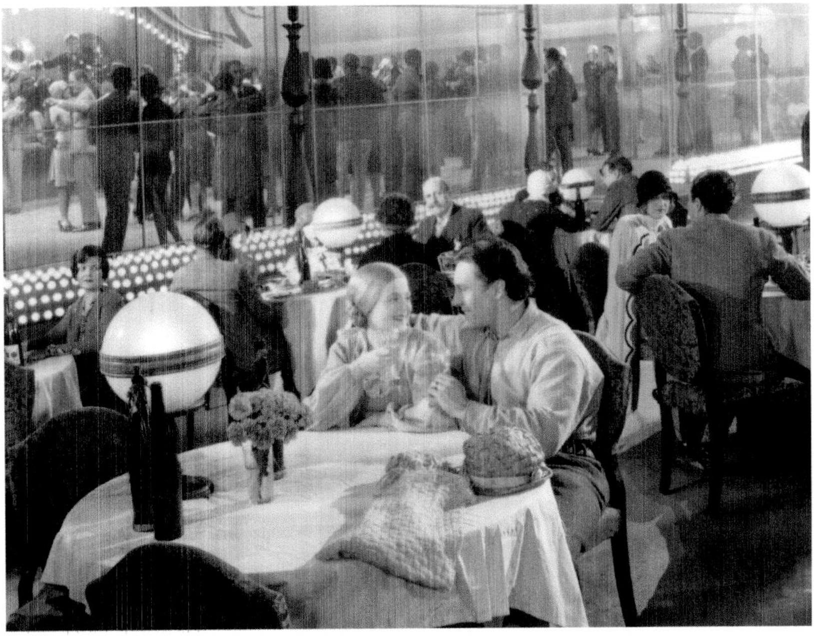

Fig. 12.14 Production still of the set of *Sunrise*.

Forced perspective was also important for very large interior sets (Fig. 12.14), at times as seemingly noticeable as in the Cézanne shot, but it should be emphasized that the departure from realistic depiction is much more pronounced in still photographs than among the myriad complexities of the film in motion: Struss described how he and Rosher photographed false perspective, using special lighting techniques as necessary, in what he called "the big set, the restaurant set, where the couple goes."[38] According to Struss,

> We'd shoot from the back of that set. There the floor came up in the back and we were up high. The floor comes up and the tables come up, and we'd shoot down on this, and get the perspective out through to the real, live exterior. Of course, we had to illuminate the interior then too, so that it would balance up with the terrific light of the daylight outside. There too we used some black bobbinets to take the place of plate glass windows. That's a form of gauze. It softened the exterior a little bit and also cut the exposure of the sunlight. And then when we made a similar shot, the reverse. We had action in the foreground, and across the street, in perspective — we couldn't use real people — we used midgets.[39]

38 Dorr, "Recollections of Karl Struss," op. cit., p. 59.

39 Ibid.

Janet Gaynor remembered how difficult it was to move as if the floors were flat:

> In some of our sets, the floors would be slanted — the whole thing was done to bring about this terrific perspective [...] I know in *Sunrise* George O'Brien and I had to do a dance. We certainly weren't professional dancers — anyway, it was supposed to be a folk-dance — but even so, to try to do it on that slanted floor was something. We had to whirl, and we'd usually whirl and whirl and end up in the camera.[40]

Effects Specialists: Williams Shots

One of the visualizations in *Sunrise* that is easiest to remember but that none of the people involved in making the film have commented on in the published record, to my knowledge, occurs just after the couple leave the church where they have witnessed a wedding. They have an emotional reunion, renewing their own wedding vows to each other, then walk down the stairs of the church with the pride and devotion of newlyweds, caught up in each other, oblivious of their surroundings, and go right into the midst of the busy traffic in the plaza. They keep on walking and walking: the traffic becomes transformed into a leafy meadow where they pause for a long kiss. Finally, they awaken to the sounds of cars honking and men yelling.[41] Cars, trucks, and horses converge on them, as drivers angrily hit their brakes.

These shots were not accomplished in-camera by the cinematographers nor through rear projection, as many have supposed. This was a Williams shot. Actually, Fox's Cutting Continuity for *Sunrise* lists two Williams shots separated by four shots photographed normally, cut-aways from the main shot in this scene.[42]

> Scene [shot] Description
> 200 Ext. [exterior] Street – WILLIAMS SHOT – General traffic – O'Brien and Gaynor enter from left, looking at each other. They walk right straight

40 "The Reminiscences of Janet Gaynor," (1958 interview), Oral History Collection, Butler Library, Columbia University, New York, p. 10.

41 The Movietone version delivers these sounds through its synchronized music and effects track; in a silent version, they could be created in the theater.

42 "*Sunrise* Cutting Continuity," shots 200–204 [June 1927]. Karl Thiede Collection.

through the dense traffic, which barely misses them. As though their thoughts were in heaven they suddenly start walking into the air. The city b.g. [background] disappears and in its place appears a meadow. They turn their heads and lovingly walk through the meadow as the camera still follows them. They stop and embrace and as they kiss, the meadow b.g. disappears and reality of the city traffic appears again. All the automobiles tied up in a traffic jam and the two kissing in the middle of the street.

[Shots 201, 201a, 202 and 203 show vehicles stopping suddenly.]

204 WILLIAMS SHOT – O'Brien and Gaynor standing kissing. Traffic in b.g. all is stopped. [changed indent on b.g.]

The magic of photographic realism used to achieve this dream-like effect was accomplished using a patented process named after its inventor, visual effects specialist Frank Williams, who worked in his own lab behind closed doors.

In his heavily annotated shooting script, Murnau jotted down notes for the Williams shot sequence on a blank page facing the typed script: "Williams matte shot until kiss / city traffic jam returns within kiss / then honking ringing ranting raving [faces?] hands braking / [people] falling over each other in bus / then (without Williams) a jolt out of the kiss, laughing onlookers running back and forth, running away, street traffic."[43] Before the film went into Williams's lab, the shots had to be taken on the set in a way that would prepare them for the specialist's work. Pictures of the setups and photography of this shot from different angles show some of the difficulties involved (Fig. 12.15). First, we see the couple leaving the church, walking down its stairway toward us. The Williams shot begins with a cut to O'Brien and Gaynor, now photographed from behind, walking away from us into the plaza traffic. It is worth quoting at length visual effects specialist Jonathan Erland's detailed elucidation of a

43 Murnau's original German handwritten text: "Williamsmatschuss bis Kuss / wiederersteht Stadt Verkehrsstörung im Kuss / dann Hupen Klingeln schimpfende / Gesichter [?] Hände Bremsen überein/ anderfallende im Autobus / dann (ohne Williams) Aufschrecken aus Kuss Hin und Herrennen Lachen der Zuschauer, weglaufen, Strassenverkehr," *Sunrise (Sonnenaufgang), Ein Drehbuch*, op. cit., p. 86 verso, on the back of the page facing the main scene. My thanks to Werner Sudendorf for his transcription and to Tessa Smith for her translation, slightly modified.

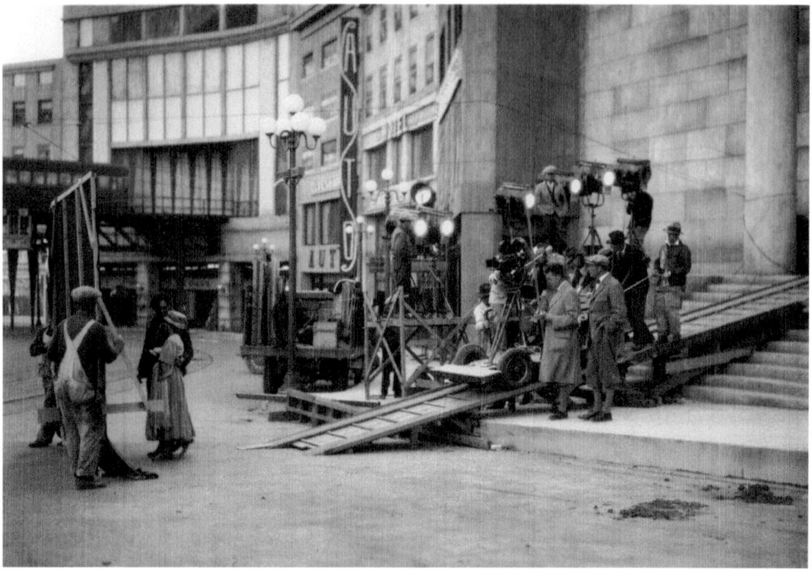

Fig. 12.15 Production still of *Sunrise*: setting-up the Williams shot.

photograph that has been referred to frequently as evidence of a process other than Williams's:[44]

> Charles Rosher is operating his Mitchell camera on a dolly moving on a ramp tilting downward starting from the height of about eight of the church's stairs from the bottom, then moving over the sidewalk and the curb to reach street level. The dolly is surrounded by men keeping its movement slow and steady. Murnau and Struss walk alongside, Struss taking notes.
>
> The actors are not walking on a treadmill after they reach level ground, as was commonly used for actors (or animals) to simulate walking or running in space when in fact they are walking or running on a treadmill in front

44 Film historian Richard Koszarski published this photo in his essay "Karl Struss: The Cinematographer", but he mistakenly believed that the special visual effects — the couple in traffic, then a meadow, then traffic, with vehicles passing in front of them in the foreground and behind them in the background — were accomplished by the cinematographers in the camera. Rosher is nearly hidden behind his camera and is not identified in Koszarski's frequently quoted caption: "Karl Struss with director F.W. Murnau and camera crew filming scene for *Sunrise*, c. 1926. Standing next to Murnau, Struss keeps a record of the exact footage shot for a complicated trick effect. Before the introduction of the optical printer, all such effects had to be done directly on the original camera negative." Richard Koszarski, in *Karl Struss: New York to Hollywood* (Fort Worth: Amon Carter Museum, 1995), p. 178.

> of an image projected behind them that shows movement in space and makes it seem that they are moving. In this picture, a black velvet screen isolates the actors from their background on the set: it is an essential element in the creation of a shot that can be used for Williams' travelling matte composite process in his lab. Williams' patent (1,273,435) covered the matte extraction process only. The patent office requires that the description has to be sufficient for one skilled in the art to accomplish the objective, and Williams did that.
>
> What can't be found in the patent is all the rest of the technique that goes into the design of a visual effect. Some of that can be readily seen from the production stills, for instance, building a ramp down the stairs. Much cannot be readily seen, such as the layout of the scene in which each step the actors take will have been planned in advance. The ground under the dolly would have been super smooth. In the stills you can see Struss with a notepad walking along with the dolly. He will have been counting off the preplanned and rehearsed steps. So many steps into the plaza up to the dissolve to the orchard, so many more to the dissolve back to the plaza, and so on. Meanwhile, the plate to be composited with the matte (the scene of the plaza with all the traffic) will have been planned, rehearsed and shot with the dolly crew following a precise path, so that all the elements — cars, horses, people, etc. — arrive at the right time and place. The camera crew would have shot the same scene without the actors but including all of the traffic, etc., that was to occur in front of the couple (who were facing away from the camera). They were then 'processed' into that scene and then the result became the background for the foreground traffic elements that were shot separately, each with the black velvet hand-held backing, resulting in the couple being 'sandwiched' amongst the traffic. The backing would have to be velvet because it needs to be jet black; ordinary black fabrics would be too gray.[45]

According to Fox's Cutting Continuity, two other Williams shots were created for *Sunrise*: both are seen in the marsh as part of the city woman's seduction of O'Brien. The first is the "illustrated title" (Fox's term) "Couldn't she get drowned?" The words slowly sink down in the middle of the frame, as if mimicking what we see next, still inside the intertitle: O'Brien pushes Gaynor from their rowboat into the water.[46] Another Williams shot comes

45 Jonathan Erland, emails to the author, 17 August and 11 September 2018.

46 "*Sunrise* Cutting Continuity," shots 38, 45, 68, 200, 204 [June 1927]. It is the second of four special titles requested by Murnau to be shipped to him in Berlin with the "'*Sunrise*' Second

just after the city woman's dialogue title "COME TO THE CITY!" It begins when the couple's heads (facing away from us) are visible at the bottom of the frame — as if they are watching fantastic images of the city playing out before them — and ends when the entire screen is taken over by a series of dynamic urban scenes.

Returning to in-camera superimpositions after O'Brien returns home, seductive apparitions of the vamp encircle him as he sits on his bed, fading in and out, tempting him. In contrast to his wife's happiness at her belief in his renewed affection, he stands up alone in semi-darkness, stares into space and pushes his fists against his temples. This triggers the reappearance of the drowning vision — the Williams shot from the intertitle that shows him pushing his wife overboard.

Frank Williams and the Travelling Matte

Frank D. Williams (21 March 1893 – 16 October 1961) started as a cinematographer in 1912 or 1913. He can be seen playing a cameraman in 1914 in Keystone's Mack-Sennett-produced *Kid Auto Races at Venice* (Henry Lehrman), the first movie released in which Chaplin appeared as the Tramp. Williams's first patent for his travelling matte process, titled "Method of Taking Motion-Pictures," was granted 23 July 1918 and was still in effect when he worked on *Sunrise.*[47] By that time, he was a considered a leader in special visual effects, drawing business to his own lab for his patented process.

In May 1926, Carl Louis Gregory presented his paper "Trick Photography Methods Summarized" before the Society for Motion Picture Engineers; it was subsequently published in both *American Cinematographer* and *Transactions of the S.M.P.E.*[48] Gregory differentiated 15 types of effects, suggesting that more specific terms than the catch-all "trick photography" should be used for distinct methods: First, [Basis] the Standard is: Straight cinematography, 16 exposures per second; Second, High Speed, Slow Motion; Third, Varied Taking Speeds; Fourth, Animated; Fifth, Reverse Order; Sixth,

Negative': 2. The Williams shot illustrated title: (Couldn't she get drowned?)", William Fox Studio memo from Don Hetrick to Mike Farley, 4 June 1927. Karl Thiede Collection. Murnau added a handwritten note to his shooting script calling this shot a trick "studio shot," with a drawing, in *Sunrise (Sonnenaufgang), Ein Drehbuch*, op. cit., p. 22 verso.

47 The application for the patent for "Method of Taking Motion-Pictures," was filed 22 May 1916, no. 1,273,435: https://patents.google.com/patent/US1273435.

48 The S.M.P.E. meeting was held in Washington DC in May 1926; Gregory's paper was published in *American Cinematographer* (June 1926) and in *Transactions of the S.M.P.E.*, 25 (1926).

Mattes; Seventh, Stop Motion; Eighth, Fade and Dissolve; Ninth, Multiple Exposure; Tenth, Glass; Eleventh, Mirrors and Prisms; Twelfth, Double Printing; Thirteenth, Travelling Matte; Fourteenth, Projection Printing; Fifteenth, Other Method.[49] According to Gregory's taxonomy, "mattes" are listed sixth but do not include "travelling mattes" which constitute a separate method, near the end:

> Thirteenth, the travelling matte. By this process figures in action may be superimposed against any background without being necessary to build any sets at all. It requires [a] very accurate mechanism to work it and is patented. It is sometimes called the Williams process from the name of the patentee, Frank Williams [...][50]

One of the ways a Williams shot can sometimes be identified is from an inexact match between the separate layers that make up the composite, resulting in a slight ghosting or halo around the edge of the image. Regardless, the Williams shot guaranteed improved image quality in this type of compositing because the photography is equally sharp from foreground to background, whereas with rear projection, the background is re-photographed and is therefore a second-generation image (a dupe).[51]

Special effects were becoming popular, a subject that could be talked about by non-specialists. In April 1926, *Photoplay*, generally thought of as a fan magazine, published "How They Do It!," an informative illustrated article that quickly moves into a rags-to-riches feature on Frank Williams, "the man who invented the new miracle travelling mat [sic] process and made possible most of the wonders which we see in motion pictures today [...]. When he comes on a set, his quiet presence is scarcely noticed."[52] Illustrations are used to explain the steps in his compositing method, and some details of his printing machine are revealed. His obstacles on the

49 Carl Louis Gregory, "'Trick' Photography Methods Summarized," in *American Cinematographer* (June 1926), pp. 9, 16–17, 20–22.

50 See Gregory's essay for his full summary of the technical details of Williams's travelling matte method. Ibid., p. 22. Both Gregory in this essay and Williams in his essay "Trick Photography," *Transactions of the S.M.P.E.* (April 1928), state that the background cloth or screen could be white or black, depending on the case at hand.

51 Other patented travelling matte compositing processes were in competition with Williams, especially the Dunning Process, patented by C. Dodge Dunning, and later that of Roy J. Pomeroy. Raymond Fielding devotes a chapter to the development and differences among "Travelling Mattes" in his classic study *The Technique of Special Effects Cinematography* (London & Boston: Focal Press, fourth edition 1985).

52 Cal York, "How They Do It!," in *Photoplay* (April 1926), p. 31.

road to success, starting from an idea he worked on back in his days with Mack Sennett, were finally overcome "through the better grade of film, a motor-cranked camera set on a solid tripod, and through his own printing machine, built according to Williams's own drawings at a cost of $18,000, one whose accuracy is to within one ten-thousandth of an inch [...]".[53]

Questions remain. Even if we surmise that Murnau probably was shown what could be done with the Williams process after he arrived in Hollywood and then decided to bring it into his production, we know that the director worked closely with his creative-technical team during lengthy pre-production meetings and then continuously during production. Williams was an outside specialist. What relationship did Williams have with Murnau, the cinematographers, or the art director? If only we could overhear their conversations! The Williams shot was worked into the overall design for forced perspectival shooting in *Sunrise* that could incorporate all the many other kinds of special effects that Murnau had the talent and the luxury of building, one way or another, into seemingly every single shot in that film without losing the intensity of his human drama. Officially, Williams worked in the background. He did not receive screen credit.

L.B. Abbott worked as an assistant on the cinematography for Fox's *What Price Glory?* (1926), directed by Raoul Walsh, which was released one year before *Sunrise*. Abbott wrote in detail about his role in preparing for using the Williams travelling matte technique for the night battle sequence in that film. This is the closest description I have seen of how studio professionals interacted with Williams's outside specialist lab.

> On July 28, 1926, we gathered up the cameras, three Bell & Howells, and left the Western [Avenue] lot for the ranch. The location was a large, level, barren area on which had been erected three parallels [wooden platforms used to elevate cameras], two of them 50 feet high and the other 72 feet. They were larger at the bottom than at the camera platform on top in order to insure a steady base for composite photography. We were to make shots of about 700 World War I soldiers marching at night through a French forest. The forest was a miniature, which already had been shot. In the final scenes, the marching men were double-printed into the miniature forest by use of a travelling matte system invented by Frank Williams. The technique was based on the density difference system.
>
> I was assigned to the 72-foot parallel, and proceeded to climb the ladder, one-handed, hauling up the tripod and setting it in place. Then I went

53 Ibid., pp. 30–31.

> back down and got the camera, hauled it up and mounted it on the tripod. Next, I mounted a film clip of the miniature forest on the pins of a prism viewing device that was fitted on the camera movement rails. The pins in the viewer then matched the location of those in the camera movement. Now a sun arc gobo (a four-by-eight foot wooden frame covered with black cloth) had to be placed in the proper position to fit the trunks of each of the miniature trees. This made it appear in the final shot that a marching man passing in the area behind a tree would disappear behind it, while a man passing in the area in front of the tree would appear to walk in front of it [...].
>
> When rehearsals were completed and the lights were all burning, the cameras rolled. The men marched and a take was made. This procedure was repeated throughout the night, with the positions of the marchers and the bomb blasts changed each time so that in the final picture it would appear that there were thousands of men. At dawn we stopped shooting and the camera assistants gathered up the exposed magazines and went to the Frank Williams Laboratory, where we unloaded the exposed film and reloaded the magazines with fresh stock. Then we went back to the location, where we spent another day of lineup, using different lineup clips. The shooting that night was similar to that of the night before, and at dawn we were finished with the sequence. We gathered up all the camera gear, stopped at the Williams lab, unloaded the exposed film, then returned the camera equipment to the Western Avenue studio.[54]

Abbott brings us to the door of the lab, even inside the door, but then what? How did the director of photography or the director of the film work with Williams and the lab? Who was in control? How were decisions made? The lab is a classic example of the proverbial black box: we know what went in and what came out, but not what happened in between.[55]

Frank Williams published a short article called "Trick Photography" in April 1928 in the *Transactions of the S.M.P.E.* He states in his first sentence: "Although I have titled this paper as dealing with trick photography, this

54 L.B. Abbott, A.S.C., *Special Effects; Wire, Tape and Rubber Band Style* (Hollywood: ASC Press, 1984), pp. 14–15.

55 Another tantalizing clue: Frank Williams advertised how his lab interfaced with a studio in *The Film Spectator* on 9 June 1929: "WILLIAMS 'SHOTS' / The success of composite scenes depends upon how well the film is taken that is to be combined. / I have just finished a complete sequence in 'The Michigan Kid', a Rex Beach picture for Universal — done entirely at the studio. It both thrilled and baffled the executives / IRVIN WILLAT personally directed these scenes and John Fulton photographed them for him. Frank Williams."

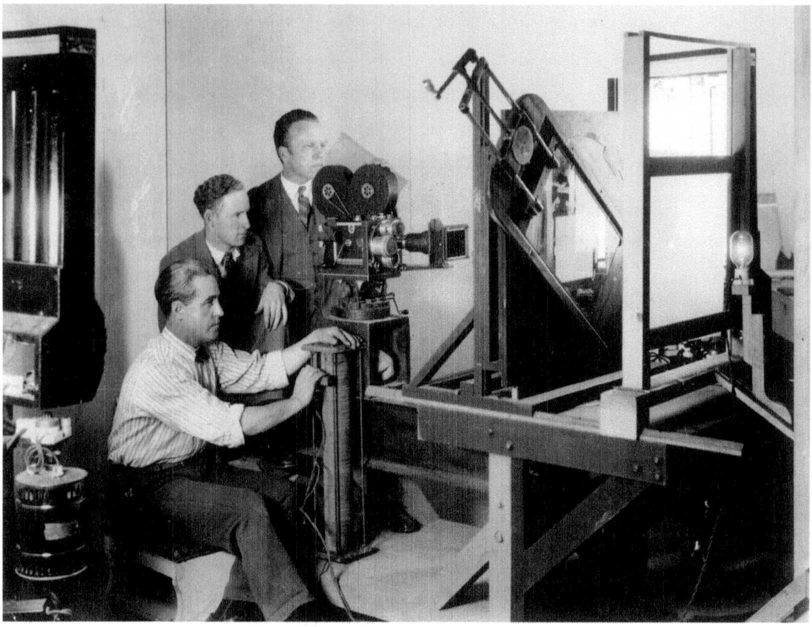

Fig. 12.16 Charles Rosher (far right) and his *Sunrise* machine (for transparency shot).

term has been discarded in Hollywood in favor of the term 'special process work'."[56] His own work, he tells us, falls into the category of "[e]ffects obtained by inserting action into backgrounds." He then outlines some of the steps involved in achieving the art of his effects, and he answers a few questions from his technical audience, perhaps giving a few clues, but he is not about to illuminate the black box.

Besides visual effects done in the camera by cinematographers and those created by outside specialists, *Sunrise* also contained shots composited (i.e., not done in-camera) by Charles Rosher (Fig. 12.16). At the end of the film, Murnau wanted the sun to rise behind the village. Rosher intended to photograph the sun rising while they were shooting at Lake Arrowhead, but he faced an impasse and needed a different kind of solution (Fig. 12.17). He wrote out what he had done on the back of a photograph of the machine that he ended up using:

> This is the same device used in *Little Lord Fauntleroy* [Alfred E. Green and Jack Pickford, 1921 — starring Mary Pickford] with many additions in order to make the sunrise effects necessary for the film *Sunrise*. I had

56 Williams, "Trick Photography," op. cit., p. 537.

> tried for over a month to make some real sunrises behind the beautiful village setting we had built at Lake Arrowhead, but clouds are very scarce in California. I employed a transparency of the village and used transparencies I made from my collection of still cloud negatives. These were moved by the rack & pinion and the moveable light was controlled by rheostat. Nowadays this could have been done in the optical department.[57]

Once again, a cinematographer who worked during the silent era and beyond it describes a one-of-a-kind artisanal solution that, in looking back from a lifetime of work, is referred to, in an off-hand manner, as something that could be handled without problem later through optical printing, yet another indication of the sea change that had taken place.

Along with praise for *Sunrise* followed by imitations of different aspects of its design and shooting techniques, some of them highly inspired, there was also a reaction against it. James Mitchell Leisen was an art director before he became a director. In April 1928, he published "Some Problems of the Art Director" in *Transactions of the S.M.P.E.* in which he wrote against the German technique of false perspective:

> In the German technique, which styles a great deal of false perspective, the cameraman is limited to one angle; otherwise, a new set is needed. A cameraman, to avoid monotony, likes to move around. We are hampered in our type of stories by the necessity of doing them in actual scale. They deal with the mental reaction of a person and express it in the story. They can use false perspectives because they limit themselves to one or two angles. It is not a practical way, because the set can be used only once. It is a saving in production cost to keep sets standing because by slight changes they can be given a totally different appearance. We did one picture entirely as an experiment [using the German technique] to see how it would work. We made cameras turn circles, run down stairs, and got an interesting result, but the cost was not commensurate with the return.[58]

Leisen's description of his exercise in what he characterized as the German perspectival method is an example of a mechanical undertaking without integrating creative-technical possibilities throughout and disregarding the role technology should play in advancing the heart of a cinematic story.

57 The photograph and its reverse side with Rosher's explanation courtesy of Kevin Brownlow.

58 James Mitchell Leisen, "Some Problems of the Art Director," *Transactions of the S.M.P.E.* (April 1928), p. 77.

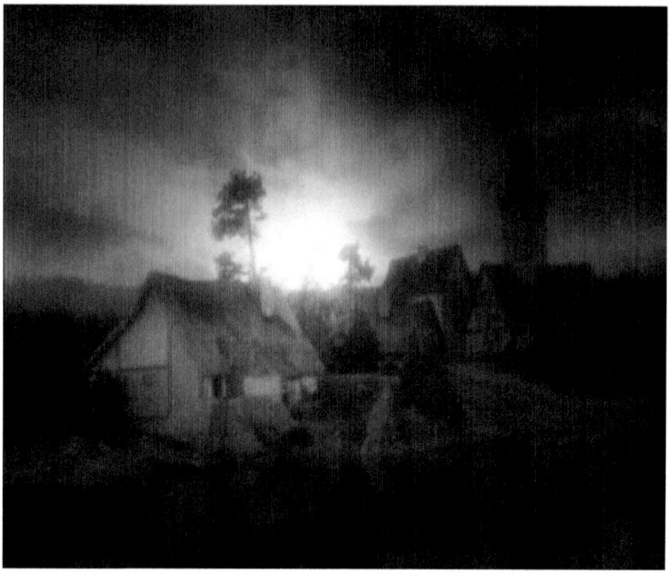

Fig. 12.17 Still from *Sunrise* (F. W. Murnau, 1927). Transparency shot.

In modern times, of course, cinematographers commenting on films like *Sunrise* — who were not professionals during the silent era — have lived through decades of doing things in other ways. Cinematographer Nestor Almendros offered an insightful lecture on *Sunrise* to students at the University of Ohio in 1984 that was published in *American Cinematographer*.[59] He pointed out the unusual importance of action in the background in Murnau's films. In this context, he referred to the Williams shot outside of the church, but he believed it was accomplished through rear projection:

> In Murnau's movies; there's always been movement in the background, as in that scene of Janet Gaynor and George O'Brien in a restaurant with people dancing in the background. [See Fig. 18] In *Sunrise* there are often two actions going on at the same time. In the dream sequence, when the man and wife are walking in the city, suddenly they seem to be in the country, then they come back to themselves because there is a traffic jam, and they again are in the city. This is how it was done: they were walking on a roller (a conveyor belt) and behind them was a projected film which included the dissolve from the city to the country, and back to the city.

59 "Sunrise" by Nestor Almendros, in *American Cinematographer* (April 1984; partial rept. June 2003) https://theasc.com/magazine/june03/sub/page2.html

He could not have seen the photo that shows that the couple were not walking on a conveyor belt.

In cinematographer John Bailey's voice-over commentary for Fox's DVD of *Sunrise*, he discusses the same shot but does not believe that rear projection was involved. In his opinion, the cinematographers created the unusual visual effects in-camera. "This car that passes in the foreground that looks matted in [...] The reason it looks so strange is because you'll see in a second there is an effect that happens here [...] Several people have said they thought this was a back projection. I don't really think so [...] It's amazing the way Rosher and Struss were able to do all these effects in-camera [...]." He goes on to explain

> why I thought that effects shot was not back projection but done as a multiple exposure. There's a very famous photograph [the same one discussed above] of the two actors walking down a ramp with the camera on a dolly track behind them and at the extreme left end of the photograph there are two grips carrying a big black gobo [the cloth background]. It suggests to me that they were shooting into black [...]. A first pass was made on black, and then another pass was made in the sylvan environment, and in the background, the city background, and then composited in the camera.[60]

Neither Almendros nor Bailey recognize that the shot had been created with an effects specialist.

This discussion of in-camera effects and effects specialists can be seen as analogous to contemporary questions about who should, and who does, control the final look of the film and also of the DVD or Blu-Ray editions, including those marketed as "restorations." The digital era has made it possible for many more kinds of people, specialists or not, to alter the look that the Director of Photography originally accomplished in concert with the director of the film. This discussion has been ongoing since the ubiquity of the digital intermediate and more recent digital "improvements" to the cinematographer's work.[61]

60 *Sunrise* DVD (20th Century Fox, 2002 or any edition), John Bailey's commentary-track for this section starts at 44:45.

61 "Originally, 'digital intermediate' referred to the electronic production of an intermediate negative for film prints that was created by color correcting the scanned camera original negative after the final editing was performed in the digital realm. In other words, you shot film, developed the negative and then scanned it to create digital files. The original negative might never be touched after this initial scan. You edited with a computer. To make a film print

In the February 2018 editorial of *American Cinematographer*, ASC President Kees van Oostrum described the loss of control by Directors of Photography over the final look of the film because of the increasing incursions made possible by digital postproduction and attempts to counter that movement:

> Today, when we create an image, we simply no longer remain its author. Instead, besieged by strong outside opinions about dynamic range, resolution, bit depth, and on and on, we seem to be becoming mere suppliers of some fundamental, base image, and we are then constantly challenged about how that image should be interpreted, and told how it should be changed.
> In reality, the parameters and capabilities of digital imagery are constantly increasing. As a matter of artistic survival, we feel the need to counter these technical achievements with heavy filtration, older and softer lenses, etc. — completely defeating the technological development because it does not represent what most of us seem to like.[62]

There can be no doubt that Murnau was in control when it came to *Sunrise*. The outside specialists must have been delighted to contribute to it, although they would not receive screen credit or be mentioned by either cinematographer, to my knowledge. *Sunrise* was, in many ways, a special case, but the transition from ingenious visual solutions through in-camera effects coordinated with the art director's forced perspective sets, to the movement toward effects specialists and Special or Visual Effects Departments in the studios was in progress. Before long — especially given the non-linear yet steady incorporation of sound recording, which made in-camera effects almost impossible — optical printing would take the place of that art, as almost every interview with a cinematographer from the silent era who continued to work with recorded sound would mention.

Just because this transition did not last a long time does not mean it was not important, like the uneven transition from silent to sound film or from orthochromatic to panchromatic film.

from the edited digital file, the Director of Photography (in most but not all cases) went to the DI facility, sat down with the colorist and timed the picture. When everyone was happy, the digital information was output onto film negative from which prints or an interpositive would be made to produce more inter-negatives for creating prints." Dave Kenig, Panavision, communication to the author.

62 Kees van Oostrum, "Inward Significance," in *American Cinematographer* (February 2018), p. 12.

Bibliography

Abbott, L.B. *Special Effects; Wire, Tape and Rubber Band Style* (Hollywood: ASC Press, 1984).

Anon. "Method of Taking Motion-Pictures." 22 May 1916, no. 1,273,435. Accessed 17 November 2021. https://patents.google.com/patent/US1273435.

Anon. "*Sunrise* Paves the Way," *Los Angeles Times* (8 December 1927): A9.

Bergstrom, Janet. "Invitation to Phantom." In *Phantom,* directed by Frederich W. Murnau. Flicker Alley, (DVD) 2003, revised for BluRay 2019.

Brownlow, Kevin. *The Parade's Gone by...* (Berkeley, CA: University of California Press, 1968).

Chambers, Gordon A. "Process Photography." In *Cinematographic Annual,* vol. 21, edited by Hal Hall (Hollywood: American Society of Cinematographers, 1931).

Clarke, Charles G. *Highlights and Shadows: The Memoirs of a Hollywood Cameraman* (Metuchen: Scarecrow Press, 1989).

Eisenschitz, Bernard. "Le voyage à Hollywood," *Cinématographe* 75 (February 1982): 16.

Eisner, Lotte H. *The Haunted Screen.* Translated by Goer Greaves (Berkeley, CA: University of California Press, 1969).

Eyman, Scott. *Five Cinematographers: Interviews with Karl Struss, Joseph Ruttenberg, James Wong Howe, Linwood Dunn and William H. Clothier* (Metuchen: Scarecrow Press, 1987).

Fielding, Raymond. *The Technique of Special Effects Cinematography* (London, Boson: Focal Press, 1985).

Freund, Karl. Interview by Frank Camino (1 June 1964). UCLA Special Collections, Collection 543, Box 2.

Gaynor, Janet. "The Reminiscences of Janet Gaynor." *Oral History Collection* (New York: Butler Library, 1958).

Gregory, Carl Louis. "'Trick' Photography Methods Summarized," *American Cinematographer* (June 1926): 9, 16–17, 20–22.

Hall, Mordaunt. "Murnau's Drums and Fifes of Life," *The New York Times* (2 October 1927): x 7.

Harvith, Susan, and John Harvith, eds. *Karl Struss: Man with a Camera* (Cranbrook: Cranbrook Academy of Art/Museum, 1976).

Kann. "*Sunrise* and Movietone," *Film Daily* 41, no. 72 (25 September 1927): 3-4.

Koszarki, Richard. *Karl Struss: New York to Hollywood* (Fort Worth, TX: Amon Carter Museum, 1995).

Leisen, James Mitchell. "Some Problems of the Art Director," *Transactions of the Society of Motion Picture Engineers* 12, no. 33 (April 9-14, 1928): 71-80.

Oostrum, Kees van. "President's Desk: Inward Significance," *American Cinematographer.* (February 2018). https://ascmag.com/blog/presidents-desk/presidents-desk-inward-significance

Schallert, Edwin. "*Sunrise* Rare Art Feature. Murnau Production Ultra in Treatment. Camera Effects Exercise Dazzling Influence. Brilliant Audience Applauds Principals," *Los Angeles Times* (1 December 1927): A9.

Struss, Karl. "Dramatic Cinematography," *Transactions of the the Society of Motion Picture Engineers* 12, no. 34 (April 9-14, 1928): 317-319.

———. "Recollections of Karl Struss: An Oral History of the Motion Picture in America." Interview by John Dorr. UCLA Special Collections 1969 (9, 24, 30 July, 13–27 August 1968).

Turner, George E., ed. *The ASC Treasury of Visual Effects* (Hollywood: American Society of Cinematographers, 1983).

Warm, Hermann. *Murnau* (Berkeley, CA: University of California Press, 1973).

Williams, Frank. "Trick Photography," *Transactions of the Society of Motion Picture Engineers* 12, no. 34 (April 9-14, 1928): 537-540.

York, Cal. "How They Do It!," *Photoplay* 29, no. 5 (April 1926): 28-31.

About the Author

Janet Bergstrom is Research Professor of cinema studies at the University of California, Los Angeles. She is primarily engaged in archivally based research on European film directors who have worked in more than one national cinema, such as Murnau, Renoir, Sternberg, Hitchcock, or Lang. She publishes visual essays along the same lines on DVD/BluRay and is writing a book on the development of Murnau's *mise-en-scène*.

13. *King Kong*, An Open Perspective

Suzanne Liandrat-Guigues

Abstract

King Kong, that masterpiece of 1930s filmmaking, is a key work for the study of special effects in cinema. Its value comes in great measure from how the film's effects move beyond that most basic function of special effects of merely giving to perception what neither eye nor camera can see. If *Kong* offers a remarkable visual and formal hybridity in the way it weaves together non-effects shots with a multitude of effects and techniques (animation, miniatures, etc.), its strength lies in how it combines and incorporates them to the film's *mise en scène* in a singular *poietic* act: the emergence of a *form* whereby multiple spatial and temporal perspectives ensure that the natural landscape of Skull Island and the urban landscape of Manhattan Island entwine in the most unexpected way.

Keywords: realism, tricks, aesthetics of perspective

Visual or audio objects on the screen and soundtrack are the result of a singular *poietic* act of re-creation or invention, which characterizes the art of film.[1] Realism, when applied to film images, only refers therefore to what can properly be called a work's final impression. The fact is that a film may involve many different operations that can produce, to varying degrees, the illusion or impression of realism. The notion of special effects concerns a series of operations that keep alive the opposition between shooting 'real' or live-action and the various other ways for creating moving images. This initial distinction is necessary to justify discussing special effects without referring to the entire enterprise of filmmaking as a whole.

Throughout the history of cinema, the filmmaking processes that make up what we call "special effects" have been diverse in form and have served

1 See Stanley Cavell, *The World Viewed* (Cambridge, MA: Harvard University Press, 1971).

Lefebvre, M. & M. Furstenau (eds.), *Special Effects on the Screen: Faking the View from Méliès to Motion Capture*. Taylor & Francis Group, 2022
DOI 10.5117/9789462980730_CH13

various purposes. These forms and purposes can be described in a non-restrictive way by terms such as "tricks," "glass shot," "process-shot," or more generally "special effects" and can be mechanical or purely visual, they may affect the image or the soundtrack, can be shot live or else obtained in the laboratory.[2] Today, after years of virtuoso developments in computer effects, the field has reached an all-time high degree of proficiency, to the point where we may be oblivious to special effects and even forget about them altogether as they blend so perfectly with recorded live action shots. But as it happens, we may wonder whether current special effects are not losing part of their own reason for existing in the first place. We know that the use of special effects is often the result of an inability to shoot and record something live. The object to be depicted may have ceased to exist or else may have never existed at all, or shooting it live may be impossible, difficult, or dangerous to achieve, etc. Yet in seeking to "erase" from sight the effects required for depicting something unavailable for live-action shooting, filmmakers and effects designers can be seen as forfeiting part of the art of film's poietic, or formative, aspect. I propose here to move against the grain of the current use of special effects by offering an analysis of the famous 1933 film *King Kong* (Merian C. Cooper and Ernest B. Schoedsack) that will take into consideration the perceivability[3] of its special effects as an aesthetic principle in its own right.

King Kong undoubtedly carved a path in the history of filmmaking, and several of the solutions it presented to various problems of representation posed by its subject matter came to be exploited later on in the century. But, above all, the film argues that special effects must not solely be "felt," but that they also require a true purpose beyond merely responding to the inability of shooting something live. In short, special effects should not solely "supplement" a film, for they can also serve to question the nature of what

2 Réjane Hamus-Vallée, *Les effets spéciaux* (Paris: Cahiers du cinéma, 2004), pp. 5–6.

3 It is well known that Christian Metz (1977) proposed a taxonomy of *trucages* that distinguished between "visible," "invisible," and "imperceptible" effects. By "*perceptible* effects," I refer here to those effects that Metz called "invisible," but in so doing I wish to emphasize the fact that we can easily perceive the effects of the technical machinations on the screen. Although the viewer may not know how they were achieved, they nonetheless are perceived as an obvious trace that some form of manipulation or trick has transpired. Metz himself recognized this when he wrote that invisible effects "[are] *perceptible*, because we perceive [their] presence, because we 'sense' [them], and because that feeling may even be indispensable, according to the codes, to an accurate appreciation of the film" (Metz 1977: 664, emphasis mine). In fact, not only are the special effects highly perceptible in *Kong* but, as we shall argue, it is important that they be so: much of the film's meaning comes from our perceiving them. Christian Metz, "*Trucage* and the Film," in *Critical Inquiry*, (3)4, 1977.

it is that they seek to depict, though this implies that that they do not mask their own nature. Analyzing special effects should therefore not limit itself to evaluating whether they are outdated or insufficiently "sophisticated" in creating realist illusions.

In *King Kong*, the special effects depict a long-past prehistoric world, which, at first glance, is the obvious reason for their presence; however, they often commingle with "real" or live-action elements in a highly suggestive *ordering* (or *dispositio*, to use a term from ancient rhetoric) — a far less obvious reason for their use. The term "ordering" here is meant to be understood as an arrangement or composition involving a combination of heterogeneous visual elements. Vital to the overall design of the film, this combination has motivated the use of hybrid imaging processes: they can be observed in the many simulacra involving the presence of real actors as well as in the dual cinematic regime of live action shooting and stop motion animation. It follows that we never lose sight of the heterogeneity of the "ordering" in this cinematic fable.

Born from the imagination of its directors, *King Kong*'s universe exists only through the cinematic process. For one thing, unlike other narratives of the era that depicted characters such as Nosferatu, Dracula, Mabuse, or the Frankenstein monster, *King Kong* was not adapted from literature. And although one can find in it a muted reference to the story of *The Beauty and the Beast* that could arguably align the film with the literary series mentioned above, the fact remains that Kong is not a literary character and that its "invention" gained its illustrious reputation from its cinematic nature. On the other hand, the narrative novelty of *King Kong* prompted the use of special effects for cinematically creating and exhibiting the film's imaginary universe.

The film was essentially conceived by Merian C. Cooper and Ernest B. Schoedsack, who were joined by a crew of technicians who remained famous in film history, including chief technician Willis O'Brien, the scenic designer who supervised the rear projection effects, stop motion special effects, as well as other animation effects and travelling mattes.[4] The

4 The idea for the film initially came from Cooper, who worked as executive assistant to David O. Selznik at RKO. Cooper, who had previously worked with Schoedsack, hired him (and Irving Pichel) to direct *The Most Dangerous Game* (also starring Fay Wray). Both Cooper and Schoedsack acted as producers on *Kong* and shared directing duties. While Schoedsack shot mostly the dialogue scenes and directed the actors, Copper worked closely with special-effects designer Willis O'Brien, who had designed and supervised the stop action effects for Harry O. Hoyt's *The Lost World* (1925, an adaptation of a story by Sir Arthur Conan Doyle), a film that also included stop motion dinosaurs. In 1915, O'Brien had directed the stop motion film *The*

brothers Mario and Juan Larrinaga and Byron L. Crabbe were also hired as miniature, matte artists and illustrators and helped shape Kong's world. Due to its impressive mix of techniques, materials, and innovations, the film was considered ahead of its time and was hailed as a wonder of filmmaking (most likely the "8th Wonder of the World" as can be seen on the banner that appears in the New York segment of the film and which can also be read as referring to both the character and the film itself — or else, to the eponymous film *qua* character!).

Offering an inventory of the different effects that were used for Kong is here less relevant[5] than wondering what expressive power this rich cinematic matter can wield. The film exhibits a world created by a cinematic machinery that continuously foregrounds itself to our attention. The stake, then, is to offer viewers a diegetic world that has no referent or "presence" in our world and which therefore can only be seen or experienced in the cinema as a pure cinematic effect. It follows that the most remarkable solutions to the problems raised by the creation of this world achieve much more than what might strike us today as merely the result of some archaic special effects technology. To put it simply, they contribute fully in expressing an aesthetic principle founded on the *mise en abyme* of the very cinematic dispositive constitutive of the world we see on the screen.[6]

The Anti 'Natural Drama'

The figure of Kong, whose name supposedly means "ape" in Indonesian, is an invention that required many ingenious tricks. This cinematic construct was the product of the imagination of two adventurers who were also documentary filmmakers, having previously made "exotic" films in Asia Minor (*Grass: A Nation's Battle for Life*, 1925) as well as in Siam and Sumatra (*Chang: A Drama of the Wilderness*, 1927). The presence of many wild animals in the latter film resulted in RKO coining a term for a new genre: the "natural drama" film. Natural dramas sought to arouse emotions, especially in the way they featured the risks faced by the production crews. In *Chang*, there is a famous scene where a wild beast rushes the tree where

Dinosaur and the Missing Link: A Prehistoric Tragedy, which he later claimed enabled him to create an early sketch of Kong made out of wood, textile, and clay.

5 Especially since such an account already exists. See Orville Goldner and George E. Turner, *The Making of King Kong: The Story Behind a Film Classic* (New York: Ballantine Books, 1976).

6 See also Hamus-Vallée, p. 50

the camera operator is hiding as he continues to shoot the frontal attack of the animal. Natural dramas showed animals in the wild, sometimes more or less mythical (such as giant lizards), though real nevertheless. Such scenes clearly fed the creation of *King Kong*, but in a sort of imaginary reversal, as the special effects and studio setting now replaced the real-life experiences and dangers of location shooting (an irony perhaps purposely pointed to by the Indonesian name of the creature).

In *Kong*, then, the idea is to show a prehistoric world. The fact that most of the beasts that are shown have once existed (e.g., the dinosaurs) serves to somewhat curtail the inventiveness of the filmmakers who did not create otherwise unimaginable monsters. This includes Kong, an "invention" entirely based on an existing model, that of the great ape. The film is famous, among other things, for its use of miniatures and animated puppets. Much ingenuity had to be deployed to create the impression of a giant-size gorilla or that of the "lost world" of Kong's island. Numerous puppets and miniatures were built, using different materials, and in different sizes, to create Kong and other oversized (or miniature!) beasts and to animate them in a mostly fictional, studio-based setting. The effects, though they largely benefitted from the specificities of filmmaking (framing, *découpage*, use of lighting, depth of field, stop motion, etc.), are not without revealing a very specific lineage.

The lost island of *Kong* (later named "Skull Island" in the sequels because its shape is reminiscent of a human skull) belongs to a European imaginary of exoticism initiated in good measure by *Robinson Crusoe* in the eighteenth century, a novel that was contemporaneous, moreover, with a certain taste for automatons (think of La Mettrie, Vaucanson, and the fascination for "animated animals"). If both these sources meet in *King Kong*, it is because the aim is no longer — as it had been in the previous documentaries of its directors — to *discover* an exotic world with its "natives" (nor, as in *The Most Dangerous Game* [Ernest B. Schoedsack and Irving Pinchel, 1932], to depict a terrifying manhunt) but rather to *construct* an imaginary exotic world with the help of "machines" or technology.

In other respects, the film follows a path that was traced in the nineteenth century with the emergence of "human zoos" or "freak shows" in carnivals, circuses, sideshows, and fairs, which exhibited natural "monstrosities," including those affecting human beings. P.T. Barnum showcased "human curiosities" in his Manhattan Museum — very tall and very short people, albinos, etc. — before embarking them on his tours of the US and Europe.[7]

7 See Raymond Bellour, *Le Corps du cinéma. Hypnoses, émotions, animalités* (POL, 2009), pp. 451ff.

Carl Denham, the fictional filmmaker in *King Kong*, clearly revives this practice when he seeks to capture and exhibit the giant gorilla — though he does it with a bit more "shine" since the film avoids detailing the squalor of the sideshows (in contrast, see David Lynch's 1980 *Elephant Man*, or, more recently, Abdellatif Kechiche's 2010 *Vénus Noire* [Black Venus]). Once captured, Kong is thus exhibited on a Broadway stage in the last part of the film, which leads to its death. If the demise of Kong is upsetting, part of the reason, I believe, rests on a certain ambivalence in the categories of "human" and "animal" that pervaded the exhibition of "human curiosities" and "freaks."[8] A distance nonetheless is maintained between the "human" and the "animal" in Kong's death scene since the ape's movements were all produced by stop motion animation and not, as would be the case today, with motion capture by some underlying use of real human movement[9] or else by the use of a gorilla suit (which is what was often believed to be the case at the time the film was first released, thus reinforcing the parallel with the sideshow "freak").

Various other references make their way into *King Kong*'s figural network, including references to visual traditions whose expressive functionality it revives: the production drawings made for the film by Willis O'Brien and the Larrinsga brothers bring to mind the engravings of Gustave Doré (especially his illustrations for *The Bible*, for Dante's *Divine Comedy*, and for Milton's *Paradise Lost*); we can also recognize the influence of Arnold Böcklin's painting, *Isle of the Dead*, whose 1880 version O'Brien had reproduced for the creation of the island in the studio with a matte painting. The production drawings on which the film's design was based also remind us of the illustrations that were made for Jules Verne's stories in the famous Hetzel illustrated edition. The importance afforded to size in the film — the contrast between Kong and the humans, the oversized sets on the island, etc. — also brings to mind narratives that emphasize variations in size such as *Gulliver's Travels* by Jonathan Swift or *Alice in Wonderland* by Lewis Carroll — the principle of which has led to the use of special effects whenever they were adapted to the screen.

If *King Kong* inaugurated a new trend in the adventure film genre, not to mention the disaster and horror genres, it also found inspiration within film history. In addition to self-parodying their own explorer/filmmaker

8 Interestingly, a similar ambivalence is thematized in *The Most Dangerous Game*, where Count Zarov, a master hunter, turns to hunting humans as game.

9 Think of Andy Serkis playing Caesar in the latest installment of the *Planet of the Apes* series, *War for the Planet of the Apes* (Matt Reeves, 2017).

past in the character of Carl Denham, or else their own adventurous First World War exploits (Schoedsack as a cameraman in the Signals Corps and Cooper as a bomber pilot), the two directors reused the same jungle studio set RKO had built for *The Most Dangerous Game*. The great wall of the village was recycled from a film by Cecil B. DeMille whose title — *The Sign of the Cross* (1932) — might even have inspired some of *Kong*'s Christlike imagery, all the while offering an homage to DeMille. As for the scenes showing the natives and their massacre, they could just as well have belonged to the series of jungle films featuring the character of Tarzan, starting with the 1918 version (*Tarzan of the Apes*, Scott Sidney) and the critically acclaimed remake of 1932 (*Tarzan the Ape Man*) directed by W.S. Van Dyke (in between them appeared the famous Hal Foster comic strip in 1929).[10]

What we find here, therefore, is a more or less active or explicit network of relations. Some of them are clearly ironic, as is the case with the indigenous ceremony celebrating King Kong, a pure fantasy ritual where the natives are seen wearing gorilla suits in a scene that mimics (or better yet "apes"!) the famous "Hot Voodoo" cabaret routine of Marlene Dietrich's character in Joseph von Sternberg's *Blonde Venus* (1932). Here a large gorilla is led in chains throughout the cabaret by a group of black women armed with shields and spears. It soon turns out that this is a gorilla suit which Dietrich sheds, "striptease" like, as she now dons a large blonde wig. The reversal act would be the one found in many a fairground attraction at the time where a girl "turns into a gorilla." The reference to *Blonde Venus* is furthered in *Kong* where Fay Wray sports blonde hair (whereas she was a brunette in *The Most Dangerous Game*, both films being shot at approximately the same time).

All these relations to other texts, films, genres, whether as homages or simple acknowledgments — a common enough trait of filmmaking — are at one here, as we shall see, with the singular aesthetic of *King Kong*.

The Machinic Fable

Throughout the film, one detects a continuous foregrounding of the act of seeing. For instance, the vegetation often offers a "frame within the frame" (Fig. 13.1). And when it recesses in the various planes of the image — creating

10 A literary character created in 1912 by Edgar Rice Burroughs, the immense success of the 1932 film, starring former US Olympian Johnny Weissmuller and Maureen O'Sullivan, led to several sequels and paved the way for future incarnations of the character in countless films and television series.

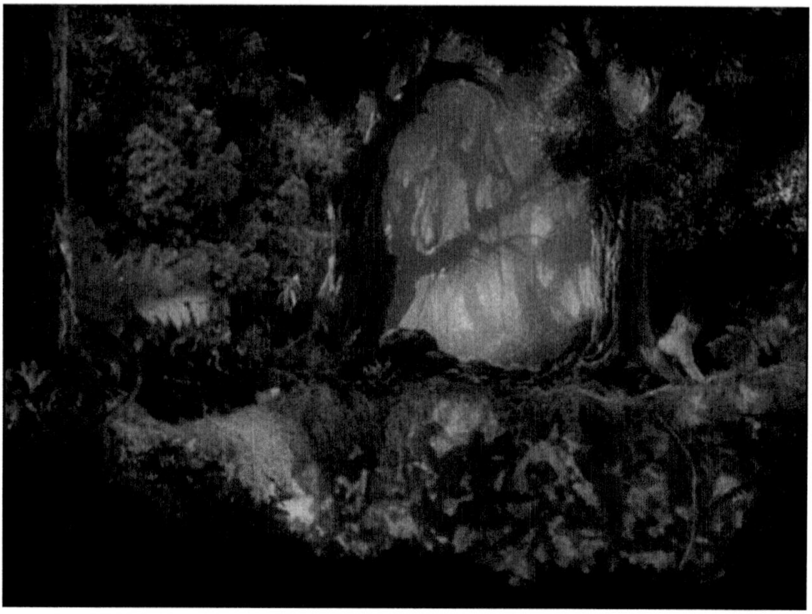

Fig. 13.1 Circular motif of vegetation in *King Kong* (Merian C. Cooper and Ernest B. Schoedsack, 1933).

a layered perspective — it forms a circular shape that figuratively stands for a point of view. If we compare this with the film's use of an iris — a common optical effect of the silent era — we can see that the composition serves a different function. Whereas the iris is used disjunctively to lead the viewer into attending to a detail in the composition of the image, thereby eliminating the background and any sense of spatial depth, the "framing within the frame" created by the recesses of the vegetation emphasizes the impression of depth and makes the viewer aware of the different layers the gaze must travel as it penetrates this dense jungle world.

This heightened awareness that the image is made up of layers is also achieved by the various effects that are used to create composites: matte paintings, rear projections, double exposures, etc. For example, after Ann has been abducted by Kong, we often see the men in the foreground of the frame as they search for her and the gorilla. This composition presides over the first appearance of a dinosaur in the background. Connecting the men in the foreground and the dinosaur in the background, arching vegetation recedes back into pictorial space from whence the well-lit dinosaur emerges, its rounded, plated-covered back reinforcing the circular motif of the vegetation. Repeatedly throughout the film, real actors and various miniatures, puppets, and paintings share the screen though they are laid out in different planes and constructed from different layers. The live actors are not the only visual element to claim the foreground, however. The layering that composes the film's space uses various combinations and thus continually reinvents itself (the use of glass shots, for instance, may situate various ornamental items in the foreground).

What is striking about this visual arrangement is the way it induces a perspectival gaze in the viewer. It isn't so much that the film seeks to directly apply any distinct set of rules inherited from various optical treatises (such as were produced during the Renaissance) but rather that it borrows from the visual "language" of perspective the art of directing the gaze. The result reminds us of pictorial compositions that were typical in the fifteenth and sixteenth centuries which emphasized depth by associating a marked foreground with an ostentatious realism (showing the patrons of the artist, easily identified by viewers, or else in the spirit of Crivelli's fly in his painting of the *Madonna and Child* [1480]) (Fig. 13.2) in such a way as to create an effect of continuity between the world of the spectator *outside* the image and the world depicted *in* the image, yet one to be understood as a series of planes leading the gaze away from the space of reality (the space of the viewer; the foreground of the painting) to that of non-reality (the furthest point in the image).

Such insistent layering, whether it results from live shooting, stop motion animation, or better yet, from a combination of both, creates a heterogeneous world that opens up on what appears to be a sort of crisscrossing through time, which is really what drives the imaginary force of the film. For not only are the special effects made to be *perceptible*, they are also seeking to *erase the separation between different* (*time-*)*worlds*. Which is why the discovery of Kong's island does not result, as it does in other some such narratives, from a breach in the Earth's crust following a natural disaster (as in Harry O. Hoyt's *The Lost World,* 1925) or an atomic blast (as in *Gojira* [*Godzilla,* Ishirō Honda, 1954]). In Kong, nothing buried in the earth's core emerges from a geological fault; on the contrary, Kong's island and the modern world are *coextensive*. Rather than proceeding from some rupture implemented by science fiction, the film's imaginary rests on a "natural drama" that stands as the fictional reversal of the real-life experiences of its explorer-filmmakers. *King Kong* thus reenacts the contradictory contemporaneity of an archaic world existing in the present and extends it to all forms of civilization.

The effect of this singular form, with its overlapping or communicating layers of time and space, repeats itself in different ways, whether it is to make us consider the "thickness" of the natives' territory which appears

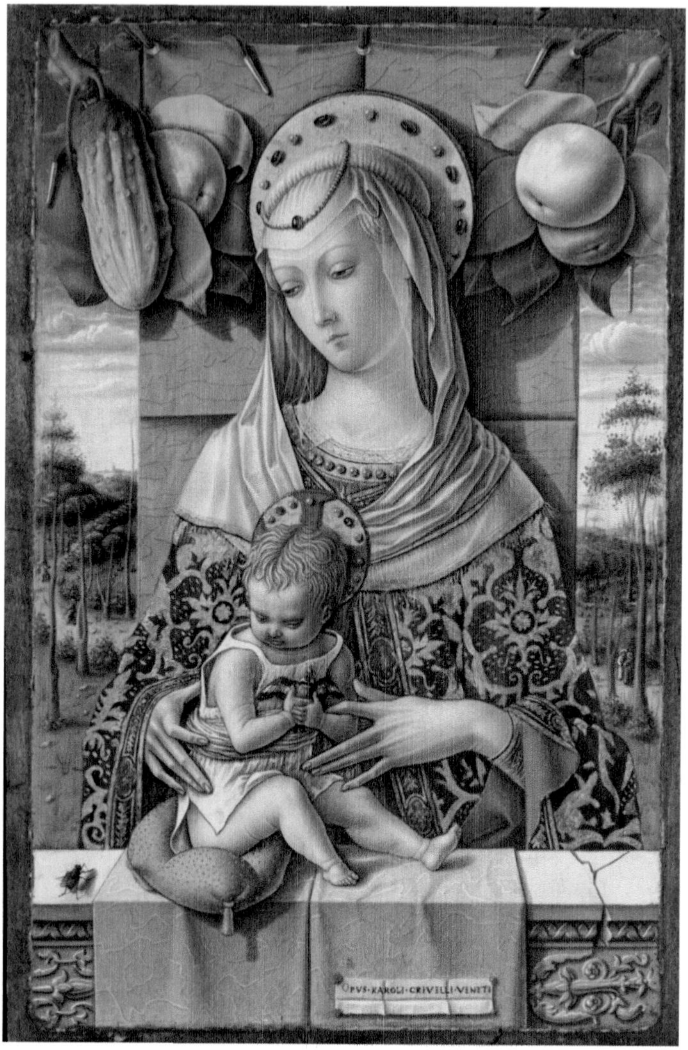

Fig. 13.2 *Madonna and Child* (Carlo Crivelli, 1480).

continuous with its primeval origins, or else to make us meditate on the continuities and the diffractions of time and space that tie Kong's island with Manhattan island at both ends of the narrative. In fact, these are two spaces and temporalities that the film seeks to connect rather than insulate, the purpose of which is rather interesting. It consists of establishing for both locales a common principle of temporal layering and hybridization that allows for the co-presence of the prehistoric past and the present on an island of the Indian Ocean as it does on Manhattan, which equally showcases the archaic alongside the futuristic.

To meet this purpose, the film had to marshal a collection of heterogeneous images such as could only be obtained by the use of special effects. These, in turn, necessitated painstaking precision and endless adjustments, often frame by frame, to integrate these very diverse materials in the image.[11] As for the images of Kong, they required numerous puppets of different sizes as well as many successive "framings within the frame" to ensure the merging of heterogeneous image elements. Thus, whether it is on the South Pacific island, where it is framed by the giant palisade built by the natives or by the pillars to which Ann is attached in sacrifice, or in New York where it is exhibited on a Broadway stage or seen through the windows of the building it climbs, Kong constantly occupies the image not only as a framed "subject" of the look but also through compositions and a *découpage* that emphasize the framing and the act of looking (what in French is called *surcadrage* ["overframing"]).

The visual *mise en scène* of Kong is also supported by the breaking down of its body parts thanks to the special-effects "stand-ins" (full-scale giant hand, arm, foot, and head/chest/shoulders, puppets of various sizes, etc.),[12] so that Kong appears on screen more often through its parts than as a whole body (Kong as synecdoche!). Similarly, Kong's many exploits, as he successively eliminates prehistoric beasts and defends himself from the attacks of the modern world, form a cumulative network of associations. The manifold visualizations and viewpoints on the island (from above, from below, on land, on water, on a mountain top, overhanging an abyss, etc.) are re-enacted in New York (the elevated railway, climbing up buildings, the upper ledge of the Empire State Building, hanging off the skyscraper, views from the foot of the building, etc.).

A sort of nesting or mirroring principle presides over the film's aesthetic. It first reveals itself with the 'film within the film' segment aboard the ship that takes the film crew to Kong's island. During the trip, Denham, the fictional filmmaker, decides to prepare Ann for her role in the film he plans to shoot on the island. The future movie 'star' must leave behind Driscoll, the second in command with whom she was chatting, and prepare herself for the screen test. She returns with her hair let loose and wearing a long medieval gown (Denham refers to it as the "Beauty and Beast costume"). Ann then follows the directions of the filmmaker, who has set up his camera on the ship's deck, executing various poses and gestures. With her head back, her bulging eyes looking up, raising her arms in horror, her hands

11 See Goldner and Turner, *The Making of King Kong*, op. cit., pp. 55–58.

12 Ibid., pp. 87–100.

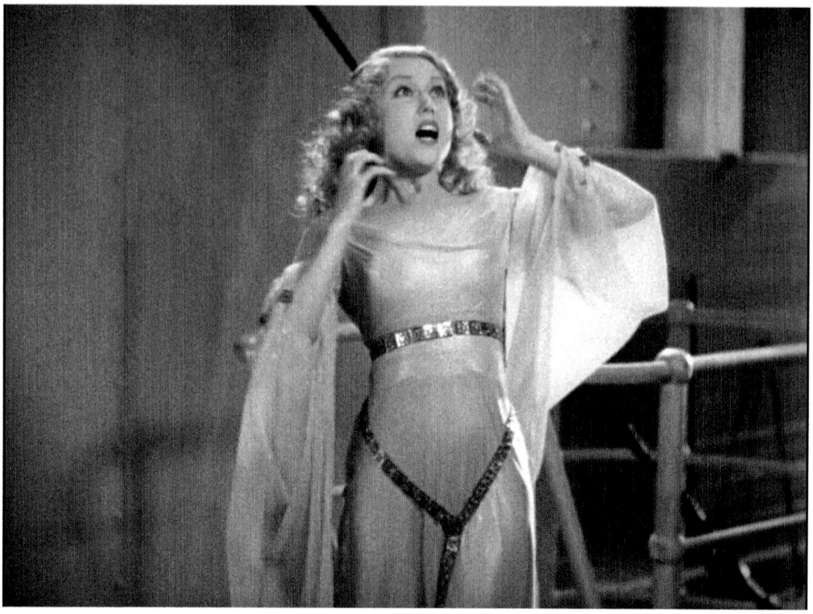

Fig. 13.3 Screaming at an imaginary vision (*King Kong*, Merian C. Cooper and Ernest B. Schoedsack, 1933).

framing her face until she closes her eyes and covers them with her arm, Ann adopts the codified gestures of someone in the grips of fear as she is invited to gaze at some imaginary thing, high up and off-screen (Fig. 13.3). Her imaginary vision pushes her to scream. This objectless, imaginary terror replays for Denham what he first saw in Ann and which had seduced him during their first meeting near the docks of Hoboken. The penniless young girl was trying to steal an apple from a shop, and her demeanor was then motivated by the miserable living conditions during the Great Depression. The distraught attitude, the gaze of someone who fears arrest, are justified by the situation. It is Fay Wray who "acts" here, not Ann Darrow. On the boat, however, the acting is "doubled": both Fay Wray and Ann are acting and simulating terror (Fay Wray simulating Ann simulating terror).

The cinematic *mise en abyme* instilled by a fictional "seeing as if" opens up a path in the film's system of representation and regulates a procedure that is initiated with the "film within the film" segment, although in this case it refrains from using special effects. To put it simply, that which cannot be seen directly necessitates either a feigned representation (as is the case with Ann simulating terror at some imaginary, off-screen presence — a process achieved through fiction but not requiring any special effects) or a different regime of visibility altogether whose fictional "seeing as if" now involves the use of special effects. It is in accordance with the latter that the special effects in *King Kong* do not efface themselves to merely serve the demands of verisimilitude but instead exhibit or display themselves in the way they allegorically interpret the fictional world that the film creates. As a simulation, or else a simulation of a simulation (as is the case with the "film within the film" segment on board the ship or with the "perspectival" compositions discussed above), this world demands the use of visible nesting-mirroring procedures.

From the moment she sets foot on the island and throughout her ordeal in the savage land, Ann constantly replays the simulated gestures of her screen test, only intensifying her scream. After Kong abducts her, he positions her, often in a seated position, as if she were in front of a screen, to face the sundry nightmarish visions that offer themselves to her gaze as she undergoes one harrowing predicament after the other. This recurring spectacle to which she must submit after arriving on the island unfurls what had been left off-screen in her screen test on the boat. Only now it calls forth the use of special effects.

During the shooting onboard the ship, when Ann screamed at Denham's request, Driscoll, who had been watching the scene with the ship's skipper, had asked: "What does he think she's really going to see?" The question was left pending, unresolved, as an enigma over the fade to black that immediately follows. It is in fact a question of great significance for the entire film. After the fade, the next segment continues to unfold this leitmotiv of visibility. This time, it is the anxious gaze of the crew as their boat, the Venture, approaches land through heavy fog until, after a cut, Kong's island finally comes into view. The shot shows the island in the background (Fig. 13.4). While strongly evocative of Arnold Böcklin's painting *The Isle of the Dead*, the foreground frames the view with ship structures and rigging (the "frame within the frame" roughly matching the 1.37:1 ratio of the film itself). It is a complex composite shot that combines live action in the studio (as first the skipper and then Denham enter the frame), a rear projection of a California beach and ocean, with a matte painting of the island in the background and superimposed gulls in the middle ground. To "see" by way of special effects is thus to see "in the imagination" as Ann did during her screen test, it is to "see as if," which is precisely what this cinematic fable invites us to do.

All this leads us to ask what is there to see in this film that is not a mental projection?[13] The very *raison d'être* of the film's special effects — which are

13 The use of the term "projection" here can also be taken to refer to cinema as a projected medium. Indeed, during the "film within the film" segment onboard the ship, as Denham

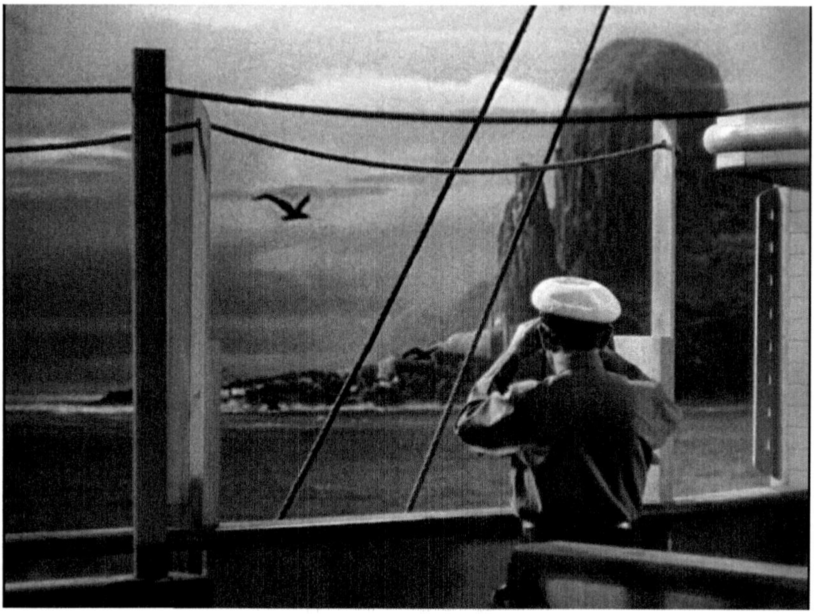

Fig. 13.4 Arrival at Skull Island: matte painting and the frame within the frame in *King Kong* (Merian C. Cooper and Ernest B. Schoedsack, 1933).

nothing if not an adventure of the gaze — is what is at stake. Their presence progressively breaks down the boundaries between the film being made (Denham's film) and the film that was shot (Cooper's and Schoedsack's); between on and off-screen space; between ancient and contemporary worlds; and between savagery and modern civilization. At each juncture, the special effects connect anachronistic co-presences (though this is also how we can make sense of Ann's somewhat unusual medieval costume, beyond the sole reference to *Beauty and the Beast*: it connects the present of the screen test with the foreshadowing of a near future marked by regression). Continuities are thus constantly established over discrepancies, uniting past and present, wilderness and civilization, Kong's island and Manhattan: the subway cars

shoots Ann, there is a constant motor-like rumble that accompanies the long-take view of her. One might first think this is the sound made by the running of the camera, yet Denham, who is shooting a silent film, is using a hand-cranked model! As one watches the film and hears this rumble (whose function is clearly to mark the shot we are seeing as a "film shot" — one simply needs to imagine the long take without the rumble to understand this), it is difficult not to think that what we are seeing is a "projected image" and that the sound we are hearing is that which is made by a projector, thus mimicking the sound that spectators of *King Kong* would have heard in the more silent passages of the film. At this point, the two meaning of "projection" — the imaginary and the cinematic — meet.

of the elevated railway are mangled much like the giant lizard; a pterodactyl is flung to the bottom of a cliff much like an airplane is hurled to crash at the base of the Empire State Building. As the film unfolds, process shots and visual effects follow one another, echoing like the alliteration of phonemes in the film's title. Kong's repeated actions and gestures, in renewing themselves, inflect the great hominid's encounter with the contemporary modern world. The layout of the skyscrapers' windows Kong uses for its final climb brings with them an inexorable geometric dimension, an abstraction that quashes nature and substitutes for the flora and fauna of Kong's island, imposing an immediate threat for Kong. Compositionally, the layers remain, but the axis has changed and the play of foreground and background now becomes "vertical": we see Ann in the foreground, wearing yet another long gown, as she lies high up on the ledge of the skyscraper, Kong is above her hanging on the mooring mast while attack planes appear in the background. At this point, the crisscrossing of temporalities comes to a resolution as abstract modernity asserts itself through architecture and technology.

All these many assorted layers, the 'framing within the frame', the 'film within the film', the endless succession of special effects — miniatures, puppets, matte paintings, travelling mattes, transparencies, rear projections, giant-size props, etc. — present us with an overwhelming flow of operations that can certainly help us appreciate how creativity and imagination may come to embody itself on film thanks to the work of dedicated specialists and various production units. Yet the real strength of the film lies less in it being a new version of *The Lost World* than in the poietic and rhetorical (in terms of *dispositio*) aspects of its *mise en scène*. In short, what is so striking about *King Kong* is the fusion it operates between the concrete texture of its images (and their effects) and the content of its discourse (its thematic or conceptual stuff).

If the film's special effects call on an assortment of technical means and knowhow to give the viewer imaginary sights to behold, they also belong, as I hope to have shown, to the *discursive form* of the film. There are therefore two parallel *historicities* that meet in *King Kong*: the historicity of techniques and of technology used to produce — to *invent* [*poiein*] — such remarkable and heretofore unparalleled images; and the historicity of the discourse — as well as the discourse on historicity — these images create. So while one may assert that an effects specialist such as Ray Harryhausen was the reverent heir of the technical innovators working on *King Kong* under the supervision of Willis O'Brien — the film having a well-defined place in the technical history of special effects — the way *Kong* layers its effects, arranges them —their *dispositio* — belongs to a history of discourse and ideas that establishes the film's unique *signature*. These ideas that the film's ordering of images and special effects express are drawn from the "natural dramas" Cooper and Schoedsack had lived and filmed in their "exotic" documentaries. Through them, they experienced the many temporalities that crisscross to form the "contemporaneous," or what Ernst Bloch referred to as the "non-simultaneity of the simultaneous."[14] In other words, what the complex visual layering and special effects of *King Kong* manage to disclose concerns the simultaneous coexistence of heterogeneous historical moments.

In the end, however, it is the very position of the spectator that these compositional and rhetorical choices underscore. For unlike realism and the illusion of immediacy it struggles to achieve, the fable of *King Kong*, the object of some monstrous bricolage (ostensibly mythical, melodramatic, erotic, psychoanalytic,[15] sociological, and political,[16] etc.), does not eliminate nor even lessen the boundary that exists between the spectator and the world the film invents. The sheer heterogeneous power the images/effects (and their *dispositio*) inherit from the techniques that are used suggests to the viewer the possibility of bridging any air-tight categories or antinomies, including those of the contemporary world which includes all of them (barbarian/civilized/savage; present/past/future; modern/archaic/contemporary, etc.) in what can nonetheless be seen as a highly problematic conceptual proposition. I am referring here to the propagation, in the US, of a vulgar "social Darwinism," as buttressed by the excesses of British philosopher Herbert Spencer. At the time of the making of *King Kong*, a long-standing tradition in the American cinema had already turned these ideas into articles of faith.[17] What is more, during the 1920s, studies seeking to compare humans and other primates either led to laws that prohibited the teaching of evolution in public schools (a famous case involving such a law was the Scopes trial in Tennessee in 1925)[18] or else revived older racist

14 Ernst Bloch, *Héritage de notre temps*, translated from German into French by Jean Lacoste (Paris: Payot, [1935] 1978), pp. 95–116.

15 As developed by Jean-Pierre Moussaron in a psychoanalytic study of the film, "*King Kong* (1933) ou la migration des sens" in *Modernités* (25), (Bordeaux: Presses Universitaires de Bordeaux, 2007). In a previous interpretation, Roger Dadoun had shown the film's skillful "synthesis" of key notions of the psychoanalytic narrative: Eros [...] Oedipus [...] The primordial Father [...] archaic Mother [...], Aggression [...], the Id. See Roger Dadoun, "*King Kong*: du monstre comme démonstration," in *Littérature* (8), (Paris: Larousse, December 1972).

16 The incredible accumulation of monsters in this period of film history has often been analyzed as an allegory for the effects of the Great Depression.

17 See Jean-Louis Leutrat, *L'alliance brisée. Le western des années 1920* (Lyon: Presses Universitaires de Lyon, 1985).

18 See Bellour, *Le Corps du cinéma*, op. cit., p. 450.

discourses. I would argue, however, that *King Kong* manages to avoid such failings. But be that as it may, *King Kong*, to our utmost delight, is perhaps best captured in a phrase by Deleuze, namely that "nothing happens in the viewer's head which does not derive from the image"[19] or, in other words, from the judicious use of such attractive special effects.

Bibliography

Bellour, Raymond. *Le corps du cinéma. Hypnoses, émotions, animalités* (Paris: POL, 2009).

Bloch, Ernst. *Héritage de notre temps*. Translated by Jean Lacoste (Paris: Pavot, 1978 [1935]).

Cavell, Stanley. *The World Viewed* (Cambridge, MA: Harvard University Press, 1971).

Dadoun, Roger. "*King Kong*: du monstre comme démonstration," *Littérature* 8 (Paris: Larousse, 1972).

Deleuze, Gilles. *Cinéma 2. The Image Time* (London: Athlone Press, 1989).

Hamus-Vallée, Réjane. *Les effets spéciaux* (Paris: Cahiers du cinéma/CNDP, 2004).

Leutrat, Jean-Louis. *L'alliance brisée. Le western des années 1920* (Lyon: Presses Universitaires de Lyon, 1985).

Metz, Christian. "*Trucages* and the Film," translated by François Metzler, *Critical Inquiry* 3, no. 4 (1977): 657–675.

Moussaron, Jean-Pierre. "*King Kong* (1933) ou la migration des sens," *Modernités* 25 (Bordeaux: Presses Universitaire de Bordeaux, 2007): 251-264.

Turner, Georges E. *The Making of* King Kong*: The Story Behind a Film Classic* (New York: Ballantine Books, 1976).

About the Author

Suzanne Liandrat-Guigues is Professor Emeritus at the University of Paris 8 — Vincennes-St-Denis. She has co-authored, with Jean-Louis Leutrat, books on Godard, Resnais, Pollet, and on the western and co-edited with him *Rivette*; *Barthélemy Amengual. Du réalisme au cinéma*; *L'essai et le cinéma*. She has also written books on modernity and on the aesthetics of movement. Her most recent works include: *Paris 1900 de Nicole Vèdres. Kaléidoscope des jours* (L'Harmattan, 2018); *Visconti* (202 Editions, 2020); and *Quatre CM de Jean Cayrol et Claude Durand* (forthcoming).

19 Gilles Deleuze, *Cinema 2. The Image Time*. (London: Athlone Press, 1989), p. 100.

14. Uncanny Visual Effects, Postwar Modernity, and *House of Wax 3D*

Kristen Whissel

Abstract
This chapter argues that *House of Wax* provides a significant example of how 3D stereo-effects were exploited in the 1950s to give expression and form to uncanny themes and optics, and, in the process, functioned as ideal visual effects for addressing the culture and experience of technological modernity in the postwar era.

Keywords: 3D cinema, *House of Wax*, postwar modernity, parallax effects

Scholars and critics have tended to analyze stereoscopic 3D cinema, its so-called waves, its modes of address, and its relation to other media and new technologies, having already taken for granted that 3D must be approached as a doomed-to-fail cinematic enterprise, cynically called forth by studios in times of intense media competition to bring audiences back to theaters.[1] To account for 3D's ostensible failure, scholars often focus on negative parallax — a stereoscopic 3D visual effect that creates the illusion that elements of the diegesis have crossed the surface of the screen and "emerged" into the space of reception. For example, focusing on *House of Wax* (André de Toth, 1953), William Paul describes 3D as an "atavism" and links 3D's mid-century demise to the tendency of the emergent image both to resist subordination to narrative and to call attention to the screen as a boundary that conventionally separates

1 See, for example, Roger Ebert, "Why I Hate 3-D (And You Should Too)," in *Newsweek*, 9 May 2010, https://www.newsweek.com/roger-ebert-why-i-hate-3d-movies-70247 (accessed 9 October 2019); and Scott Higgins "3D in depth: *Coraline, Hugo* and a sustainable aesthetic," in *Film History* 16(3) (2012), pp. 169–209.

Lefebvre, M. & M. Furstenau (eds.), *Special Effects on the Screen: Faking the View from Méliès to Motion Capture*. Taylor & Francis Group, 2022
DOI 10.5117/9789462980730_CH14

diegetic space from the space of reception.[2] Philip Sandifer[3] has echoed this assessment and argues that the potential for negative parallax to produce shock and surprise links 3D film to the silent era's "cinema of attractions" (a term coined, of course, by Tom Gunning[4]), contending that the 3D format suffers from a type of arrested development that has prevented it from moving beyond what he calls the "demo" stage.[5] In contrast to scholarship that approaches parallax effects as historically regressive gimmicks or as attractions that are extraneous to a 3D film's narrative and thematic concerns,[6] this essay is grounded in an effort to answer the following questions: to what aesthetic, narrative, and ideological ends did 3D films of the 1950s so insistently use parallax effects to call attention to the act of looking, to embodied vision, and to the imbrication of the spectator's vision with 3D technologies? Given the emergent image's direct and even (playfully) aggressive address to the spectator's technologically enhanced vision and its illusory violation of the screen and, with it, other categorical boundaries, how might parallax effects be productively analyzed through the aesthetic category of the uncanny? How do 3D films of the 1950s exploit a range of "stereo-effects"[7]

2 Paul argues that "3D is a kind of sport, an unexpected and always doomed mutation that by its very perversions defines the norms of the normative Hollywood style." See William Paul, "The Aesthetics of Emergence," in *Film History*, (5)3 (1993), p. 321.

3 Philip Sandifer, "Out of the Screen and into the Theater: 3-D Film as Demo," in *Cinema Journal* (50)3 (2011), pp. 62–78.

4 Tom Gunning, "The Cinema of Attractions: Early Film, its Spectator, and the Avant-Garde," in *Wide Angle*, (8) 3–4 (1986), pp. 63–70.

5 Sandifer, "Out of the Screen," op. cit., p. 64.

6 For detailed and insightful analyses of critical and scholarly debates surrounding the exploitation of positive and negative parallax in 3D film, see Barbara Klinger, "Beyond Cheap Thrills: 3D Cinema Today, the Parallax Debates and the 'Pop-Out'," in *Public: Art/Culture/Ideas*, (24)47 (2013), pp. 186–189; see also Miriam Ross, *3D Cinema: Optical Illusions and Tactile Experiences*, (New York: Palgrave MacMillan, 2015) and, especially, Ariel Rogers, *Cinematic Appeals: The Experience of New Movie Technologies*. (New York: Columbia University Press, 2013). Rogers provides the most detailed assessment of the 3D debates surrounding films of the 1950s onward.

7 Eisenstein (2013[1947]) applied the phrase "stereo-effects" to positive parallax (whereby "the image plunges deep inside the screen, drawing the spectator along into unprecedented depths"), negative parallax (whereby "the image, palpably three-dimensional, 'tumbles out' of the screen into the auditorium"), and images that remain "within the boundaries of ordinary cinema as a kind of flat *haut-relief*, poised somewhere within the plane of the reflecting screen." Sergei Eisenstein, *On Stereocinema*, S. Levchin, trans., *Public: Art/Culture/Ideas*, (24)47 (2013), p. 22. I would add to these stereo-effects the seeming solidity and "planar" quality of the 3D image analyzed in detail by Oliver Wendell Holmes in both "The Stereoscope and the Stereograph" and "Sun-Painting and Sun-Sculpture: With a Stereoscopic Trip Across the Atlantic," in *Soundings*

as uncanny visual effects in order to address the socio-historical postwar context into which *House of Wax* and other 3D Hollywood films emerged?

Such questions are imperative for understanding 3D cinema's intermedial history, its relationship to nineteenth-century stereoscopy, and its exploitation of parallax effects as visual effects capable of promoting uncanny affects. Elsewhere I have argued that certain types of spectacular visual effects function emblematically in narrative cinema to give meaningful expression to a film's key themes, anxieties, and obsessions — particularly those that are linked to the historical conditions and contexts in which a film was produced and received.[8] *House of Wax* provides a significant example of how 3D stereo-effects were exploited in the 1950s to give expression and form to uncanny themes and optics, and, in the process, functioned as ideal visual effects for addressing the culture and experience of technological modernity in the postwar era. If, as Anthony Vidler[9] argues, "the vicarious taste for the uncanny has been a constant in modern culture, only intensified by shifts in media," then it makes sense that 3D media's (potentially) uncanny stereo-effects and aesthetics would enjoy a resurgence in an era defined by the shock of a second World War and its aftermath, Hollywood's intense competition with television, and the industry-wide technological change that ensued.[10] Indeed, the cinematic resurgence of stereoscopic 3D during the postwar era asks that we think of 3D not in terms of its serial failure but instead in terms of its odd historical persistence — its rather uncanny status as an often forgotten but nevertheless old and familiar technology (with origins as much in the stereoscope and photography as in early cinema) that seems to return in periods of intense social, historical, and technological change.[11]

from the Atlantic, (Boston: Ticknor and Sons, 1864), pp. 124–165 and 166–227; as well as by Jonathan Crary in *Techniques of the Observer: On Vision and Modernity in the Nineteenth Century* (Cambridge, MA: MIT Press. 1990), pp. 125–126.

8 Kristen Whissel, *Spectacular Digital Effects. CGI and Contemporary Cinema* (Durham, NC: Duke University Press, 2014).

9 Anthony Vidler, *The Architectural Uncanny. Essays in the Modern Unhomely*, (Cambridge, MA: MIT Press, 1992), p. 4.

10 On media and technological change during this era, see John Belton, *Widescreen Cinema*, (Cambridge, MA: Harvard University Press, 1992); Kevin Heffernan, *Ghouls, Gimmicks, and Gold: Horror Movies and the American Movie Business, 1953–1968* (Durham, NC: Duke University Press, 2004); and Rogers, *Cinematic Appeals*, op. cit.

11 For a discussion of the forgotten history of stereoscopic 3D media beyond 3D film's four waves, see Leon Gurevitch and Miriam Ross, "Stereoscopic Media: Beyond Booms and Busts," in *Public*, (24)47 (2013), pp. 83–92.

Freud, Holmes, and the Optical Pleasures of Parallax Effects

Published in the aftermath of WWI, Freud's essay on the uncanny provides an ideal conceptual framework for situating 3D cinema of the 1950s within its own postwar era (which included the rise of the Cold War and US involvement in the Korean War [1950–1953]) and for analyzing various boundary violations that define stereoscopic 3D aesthetics. To be sure, the uncanny has provided media historians with a conceptual framework for theorizing (the experience of) a range of media and the sensations they generate — from photography,[12] radio,[13] and phonography[14] to other forms of entertainment such as magic shows and phantasmagoria[15] and early synchronous sound cinema.[16] As scholars such as Mark Sandberg,[17] Laura Mulvey,[18] and Tom Gunning[19] have emphasized, media can generate uncanny experiences by fostering intellectual doubt around the stability of categorical boundaries and differences.[20] As I argue elsewhere,[21] 3D cinema's stereo-effects can be successfully harnessed to uncanny ends precisely because they seem to

12 Laura Mulvey, *Death 24x a Second: Stillness and the Moving Image* (London: Reaktion Books, 2006).

13 Jeffrey Sconce, *Haunted Media: Electronic Presence from Telegraphy to Television* (Durham, NC: Duke University Press, 2000).

14 Tom Gunning, "ReNewing Old Technologies: Astonishment, Second Nature, and the Uncanny in Technology from the Previous Turn-of-the Century," In D. Thorburn and H. Jenkins, eds., *Rethinking Media Change: The Aesthetics of Transition* (Cambridge, MA: MIT Press, 2004), pp. 39–59.

15 Tom Gunning, "Uncanny Reflections, Modern Illusions: Sighting the Modern Optical Uncanny," in J. Collins and J. Jervis, eds., *Uncanny Modernity: Cultural Theories, Modern Anxieties* (London and New York: Palgrave MacMillan, 2006), pp. 68–89.

16 Robert Spadoni, *Uncanny Bodies: The Coming of Sound Film and the Origins of the Horror Genre* (Berkeley, CA: University of California Press, 2007).

17 Mark Sandberg, *Living Pictures, Missing Persons: Mannequins, Museums, and Modernity*, (Princeton, NJ: Princeton University Press, 2003).

18 Mulvey, *Death 24x a Second*, op. cit.

19 Gunning, "Uncanny Reflections," op. cit.

20 Mulvey is particularly eloquent on this point, explaining that, "In the cinema, organic movement is transformed into its inorganic replica, a series of static, inanimate, images, which, once projected, then become animated to blur the distinctions between the oppositions. The homologies extend: on the one hand, the inanimate, inorganic, still, dead; on the other, organic, animate, moving, alive. It is here with the blurring of these boundaries, that the uncanny nature of the cinematic image returns most forcefully and, with it, the conceptual space of uncertainty: that is, the difficulty of understanding time and the presence of death in life." See Laura Mulvey, *Death 24x a Second*, op. cit., pp. 52–53.

21 See Kristen Whissel, "*Dial M for Murder*: The Detective Thriller, the Postwar Uncanny, and 3D Cinema," *New Review of Film and Television Studies*, (20)1 (Winter 2022): 49-62; and Kristen Whissel, "Digital 3D, Parallax Effects, and the Construction of Film Space in *Tangled* 3D and

efface the boundaries separating sight from touch, the immaterial from the material, and, in the case of negative parallax, diegetic space from the space of exhibition. As Mulvey in particular emphasizes, the production of such categorical ambivalence is a defining feature of the uncanny, particularly concerning "confusion between the animate and the inanimate…associated with death and the return of the dead,"[22] and one that is at the core of any recording media's indexicality as "a trace of the past that persists into the present."[23] *House of Wax* harnesses 3D stereo-effects to uncanny ends, not only by remediating (to borrow a phrase from David Jay Bolter and Richard Grusin)[24] waxworks in ways that further undermine categorical differences but also by offering perceptual shocks able to address the experience of postwar modernity and by producing a familiar-but-strange spatiality that seems to confirm Adorno's claim that, in the middle of the twentieth century, 'Dwelling, in the proper sense, is now impossible'.[25]

In his essay, "The Uncanny," Freud reworks Ernst Jentsch's definition of "*unheimlich*"[26] through an analysis of E.T.A. Hoffmann's short story, "The Sand-Man." Freud builds upon Schelling's assertion that "*Unheimlich* is the name for everything that ought to have remained… secret and hidden but has come to light,"[27] and explains that the uncanny affects can arise from the return of "something which is familiar and old-established in the mind and which has become alienated from it."[28] This can include the reoccurrence of "surmounted" beliefs — in animism, for example, or the possibility of the return of the dead. In turn, familiar-but-alienated material might return as repressed psychological content, especially the castration complex, as figured through dismemberment or the "idea of being robbed of one's eyes,"[29] as well as the female genitals (for Freud, "This

Cave of Forgotten Dreams 3D," in *Screen Space Reconfigured*, S. Ø. Saether and S.T. Bull, eds., (Amsterdam: Amsterdam University Press, 2020) pp. 77–103.

22 Laura Mulvey, *Death 24x a Second*, op. cit., pp. 60–61.

23 Ibid., p. 31

24 Jay David Bolter and Richard Grusin, *Remediation: Understanding New Media* (Cambridge, MA: MIT Press, 2000).

25 Theodor Adorno, *Minima Moralia: Reflections from a Damaged Life* (London: Verso, 2005[1951]), p. 38.

26 Ernst Jentsch, "On the Psychology of the Uncanny," in *Uncanny Modernity: Cultural Theories, Modern Anxieties*, J. Collins and J. Jervis, eds., op. cit., pp. 216–227.

27 Sigmund Freud, "The Uncanny," In A. Dickson and J. Strachey, eds., *The Penguin Freud Library: Art and Literature*. Volume 14. (London: Penguin Books, 1985[1919]). The quote from Schelling is found on p. 345.

28 Ibid., pp. 363–364.

29 Ibid., p. 351.

unheimlich place [...] is the entrance to the former *Heim* [home] of all human beings" and constitutes "what was once *heimisch*, familiar; the prefix '*un*' is the token of repression").[30] Uncanny affects (such as disquietude and dread) can arise from the experience of doubling and doppelgangers, the inexplicable recurrence (of names, crimes, numbers, events, etc.), as well as from uncertainty as to whether something is alive or dead.[31] At the level of content and theme, *House of Wax* takes up several features of the uncanny that also structure the waxwork, including doubling and doppelgangers, the repetition and return of the past in the present, and a blurring of the boundaries separating the animate from the inanimate, the living and the dead — and I will return to these later. As important are the various settings in the film that serve as relatively *unheimlich* homes and houses and the optical means by which audiences are encouraged to investigate them and other venues of commercialized leisure as (alienated) spaces of modernity.

Both the wax museum and the House of Wax that serve as the film's main settings are decidedly unhomely 3D spaces that demand critical analysis through Freud's overlapping definitions of *heimlich* and *unheimlich*. As Vidler explains, "Freud deliberately approached the definition of *unheimlich* by way of that of its apparent opposite, *heimlich*, thereby exposing the disturbing affiliations between the two and constituting the one as a direct outgrowth of the other."[32] Freud cites a definition of *heimlich* as "belonging to the house, not strange, familiar" and as that which is "intimate, friendlily comfortable; the enjoyment of quiet content, etc., arousing a sense of agreeable restfulness and security as in one within the four walls of [a] house."[33] Yet, as Vidler notes, Freud goes on to cite definitions that associate *heimlich* with that which is "concealed, kept from sight, so that others do not get to know of or about it, withheld from others"[34] — bringing it closer to Schelling's definition of *unheimlich* as "everything that ought to have remained... secret and hidden but has come to light." In *House of Wax*, stereo-effects expose "disturbing affiliations" between the apparently opposed categories of concealment and display, past and present, and private and public that prevail within the all-too familiar spaces of commercialized leisure that the film's antagonist calls home. They thereby confirm Freud's conclusion that "*heimlich* is a word the meaning of which develops in the direction of ambivalence, until

30 Ibid., p. 368.
31 Ibid., p. 364–365.
32 Vidler, *The Architectural Uncanny*, op. cit., p. 23
33 Freud, "The Uncanny," op. cit., p. 342.
34 Ibid., p. 344.

it finally coincides with its opposite, *unheimlich*. *Unheimlich* is in some way or other a sub-species of *heimlich*."[35] Such categorical uncertainty concerning space — whereby the familiar, comfortable, and homely may become unnervingly indistinguishable from the secret, the hidden, and the unhomely — is central to the perceptual experience, thematic obsessions, and visual pleasures manufactured by a number of stereoscopic 3D films of the 1950s. This uncertainty, moreover, extends to the conventional separation of exhibition space from diegetic space, spectator from spectacle, as these films use positive and negative parallax to create ambivalence around 3D's technological enhancement of vision, on one hand, and its facilitation of (playful) aggression towards the act of looking through emergence effects, on the other.

Freud's analysis of "The Sand-Man" and the story's eponymous figure who "tears out" children's eyes are helpful for understanding the uncanny visual effects of *House of Wax*, their relation to 3D spectatorship, and their centrality to the film's investigation of modern homelessness, the return of the past in the present, and *(un)heimlich* spaces of commercialized leisure. Hoffmann's story turns on the protagonist Nathaniel's dread of his father's colleague, a lawyer named Coppelius. As a child, Nathaniel suspects that Coppelius is the Sand-Man; later in life, Coppelius returns in the guise of the optician and barometer salesman, Coppola. Nathaniel's initial dread of Coppelius/Coppola/the Sand-Man is linked to the act of concealed looking: as a child, he is caught spying on his father and Coppelius as they work together and is punished by the latter. As Vidler notes, there are numerous references to the optical devices (spectacles, telescopes, spyglasses) used by Nathaniel as he spies on others.[36] Significantly, Hoffmann's emphasis on the technological enhancement of vision is accompanied by references to alarming aggression towards it. Freud analyzes a moment in the story which confirms that "the feeling of something uncanny is directly attached to the figure of the Sand-Man, that is, to the idea of being robbed of one's eyes": the adult Nathaniel encounters the Sand-Man/Coppelius in the guise of the optician Coppola, who tries to sell him some "weather glasses," which Nathaniel refuses.[37] Freud quotes Coppola's response, "Not weather glasses, not weather glasses? Also got fine eyes, fine eyes," which the horrified

35 Ibid., pp. 346–347. Elsewhere I discuss the (un)heimlich home in relation to *Dial M for Murder*'s elaboration of a postwar uncanny through 3D stereo-effects. See Kristen Whissel, "*Dial M for Murder*: The Detective Thriller, the Postwar Uncanny, and 3D Cinema," op., cit.

36 Vidler, *The Architectural Uncanny*, op. cit., p. 34.

37 Freud, "The Uncanny," op. cit., p. 351.

Nathaniel assumes are children's eyes torn out by the Sand-Man.[38] Linking the enhancement of vision to aggression against it, Freud explains that "The student's terror is allayed when he finds that the proffered eyes are only harmless spectacles, and he buys a pocket glass from Coppola."[39]

Much like the glasses audiences wear to watch a 3D film or the eyepieces of a stereoscope, the optical devices that proliferate throughout the story are, as Vidler argues, "fabricated either to imitate real eyes…or to extend the powers of real eyes" and therefore "take on uncanny roles": "They add to the already formidable powers of the natural eye, and more often than not they trick it."[40] I would take Vidler's claim further, however: such devices participate in the story's articulation of the uncanny because their ability to double up, enhance, and/or trick vision is never entirely separated from the Sand-Man's aggression towards looking and the (pleasurable) dread he creates for the reader around the vulnerability of embodied sight — hence Nathaniel's momentary confusion between spectacles and torn-out eyes. 3D technologies have historically exploited their peculiar ability to enact this dual address to the observer, both in their simulation of the perception of three-dimensional objects and space and in their cultivation of negative parallax to playfully and (sometimes) aggressively address the act of looking itself. That is, the close relationship between the technological enhancement of vision (materialized by the 3D glasses worn by spectators)[41] that satisfies a desire to see and know, on one hand, and an affectively charged direct address to embodied vision, on the other, has been central to the distinct perceptual pleasures of 3D since the invention and commercialization of the stereoscope.

Nineteenth-century writings on the aesthetics of stereoscopic 3D foreground precisely the ability of the latter simultaneously to enhance vision while facilitating playful aggression towards it. In his published essays on the stereoscope and stereograph, Oliver Wendell Holmes (inventor of the hand-held Holmes stereoscope), describes these optical principles and goes on to detail the various ways in which stereographic images provided observers with "a surprise such as no painting ever produced. The mind feels its way into the very depths of a picture. The scraggy branches of a tree

38 Ibid., p. 350.

39 Ibid.

40 Vidler, *Architectural Uncanny*, op. cit., p. 34.

41 Simon Lefebvre argues that 3D glasses "have the ambivalence of the window and the cover, of what opens and what confronts, they are a kind of mask, a veil thrown over the eyes, and the visibility tool of an image that would otherwise be blurred." "The Disappearance of the Surface," in *Screens*, D. Chateau and J. Moure, eds. (Amsterdam: Amsterdam University Press, 2016), p. 105.

in the foreground run out at us as if they would scratch our eyes out. The elbow of a figure stands forth so as to make us almost uncomfortable."[42] Here, binocular vision is replicated by an optical technology that uses positive parallax to facilitate curiosity such that "the mind feels its way into the picture" to investigate the exaggerated depth of a space unfolding along the z-axis. In turn, negative parallax promotes what David Trotter[43] describes as the stereoscope's ability to promote a "haptic look," not only by activating the sense of touch through sight but also by facilitating a disquieting awareness of embodied vision through foregrounded objects that can "make us almost uncomfortable" and, in some cases, seem "as if they would scratch our eyes out."[44] This conjoining of vision's technological enhancement with a playful and sensational aggression towards it, along with the tendency of parallax effects to (momentarily) trick the senses by producing the illusory materiality and persuasive presence of objects depicted in 3D, place the visual pleasures of stereoscopic 3D at least partially within the domain of the uncanny.

As Trotter's reference to the haptic quality of the 3D image suggests, the dual address of the stereoscopic image (to curiosity and embodied sensation) depends in part upon the exaggerated relief and volumetric solidity that the stereoscope gives to depicted objects. In his well-known formulation, Holmes describes the stereoscope as "an instrument which makes surfaces look solid. All pictures in which perspective and light and shade are properly managed, have more or less the effect of solidity; but by this instrument that effect is so heightened as to produce an appearance of reality which cheats the senses with its seeming truth."[45] So persuasive was this "seeming truth" that stereoscopic images created, Crary explains, "not simply likeness, but immediate, apparent *tangibility* [...] transformed into a purely visual

42 Holmes, "The Stereoscope and the Stereograph," op. cit., p. 148.

43 David Trotter, "Stereoscopy: Modernism and the 'Haptic'," in *Critical Quarterly*, (46)4 (2004): pp. 39–42.

44 For detailed analyses of embodiment and the haptic quality of the moving image, see Vivian Sobchack *The Address of the Eye: A Phenomenology of Film Experience* (Princeton: Princeton University Press, 1992); Laura U. Marks, *The Skin of the Film: Intercultural Cinema, Embodiment, and the Senses* (Durham, NC: Duke University Press, 2000); and Jennifer M. Barker *The Tactile Eye: Touch and Cinematic Experience* (Berkeley, CA: University of California Press, 2009). For discussions of the haptic quality of the digital 3D image, see especially Ariel Rogers *Cinematic Appeals,* op. cit.; Miriam Ross, "The 3D Aesthetic: *Avatar* and Hyper-Haptic Visuality," in *Screen*, (53)4 (2012), pp. 381–397, and, also by Ross, *3D Cinema: Optical Illusions and Tactile Experiences* (New York: Palgrave MacMillan, 2015).

45 Holmes, "The Stereoscope and the Stereograph," op. cit., p. 140.

experience."[46] Hence, when describing a stereoscopic 3D image of the full moon, Holmes claimed that "the sphere rounds itself out so perfectly to the eye that it seems as if we could grasp it like an orange."[47] Aided by a wealth of photographic detail, stereoscopic 3D images produce a moment of uncertainty in the viewer that demonstrates how "an uncanny effect is often and easily produced when the distinction between imagination and reality is effaced, as when something we have hitherto regarded as imaginary appears before us in reality, or when a symbol takes over the full functions of the thing it symbolizes, and so on."[48]

This latter point is significant to any discussion of *House of Wax*, for the uncanny quality of the 3D cinematic image derives in part from the ambivalence surrounding the sculptural quality and apparent tangibility with which 3D endows objects, on one hand, and the virtuality and immateriality of the stereoscopic 3D image, on the other, such that one of these qualities may seem to give way to the other, in the process creating disquieting ambiguity around the boundaries separating categorical differences and individual media. For example, upon viewing stereographs of Italian street scenes through a Kaiserpanorama in 1911, Franz Kafka described the "people inside" as appearing "like wax figures, their soles fixed to the ground on the sidewalk."[49] Here, one medium (the waxwork) seems to haunt another (the stereograph) in ways that foreground the uncanny qualities of both, not only in terms of their reality effects and their tendency to undermine differences between stillness and motion[50] but also in terms of the familiar-yet-strange dimensionality of the stereoscopic 3D image, which endows the flat, 2D photographic image of the human figure with exaggerated relief such that

46 Crary, *Techniques of the Observer*, op. cit., pp. 122–124.

47 Holmes, "Sun-Painting and Sun-Sculpture: With a Stereoscopic Trip Across the Atlantic," op. cit., p. 224.

48 Freud, "The Uncanny," op. cit., p. 367. Elsewhere, I discuss Werner Herzog's exploitation of the haptic quality, illusory tangibility, and dreamlike irreality of the digital 3D image to articulate the historical uncanny in *Cave of Forgotten Dreams* (2010). See, Kristen Whissel, "Digital 3D, Parallax Effects, and the Construction of Film Space in *Tangled* 3D and *Cave of Forgotten Dreams* 3D," in *Screen Space Reconfigured*, S.Ø. Saether and S.T. Bull, eds., (Amsterdam: Amsterdam University Press, 2020) pp. 77–103.

49 Quoted in Trotter, "Stereoscopy: Modernism and the 'Haptic,'" op. cit., p. 43.

50 Mulvey theorizes photography's uncanny ability to undermine categorical differences between stillness and motion, the past and the present in *Death 24x a Second*, op. cit., pp. 17–84. Mark Sandberg and Marina Warner analyze the reality effects and uncanny lifelikeness of the waxwork as well as its relationship to time. See Mark Sandberg, *Living Pictures, Missing Persons: Mannequins, Museums, and Modernity*, op. cit; and Marina Warner, *Phantasmagoria: Spirit Visions, Metaphor, and Media into the Twenty-First Century* (Oxford: Oxford University Press, 2006).

it appears oddly sculptural and solid. As Sir David Brewster explained to readers in his treatise on the stereoscope, the solidity of stereographed figures could be exploited to contrast with relatively transparent, ghostly figures created by trick effects that "carry" viewers "even into the realm of the supernatural." The stereographer's "art," Brewster explained,

> enables him to give a spiritual appearance to one or more of his figures and to exhibit them as "thin air" amid the solid realities of the stereoscopic picture. While a party is engaged with their whist or their gossip, a female figure appears in the midst of them with all the attributes of the supernatural. Her form is transparent, every object or person beyond her being seen in shadowy but distinct outline. She may occupy more than one place in the scene, and different portions of the group might be made to gaze upon one or other of the visions before them.[51]

Here, the optical effect of underexposing a figure creates an "aerial personage"[52] that contrasts with the "solid realities of the stereoscopic picture."[53] As Brooke Belisle shows, Holmes, too, was drawn to the haunting ethereality of images created when a figure moved during a stereograph's exposure, or when a figure moved out of the frame during the interval between the taking of the first and second image, which, Belisle explains, introduced a temporal element into the finished stereograph.[54] For example, describing a stereograph of Temple Bar, Holmes mentions, "Ghost of a boy with a bundle, — seen with right eye only. Other ghosts of passers or loiterers."[55] From the outset, the stereoscopic 3D image introduced doubt around the ostensible materiality or immateriality of different objects

51 David Brewster, *The Stereoscope: Its History, Theory, and Construction, With Its Application to the Fine and Useful Arts and to Education* (London: John Murray, 2012[1856]), p. 205.

52 Ibid., p. 206.

53 Brewster describes the method for creating this trick effect in the following way: "In order to produce such a scene, the parties which are to compose the group must have their portraits nearly finished in the binocular camera, in the attitude which they may be supposed to take, and with the expression which they may be supposed to assume, if the vision were real. When the party have nearly sat the proper length of time, the female figure, suitably attired, walks quickly into the place assigned her, and after standing a few seconds in the proper attitude, retires quickly, or takes as quickly, a second or even a third place in the picture if it is required, in each of which she remains a few seconds, so that her picture in these different positions may be taken with sufficient distinctness in the negative photograph." Ibid., p. 205.

54 See Brooke Belisle, "The Dimensional Image: Overlaps in Stereoscopic, Cinematic, and Digital Depth", in *Film Criticism*, (37)13/(38)1, 2013, esp. pp. 119–120.

55 Holmes, "Sun-Painting and Sun-Sculpture: With a Stereoscopic Trip Across the Atlantic," op. cit., pp. 187–188.

depicted within a stereograph to uncanny ends — from the illusory sense of the vulnerability of the eyes, to delight in the contrast between the stereoscope's "solid realities" and "aerial personages" — and it continued to do so in the second half of the twentieth century with the popularization of 3D cinema. For André Bazin, the virtual, ethereal nature of the 3D film image seemed to overwhelm the impression of sculptural solidity that was so striking to Kafka's and Holmes's reception of the stereoscopic image. Bazin notes that 3D cinema "gives an effective impression that the objects are in space but inhabiting it in the state of impalpable phantoms," resulting in "an impression of unreality far more perceptible than standard black and white, flat cinema."[56] This description seems to link 3D cinema more to older forms of virtual images, such as phantasmagoria or Pepper's Ghost, than to waxworks or 2D cinema.[57] Regardless of whether (still or moving) stereoscopic 3D images introduce phantom impalpability or a seemingly palpable sculptural solidity into the image, they appear — throughout their history — to have been haunted by the uncanny.[58]

The ghostly effect Bazin describes and his reference to "impalpable phantoms" returns us to the postwar context in which his description was written. Vidler links the persistence of the uncanny in literature, media, and architecture in the second half of the twentieth century to an experience of global warfare "compounded by its repetition on an even more terrible scale during World War II" and argues that the uncanny must be thought of as "a significant psychoanalytic and aesthetic response to the real shock of the modern."[59] Bazin's emphasis on "ghostly immateriality" is telling in this respect, given the brief rise of 3D cinema in the post-World War II era and, moreover, the association between Freud's essay on the uncanny and World War I. Published in 1919, Freud's essay returns repeatedly to the relationship between the uncanny and "man's attitude toward death," and he notes that "Many people experience the feeling in the highest degree in relation to death and dead bodies, to the return of the dead, and to spirits

56 André Bazin, "Will CinemaScope Save the Cinema?" In *André Bazin's New Media*, trans. and ed., Dudley Andrew (Berkeley: University of California Press, 2014[1953]), pp. 282–283.

57 For a fascinating history of Pepper's Ghost and other 3D virtual images, see Katharina Loew, "Tangible Spectres: 3D Cinema in the 1910s," in *Film Criticism*, (37)3/(38)1 (2013), pp. 87–115.

58 It is important to note that Holmes's most poignant discussions of the uncanny effects of individual stereographs, and the uncertainty they create around the line separating the living and dead, focus on stereographs that depict the bodies of dead soldiers awaiting mass burial: "two youths... lie in the foreground, so simple-looking, so like boys who had been over-worked and were lying down to sleep." Holmes, "Sun-Painting and Sun-Sculpture: With a Stereoscopic Trip Across the Atlantic," op. cit., p. 219.

59 Vidler, *The Architectural Uncanny*, op. cit., p. 9.

and ghosts."[60] To be sure, the (violent) return of the past (and the dead) in the present is a theme taken up by a number of 3D films of the 1950s and elaborated through their parallax effects and optics. It makes sense, then, that stereoscopic 3D found a (temporary) home in popular culture of the early 1950s as it responded to "the real shock of the modern." Not only do 3D films of the postwar era exploit stereo-effects to undermine the boundaries that separate touch and sight, the material and immaterial, and diegetic space from the space of exhibition, but they also tie 3D spectatorship to a curious look that is playfully assailed for its desire to see and know the secrets concealed or hidden beneath the surface appearances of modern life — not only those found in venues of commercialized leisure but also in a variety of *unheimlich* homes (from the Wendice flat in *Dial M for Murder* 3D [Alfred Hitchcock, 1954] to the creature's dark cave in *Creature from the Black Lagoon* 3D [Jack Arnold, 1954] and the historically pastiched version of small town USA in *The Bubble* 3D [Arch Oboler, 1966]). This is an important point: given the uncanny's origins in the secret and the hidden brought to light, it is inseparable from questions of visibility and knowability. If, as Jens Schröter argues, the transplane (3D) image's defining feature is its ability "to provide a heightened knowledge of space,"[61] then it would have made sense for 3D filmmakers of the postwar era to deploy parallax effects in a manner that transforms diegetic space and *mise en scène* into tools of heightened spatial expression capable of both "choreographing the z-axis as a storytelling device"[62] and of staging an investigation of fictional spaces of modernity organized around, and defined by, the postwar uncanny and its affects.

House of Wax

In *House of Wax*, parallax and other stereo-effects remediate the waxwork and the 2D moving image to give expression to a version of the historical uncanny in which the past persists in the present through recurring cycles of catastrophic violence. As part of this project, the film mobilizes numerous fantasies around disquieting doublings and the return of the dead, and,

60 Freud, "The Uncanny," op. cit., p. 364

61 Jens Schröter, *3D: History, Theory and Aesthetics of the Transplane Image*, trans., B. Pichon and D. Rudytsky (London: Bloomsbury, 2014), pp. 40–41.

62 Stephen Prince, *Digital Visual Effects in Cinema: The Seduction of Reality* (New Brunswick, NJ: Rutgers University Press, 2012), p. 202.

moreover, undermines the boundaries that separate animate and inanimate, the living and the (merely) lifelike, spectator and spectacle. The sculptural solidity that 3D stereo-effects give to foregrounded objects in the *mise en scène* contributes, moreover, to the generation of uncertainty around the opposition of stillness and motion and the tangible and intangible. At the center of the film's elaboration of the uncanny is the sculptor and wax museum owner, Henri Jarrod (Vincent Price). He is horribly disfigured in a fire set by his business partner, Matthew Burke (Roy Roberts), who destroys their wax museum and leaves Jarrod for dead in order to collect an insurance payout. Burke and his girlfriend, Cathy (Carolyn Jones), are murdered shortly after by a disfigured killer. Though missing since the fire and presumed dead, Jarrod returns wheelchair-bound, with hands twisted and scarred (an injury that stands in for castration)[63] — to open a new House of Wax. The latter features a Chamber of Horrors and a waxwork Joan of Arc bearing a disquieting resemblance to Cathy (whose corpse has been stolen from the morgue by the killer) and a tableau scene depicting Burke's own mysterious death by hanging. Unnerved by the astonishing resemblance of Joan of Arc to Cathy, Cathy's friend Sue (Phyllis Kirk) investigates the source of the uncanny dread the wax figure inspires in her. She discovers that Cathy's missing corpse lies beneath Joan of Arc's "skin of wax," revealing Jarrod as the disfigured murderer whose identity has been concealed beneath his own wax mask.

The film's obsessive focus on both the return of the dead in various guises (as grotesquely scarred monsters, mummified corpses, and as lifelike wax figures of martyred saints, beheaded queens, and assassinated presidents) and the violent return of the past in the present is inseparable from the post-war (and wartime) historical context in which *House of Wax* was produced and exhibited. *House of Wax* marshals up the uncanny reality effects of a range of old and new media in order to rehearse fictional scenarios that seem "to confirm the old, discarded belief" that "the dead *do* live on and appear on the scene of their former activities."[64] In the process, as Paul[65]

63 As William Paul has noted, the film foregrounds the theme of castration with almost gleeful hilarity. For example, following the fire, Jarrod tells his future business partner, Wallace (Paul Cavanagh), "As for my hands, they are useless to me now. As a sculptor I can't control them. But they serve for ordinary functions." Later, after asking to see the hands of the young sculptor Scott (Paul Picerni), Jarrod wistfully explains, "Yes, I used to have hands like that. How I envy you." In order to bring his destroyed (wax) family back to life, Jarrod murders and coats in wax individuals who look like living doppelgangers of his "dead" wax figures. See Paul, "The Aesthetics of Emergence," op. cit., p. 343.

64 Freud, "The Uncanny," op. cit., p. 371; emphasis in original.

65 Paul, "The Aesthetics of Emergence," op. cit., p. 342.

notes, the film dramatizes — and, I would add, charges with uncanny affect — the "mummy complex" that Bazin[66] defined as central to the desires structuring the "plastic arts" and indexical recording media. The film's own postwar setting (New York in 1902, just following the Spanish-American and Philippine-American wars, both of which — as *This is Cinerama* [Mike Todd, Michael Todd Jr., Walter A. Thompson, Fred Rickey, 1952] had recently detailed for its audiences — were thoroughly mediated by early cinema) is salient in this respect and, importantly, bears some relation to the simultaneously postwar (and wartime) context in which *House of Wax* was released. Through its protagonist, Sue, the film references the early twentieth-century migration of populations from rural areas and small towns to cities: she has come to New York from small-town Massachusetts to find work and, unsuccessful in the latter, has fallen behind in the rent she pays for the rooms she shares with Cathy in a boarding house and faces eviction. Such urban migrations and their resulting dislocations helped spur an intensely competitive, burgeoning culture of commercialized leisure as well as a housing crisis and a pervasive sense of alienation and uprootedness — all of which helped make "estrangement and unhomeliness" "intellectual watchwords" of the twentieth century.[67] Both watchwords apply to the wartime/postwar context in which the film was released, which was itself marked by another era of US imperialism in East Asia, of rapid media change and competition (between cinema and television, and between rival studios and new formats),[68] of housing shortages and, along with the latter, another alienating mass migration, this time away from city centers to suburbs, "bedroom communities," and small towns.[69]

House of Wax takes up the trope of modern homelessness in its opening sequence, which remediates the waxwork through 3D stereo-effects to provide the past with an *unheimlich* home in the present. Deriving its claims to authenticity from the death (or life) mask molded directly from the face of the "original," the waxwork is often characterized by a hyperreal lifelikeness

66 André Bazin, "The Ontology of the Photographic Image," In *What is Cinema?*, vol. 1, trans. Hugh Gray (Berkeley, CA: University of California Press, (1967), p. 9.

67 Vidler, *The Architectural Uncanny*, op. cit., p .9.

68 Lynn Spigel, *Make Room for Television: Television and the Family Ideal in Postwar America.* (Chicago: University of Chicago Press, 1992); John Belton, *Widescreen Cinema*, op. cit.; and Ariel Rogers, *Cinematic Appeals*, op. cit.

69 Kenneth Jackson, *Crabgrass Frontier: The Suburbanization of the United States* (London: Oxford University Press, 1987); Lynn Spigel, *Make Room for Television*, op. cit.; Eric Avila, *Popular Culture in the Age of White Flight: Fear and Fantasy in Suburban Los Angeles* (Berkeley, CA: University of California Press, 2006).

that blurs the boundary between the living and the dead, the animate and the inanimate, in order to create an overwhelming (yet fleeting) sense of the embodied presence of the depicted figure.[70] Jarrod revives dead historical figures in his waxworks and houses them in a museum that shelters his creations from what the artist regards as the vagaries and vulgarity of urban modernity in general and more commercialized forms of leisure and their mass audiences in particular. Indeed, as Paul[71] and Kevin Heffernan[72] note, the film activates an opposition between art/beauty and sensational commercialized leisure in the violent confrontation between Jarrod and Burke, who wants to add a profitable chamber of horrors to the museum that depicts recent crimes of a violent and sordid nature. Jarrod refuses — he wishes to create "beauty" through startlingly lifelike wax figures that appear to preserve the historical past in the present and have the status of art (while touring the museum, a prospective investor, Wallace [Paul Cavanagh], compares Jarrod's tableaux to "the dimensional paintings of the Masters"). Despite their beauty, however, the waxworks produce an uncanny historicity that creates ambivalence around the boundaries separating the past from the present and the living from the dead (not only does Jarrod refer to the figures as his "friends," "children," and "people," he also lives in a room above the museum's exhibition space). In the process, the stereoscopically rendered waxworks demonstrate how the uncanny functions as what Jo Collins and John Jervis call "a figure for the simultaneous homelessness of the present, and haunting by the past, that has been associated with modernity since at least the time of Baudelaire."[73]

The film's crucial but much overlooked opening scene suggests that 3D stereo-effects will be deployed throughout the film in service of the uncanny. The scene commences with a tracking shot that moves fluidly from an empty, rainy city street (shot in perspectival depth) through a window and into the display space of the wax museum, which is populated with individual waxworks and group tableau scenes — some behind proscenia, others staged on the museum floor. The camera glides past the eerily still and silent wax displays until it reaches the sculptor at work in the studio adjacent to the museum, where he creates the lifelike figures that allow

70 Sandberg, *Living Pictures*, op. cit., pp. 37–67; and Warner, *Phantasmagoria*, op. cit., pp. 21–57.

71 Paul, "The Aesthetics of Emergence," op. cit., pp. 342–343.

72 Kevin Heffernan, *Ghouls, Gimmicks, and Gold: Horror Movies and the American Movie Business, 1953–1968*, op. cit., p. 27.

73 Jo Collins and John Jervis, "Introduction," In Collins, J. and Jervis, J., eds., *Uncanny Modernity: Cultural Theories, Modern Anxieties* (London and New York: Palgrave MacMillan, 1–9, 2006), p. 2.

him to dwell in the past. In the process of moving through the displays, the camera discloses the highly mediated form of uncanny historicity that emerges from the imbrication of 3D aesthetics with waxwork figures. The displays depict historical scenes populated by wax figures fixed in a state of suspended animation — most often in the moments immediately preceding or following their executions or assassinations. Tableaux scenes depict Cleopatra and Mark Antony in their final meeting before they die; Lincoln, collapsed in a chair moments after being shot by John Wilkes Booth; Anne Boleyn, kneeling before the executioner's block; Joan of Arc, about to be burned at the stake; and an emergent Marie-Antoinette (the historical figure most closely associated with the waxwork), her head still affixed to her body, but not for long. Like the wax displays of the Musée Grévin analyzed in detail by Vanessa Schwartz, Jerrod's waxworks depict historical figures suspended "somewhere on the threshold between life and death" and situated in tableaux settings that seem to "beat back death by halting a narrative before or at the moment of death."[74]

Hence, the figures inhabit a complex and temporally ambivalent moment in time. While their uncanny lifelikeness gives them the appearance of having transcended death and of resurrecting the past in the present, it is death that has created the condition of possibility for their lifelikeness: the authenticity and iconicity of the waxwork is often an effect of indexicality — of the death masks used to create them. As Sandberg explains, the indexical relationship between the waxwork and the model gives the figure a material/physical connection to the past, to the moment when the death masks were made from a corpse.[75] In Warner's formulation, waxworks "proceed from an original to a sequence of replicas, each of which partakes of the essence of the progenitor or model. That original contact with the subject's vanished physical being intensifies the hallucinatory effect of presence in absence, of ubiquity and deathlessness in the waxwork — as process and artefact."[76] Jarrod refers to this process when he tells the crowd at his new House of Wax that, with the exception of Joan of Arc, "The historic figures I'm about to show you will be more interesting when I tell you that their faces were molded from the original death masks now in the possession of certain European governments." Given that so many of Jarrod's waxworks are of historical figures who were assassinated or executed, their historicity

74 Vanessa Schwartz, *Spectacular Realities: Early Mass Culture in Fin-de-Siecle Paris* (Berkeley, CA: University of California Press, 1998), p. 140.

75 Sandberg, *Living Pictures*, op. cit., pp. 47–52.

76 Warner, *Phantasmagoria*, op. cit., p. 38.

is equally complex, recursive, and ambivalent: not only do the waxworks revive the dead only to return them to the moment of, or immediately preceding, death, but many of them are associated with war, revolution, and violent historical change. Depending upon death for their lifelikeness and reviving the past in the present in a manner that, Warner explains, "perpetuate[s] infinite stasis, time standing still,"[77] the film's waxworks materialize a postwar historical uncanny.

The opening sequence exploits parallax effects to play upon the waxwork's ability to blur the line between past and present, absence and presence, by allowing Jarrod's most prized figure — Marie-Antoinette — to reach past the threshold of the screen into the space of exhibition, foregrounding the illusory solidity and phantom intangibility of the 3D image. The sculptural solidity of the 3D image enhances its "haptic" quality, tricking the senses by seeming to offer to the spectator a tangible connection to the past that is ultimately phantasmatic while playing upon the compulsion to confirm by touch that a wax figure is not truly alive.[78] Such contact is, of course, prohibited by the conventions of the museum display (a waxwork policeman at the entrance of the House of Wax stands next to a placard that exhorts, "Please do not touch the displays"); in turn, the 3D image invites the spectator to meet the grasp of a phantom figure whose outstretched arms only confirm 3D's uncanny ability to (playfully) "cheat" the senses, creating, as Paul argues, "a seeming reality that is in fact an illusion thinner than the air through which it moves."[79] Much as 3D enhances the waxworks' lifelikeness by bringing their sculptural solidity into relief, so, too, does the waxwork's sculptural three-dimensionality contribute to and thematize 3D's familiar-but-strange in-between-ness, its imperfect, uncanny replication of the perceptual experience of three-dimensional objects and space.[80]

The particular selection of past historical moments accumulated within the museum links the waxworks' uncanny historicity to the experience of modern life. The small space of the museum is crowded with figures and tableaux arranged in the foreground, midfield, and background of most

77 Ibid., p. 43.

78 For an extensive analysis of outstretched arms and touch in 3D cinema and *Creature from the Black Lagoon*, see Rogers, *Cinematic Appeals*, op. cit., pp. 179–210.

79 Paul, "The Aesthetics of Emergence," op. cit., p. 345.

80 On the strangeness of 3D spatiality in digital 3D cinema, see: Nick Jones, *Spaces Mapped and Monstrous: Digital 3D Cinema and Visual Culture* (New York: Columbia University Press, 2020); Miriam Ross, *3D Cinema*; and Akira Mizuta Lippit, "Three Phantasies of Cinema: Reproduction, Mimesis, Annihilation," *Paragraph*, (22)3, pp. 213–27.

shots in the opening scenes, creating a dense visual field that is evocative of nineteenth-century stereoscopic images which, Crary argues, offered the most intense experience of three-dimensionality when depicting "an object-filled space, with a material plenitude that bespeaks a nineteenth-century bourgeois horror of the void."[81] The museum's own object-filled space suggests a horror of the historical void created by modernity's dissolution of tradition and rejection of the past, its alienating (and, sometimes, forced) migrations, its demand for the always new, and its imbrication of everyday life with capitalism's inexorable futurity and demand for profitability.[82] Moreover, Jarrod's tendency to refer to the wax figures as his "children" when showing them to Wallace defines the museum as a simultaneously *heimlich* (for Jarrod) and *unheimlich* (for the spectator) "home" for the abandoned past, an idea underscored by a wax display featuring a woman surrounded by children and labeled with a placard reading "Mother Love." In this respect, the museum functions as a shelter from modern homelessness that pits "the homely, the domestic, the nostalgic, against their ever-threatening, always invading, and often subversive opposites."[83]

Much as the wax displays suggest a nostalgia-driven need to retreat into the past and to transcend the effects of (historical) time, they also materialize, as a form of "vernacular modernism,"[84] a conceptualization of history as a destructive force of violence and ruin. In terms of their content and their organization in stereoscopic three-dimensional space, the waxworks reject an understanding of the historical past in terms of linear progress to suggest instead a Benjaminian model of history as "wreckage upon wreckage;"[85] the lifelikeness and beauty of the waxworks are the *unheimlich* outcome of a drive to "waken the dead, make whole what has been smashed."[86] 3D's tendency to produce, in Crary's words, a "derangement of the conventional

81 Crary, *Techniques of the Observer*, op. cit., p. 125.

82 On these and other aspects of modernity, see Marshall Berman, *All That is Solid Melts Into Air* (New York: Penguin Books, 1982); Anthony Giddens, *The Consequences of Modernity* (Palo Alto: Stanford University Press, 1990); and David Frisby and Mike Featherstone, eds., *Simmel on Culture: Selected Writings* (London: Sage, 1997). Sandberg argues that wax effigies and museums at the turn of the century functioned as a "compensatory project of *mise en scène* that gave" modernity's "displaced objects and bodies a new kind of scenic home." See Sandberg, *Missing Persons*, op. cit., p. 8.

83 Vidler, *The Architectural Uncanny*, op. cit., p. 13.

84 Miriam Hansen, "The Mass Production of the Senses: Classical Cinema as Vernacular Modernism," *Modernism/Modernity*, (6)2 (1999): p. 59.

85 Walter Benjamin, *Illuminations: Essays and Reflections*, Hannah Arendt, ed. and H. Zohn, trans. (New York: Schocken Books, 2007[1950]), p. 257.

86 Ibid.

functioning of optical cues"[87] is key to the film's articulation of an uncanny historicity. As Crary explains, within the stereoscopic 3D image, "Certain planes or surfaces, even though composed of indications of light or shade that normally designate volume, are perceived as flat," creating the effect of a cardboard cut-out, while foregrounded objects "seem to occupy space aggressively."[88] This "planar" quality of the 3D image distorts space and exacerbates uncertainty about the size of the museum's display space and the distance between the various figures and tableaux, thereby undermining linear perspective and refusing to give the space of the museum "a unifying logic or order."[89] In this respect, the 3D space of the museum contrasts sharply with orderly shots of empty city streets, which are organized around linear perspective and staged in depth. 3D stereo-effects ensure that we perceive the space of the museum as "an assemblage of local zones of three-dimensionality" that does not "coalesce into a homogeneous field."[90] This is important, for if linear perspective implies, as Crary explains, a unified, "homogeneous and potentially metric space," then 3D gives the museum the appearance of "a fundamentally disunified and aggregate field of disjunct elements,"[91] creating a spatial aesthetic that corresponds to the waxwork displays' uncanny historicity and their rejection of notions of linear progress.

The use of 3D stereo-effects and waxworks to articulate an understanding of history as the (uncanny) repetition and return of the violent past in the present is not surprising, for as mentioned above, not only was the film produced in the aftermath of World War II and set just after the Spanish-American and Philippine-American wars, it also evidences a morbid fascination with violent conflict by evoking the Civil War, the French Revolution, and the Hundred Years War through its wax displays. If, as Sandberg argues, the waxwork is a technology of revivification that attempts to fix time and forestall decay,[92] then once framed by the film's representation of history as the uncanny repetition of violent conflict, the waxwork figures are doomed to "die" violently — and return from the dead — once again. After Burke sets the museum on fire, Joan of Arc and John Wilkes Booth are both burned "alive" again as they were in the past, Anne Boleyn loses her head once more, and the others melt away as their 3D sculptural solidity is liquefied by heat and fire (Fig. 14.1). The fire is an incendiary imposition of

87 Crary, *Techniques of the Observer*, op. cit., p. 125.
88 Ibid.
89 Ibid.
90 Ibid., p. 126.
91 Ibid., p. 125.
92 Sandberg, *Living Pictures*, op. cit., p. 37.

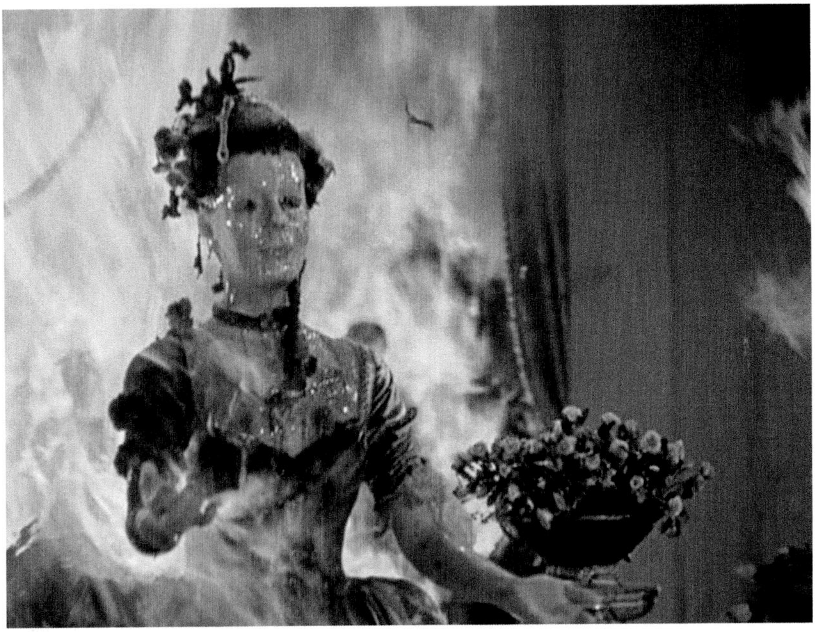

Fig. 14.1 *House of Wax*: Marie-Antoinette melts and burns in the gas-fed fire.

modernity's insatiable demand for the new and urban capital's tendency to dissolve the past, immolating it in a grotesque demonstration of what Charles Baudelaire called the "ephemerality" of modern life.[93] In turn, images of figures from the past going up in flames, along with the (gas-fed) explosion that destroys the museum (including Jarrod's "friends," family," and people") and leaves Jarrod missing and presumed dead, seem to address, in a displaced manner, not only the mass death of the most recent World War (including the nuclear bombings of Hiroshima and Nagasaki, which left victims severely burned)[94] but also the war's violent global destruction of material traces of humanity's shared historical past by technological means.

93 Charles Baudelaire, "The Painter of Modern Life," in *The Painter of Modern Life and Other Essays*, trans. and ed., J. Mayne (London: Phaidon Press, 1965 [1863]), p. 13. Vanessa Schwartz shows that the Musée Grévin's focus on current events and its reproduction of sensational newspaper stories in wax tableaux led the journalist Jules Claretie to celebrate "both the museum's commitment to rapid change and its embrace of ephemerality" by praising its "glorious transitory wax figures! Celebrities of the day! Pantheon of the moment!" See Schwartz, *Spectacular Realities*, op. cit., p. 110.

94 My thanks to Alexandra Bush for this point. André de Toth was born in Hungary and emigrated to the United States during WWII. In his autobiography (*Fragments – Portraits from the Inside* [London: Faber and Faber, 1994]), de Toth wrote of WWII that "Hitler died in the fire he lit himself. Unfortunately, and to our everlasting shame, we just stood by till it was too late and

Given the museum's emphasis on the past and reference to modern homelessness, the inclusion of the "Mother Love" display in the opening scene and its fiery demise[95] suggest that, along with the other historical figures, this nostalgic patriarchal image of maternal femininity has also been lost to the past, displaced by processes of urbanization and the arrival of the New Woman, a modern type of femininity that was not only associated with financial independence and urban life but was also conflated, as Shelley Stamp argues, with the burgeoning culture of sensational commercialized leisure and urbanization in the early twentieth century.[96] It is telling, in this respect, that the working-class Cathy is made to embody the social change associated with urban modernity, which the film associates with the increased sexual availability of women and the commodification of heterosexual romance and marriage. Cathy tracks her rise up the social ladder according to the expense and location of the forms of commercialized leisure to which her increasingly wealthy dates treat her. On the night of her murder, she describes her latest suitor to Sue as "a swell who spends freely" and explains that he is taking her to Tony Pastor's to "see the vaudeville show." Cathy explains, "I'm moving up in the world, honey. A girl never hits the jackpot until after she passes 14th Street… Not too long ago I was down on Delancey Street and tonight I'll be right up there among the bright lights on 23rd." Yet like Jarrod and Sue, Cathy suffers from modern homelessness; eager to leave the boarding house and to make her relationship with Burke "legitimate," she hints at her desire to marry Burke after learning of the insurance payout from the museum fire and Jarrod's presumed death. Shortly after, Jarrod murders her and uses her corpse as a prop to "revivify" in wax the saint and martyr, Joan of Arc. In other words, once embalmed,

millions went with him in the inferno. We were burned too before the international fire brigade was able to quell the fire," p. 210. I would also include in the reference the war's high-tech mass destruction not only the concentration camps (the sequence focuses intensely on the release of gas from their lamps) but also Germany's bombing of London from 7 September 1940, until 10 May 1941, which left many homeless. Stansky notes that "Ultimately in the course of the war 40 per cent of London's housing stock was made uninhabitable" by the German *blitzkrieg*. See Peter Stansky, *The First Day of the Blitz: September 7, 1940* (Yale University Press, 2007), p. 132. On the destruction of Hiroshima and Nagasaki, see Akira Mizuta Lippit, *Atomic Light (Shadow Optics)* (Minneapolis: University of Minnesota Press, 2005).

95 One shot shows Jarrod lying on the floor with the museum in flames around him and the "Mother Love" placard fully visible behind him. As Bob Furmanek and Greg Kintz have shown, this shot was featured in a View-Master reel used as a promotional tie-in for *House of Wax* and other 3D features. http://www.3dfilmarchive.com/House-of-Wax (accessed 30 June 2019).

96 Shelley Stamp, *Movie-Struck Girls: Women and Motion Picture Culture After the Nickelodeon* (Princeton: Princeton University Press, 2000).

Cathy-as-Joan-of-Arc represents a perversely gendered literalization of the Bazinian notion of "change mummified,"[97] in which wax does not simply preserve her dead body against the "ravages of time" but aids in the disavowal of the socio-cultural change she represents. By charging with uncanny affect Jarrod's efforts to kill off modern femininity and re-dress it in the guise of the past, the film (perhaps inadvertently) adds to its conceptualization of history as "wreckage upon wreckage" the violence enacted against (working and poor) women as a means for preserving patriarchal privilege in the face of rapid social change.[98]

In keeping with the film's emphasis on uncanny historicity, the return of the (violent) past in the present, and uncanny doublings, Jarrod and his family of waxwork revenants return from the dead following the fire to once again "appear on the scene of their former activities."[99] When Wallace exclaims, "I thought you were dead," Jarrod replies, "Jarrod is dead; I am a reincarnation." The fire has transformed him into a showman ready to give the people what they want ("sensation, shock, horror") and into the grotesque killer who procures corpses for some of the new museum's waxworks displays. The reality effects of the new wax displays are organized around an aesthetic principle that intensifies the waxwork's uncanniness. Jarrod explains to Wallace that "each subject must be taken from life. How can I convince my audience they're alive if I don't believe it myself?" Jarrod's implementation of the principle "taken from life" is as Sandberg notes "too literal and too complete" — some of his wax figures conceal the lifeless bodies of his murder victims — thereby exacerbating the categorical ambivalence around presence and absence, living and dead, original and copy.[100]

Whereas within the domain of the first wax museum, wax functions as a medium that produces startling lifelikeness and the illusion of animacy in the inanimate wax figure, within the new House of Wax it is returned to its (perhaps "old and forgotten") function as a medium for mummifying corpses. The bodies beneath the wax, the concealment of the murderer's disfigured

97 Bazin, "The Ontology of the Photographic Image," op. cit., p. 15.

98 For analyses of women and urban forms of commercialized leisure and labor, see Stamp, *Movie-Struck Girls,* op. cit.; and Maggie Hennefeld, *Specters of Slapstick and Silent Film Comediennes* (New York: Columbia University Press, 2018).

99 Freud, "The Uncanny," op. cit., p. 371.

100 See Sandberg, *Living Pictures*, op. cit., pp. 44. In turn, the use of wax to embalm the bodies passed off as effigies brings them closer to the Musée Grévin's first wax figures, which, as Schwartz explains, represented corpses from the morgue. See Schwartz, *Spectacular Realities*, op. cit., p. 88.

face beneath Jarrod's own life mask, and the process by which Jarrod coats the bodies of his victims in "a skin of wax" in his basement studio constitute the hidden secrets of the new House of Wax that Sue will "bring to light." Within the fictional world of the film, the displays provoke dread in Sue (or, what she describes as "a fear, an intuition") because of an *unheimlich* presence haunting the absence that the wax figure conventionally and ultimately signifies;[101] that is, Joan of Arc is uncanny for Sue not because the figure seems so lifelike but because she fears it is dead.[102] Significantly, the organization of the House of Wax around Jarrod's attempt to pass off the "original" (body, corpse) as a hyperreal, lifelike "copy" not only plays upon the overly close association of waxworks with corpses but also informs de Toth's use of positive and negative parallax as uncanny visual effects organized around a broader investigation into the experience of modern life in the postwar era.

The film's infamous paddleball sequences provide ideal examples of how parallax effects engage the spectator in an investigation of the spaces and experience of urban modernity while elaborating in spectacular fashion not only the film's thematic concerns but also the stakes of its protagonist's disquietude and, with it, the solution to the film's narrative enigma. Significantly, the two sequences bookend the protagonists' introduction to the new House of Wax and draw, as Bob Furmanek and Greg Kintz note, diegetic passers-by into the new establishment *and* the spectator back into the fictional world of the film following its intermission.[103] The barker's (Reggie Rymal) paddleball antics define the House of Wax as a space where categorical boundaries are effaced — not only those that separate the present from the past and the living from the dead but also those that separate original and copy, spectator and spectacle. Positive and negative parallax, in turn, create ambivalence around the boundaries separating theatrical space and the fictional world of the film, the tangible and intangible. The sequences exaggerate stereoscopic 3D's technological enhancement of vision and actively solicit the look of the film spectator in order to (playfully) assail it through emergence effects. Describing the House of Wax as "a cultural exhibition that will enlighten and amaze you" while offering a lurid glimpse

101 This corresponds to what Sandberg refers to as the waxwork's "missing person." ibid., pp. 1–17.
102 According to Sandberg, "One might say that the corpse is the hidden secret of the wax museum" because the wax museum's "bid for legitimacy depended on constructing vivacity, on distracting from the fact that all of the bodies on display were 'corpses,' that is, dead matter." *Living Pictures*, op. cit. p. 22.
103 Bob Furmanek and Greg Kintz, "Top 10 3-D myths," in http://www.3dfilmarchive.com/home/top-10-3-d-myths (accessed 26 March 2016).

of "beauties who died in torture and on the block," the barker foregrounds the function of the waxworks in provoking and satisfying curiosity while providing spectators with a pleasurable affective charge organized around (profoundly gendered forms of) sensation and dread. During his spiel, the barker hits the ball into the depths of the image, towards the faces of (mostly female) members of the crowd gathered around him and then, with the help of negative parallax, across the surface of the screen and (seemingly) into reception space. On several occasions, the barker hits the ball so closely to the eyes of onlookers in the diegetic crowd (first, a trio of young women and then, a respectable-looking couple) that they flinch reflexively — mirroring the somatic response of the theatrical audience as the barker then turns to the camera and propels the (phantom) ball across the screen directly towards the camera and the 3D glasses worn by spectators. Shot composition and parallax effects combine to create a mirroring effect, such that two audiences appear to face one another, each watching and then repeating the other's somatic and affective response to the ball's playful aggression towards (technologically enhanced) vision. This sequence foregrounds the ability of 3D stereo-effects to playfully confuse or "cheat" the senses by emphasizing the illusory tangibility of the emergent 3D image, creating the spectator's Holmesian awareness of embodied vision and the vulnerability of the eyes through an emergent, but phantom, projectile. These uncanny visual effects, which trick the eye into momentarily mistaking a representation for the real, correspond to those of the waxwork, which aim to accomplish the same through its own astonishing reality effects (directing the crowd's attention to Little Egypt, the barker asks, "Is she wax or is she flesh and blood?") that tempt amusement-seekers to violate the prohibition against touching the figures.[104]

The fact that this use of parallax effects creates a correspondence between the diegetic spectator of the House of Wax and the theatrical spectator of *House of Wax* is quite important, for it binds the paddleball display to the narrative by gesturing towards the hidden truth behind Jarrod's new waxworks. At one point during his spiel, the barker faces the camera with the diegetic audience arrayed behind him. He creates a triangle from the paddle's elastic and propels the ball through it, mimicking before enacting its passage across the permeable surface of the screen and back again into the depths of the diegesis (Fig. 14.2). Together, the organization of looks

104 Paul takes this further by arguing that "*House of Wax* philosophically explores the play between illusion and reality central to the experience of stereoscopic films by using the waxworks as a kind of metaphor for 3D itself." "Aesthetics of Emergence," op. cit. p. 341.

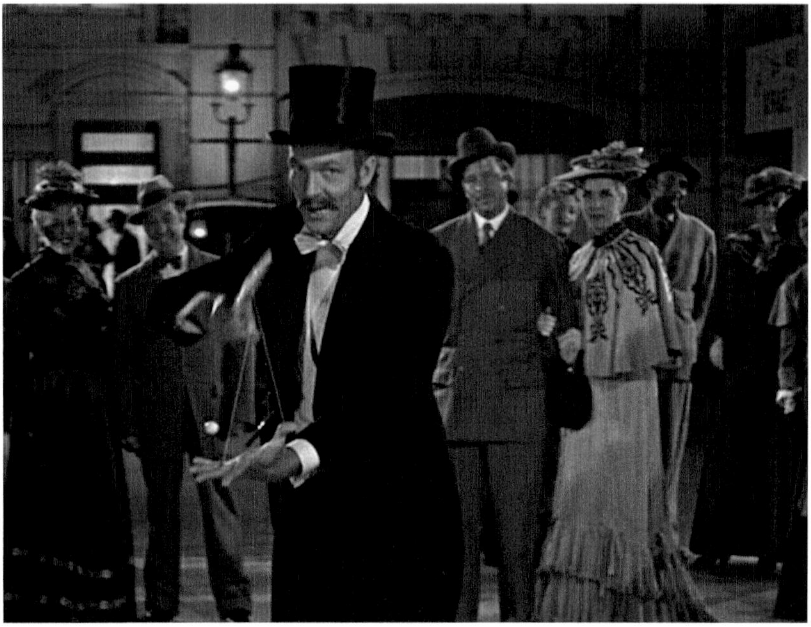

Fig. 14.2 *House of Wax*: The ball's passage through the elastic triangle mimics its illusory passage across the screen's threshold while the diegetic audience watches.

and the use of parallax effects once again doubles the barker's audience, such that the spectator appears to become spectacle (for the diegetic audience) and vice versa, each time the ball passes across the threshold of the screen. Parallax effects join diegetic and theatrical spectators along the z-axis traced by the paddleball to hint at the hidden truth behind the wax displays inside: some are made from spectators (Cathy and Burke) who were once part of the sensation- and spectacle-seeking public that flocks to new venues of commercialized leisure such as the House of Wax. Like the "audience" arrayed behind the barker, they have been transformed into lifelike spectacles that (only) appear to look back at their audience. Within the paddleball sequences, depth and emergence effects function as uncanny visual effects that identify the House of Wax as an *unheimlich* space where the boundary between spectacle and spectator, representation and the real, has been violently compromised. Moreover, the continuum traced along the z-axis by the ball becomes a metaphor for a connection forged by the film between two postwar audiences in search of sensational forms of visual culture capable of addressing the "shock of the modern" — including high-tech (overseas) warfare and alienating economic, technological, and social change. Even as it participates in the culture of modernity by foregrounding its status as a form of new media, *House of Wax* deploys parallax effects in order to identify a defining feature of modern life: the wartime violence endured by an audience in one era will inevitably return from the past as an uncanny spectacle for the visual pleasure of a subsequent (postwar) audience, *ad infinitum.*

The film's final scenes demonstrate the tendency of *House of Wax* to harness positive parallax to a curious, investigative look that penetrates depth in order to see, know, and "bring to light" that which is secret and hidden, old and forgotten, while simultaneously playing upon the limits to knowability and the epistemological uncertainty around categorical boundaries provoked by the waxwork and stereoscopic 3D. In the film's third act, Sue moves closer to uncovering the secrets behind Joan of Arc's unnerving similarity to Cathy one evening as she searches for her friend, Scott (Paul Picerni), after hours in the House of Wax. As she moves through the museum's darkened, cavernous chambers, positive parallax gives the space of the museum exaggerated depth while pale pools of light illuminate the wax figures and tableaux scenes scattered throughout. In these shots, lighting, *mise en scène*, and exaggerated depth effects once again emphasize the planar quality of the stereoscopic 3D image — that "insistent sense of 'in front of' and 'in back of'"[105] given to objects in the midfield and depths of the frame. An effect of positive parallax, the planar quality of the displays not only accomplishes a "derangement of the conventional functioning of optical cues" as it did in the film's opening sequence, it also combines with lighting and composition to suggest that each tableau might represent two acts of violence — the violence depicted by the tableau and the murder needed to procure the corpse beneath the wax — and might therefore conceal a second horror equal to the one it displays. In the second half of the film, *House of Wax* employs positive parallax to create a scopic field linked to seeing and knowing (as in Holmes's description of the stereoscope as an instrument that allows "the mind" to "feel its way into the depths of the picture") and to the investigation of uncanny media, of the "secret and hidden," and of the horrifying tendency of the past to repeat itself, through cyclical violence, in the present.

The logic of the "display that conceals" derives not only from the planar quality of the 3D image but also from a range of stereo-effects provoked by the film's remediation of the waxwork, stereoscopy, and photography, all of which allows *House of Wax* to exploit both the uncanny stillness that, Mulvey argues, haunts the moving image[106] and the uncanny movement that haunts

105 Crary, *Techniques of the Observer,* op. cit., p. 125.
106 Mulvey, *Death 24x a Second,* op. cit., p. 67.

the immobile waxwork.[107] As Sue enters the darkened Chamber of Horrors, she passes by a palely lit display case lined with a row of disembodied heads, one of which looks like Jarrod's assistant, Igor (Charles Bronson). Identical sculptures of Igor's disembodied head proliferate throughout the *mise en scène* in the second half of the film (Igor has sculpted numerous figures to look like himself — including a waxwork William Kemmler in the electric chair) contributing to doubt over the head's status as embodied or disembodied, animate or inanimate. Sue regards the display while passing by and, once she is out of the frame, the camera cuts to a close-up; the head remains perfectly still for several beats until finally its eyes shift slightly as Igor tries to keep Sue in sight. Moments later, the head disappears behind the display case and Igor emerges from behind it in pursuit of Sue, exaggerating and making potentially lethal the degree to which waxwork tableaux settings depended upon, as Schwartz explains, the curious viewer's mobilization "across, through and even into the display" in order to "literally dissolve the boundaries between self and spectacle —to blur reality and representation."[108]

Rather than simply rehearsing the cliché (staged several times in *House of Wax*) of the naïve spectator (usually a woman), this sequence exploits the uncertainty at stake in the uncanny reality effects of the various media — film, the waxwork, and the stereoscopic 3D image — that intersect in this film. (As Sandberg explains, uncertainty about the [in]animacy of the waxwork generates confusion around the ontological status of wax figures, on one hand, and museum visitors or staff, on the other.)[109] In the process of exploiting the disquieting effect of a figure's protracted motionlessness, this shot invokes the cinema's own potential for producing disquieting uncertainty around what Mulvey describes as "the fundamental, and irreconcilable, opposition between stillness and movement that reverberates across the aesthetics of cinema"[110] as the projector reconstitutes still frames of the filmstrip into an illusion of motion. Mulvey argues that

> Although the projector reconciles the opposition and the still frames come to life, this underlying stillness provides cinema with a secret, with a hidden past that might or might not find its way to the surface. The

107 Warner refers to this stilled movement as the waxwork's "creepy inertia." See Marina Warner, *Phantasmagoria,* op. cit., p. 47.

108 Schwartz, *Spectacular Realities*, op. cit., p. 132.

109 Sandberg, *Living Pictures*, op. cit., pp. 104–107.

110 Mulvey, *Death 24x a Second,* op. cit., p. 67.

> inanimate frames come to life, the unglamorous mechanics are covered over and the entrancing illusion fills the screen. But like the beautiful automaton, a residual trace of stillness, or the hint of stillness within movement, survives, sometimes enhancing, sometimes threatening.[111]

Scenarios of misperception in *House of Wax* that confuse the animacy of the human with the inanimacy of the waxwork play upon this "hint of stillness" (and the death it references), bringing the cinema's secret "to the surface" when disquietingly still bodies are restored to life through human motion or when a previously animate character is transformed into a wax figure and fixed in a state of suspended animation.

For example, the shot of Igor recalls an earlier one in the film that similarly foregrounds the "hint of stillness within movement" that haunts the cinema and the waxwork and, moreover, suggests the true cause of Joan of Arc's disquieting similarity to Cathy. On the opening night of the House of Wax, Jarrod, Scott, and Wallace discuss Sue's striking resemblance to the wax figure of Marie-Antoinette that was destroyed in the fire.

> Wallace: I haven't known Miss Allen for more than ten minutes, but there is something about her...
> Jarrod: ...face that haunts you as the face of my Marie Antoinette haunts me?
> Wallace: Marie-Antoinette ? Of course — I should have seen it at once! A figure in wax — Mr. Jarrod's greatest work.
> Jarrod: More than wax — she lived!
> Sue: You mean I look like she did?
> Wallace: Exactly as she did!
> Jarrod: Once in his lifetime, every artist feels the hand of God and creates something that comes alive. So it was with my Marie Antoinette and I loved her. But she's gone now, horribly destroyed. Perhaps you will help me bring her back? You will come to see me, soon?

As Wallace and Jarrod discuss Sue's resemblance to the destroyed wax figure, a close-up frames her from the shoulders up, then dissolves into a shot of Sue dressed as Marie Antoinette and then back again to Sue. She holds her head perfectly still at the beginning and end of the dissolve, moving her eyes slightly only after she transforms completely into Marie Antoinette. Taking advantage of the sculptural solidity that 3D gives to the image,

111 Ibid.

the dissolve from Sue to Marie Antoinette makes Sue appear inanimate, statuesque, in the process evoking the stillness that haunts the moving image; in turn, once transformed into Marie-Antoinette , Sue's eyes animate a figure we momentarily mistake for Jarrod's destroyed wax figure, thereby calling forth the illusory animacy ("she lived!") that haunts the inanimate waxwork. Indeed, the restriction of movement to a slightly shifting glance momentarily creates the illusion that Sue is peering out from behind Marie Antoinette's "skin of wax" and costume. The momentary hesitation created by this dissolve and its play upon stillness and motion, model and copy, momentarily brings "to light" the museum's hidden/*heimlich* secrets. This sequence of still and moving images (joined together by the overlapping dissolve — one of the oldest types of optical effect) not only foreshadows Jarrod's plans for Sue (he will attempt to transform her into a spectacle that cannot, but seems convincingly to, look back at the spectator) but also hints at the location of Cathy's missing corpse. In this way, *House of Wax* marshals up the uncanny effects and affects of a broad range of old and new media — and old and new visual effects — in order to return us to the desire that, Bazin argued, informs them all (the desire to embalm time and cheat death) — and, additionally, in order to forestall a range of socio-historical transformations (particularly around gender) wrought by modern life.[112]

Given this emphasis on 3D's remediation of the waxwork and the articulation of the film's uncanny themes through parallax effects, it is significant that the hidden secret behind Joan of Arc's uncanny resemblance to Cathy "comes to light" through the violation of the prohibition against touch that informs the reception of the 3D image and the waxwork. After scratching the figure's skin of wax, Sue reaches up and pulls back Joan's black wig to reveal Cathy's blonde hair hidden beneath. When Jarrod catches her in the act, Sue exclaims, "It *is* Cathy! It's Cathy's body under the wax. I *knew* it! I knew it all the time!" Touch becomes the means for dispelling the uncertainty linked to an optical encounter with the uncanny.[113] The

112 Bazin, "The Ontology of the Photographic Image," op. cit. Tom Gunning analyzes the uncanny effects of technologies of reproduction and the defense against death in "Re-Newing Old Technologies," op. cit.

113 Gunning discusses the "optical uncanny" of magic tricks and literature of the fantastic in "Uncanny Reflections." He summarizes the history of the optical uncanny as follows: "modern skepticism about supernatural causes and mastering of the principles of optics through modern science and mechanics gave birth to a modern uncanny in Freud's sense: an encounter with a mode of thought seemingly surmounted whose apparent recrudescence causes an unsettling yet fascinating sensation. The uncanny gazes across an abyss at something left behind, and

embalmed corpse-as-waxwork not only returns us to the past by referencing the use of wax to mummify the dead but also foregrounds the ambivalence around the various boundaries separating old and new technologies and media by sustaining the film's intermedial investigation into the various ways that the past returns in the present (often in the guise of the new) through the exploitation and thematization of the sculptural solidity and illusory tangibility of 3D image. With its emphasis on the cyclical return of violence and death, *House of Wax* utilizes 3D's visual effects in order to dramatize the various ways the past, the old and forgotten, and the dead haunt the new and the living by reviving and remediating old media to uncanny ends.

Conclusion

House of Wax provides a compelling example of how 3D cinema's various stereo-effects — positive and negative parallax, the illusory solidity and seeming tangibility of foregrounded objects, the planar quality of the image — can function as spectacular visual effects that emblematize a film's central concepts, themes, and anxieties. Given the film's emphasis on the violent resurgence of the past in the present, the cyclical repetition of violent conflict, the return of the dead, its mediation of the experience of technological modernity and media change, and its dramatization of various forms of modern homelessness, *House of Wax*'s insistent elaboration of uncanny themes and optics through stereoscopic 3D seems inseparable from the postwar era in which it was produced and exhibited. And while *House of Wax* might seem like an overdetermined example of how 3D cinema might deploy its stereo-effects to uncanny ends, it is important to keep in mind that a number of important 3D films from the era use positive and negative parallax, along with other stereo-effects, in a similar fashion. I will note briefly, for example, that *Creature from the Black Lagoon* (Jack Arnold, 1954) deploys parallax effects as uncanny visual effects in ways that resonate with Schelling's argument that "everything is *unheimlich* that ought to have remained secret and hidden but has come to light": the

yet suddenly recalled. With the modern science of optics the possibility of generating such sensations directly and intentionally gave birth to a modern optical conjuring, which dates back to at least the seventeenth-century catoptric mirror tricks of Kircher and Della Porta and gains impetus in the next century with the technical perfection of projected and reflected images." Tom Gunning, "Uncanny Reflections," op. cit., p. 85.

creature's existence comes "to light" through an archaeological dig that exposes the webbed claw of one of its predecessors. Initially dismissed, as Kevin Heffernan notes, as a "'native' superstition,"[114] the creature's existence gives rise to an uncanny temporality in which the distant evolutionary past erupts in the present in the shape of the creature, a biological anomaly (with origins in the Devonian era) that blurs the boundaries between human/nonhuman (the "gill-man" is neither entirely "lungfish" nor "human"). While the thoroughly repeated trope of the creature's emergent claw (which it rakes across the face of one character, blinding him) suggests that the creature should have remained hidden from scientific investigation, underwater shots exploit positive parallax to create an opaque scopic field in which darkness, seaweed, debris, and rock formations limit visibility and therefore knowability to suggest that the creature will ultimately remain unknowable. And while one of the motivations behind the scientists' pursuit of the creature is the need to learn from it in order to "teach men to adapt themselves to some new world of the future," there is, as Ariel Rogers argues, an overt association between the female scientist, Kay, and the creature, which suggests that the "affective appeal of [their] presentation in 3D...derived from that the promise that the new technology would enhance the intimacy with which the viewer could encounter a spectacle of alterity."[115]

In *House of Wax*, 3D cinema's "stereo-effects" function as spectacular visual effects that give expression to uncanny themes and optics that would have resonated with audiences in the postwar era, in the process demonstrating how 3D cinema of the 1950s might be situated within its broader historical context and how emergence effects function emblematically, to give expression to a film's broader narrative and thematic concerns. *House of Wax* suggests that, when approached less as a gimmick for luring spectators back to the cinema and more as an "aesthetic response to the real shock of the modern"[116] that made the familiar experience of narrative cinema strange, parallax effects and the 3D films in which they appear open up new ways of "thinking stereoscopically" (to borrow a phrase from Trotter),[117] about the place of visual effects within American film history.

114 Heffernan, *Ghouls, Gimmicks, and Gold*, op. cit., p. 40.

115 Rogers, *Cinematic Appeals*, op. cit., p. 202.

116 Vidler, *The Architectural Uncanny*, op. cit., p. 9.

117 Trotter, "Stereoscopy: Modernism and the 'Haptic,'" op. cit., p. 39.

Bibliography

Adorno, Theodor. *Minima Moralia: Reflections from a Damaged Life* (London: Verso, 2005 [1951]).

Avila, Eric. *Popular Culture in the Age of White Flight: Fear and Fantasy in Suburban Los Angeles*. (Berkeley, CA: University of California Press, 2006).

Barker, Jennifer M. *The Tactile Eye: Touch and Cinematic Experience* (Berkeley, CA: University of California Press, 2009).

Bazin, André. "The Ontology of the Photographic Image." In *What is Cinema?*, vol. 1, translated by Hugh Gray (Berkeley, CA: University of California Press, 1967).

———. "Will CinemaScope Save the Cinema?" In *André Bazin's New Media*, edited and translated by Dudley Andrew (Berkeley, CA: University of California Press, 2014 [1953]): 267–287.

Beaudelaire, Charles. "The Painter of Modern Life." In *The Painter of Modern Life and Other Essays*, translated and edited by Jonathan Mayne (London: Phaidon Press, 1965 [1863]).

Belisle, Brooke. "The Dimensional Image: Overlaps in Stereoscopic, Cinematic, and Digital Depth," *Film Criticism* 37-38, no. 3-1 (2013): 117–137.

Belton, John. *Widescreen Cinema* (Cambridge, MA: Harvard University Press, 1992).

Benjamin, Walter. *Illuminations: Essays and Reflections*, translated by H. Zohn, edited by Hannah Arendt (New York: Schocken Books, 2007 [1950]).

Berman, Marshall. *All That is Solid Melts Into Air* (New York: Penguin Books, 1982).

Bolter, Jay David, and Richard Grusin. *Remediation: Understanding New Media* (Cambridge, MA: MIT Press, 2000).

Brewster, David. *The Stereoscope: Its History, Theory, and Construction, With Its Application to the Fine and Useful Arts and to Education* (London: John Murry, 2012 [1856]).

Collins, Jo, and John Jervis. "Introduction." In *Uncanny Modernity: Cultural Theories, Modern Anxieties*, edited by Jo Collins and John Jervis (London and New York: Palgrave MacMillan, 2006): 1–9.

Crary, Jonathan. *Techniques of the Observer: On Vision and Modernity in the Nineteenth Century* (Cambridge, MA: MIT Press, 1990).

de Toth, André. *Fragments – Portraits from the Inside* (London: Faber and Faber, 1994).

Ebert, Roger. "Why I Hate 3-D (And You Should Too)," *Newsweek*, (9 May 2010). https://www.newsweek.com/roger-ebert-why-i-hate-3d-movies-70247.

Eisenstein, Sergei. "On Stereocinema," translated by Sergey Levchin, *Public: Art/Culture/Ideas* 24, no. 47 (2013): 22.

Freud, Sigmund. "The Uncanny," In *The Penguin Freud Library: Art and Literature. Vol. 14*, edited by Albert Dickson, translated by James Strachey (London: Penguin Books, 1985 [1919]).

Frisby, David, and Mike Featherstone, eds. *Simmel on Culture: Selected Writings* (London: Sage, 1997).

Furmanek, Bob, and Greg Kintz. "An In-Depth Look at *House of Wax*." 3-D Film Archive. Accessed 16 November 2021. http://www.3dfilmarchive.com/House-of-Wax.

Giddens, Anthony. *The Consequences of Modernity* (Palo Alto: Stanford University Press, 1990).

Gunning, Tom. "The Cinema of Attractions: Early Film, its Spectator, and the Avant-Garde," *Wide Angle* 8, nos. 3-4 (1986): 63–70.

———. "ReNewing Old Technologies: Astonishment, Second Nature, and the Uncanny in Technology from the Previous Turn-of-the Century." In *Rethinking Media Change: The Aesthetics of Transition*, edited by David Thorburn and Henry Jenkins (Cambridge, MA: MIT Press, 2004): 39–59.

———. "Uncanny Reflections, Modern Illusions: Sighting the Modern Optical Uncanny." In *Uncanny Modernity: Cultural Theories, Modern Anxieties*, edited by Jo Collins and John Jervis (London and New York: Palgrave MacMillan, 2006): 68–89.

Gurevitch, Leon, and Miriam Ross. "Stereoscopic Media: Beyond Booms and Busts," *Public* 24, no. 47 (2013): 83–92.

Hansen, Miriam. "The Mass Production of the Senses: Classical Cinema as Vernacular Modernism," *Modernism/Modernity* 6, no. 2 (1999): 59–77.

Heffernan, Kevin. *Ghouls, Gimmicks, and Gold: Horror Movies and the American Movie Business, 1953–1968* (Durham, NC: Duke University Press, 2004).

Hennefeld, Maggie. *Specters of Slapstick and Silent Film Comediennes* (New York: Columbia University Press, 2018).

Higgins, Scott. "3D in depth: *Coraline, Hugo* and a sustainable aesthetic," *Film History* 16, no. 3 (2012): 196-209.

Holmes Jr., Oliver Wendell. "The Stereoscope and the Stereograph." In *Soundings from the Atlantic* (Boston: Ticknor and Sons, 1864): 124–165.

———. "Sun-Painting and Sun-Sculpture: With a Stereoscopic Trip Across the Atlantic." In *Soundings from the Atlantic* (Boston: Ticknor and Sons, 1864): 166–227.

Jackson, Kenneth. *Crabgrass Frontier: The Suburbanization of the United States* (London: Oxford University Press, 1987).

Jentsch, Ernst. "On the Psychology of the Uncanny." In *Uncanny Modernity: Cultural Theories, Modern Anxieties*, edited by Jo Collins and John Jervis (London and New York: Palgrave MacMillan, 2006): 216–227.

Klinger, Barbara. "Beyond Cheap Thrills: 3D Cinema Today, the Parallax Debates and the 'Pop-Out'," *Public: Art/Culture/Ideas* 24, no. 47 (2013): 186–189.

Lefebvre, Simon. "The Disappearance of the Surface." In *Screens*, edited by Dominique Chateau and José Moure (Amsterdam: Amsterdam University Press, 2016): 97–106.

Lippit, Akira Mizuta. *Atomic Light (Shadow Optics.)* (Minneapolis, MN: University of Minnesota Press, 2005).

Loew, Katharina. "Tangible Spectres: 3D Cinema in the 1910s," *Film Criticism* 37–38, nos. 3-1 (2013): 87–115.

Marks, Laura U. *The Skin of the Film: Interculture Cinema, Embodiment, and the Senses* (Durham, NC: Duke University Press, 2000).

Mulvey, Laura. *Death 24x a Second: Stillness and the Moving Image* (London: Reaktion Books, 2006).

Paul, William. "The Aesthetics of Emergence," *Film History* 5, no. 3 (1993): 321–355.

Prince, Stephen. *Digital Visual Effects in Cinema: The Seduction of Reality* (New Brunswick, NJ: Rutgers University Press, 2012).

Rhodes, John David. *Spectacle of Property: The House in American Film* (Minneapolis, MN: University of Minnesota Press, 2017).

Rogers, Ariel. *Cinematic Appeals: The Experience of New Movie Technologies* (New York: Columbia University Press, 2013).

Ross, Miriam. "The 3D Aesthetic: *Avatar* and Hyper-Haptic Visuality," *Screen* 53, no. 4 (2012): 381–397.

———. *3D Cinema: Optical Illusions and Tactile Experiences* (New York: Palgrave MacMillan, 2015).

Sandberg, Mark. *Living Pictures, Missing Persons: Mannequins, Museums, and Modernity* (Princeton, NJ: Princeton University Press, 2003).

Sandifer, Philip. "Out of the Screen and Into the Theater: 3-D Film as Demo," *Cinema Journal* 50 no. 3 (2011): 62–78.

Schroter, Jens. *3D: History, Theory and Aesthetics of the Transplane Image*, translated by B. Pichon and D. Rudvtsky (London: Bloomsbury, 2014).

Schwartz, Vanessa. *Spectacular Realities: Early Mass Culture in Fin-de-Siècle Paris* (Berkeley, CA: University of California Press, 1998).

Sconce, Jeffrey. *Haunted Media: Electronic Presence from Telegraphy to Television* (Durham, NC: Duke University Press, 2000).

Sobchack, Vivan. *The Address of the Eye: A Phenomenology of Film Experience* (Princeton, NJ: Princeton University Press, 1992).

Spadoni, Robert. *Uncanny Bodies: The Coming of Sound Film and the Origins of the Horror Genre* (Berkeley, CA: University of California Press, 2007).

Spigel, Lynn. *Make Room for Television: Television and the Family Ideal in Postwar American* (Chicago: University of Chicago Press, 1992).

Stamp, Shelley. *Movie-Struck Girls: Women and Motion Picture Culture after the Nickelodeon* (Princeton, NJ: Princeton University Press, 2000).

Stansky, Peter. *The First Day of the Blitz: September 7, 1940* (New Haven, CT: Yale University Press, 2007).

Trotter, David. "Stereoscopy: Modernism and the 'Haptic'," *Critical Quarterly* 46, no. 4 (2004): 39–42.
Vilder, Anthony. *The Architectural Uncanny. Essays in the Modern Unhomely* (Cambridge, MA: MIT Press, 1992).
Warner, Marina. *Phantasmagoria: Spirit Visions, Metaphor, and Media into the Twenty-First Century* (Oxford: Oxford University Press, 2006).
Whissel, Kristen. *Spectacular Digital Effects. CGI and Contemporary Cinema* (Durham, NC: Duke University Press, 2014).
———. "Digital 3D, Parallax Effects, and the Construction of Film Space in *Tangled* 3D and *Cave of Forgotten Dreams* 3D." In *Screen Space Reconfigured*, edited by Susanne Ø. Saether and Synne Tollerud Bull (Amsterdam: Amsterdam University Press, 2020): 77–103.
———. "*Dial M for Murder*: The Detective Thriller, the Postwar Uncanny, and 3D Cinema," *New Review of Film and Television Studies* 20, no. 1 (Winter 2022): 49-62

About the Author

Kristen Whissel is Professor of Film & Media at the University of California, Berkeley. She is the author of *Picturing American Modernity: Traffic, Technology, and the Silent Cinema* (2008) and *Spectacular Visual Effects: CGI and Contemporary Cinema* (2014) and is a co-editor of the volume *Editing and Special/Visual Effects* (2016).

15. *Oblivion*: Of Time and Special Effects

Sean Cubitt

Abstract

The opening scene of *Oblivion* (Joseph Kosinski, 2013) melds photogrammetry (including texture maps of actual spacecraft), location reference footage, back-projection, light detection and ranging (lidar), and an elaborate physical set in which actors move but are also matched to digital doubles on a digital set. These processes match a narrative concerning clones on a post-apocalyptic Earth. These spatial, architectural, and digital-analog doublings along with the cloning and the trope of the 'post' invite questions about the temporalities deployed in the film, questions of whether identity and non-identity can be held in balance in a single presence or require a cluster of time-based adjustments (narrative, fantasy, co-temporality) that destabilize the illusion and effectively demand a story-arc of remarkable complexity for a relatively low-budget and unambitious project.

Keywords: special effects, *Oblivion, déjà vu,* digital double, time

This paper stems from a concern with the lack of identity that permeates the cinema of special effects. I understand "identity" here both as a question of the persons depicted — can such cyborgs have an identity? — and as an ontological question following Aristotle's premise that everything that exists is identical to itself (A=A).[1]

1 The law of identity, commonly held as one of the three founding laws of logic, states that any term is identical to itself and is widely held to derive from Aristotle's *Prior Analytics* Book II, trans. by Robin Smith, (Indianapolis: Hackett, 1989, Part 22), p. 68a. The point is reasserted in Leibniz's *New Essays* IV, 2, § i (1896) in the form "Everything is what it is," and in modified form in Frege's *Foundations* III, ix, § 38, (1953), where it is restricted to the number 1, where Frege also determines 0 as the non-identical. Frege, Gottlob, *The Foundations of Arithmetic,* trans. by J.L. Austin, second revised edition (New York: Harper, 1953).

Lefebvre, M. & M. Furstenau (eds.), *Special Effects on the Screen: Faking the View from Méliès to Motion Capture*. Taylor & Francis Group, 2022
DOI 10.5117/9789462980730_CH15

As a time-based art, all cinema engages with the instabilities of self-identity and whether it is consistent during change; but in the special-effects film, we encounter these instabilities in a particularly poignant form. All film speaks from the past to the present; some films speak from multiple pasts and towards multiple futures. But in all of these cases, we experience the film itself, during a screening, as present. That present is not always so coherent in the case of the special-effects film. The disruption of its present has implications for all its temporal relationships. Some years ago, I argued for a distinction between the magical effects in Méliès and his near contemporary Dadasaheb Phalke, the founder of Indian cinema.[2] Phalke, I argued, was coping with the incommensurability of historical and mythical time, the coincidence of an immanent plane of divinities with the sublunar world of human action; while Méliès confronted the broken temporalities of the commodity form as it emerged in the Belle Epoque, whose glittering surfaces compressed production into retailing and displaced desire and mortality in forms at once alluring and ridiculous. Today, special effects respond to other temporalities, other configurations.

Take a perfectly ordinary effects movie of 2013: *Oblivion* (Joseph Kosinski). One of several summer season box-office disappointments (it took 89 million dollars at the domestic box office), *Oblivion* was nonetheless blessed with highly attractive design work. Unusually, a great deal of what we see — notably the 'skytower' house in the clouds, the bubbleship aircraft, and the attack drones — are physical effects, and the cloudscapes (filmed from the peak of Haleakala in Hawai'i), locations (for example, a deserted 1904 power station in Baton Rouge), and landscapes (mainly Icelandic) picture physical places. The story demands some digitization, but much of it is derived from lidar recordings, photogrammetry, and texture mapping from the actors and props. The indexical quality is perhaps most striking in the battered exterior of the returning spacecraft *Odyssey*, which is modelled on texture maps and reference photographs of the space shuttle *Endeavour*, which had just gone on display in Los Angeles as production geared up. Few films can boast having textures mapped from surfaces that have genuinely left the earth's atmosphere. This does not remove the question of the fidelity of cinema to its world: it helps set up the terms on which we might enquire into how not only the practice of fidelity but the nature of the world may have changed.

2 Sean Cubitt, "Phalke, Méliès and Special Effects Today," in *Wide Angle*, (21)1, (January 2001), pp. 114–130.

Particularly innovative is the use of front projection, where an array of 21 synchronized high-resolution projectors mounted under the skytower set beam images of the cloudscape onto a curved screen covering the studio and embracing the set, giving the actors a far stronger sense of location than greenscreen recording, and making the reflective and refractive surfaces of the set sparkle with ambient colors, a process that 'probably saved us 500 visual effects shots' and the time-consuming work of removing blue- or greenscreen artefacts.[3] The verisimilitude of the scene is helped along, it must be added, by complex mathematical adjustments to make up for the reflection and deformation of the image array on the curved screen, using paint programs to fill gaps, boosting volumes to counter parallax effects, mildly speeding up the movement of clouds to fit the drama (not unlike undercranking chase sequences in the old days perhaps), and careful set lighting to avoid cast shadows. Almost all of these techniques, bar the volume boost, are of a kind with shooting on sets with painted backdrops. The major difference between these projected cycloramas and the use of mattes and painted backdrops in spectaculars like *Ben Hur* (William Wyler, 1959) is that where filters, tripack chemistry, grading, and printing were previously in the hands of lab technicians, they are now in the hands of the director, the director of photography, and the editor. Yet there is a particular falsity involved that demands investigation.

Critics liked Tom Cruise's energy in the star role of Jack Harper, while fans remarked that it was the better for not showcasing his 'shit-eating grin'. The performance, beyond the action sequences, is somewhat wooden. His director no doubt felt such performance style was apposite for the story of a clone dredging through his unconscious trying to recall the lost shreds of his humanity — surely a perfect working definition of a Hollywood star. Given the film's themes of falsehood and inauthenticity, it is proper that it should indeed be formally inauthentic, from its reprises of *Top Gun*'s (Tony Scott, 1986) comic book iconography to the mischievous presentation of physical sets as if they were digitally generated.

The film's double-bluff, echoed in its plot, pursues one of science fiction's great themes, the question of what constitutes the human. The topic was constantly broached in the original and *Next Generation* iterations of *Star Trek*, for instance, as Spock and Data were asked in numerous episodes to encounter another distinguishing feature of humanity. The question returned as something between epistemology and ethics in debates of the 2000s over the 'death of cinema' when major critics including Mary Ann

3 Joe Fordham, "Last Man Standing," in *Cinefex*, 134 (digital edition, 2013): 94-116.

Doane[4] and D.N. Rodowick[5] lamented the loss of what they saw as the privileged indexical relation of classical cinematography. The challenge was already enacted in *The Matrix* (Wachowskis, 1999), as effect and as storyline — how many viewers shared Cypher's feeling that it would be better to luxuriate in the illusion than to accept the grimness of reality?

It is by now a truism of environmental criticism that the term 'nature' acts somewhat like a floating signifier, and is 'perhaps the most complex word in the language' according to Raymond Williams.[6] The field we designate as nature has changed over history, and the corollary is that the meaning of the word 'human' also changes, most dramatically in the transition from religious to secular worldviews, but in more nuanced ways subsequently. On the surface, the central concept of humanity in *Oblivion* is that of the citizen, as evinced by the repeated citation of a line from Macauley's poem *Horatius*, a Victorian verse retelling the sacrifice of a Roman soldier holding back an attacking horde on the Sublician Bridge across the Tiber. Under that trope of the citizen-soldier, with its imperial ethos and its pre-echo of fatalist heroism in wartime Nazi cinema,[7] there lies another more complex and more time-bound sense, not only of shared mortality but of what constitutes the present of a human being.

At the 3'45" mark, after a voice-over exposition of the narrative set-up, Jack emerges into the skytower set, seen through its extensive windows that reflect both the physical bubbleship and the front-projected cyclorama of Hawaiian clouds. As he exits towards the camera, we cut to a dramatic crane shot, punctuated by a brief lens-flare, moving up from below the launch platform to a position high above it as Jack walks out to his aircraft, revealing the clouds below. 'It was impossible to shoot that with front projection' [because of the parallax effects and the missing extension of the cloudscape above and below the set], according to visual effects supervisor Bjorn Mayer, 'so we shot reference of Tom with a low camera, a medium camera and a high camera, and then we used those to re-project textures onto a digital double.'[8] The physical set itself had been built from a model constructed

4 Mary Ann Doane, "The Indexical and the Concept of Medium Specificity," *Differences: A Journal of Feminist Cultural Studies*, (18)1 (2007), pp. 128–152.

5 David Norman Rodowick, *The Virtual Life of Film* (Cambridge, MA: Harvard University Press, 2007).

6 Raymond Williams, *Keywords: A Vocabulary of Culture and Society*, rev. ed. (London: Fontana, 1983), p. 219.

7 Julian Petley, *Capital and Culture: German Cinema 1933–45* (London: British Film Institute, 1979).

8 Fordham, "Last Man Standing," op. cit., p. 7.

in Autodesk 3ds Max, which also provided the basis for the digital set in the shot, with reference footage to match the textures and reflections of the physical set. The diegesis is thus constructed in continuity between physical and digital elements, but what are the implications of constructing continuity between the actor Tom Cruise and the synthespian required for the effects shot? We have grown accustomed to the idea that humans are distinguished by being unique. As the plot unfolds, we will learn that this is not the case with Jack. The narrative offers us a resolution of the quandary this poses in its final act, but as in so many narratives, the fact that the quandary demands to be raised is more historically significant than that it can be given some more or less ideological resolution.

In a 2014 essay, Thomas Elsaesser argues that "among the specific *cinema effects*, one would first list the *impression of reality* [which] is also a consequence of the *impression of movement,* which in turn is complemented by the *impression of presence*."[9] This presence is construed within a logic of perspective that implies the co-presence of a human observer, while the digital, in contrast, "is neither visual-representational nor geometrically centered on a perceiving subject" (ibid: 34). Writing of the opening shot of *Fight Club* (David Fincher, 1999) in a 2009 essay, William Brown adds that in 'posthuman' digital cinematography, not just physical objects have been passed through, but so has an entire human being so "that human being is [...] reduced to the equivalent of 'thin air'."[10] Emphasizing the floating, subjectless continuity of digital process shots similar to the pre-title sequence of *Oblivion,* both authors point to a nexus in which both what is observed and the medium of observation are other than human. In *Fight Club,* this results in an effect of the uncanny; in *Oblivion* what is striking is the continuity between the human and non-human objects of vision, and between the physical and virtual cameras.

To take the latter part of the sequence first, the synthespian is in technical terms an object, parsed as such in the memory of the computer and treated as such in the construction of the image. But we are aware that this is an effects shot, even if we are not party to how it was achieved. This awareness is integral to the experience of the film as aesthetic — not necessarily 'aesthetic' meaning valued as artwork but 'aesthetic' in terms of belonging to the sphere

9 Thomas Elsaesser, "Digital Cinema: Convergence or Contradiction?," in *The Oxford Handbook of Sound and Image in Digital Media,* Amy Herzog, John Richardson, and Carol Vernallis, eds. (Oxford: Oxford University Press, 2014), p. 32.

10 William Brown, "Man Without a Movie Camera – Movies Without Men: Towards a Posthumanist Cinema?," in *Film Theory and Contemporary Hollywood Movies,* Warren Buckland, ed. (New York: Routledge, 2009), p. 75.

of formal presentations with some kind of claim to conveying a content, and in some more or less ideological manner claiming for that content a degree of truth. This is the purpose of Adorno's critique of the phenomenological precept that immediate sense perception is the necessary grounding of all meaningful experience: "Not knowing what one sees or hears bestows no privileged direct relation to works but instead makes their perception impossible."[11] As Barry Salt has not ceased to argue, we must understand the norms of a cinematic practice before we can assess what goes beyond them.[12] In the case of this flamboyant camera movement, the liberation from a terrestrial viewpoint immediately (or just after the titles) precedes a flying sequence. We are being introduced to a gravity-less perception but also to an inhuman one, or rather to a perspective that implies the thorough imbrication of human and machine, whether helicopter or virtual camera. The synthespian object in this shot then takes on a dual function, as a non-identical human-digital composite of a non-identical human-digital view. My claim is that the audience are party to this, as connoisseurs of compositing.

I presented an analysis of the second shot in the sequence first because it depends for its effect on a retention, to use the phenomenological expression, of its preceding shot. "The understanding of the meaning of a fleeting musical passage often depends," Adorno continues, "on the intellective comprehension of its function in a whole that is not present: the purportedly immediate experience itself depends on what goes beyond pure immediacy."[13] The experience of non-identity in the second shot depends for its effectivity on the prior identification of Jack as actual, in an actual set. At the same time, as connoisseurs of cinema, the audience is also aware that Jack is also Tom Cruise, and that the skytower is indeed a set — that is, that there is a prior non-identity, grounded in the experience of cinema as cultural form, that shapes the transgression of the ascription of identity in the first shot in its deconstruction in the second.

The principle of non-identity, to stay with Adorno, is integral to the aesthetic. Adorno is speaking specifically of 'high' art when he argues a few pages earlier that "art is an entity that is not identical with its empiria. What is essential to art is that which is not the case,"[14] but the principle can

11 Theodor W. Adorno, *Aesthetic Theory*. Gretel Adorno and Rolf Tiedemann, eds., trans. Robert Hullot-Kentor (London: Athlone Press, 1997), p. 338.

12 Cf. for example: Barry Salt, *Film Style and Technology: History and Analysis* (London: Starword, 1983); and also Barry Salt, "The Shape of 1999: The Stylistics of American Movies at the End of the Century," in *New Review of Film and Television Studies*, (2)1, May (2004), pp. 61–86.

13 Adorno, *Aesthetic Theory*, op. cit., p. 338.

14 Ibid., p. 335.

be extended to any aesthetic formation, including that through which we perceive and construe the existence of other human beings. The Aristotelean principle A=A, according to which anything that exists is identical to itself, does not necessarily hold for human beings, of whom we use the term 'identity' in order to produce certain senses of continuity for legal or moral reasons, but who we expect to be changeable, evolving, and unstable. This is, of course, doubly true of actors, in whom we admire the ability to step into and out of roles, roles that themselves often, and very much so in *Oblivion*, demonstrate the principle of instability.

Adorno's argument against the phenomenological reduction in the case of music holds also for film: we have to acquire the skill of seeing and hearing film. By the same token, film must school us in perception if there is to be a fit between the medium and its audience. The ancient and constitutive instability of humanity is perpetually re-enacted and repositioned in the arts because it is itself a mutable condition, historically contingent. It is certainly true that digital tools can be understood economically: extending the mechanical principle of automation is integral to the process of abstracting and commodifying skills and knowledges that Marx identified in the *Grundrisse*,[15] but posing economics as sole driver is unconvincing when we observe the mounting production costs of effects movies[16] and the increasing reliance on franchise and formula[17] to minimize the kind of risks that cost the studios so dearly in the summer of 2013 when *Elysium* (Neill Blomkamp) and *The Lone Ranger* (Gore Verbinski) were among the big budget movies that fared worse than anticipated at the box office. The economic determinist thesis can indeed account for some technological advances. The industry's interest in 3D in the later 2000s can be seen as an attempt both to compete with internet delivery of film content and to develop a new sales drive for high-definition domestic 3D TVs. But when the product and its mode of delivery remains essentially the same, as is the case with the move towards digital effects that have not altered theatrical screening of the 120-minute narrative feature or its delivery to other platforms, we need to investigate other and more specifically aesthetic rationales. The competition with games in a period when games design has become increasingly cinematic might be such an economic criterion, but that does not account for the longer history of digitization in production

15 Karl Marx, *Grundrisse*, trans. Martin Nicolaus (London: Penguin/New Left Books, 1973), pp. 694ff.

16 Tino Balio, *Hollywood in the New Millennium* (London: British Film Institute, 2013).

17 Dennis Schatz, *World of Inventors: Thomas Edison* (Silver Dolphin Books, 2009).

and postproduction. There is also a case to be made for increased control over workflow as a form of advanced managerialism. In a forthcoming book, I argue that this administrative impulse now extends to the levels of chip design, screen architectures, color management, and codecs. While there are economic drivers in this phenomenon, there are also issues of governance and procedure that are more properly political in the sense that they concern the administrative architectures of power, leading to a widely felt need to remarry the torn halves of political economy.

The putative tools of a new political economy informed by the Marxist and Foucaultian traditions, tools that are adumbrated in Stephen Prince's *Digital Visual Effects*[18] and Julie Turnock's *Plastic Reality*,[19] await their full deployment. In the specific case of effects cinema, there are reasons, as Prince suggests, to begin the investigation from the aesthetic and specifically from the isomorphism between what Adorno calls the empiria — especially the precise techniques of particular effects and their compositing and assembly in final cut — and the extended moment of reception by an audience. This is not to overemphasize a causal relation, still less a determining one, between effect as technique and affect as response but rather to observe that they share formal elements, most particularly their duration. *Oblivion* gives us a chance to look at the particular modes of continuity involved in transitioning from actual to virtual sets and actors and back. There is clearly an equivalence in operation: the synthespian 'is' Jack in the same way that Tom Cruise 'is' his character, even though the audience knows that the meanings of 'is' are different in each instance and constantly changing across edits. What remains continuous across the cut is 'Jack': the name and, compositionally, the deixis that points to actor and synthespian alike as object of the proper noun. Like most big budget films, *Oblivion* went through an extensive period of previsualization. In this case, Cruise was aboard the project from early on, meaning that his likeness could be used to fill in the space labelled Jack in the storyboards. I believe the key term here is 'modelling'.

I noted above that the 3ds Max model of the skytower was used both for construction (recall that the proprietor of 3ds Max, Autodesk, is dominant in the architectural and CADCAM software sectors) and for CGI. To that extent, both the CGI and the physical set are projections of the same model. We have seen that the CGI Jack is generated from a model derived from Cruise;

18 Stephen Prince, *Digital Visual Effects in Cinema: The Seduction of Reality* (New Brunswick, NJ: Rutgers University Press, 2012).

19 Julie Turnock, A *Plastic Reality: Special Effects, Technology, and the Emergence of 1970s Blockbuster Aesthetics* (New York: Columbia University Press, 2015).

and we know by the end of the film that 'our' Jack is not the only Jack, that he is replicable and indeed a replicant of a lost original from which these identical variants derive. The resolution of the final scene notwithstanding, the question raised concerns where the identity of these identical iterations derives from, and the answer must be: from the model. What does this tell us about the status of the human in contemporary effects movies?

One way to find out is to look at related effects in earlier periods. At the dramatic peak of *This Island Earth* (Joseph M. Newman, 1955), doctors Cal Meacham and Ruth Adams arrive at the underground city on the planet Metaluna by flying saucer (1'15"). Having flown over a miniature set of some complexity in the establishing shot, the slightly translucent animated saucer, along with smoke effects and miniature pyrotechnics, is superimposed on a painted matte. In the next shot, the characters traverse a set where it is frankly impossible to discern whether the backgrounds in gaps between set furnishings are painted backdrops or rear-projected animations. In the third shot, an animated element of the painted matte shows them descend to the surface, where a practical element is matted in, through which the actors appear, traverse the lower part of the screen, and begin to pass behind the painted area when a fourth cut takes them back into a physical set. As in *Oblivion*, the saucer appears as miniature, as animation, and as practical set, but unlike *Oblivion*, the human figures remain themselves. This is the case too during an earlier sequence aboard the saucer, when the Earth scientists are transformed to allow them to breath on the alien planet. Here the actors remain in costume while images of musculature, blood vessels and skeletons are superimposed on them. Though threatened with a brain-washing device, they remain entirely themselves throughout the adventure.

Freedom of thought was a major ideological structure of Cold War America, a war evoked by the central plot device of atomic power for energy and weaponry (I am grateful to James Williamson for this observation). While self-identity is sacrosanct in American individualism, the alien artefacts and alien planet can be presented as alien not only in design but in the radically different dimensionality of mattes and cell animation. Yet as J.P Telotte suggests, flying saucer invasion movies have the unsettling characteristic of "envisioning *us* as versions of the alien other that our SF films were at the time presenting as so threatening."[20] The explicit message of these films concerns the right to defense; less explicit is the anxiety over

20 Jay P. Telotte, "Animating Space: Disney, Science, and Empowerment," in *Science Fiction Studies*, (35)1, (February 2008), p. 104; http://www.depauw.edu/sfs/backissues/104/telotte104.htm (accessed 9 October 2019).

the conduct of the Cold War and the apocalyptic potential of nuclear energy. Beyond that again lies the fear that human exteriors might house the alien other, the tactic of the Metalunan Exeter. To be as you seem is important in this paranoid moment in US history. In certain respects, this trope, whose classical expression is *The Manchurian Candidate* (Frankenheimer, 1962), rewrites the 'passing for white' narrative and the less explicit beginnings of border paranoia focused on the illegal alien.

The coincidence of appearance and identity is then both axiomatic and the object of paranoia in the mid-1950s. It is the internal, however, that matters in *Oblivion*, as it does in a series of recent films I have taken to calling the irreality sub-genre, among them *Deja Vu* (Tony Scott, 2005), *Next* (Lee Tamahori, 2007), *Knowing* (Alex Proyas, 2009), *Inception* (Christopher Nolan, 2010), *The Adjustment Bureau* (Georges Nolfi, 2011), and *Source Code* (Duncan Jones, 2011). In *Source Code*, for example, protagonist Colter Stevens repeatedly awakes on a train where his task, it transpires, is to rewrite history while wooing the girl, his mind housed in the borrowed body of a deceased passenger. The actual Colter Stevens meanwhile exists in what becomes a wholly different timeline. The world is in effect reprogrammable. In *Oblivion*, Jack has effectively taken on the task of reprogramming himself in order to release an authentic self that, however, turns out to be anything but his unique possession. Is it not then quite correct that he should be supplanted by a digital double, even before the titles have rolled?

If the various Jacks are each an iteration of a single model or, as Paul Virilio might put it, an accidence of a single substance, then we are in the presence of a General Accident occurring at the level of humanity, once it has been reconceived as the serial product of codes that, while the efficient cause of human being, are extraneous to its instantiations. Whether those codes are genetic, as in the case of clones, or digital, as in the case of synthespians, is of marginal importance to the principle of code as substance and empiria as accident once it has been ascertained that the code is primary and the extant a derivation from it. To this extent, Adorno is right to assert the excess of the truth of art to its empiria but wrong in that, since he wrote in the 1960s, the relation of what presents itself in art to what is discoverable in it has been reversed. The apparent is mere epiphenomenon, a shape-shifting phenotext of non-identity beneath which lies a genotext comprised of symbols, organized according to a logic that supervenes over the truth-claims of the actually existing. The scriptwriter's axiom, that character generates story, is no longer true of the high-concept film: story generates character; not only as a heuristic device for maximally efficient entertainment but as ontological axiom of its account of human being. To be human is to be narrated, from a source — DNA or other determinations — beyond the human.

This rather different mode of determinism expresses itself in an ideology of freedom: Jack liberates himself from the lie of subordination to his authors in the *dénouement*. But this ideology masks the true fatalism of the narrative structure. It is as if we were returning to the debate between predestination and free will in mediaeval theology, with only the grave of chaos and emergence to protect us from the determinisms of scientific rationalism and bio-genetic, digital, and bio-political or protological programming. Against them, we have only the individual celebrated in *Oblivion*. This individual is not only challenged by lack of uniqueness — permitted by a flaw in programming that allows the various Jacks to recover his repressed memories — nor even by the half-exposed truth that all individuality is a social construct but by the incipient failure of the individual as social model. Cited favorably in Berardi's *Soul at Work*, Alain Ehrenberg argues that the slow collapse of disciplinary and prohibitive regimes forced individuals to take on responsibility for their lives, the etiology of the modern form of depression that "manifests itself as a pathology of responsibility, dominated by the feeling of inadequateness. The depressed individuals are not up to the task, they are tired of having to become themselves."[21] Thus the Horatian gesture of self-sacrifice that closes the second act of the film not only accomplishes the ideological task of shoring up the individual as sociological foundation, in accordance with Hollywood ideology, but also offers the individual an exit route from the unbearably compulsory nature of individuality. At the same time, the third act's revelation of another Jack exposes the realist narrative that has just concluded as the mask of another narrative whose drivers are no longer realist but naturalist: biological inheritance as in Emile Zola, or social programming as in Frank Norris.

The questions of identity and the non-identical that formed the initial problematic of this paper thus recur in a rather different form in contemporary effects movies like *Oblivion* when compared to the production of individuality in early effects films like *This Island Earth*. This is not, I believe, a function of the transition from analog to digital production. Rather, that transition itself is premised on a worldview according to which the mathematical logic of *nature* revealed by scientific enquiry has been transposed into an account of what it is to be *human*. The statistical method of what was already being called social physics in the era of Quetelet's *Sur l'homme*

21 Alain Ehrenberg quoted in Franco "Bifo" Berardi's *The Soul at Work: From Alienation to Autonomy*. Preface Jackson E. Smith (Cambridge, MA and London: MIT Press, 2009), p. 99.

et le développement de ses facultés, ou Essai de physique sociale of 1835 moved swiftly from disinterested anthropometry, the ancestor of motion capture and the modelling techniques employed in *Oblivion*, to become the basis for the new managerial and probabilistic governmentality of the twentieth century and the societies of control and protocol evoked by Deleuze[22] and Galloway.[23] To the extent that generality or law — and as a consequence being human on the contemporary model of individuality — is premised on this statistical, numerically encoded formation, the near-seamless digitization of Jack/Cruise is the opposite of an illusion. It is strictly the appropriate aesthetic for depicting a state of being that is indistinguishable from arithmetic. It is the realist claims of films like Haneke's *Caché* (2005) that offer illusions, such as that individual racism is shocking in a France programmed for structural racism.

Jack's discovery that he is not only not unique but programmed is the kind of formal freedom addressed in the Marcusean slogan 'Freedom is the condition of liberation'.[24] To know that you are programmed is not part of the program and therefore forms the formal grounds on which liberation from the program can be launched. We are not yet out of the woods. And — at the 32-minute 37 mark — nor is Jack, who has snuck off radar to take a little R&R at a lakeside cabin (at a physical location in Iceland). Just as he is called away, he splashes a few drops of water towards something unseen, asking 'You gonna miss me?'. The reverse angle shows us a brown trout flicking its tail over the sandy lake bed. With the exception of a stray dog that Jack scares away from a downed clone, this is the only animal we see in the film. The shot is doubly magical: returning the angle of Jack's point of view, his look arrives at the very spot where, in the previous shot, the camera must have been. There is, in the cabin sequence (and its reprise in the third act), absolutely nothing of the bravura camera movement we observed in the 180-degree vertical rotation displaying the skytower's aerial location.

22 Gilles Deleuze, "Postscript on the Societies of Control," in *October: The Second Decade, 1986-1996*, Rosalind Krauss, Annette Michelson, Yves-Alain Bois, Benjamin H.D. Buchloh, Hal Foster, Denis Hollier, and Sylvia Kolbowski, eds, (Cambridge, MA: MIT Press, 1997), pp. 443–447.
23 Alexander R. Galloway, *Protocol: How Control Exists After Decentralization* (Cambridge, MA: MIT Press, 2004).
24 This expression of Marcuse is taken from: Slavoj Žižek's *In Defence of Lost Causes* (London: Verso, 2008), p. 196. It conforms to the argument made in the second chapter of *One Dimensional Man*, notably where Marcuse argues that "the dialectical logic insists, against the language of brute facts and ideology, that the slaves must be free for their liberation before they can become free, and that the end must be operative in the means to attain it." Herbert Marcuse, *One-Dimensional Man Studies in the Ideology of Advanced Industrial Society*. Second Edition. With an Introduction by Douglas Kellner (London: Routledge, 2002), p. 44.

Instead, we are given a reduction to the humanist camerawork of classical *découpage* but with one magical difference: there remains the implication that we have just been observing Jack from the standpoint of the fish.

What exactly is the status of the animal here? The shot is odd enough to recall that extraordinary moment in *La Règle du jeu* (*The Rules of the Game*, Jean Renoir, 1939) when, just before the fatal final act, the camera pans down a classical statue in the gardens to single out a croaking frog. Like Renoir's image, the shot of the trout confronts human with inhuman. The difference is that Renoir's shot is unmotivated by any human observer: the frog acts as chorus to the unfolding tragedy, an external voice of unreason beyond the patterned doings of the *château*, providing a satirical commentary on the many statues and automata that populate Renoir's film but from a source outside it. In *Oblivion*, the trout is the guarantee of Jack's reality, up to this juncture accompanied only by Victoria, his comms officer and lover, with her own secrets to hide, and the distant, heavily mediated voice of their mission controller. If, as we discover, these others are also not what they seem, then this encounter is all there is to mark the authenticity of Jack's existence.

It is, to that extent, a sentimental trope, not only for its echoes of a depoliticized environmentalism but in the terms that James Joyce's Stephen Dedalus purloined from George Meredith: "the sentimentalist is he who would enjoy without incurring the immense debtorship for a thing done."[25] The diegesis has all but obliterated the human race: this little fish is to restore humanity after the apocalypse. The term 'debt,' however, opens the true abyss of temporality forgotten in *Oblivion*. In Benjamin Franklin's famous phrase, the only certain things we know about the future are death and taxes. But in the irreality film, to which this film belongs, death is not certain and not certainly future; and while taxes can be legislated away, what cannot is the engine of the digital finance economy: debt. Having mortgaged our future earnings to pay for present possessions, individually and socially, we owe our existence to a future whose bills we have no means or intention to pay. And yet the machinery of debt ensures that we pursue our employments, beyond depression, even to the suicidal point of sacrificing the planetary future, with an eye to that immanent yet perpetually deferred lonely moment of the last instance when all debts will be cancelled. Such is the messiah of fiscal prudence.

Because it does not acknowledge the billion dead of its fictional universe, this scene, which might have had the absurdist, existential profundity

25 James Joyce, *Ulysses* (Harmondsworth: Penguin, 1968 [1922]), p. 409.

of Renoir's frog, invites us to consider not what apocalypse might mean but — sentimentally — how it would feel to survive it. To its credit, it reveals that all our debts are payable to the Last Man but that the Last Man is a fiction and a con. *Oblivion* speaks from a true account of the human condition and its temporalities in the shadow of the global financial crisis five years after it began in 2007, at a moment then when the state of crisis had begun to appear as a permanent condition. The film's weakness is that it does not speak to crisis but only beyond it. Its fatalism is heartfelt where it should be cold, and its optimism saccharine where it should be passionate.

Jack's situation as clone or replicant of a model that only recurs as a form of *déjà vu*, an irruption of self-consistency of action and its trace where no consistency should be, opens a gulf between code and its expression, substance and accidence. *Oblivion* is not an allegory of the situation that gives it birth but, like Jack, an accident of its transmission. The world of the early twenty-first century is a world run by cyborg corporations whose sole motivation is profit, who lack all sense of communality and shame, and who are prepared to sacrifice the happiness of the living and the lives of the unborn. It is not a world for heroes but for commoners. If, as Gramsci believed, factory discipline was responsible for evolving a new communal, industrial humanity; perhaps our digital protocols may be the source of a posthuman commons. The opportunity for *all* the Jacks to coordinate is lost in the narrative drive here, and with it the reverse apocalypticism proposed in Walter Benjamin's theses on the philosophy of history, where he asserts the duty of those who live to be the posterity to whom those who are dead looked for their justification. In accordance with Hollywood's ideological individualism, Jack's narrative task is to save — and in so doing to realize — himself. As creditor of all the debts we owe to the future, he is in a position to pay back the dead and fails not only to do so but even to recognize that he has that responsibility.

The saving grace of the film is that it does establish the hollowness of this ungenerous solipsism. By confounding practical and digital elements as modelled by the same underpinning codes, it cedes the programmed world not as sociological satire but as anthropological situation. At the same time, it offers, in the narrative device of a glitch in programming, the vision of an identity founded in the mistaken recognition we experience in underwriting Jack's existence from the inhuman perspective of the fish. To the extent that Jack is posthuman, it is because he is an intricate oscillation between the non-human and the inhuman. On its own, this oscillation would be merely cynical, in Sloterdijk's sense. Establishing his diegetic world as doubly immaterial, and establishing Jack, Cruise, and all their avatars as accidents, the special effects, in their internal contradiction, do not indicate how to be posthuman – a merely chronological difference. The temporality of these effects is not Bloch's[26] not-yet so much as it is the not-now of potential[27] undercutting the present on which self-identity depends. Not 'post,' therefore, the otherwise-than-human is a permanent potentiality opened up in the moment when the human appears to be most fully subsumed within its programming.

Bibliography

Adorno, Theodor W. *Aesthetic Theory*, edited by Gretel Adorno and Rolf Tiedermann, translated by Robert Hullot-Kentor (London: Athlone Press, 1997).

Aristotle. *Analytics. Book II*, translated by Robin Smith (Indianapolis: Hackett, 1989).

Balio, Tino. *Hollywood in the New Millennium* (London: British Film Institute, 2013).

Berardi, Franco "Bifo." *The Soul at Work: From Alienation to Autonomy* (Cambridge, MA and London: MIT Press, 2009).

Bloch, Ernst. *The Principle of Hope*, translated by Neville Plaice, Stephen Plaice, and Paul Knight (Cambridge, MA: MIT Press, 1986).

Brown, William. "Man Without a Movie Camera – Movies Without Men: Towards a Posthumanist Cinema?" In *Film Theory and Contemporary Hollywood Movies*, edited by Warren Buckland (New York: Routledge, 2009): 66–85.

Cubitt, Sean. "Phalke, Méliès and Special Effects Today," *Wide Angle* 21, no. 1 (January 2001): 114–130.

Deleuze, Gilles. "Postscript on the Societies of Control." In *October: The Second Decade, 1986–1996*, edited by Rosalind Krauss et al. (Cambridge, MA: MIT Press, 1997): 443–447.

Doane, Mary Ann. "The Indexical and the Concept of Medium Specificity," *Difference: A Journal of Feminist Cultural Studies* 18, no. 1 (2007): 128–152.

Fordham, Joe. "Last Man Standing," *Cinefex* 134 (July 2013): 94-116.

Frege, Gottlob. *The Foundations of Arithmetic*, translated by John L. Austin (New York: Harper, 1953).

Galloway, Alexander R. *Protocol: How Control Exists After Decentralization* (Cambridge, MA: MIT Press, 2004).

Herzog, Amy, John Richardson, and Carol Vernallis, eds. *The Oxford Handbook of Sound and Image in Digital Media* (Oxford: Oxford University Press, 2014).

26 Ernst Bloch, *The Principle of Hope*. 3 vols, trans. Neville Plaice, Stephen Plaice, and Paul Knight (Cambridge, MA: MIT Press, 1986).

27 Paolo Virno, *Déjà Vu and the End of History*, trans. David Broder (London: Verso, 2015).

Joyce, James. *Ulysses* (Marmondsworth: Penguin, 1968 [1992]).
Marcuse, Herbert. *One-Dimensional Man Studies in the Ideology of Advanced Industrial Society* (London: Routledge, 2002).
Petley, Julian. *Capital and Culture: German Cinema 1933-45* (London: British Film Institute, 1979).
Prince, Stephen. *Digital Visual Effects in Cinema: The Seduction of Reality* (New Brunswick, NJ: Rutgers University Press, 2012).
Rodowick, David Norman. *The Virtual Life of Film* (Cambridge, MA: Harvard University Press, 2007).
Salt, Barry. *Film Style and Technology: History and Analysis* (London: Starword, 1983).
———. "The Shape of 1999: The Stylistics of American Movies at the End of the Century," *New Review of Film and Television Studies* 2, no. 1 (May 2004): 61–86.
Schatz, Dennis. *World of Inventors: Thomas Edison* (San Diego: Silver Dolphin Books, 2009).
Telotte, Jay P. "Animating Space: Disney, Science, and Empowerment," *Science Fiction Studies* 35, no. 104 (February 2008): 48-59.
Turnock, Julie. *Plastic Reality: Special Effects, Technology, and the Emergence of 1970s Blockbuster Aesthetics* (New York: Columbia University Press, 2015).
Virno, Paolo. *Déjà Vu and the End of History*, translated by David Broder (London: Verso, 2015).
Williams, Raymond. *Keywords: A Vocabulary of Culture and Society* (London: Fontana, 1983).
Žižek, Slavoj. *In Defence of Lost Causes* (London: Verso, 2008).

About the Author

Sean Cubitt is Professor of Screen Studies at the University of Melbourne. His publications include *The Cinema Effect*, *Ecomedia*, *The Practice of Light*, *Finite Media*, and *Anecdotal Evidence*. Series editor for Leonardo Books at MIT Press, his current research is on political aesthetics, ecocritique, and practices of truth.

Envoi

16. The Effect of Miracles and the Miracle of Effects: Bazin's Faith in Evolution

Dudley Andrew

Abstract

Special camera effects such as superimpositions, by their absence in Dreyer's *Ordet* as much as by their bold presence in Hitchcock's *The Wrong Man*, crystallize the chemical constitution, as it were, of film, making it clear that in its molecular makeup and in varying ratios, cinema binds realism to magic, evidence to illusion, and document to poetry. There must be a 'process' by which the standard registration of straight cinematography is transcended by the apparent miracle achieved by the effect. A materialist approach to film history would calculate its evolution by tracing the improvements in both realism and effects realized progressively film after film, year after year. But an aesthetic attitude can discount this linear progression and see in a miraculous film like *Ordet* a throwback strategy that insinuates miracles by primitive techniques of lighting and sound. Comparing *Ordet, The Wrong Man*, and Bresson's *A Man Escaped*, all made within a year of each other and all discussed by Bazin and Godard, one can recognize a spectrum of 'faiths' in the image and in art. Decades later, Godard's *Histoire(s) du Cinéma* subsumes such films and the beliefs they engendered within a videographic essay where superimposition is omnipresent and not special at all.

Keywords: miracle, faith, superimposition, video, evolution, reception, Godard, Bazin, Cocteau

What was it that drew André Bazin and Jean Cocteau together in the late 1940s, the one devoted to the magic of poetry and to astonishing theatrical effects, and the other an advocate of neorealism? They must have met at the first Cannes festival where Bazin wrote in praise of *La belle et la Bête*

Lefebvre, M. & M. Furstenau (eds.), *Special Effects on the Screen: Faking the View from Méliès to Motion Capture*. Taylor & Francis Group, 2022
DOI 10.5117/9789462980730_CH16

(Beauty and the Beast, Jean Cocteau, 1946). Reviewing it, he mentioned *Le sang d'un poète* (*The Blood of a Poet*, Jean Cocteau, 1932), a film he had already admired "because Cocteau only made use of tricks directly coming from the shooting," not from technological add-ons.[1] Nearly all the tricks of *La belle et la Bête* (and some are required in every fairy tale) were profilmic, made neither in the lab nor in the camera but within the photographed space and event. (Recall those arms holding torches that extend from the walls and coax both Belle and the viewer to enter.) Still, compared to the 'icy patina' of this beautiful adult fantasy, Bazin preferred the "slightly clumsy" effects of Cocteau's childlike first film. The rough edges around its trick shots document rather than hide the effort to bring off something "special."[2]

As for magic, Cocteau downplayed it. Of *La belle et la bête*, he said:

> I chose the least magical of all fairy tales — by which I mean the one that would take least advantage of the opportunities provided by modern cinematographic techniques [...]. In place of what people usually consider supernatural and poetic, I would adopt a kind of realistic style, avoiding imprecision and effects of mist, *superimposition* and other *outmoded* techniques.[3]

Later, in writing about *Orphée* (*Orpheus*, Jean Cocteau, 1950), he said "The closer you get to a mystery, the more important it is to be realistic." Had he been listening to Bazin?

Or was it vice versa? In 1955, Bazin laid out a brief evolution of the science-fiction film, which was just then undergoing something of a renaissance.[4] This genre had slumped after a highpoint with *Frankenstein* (James Whale, 1931) and *King Kong* (Merian Cooper & Ernest Schoedsack, 1933). He argues that since the mid-1930s, the realist direction of cinema as a whole had not been beneficial to traditional science fiction, which had always been based on visual effects. By the 1950s, the most mature examples of science fiction had begun to appeal more to the mind than to the eyes. Bazin links Howard Hawks's *The Thing* (1951) and Robert Wise's *The Day the Earth Stood Still* (1951)

1 André Bazin, "Vie et mort de la surimpression," part 2, in *L'Ecran Français*, (9) (29 August 1945); reprinted in André Bazin, *Écrits complets*, vol. I (Paris: Macula, 2018), p. 128.

2 André Bazin, "*La belle et la bête*," in *Le Parisien Libéré*, (691) (1 November 1946); reprinted in André Bazin, *Écrits complets*, vol. I (Paris: Macula, 2018), p. 207.

3 Jean Cocteau, *The Art of Cinema* (London: Marion Boyars 1992), p. 141.

4 André Bazin, "La science-fiction au cinéma doit faire appel au 'fantastique mental'," in *Radio-Cinéma-Télévision* (284) (26 June 1955); reprinted in André Bazin, *Écrits complets*, vol. II (Paris: Macula, 2018), pp. 1752–1753.

as offshoots of a post-Atomic bomb literary trend, headed by Ray Bradbury's *The Martian Chronicles* (1950).[5] He approves the way the "marvelous" can now be generated by strategies of "omission which are more effective than any description." The most thrilling effects are those provoked by our minds in confrontation with what is limitless and unimaginable, with the sublime. Watching *Them*! (Douglas Gordon, 1954), he found himself spooked by the single still photograph that introduces such menacing creatures, but when they are finally displayed in their "concrete realization," it is highly "disappointing."[6] This does not surprise Bazin, because, as audiences grow increasingly sophisticated, contemporary special effects need to become *spectacularly special* to do for us what *Le voyage dans la lune* (*The Trip to the Moon*, Georges Méliès) did for its 1902 audience or what *King Kong* pulled off in 1933. The visual effects of *Them!* were not up to its quite effective script.

Whereas genres have often been treated like flowers to be identified and catalogued by the botanist, in fact they develop in tandem with viewer competence. An optimist, Bazin believed audiences of the 1950s had surely become savvy enough to recognize drama manufactured at the editing table or in the animation studio rather than captured on location. But do film audiences actually evolve? Their needs generally have not. For instance, spectators of sci-fi films in the twenty-first century, just as those in the 1950s or in 1898, demand both verisimilitude and thrills. But clearly *genres* evolve. Bazin treated them as organisms that grow, mutate, cross-pollinate, and sometimes die. And they do so in relation to growing audience proclivities and skills, these being distinct from immutable needs. What counts as "realistic" and what provokes credible illusions should change with each subsequent generation, even if the demands for both realism and for effects remain the same.

A bit like addicts habituated to their current dosage of some drug and requiring more of it just so the same effects continue, spectators hunger for enhancements to be added to whatever techniques of illusion are current. Science fiction and horror films must startle viewers in novel ways just so they can experience the same level of shock these genres have always delivered. Hence the recurring quest for "greater realism" in technology (depth, sound, color, three-dimensions). And hence the research and development units that studios maintain to come up with techniques capable of producing

5 Ibid.

6 André Bazin, "Les Monstres attaquent la ville (*Them!*)," in *Radio-Cinéma-Télévision* (284) (26 June 1955); reprinted in ibid., pp. 1753–1754.

effects that, as the phrase connotes, do something "special," something marvelous. Reviewing the 1955 film *It Came from Beneath the Sea* (Robert Gordon) — which he found ludicrous — Bazin noted that "recognizing the obsolescence or insufficiency today of the traditional giant rubber octopus that has been used and reused for twenty years, the Hollywood studios this time turned principally to models and to frame-by-frame animation. But the feebleness of this technique will strike even the least informed viewer."[7] Evidently, the "*trucage*" of Méliès and even of *King Kong* was "insufficient" for spectators in the mid-1950s. And even the best effects of the 1950s would need to be surpassed. Whatever we might think of the 2005 *King Kong* (Peter Jackson), today, more than a decade later, it is no longer quite realistic or spectacular enough; were he to update this classic, Peter Jackson would want to reboot it with his 48fps delivery system just for starters.

We can get some sense of Bazin's view of audience maturation by turning to one of his earliest pieces, "Vie et mort de la surimpression."[8] Examining a particular special effect — the portrayal of ghosts and phantoms — across a few decades in film history, he writes rather like an art historian, someone like Ernst Gombrich or Michael Baxandall. In films just as in paintings, one can observe a constant drive for increased realism (Gombrich's main concern), but one notes as well the more variable drive of ambitious artists to flatter or occasionally befuddle the newest generation of sophisticated viewers (this has been Michael Baxandall's project). Bazin was sensitive to both drives when he reacted to the clear "evolution" of filmmaking that was exposed suddenly in 1945 when all at once hundreds of Hollywood films made during the war were finally available on Paris's screens. In the interim, while occupied by the Nazis, France had produced a number of intelligent films venturing into the realm of dreams and the marvelous (*La main du diable* [*The Devil's Hand*, Maurice Tourneur, 1943], *La nuit fantastique* [*The Fantastic Night*, Marcel L'Herbier, 1942], *Les visiteurs du soir* [*The Devil's Envoys*, Marcel Carné, 1942], and *Sylvie et le fantôme* [*Sylvie and the Ghost*, Claude Autant-Lara, 1946],which was in production as he wrote). However, their manner of evoking this realm seemed to him a crude throwback next to *Here Comes Mr. Jordan* (Alexander Hall, 1941) and *Our Town* (Sam Wood, 1940), which he found decidedly more subtle.[9]

7 André Bazin, "*Le Monstre vient de la mer*: qu'il y retourne!" in *Le Parisien Libéré* (3648) (2 June 1956); reprinted in ibid., pp. 1973–1974.

8 See especially Michael Baxandall, *Patterns of Intention: On the Historical Explanation of Pictures* (New Haven, CT: Yale University Press, 1985).

9 André Bazin, *Qu'est-ce que le cinéma?* vol. 1. *Ontologie et langage* (Paris: Éditions du Cerf, 1958), pp. 27–30.

But to what did he consider current French examples to be a throwback? To the German and Swedish silent films, praised in their day for expanding cinema's purview beyond the ordinarily visible and into the fantastic, the supernatural, the oneiric. For in the silent era, such films did exactly that, Bazin claims. But viewers soon became accustomed to the simple superimposition of a translucent figure floating atop the more solid image of the character being haunted (*Körkarlen* [*The Phantom Carriage*, Victor Sjöström, 1921] is the titular example he cites). Silent filmmakers had recourse as well to slow motion and occasionally to negative footage in order to "*signify*" dreams rather than to "represent" them (his italics).[10] This was consistent with the pictorialism of much European art cinema in the 1920s; but after sound had brought the floating image down to earth, so to speak, such techniques would appear anachronistic. In Bazin's view, following Roger Leenhardt, after sound, there would be a retreat of pictorialism and a more general emphasis on *mise en scène*, altering the constitution of every film and of the cinema *in toto*.

To adjust to this overall evolution of film language toward a more natural look, special effects needed to be scaled back, to become less obtrusive. This is what his article claims to observe in the Hollywood films suddenly on view after the war. Curiously, he speculates that this naturalization of the oneiric occurred first in Hollywood, thanks to America's adoption of popular Freudianism which encouraged more psychologically plausible representations of dreams. Grotesque distortions on the screen's surface were replaced by scenes suggesting the supernatural with a certain amount of visual finesse. He believed that Hollywood responded to, and abetted, cinema's generally more mature relation to the spiritual. Figures emerging out of the unknown became "interiorized" and shed the spectacular appearance of their 1920s counterparts. Less literally sensational, modern films that involved the spiritual dimension, if not including actual 'spirits', had become speculative and thoughtful. In a particularly telling parenthesis, he quipped: "The scenario of *Here Comes Mr. Jordan* possesses a subtlety which a critic of the existentialist stripe at the café de Flore would not be ashamed to claim."[11] Cinema had matured to attain the stature of philosophy, or at least of a popular brand of philosophizing.

10 The italics were dropped when the essay was reprinted (in shortened form) in the first edition of *Qu'est-ce que le cinéma*? vol. 1. They have been reintroduced for the *Écrits complets*. See Bazin, *Écrits complets*, vol. I, op. cit., p. 127.

11 Bazin excised this parenthesis when he reprinted the essay in vol. 1 of *Qu'est-ce que le cinema?* op. cit., p. 112.

While a more naturalized supernatural may have become its norm, Hollywood still often needed to go beyond suggestion; certain genres required that hypnogogic illusions and dreams be displayed. This is when Bazin notes that the old standby methods had to be superseded, for audiences always look for the *tricks* in "*trucage*." Hence new tricks must be continually concocted. In the 1940s, the Dunning Process was the most advanced way to lay two images onto the same frame while adjusting the transparency of one to the other at will. Bazin delights in explaining the apparatus that produces such effects,[12] but his key point is that such advances in effects are demanded by the equally important gains in cinematographic realism (faster film stock, better lenses, sync sound) evident in the films of Wyler and Welles, his touchstone examples. This new baseline for realism, when tied to an across-the-board increase in the public's knowledge of how cinema is fabricated, dooms traditional superimposition to quaintness, no matter how beautifully it functioned in older movies. Bazin takes superimpositions to be rather like haloes in medieval and early renaissance religious paintings. Haloes lost their appeal after the arrival of perspective and were eventually abandoned. Although they persist in certain nineteenth-century paintings and can even be found today, they appear anachronistic.

The disappearance of haloes, does not, however, mean the loss of a belief in saints, which is tied to the myth of judgment after death. Many desires, needs, and intractable conundrums are constants across human experience and are expressed via "myths." We know how much this meant to Cocteau (*L'éternel Retour* [Cocteau script directed by Jean Delanney, 1943], *Orphée, Les chevaliers de la table ronde* [*Knights of the Round Table*, theatrical play of 1937], etc). While I doubt that Bazin would have agreed with the director's famous quip "I always prefer myth to history," he certainly adhered to Cocteau's distinction between these. He deliberately deployed the term "myth" in the titles of three of his most famous pieces, all of them dealing with history: "Le mythe du cinema total,"[13] "Le mythe de Monsieur Verdoux,"[14] and "Le cinéma soviétique et le mythe de Staline."[15] While myths may come into existence at a precise moment, they are

12 Julie Turnock at the University of Illinois, a specialist in Hollywood effects, believes Bazin may have been misinformed about the process used in *Our Town* and may not have fully understood what he terms the Dunning process. But she notes that his point stands nonetheless.

13 Bazin, "Le mythe du cinema total et les origins du cinématographe" in *Critique* (6) (Nov. 1946); reprinted in Bazin, *Écrits complets,* vol. I, op. cit., pp. 202–204.

14 Bazin, "Le mythe de Monsieur Verdoux" (1948) in *La Revue du Cinéma* 19 (Jan. 1948); reprinted in André Bazin. *Écrits complets,* vol. I, op., cit., pp. 332–341.

15 Bazin, "Le cinéma soviétique et le mythe de Staline," in *Esprit*. Nouvelle série (170) (August1950), pp. 210–235; reprinted in André Bazin. *Écrits complets,* vol. I, op., cit., pp. 665-676.

effectively immutable. Charlie Chaplin understood this perfectly when he put the mythical "Charlot" into play in distinct ways from one decade to the next. While a myth remains universal, the discourses within which it circulates and the techniques that express it have lifespans. These spring to life, grow decrepit, and may expire. Genres provide obvious cases of the historical cycles of an overriding myth. Bazin traced the evolution of the Western myth (the desert and the garden, the sacrifice of the lone hero for the community, the birth of law in savage space) as it moved first from popular literature to early cinema and then as it went through a series of recognizable phases rising to a classical form in *Stagecoach* (John Ford, 1939) and moving toward a baroque form that culminated in the decadence of some of its 1950s expressions. He would have been saddened but not shocked to see its steep decline after his death but elated by its intermittent and often flamboyant appearance in Italy, Spain, Germany, India, and other national cinemas.

Now, what is true of genres is even more true of techniques that convey elements of myth, such as rear projection used during chases on horseback, to stick with the Western genre and myth. Indeed, because techniques are mechanical, they are even more susceptible to transmutation and supersession than discursive phenomena like genres and styles. Lighter cameras, improved crab-dollies, mobile units (including helicopters), and a generation of steady-cams would come to make rear projections anachronistic, at least until digital compositing resurrected the technique of layering images to produce the illusion of homogeneous space, no matter how fragmented the recording process may have been.

"Resurrected." The word defies evolution, for strictly speaking, a species, once defunct, does not return. Its function may *devolve* to some later species that develops in the place it vacated, but the earlier species itself lies buried, becomes fossilized. This is what Bazin claimed about the superimposition. And this is what, in a clever twist to the argument, Daniel Morgan has challenged. If any technique might conceivably defy evolution and be resurrected, it ought to be one whose function has always been to figure resurrection itself by bringing back on-screen phenomena such as ghosts, spirits, and the undead. Morgan would "open Bazin" to the digital era and to new technologies, some of whose techniques are indeed resurrections of what Bazin's evolutionary account had foreclosed. This is the thrust of the cleverly titled "The Afterlife of Superimposition," Morgan's extension of Bazin's original Darwinian essay.[16]

16 Daniel Morgan, "The Afterlife of Superimposition," in Dudley Andrew, ed., *Opening Bazin* (New York: Oxford University Press, 2012), pp. 127–141.

Morgan justifies updating Bazin because the latter's flexible realism allowed him to appreciate numerous genres, each of which maintains some rapport to the medium's photographic ontology, but each of which takes this in its own particular direction. Of course, Bazin had personal tastes, but he normally did not rank genres or styles.

In fact, he holds a firm belief that all styles are historically contingent, that cinema (even realist cinema) is continually being transformed in response to its circumstances. Rather than advocating a single kind of realist cinema (based on a long-take, deep-space aesthetic), he sees danger when filmmakers cling to or fetishize specific techniques. Each realist style will be superseded by subsequent developments both within and outside of cinema. The basic feature of Bazin's account of realism is to treat it not as a doctrine but as a way of understanding the interaction between style and medium.[17]

Morgan looks for the rebirth of the superimposition in the 1990s when circumstances have indeed changed thanks to the digital; the technique is fundamental to Godard's *Histoire(s) du cinema* (1988–1998), a key section of which Morgan analyzes. In Section 4a (called "Control of the Universe"), the British voice of Alfred Hitchcock runs alongside Godard's gravelly monotone and the melancholic strains of Giya Kancheli's score. These three layers of sound accompany images that are even more densely layered: Hitchcock's face is superimposed atop scenes from his films that sometimes blink on and off as if being pulled slowly through a moviescope. Crucial for Morgan and for me, Godard selects a couple scenes in which Hitchcock himself aimed to superimpose something from "the beyond" onto a realistic drama of the present. Godard clinches this section with the denouement of *The Wrong Man* (Alfred Hitchcock, 1956), in which the image of the veritable killer gradually emerges on the screen while Manny (Henry Fonda) prays. Godard's voice suggests that the superimposition here answers the prayer in the same way that a prayer is answered in Carl Dreyer's *Ordet* (1955). *Ordet,* by the way, was made just a year before *The Wrong Man*; it baldly concludes on a miracle.

As is often the case, Godard here takes his cue from Bazin who extolled *Ordet* in January 1956, comparing it to the very best Hitchcock.[18] A year

17 Ibid., p. 133.

18 Bazin reviewed *Ordet* for three different journals. The longest appeared in *France Observateur* (5 January 1956) and has been translated in *Cinema of Cruelty* (New York: Seaver Books, 1981), pp. 26–30. But the one I am relying on here came out a few days earlier, in *Radio-Cinéma-Télévision* (313) (1 January 1956), p. 42.

later, he would flip the comparison, extolling *The Wrong Man* as the purest Hitchcock of all, stripped down, like *Ordet*, to the essence of cinema: rigorous and precise realism, pure black and white cinematography, and helpless waiting, i.e., cinema as the sheer unfolding of time.[19] In writing about *Ordet*, Bazin assures us that Dreyer is like a grandmaster who dominates the chessboard of the screen, not a square millimeter of which he does not control. Dreyer's minute realism exposes the precise rules of the Protestant society in *Ordet*, so that each gesture portends significance.

Yet, Bazin says, "this most perfectly planned of films still leaves one part open to chance," and because of this, the film's "precise — that is to say, violent — realism enters into a necessary and dialectical rapport with the spiritual universe."[20] Now this is just what Godard says of *The Wrong Man*. And Godard would have remembered that Bazin opened his review by comparing this Hitchcock film to Bresson's *Un condamné à mort s'est échappé* (*A Man Escaped*, 1956), another film of unstinting realism that leaves a tiny opening for the spiritual world to enter so as ultimately to liberate the prisoner from the plot as if by miracle. Godard's article on *The Wrong Man* appeared in *Cahiers du Cinéma* in June 1957, presumably after he had spoken (and argued) with Bazin, whose own review had come out a month earlier. For Godard, like Bazin, characterizes the plot as a "machine inexorably grinding on"; however, 40 years later, in the *Histoire(s) du cinéma*, he emphasizes instead the miracle that defies narrative logic. Now this miracle is nothing other than that of the superimposition grafting a divine plot onto the fatal plot that is working itself out on earth. After all, in David Hume's notorious refutation, and in the words of the prelates in *Ordet*, a miracle breaks the order of nature, rupturing its rules and its timeline, thereby rerouting nature's plan.

According to Godard, Hitchcock does nothing special here (i.e., this is not a special effect); he is simply doing what he is always capable of doing; like God ("taking control of the universe"), he uses superimposition to interrupt Henry Fonda's fate by putting the errant judicial process back on track. Hitchcock the director stands above all worlds, mundane and spiritual alike, manipulating everything for his (and our) pleasure and instruction.

19 André Bazin, "Le Faux Coupable," in *France Observateur* (366) (16 May 1957). Morgan (ibid.) correctly points out in a footnote to his own essay that Bazin's review of *The Wrong Man* makes no mention of this extraordinary superimposition. Perhaps this is because Bazin wanted to emphasize the "pure Hitchcock" style which in this case "delivers 'suspense' evacuated of all 'suspense', a pure waiting without catharsis."

20 André Bazin, *Ordet*, in *Radio-Cinéma-Télévision* (313) (1 January 1956), p. 42; reprinted in André Bazin, *Écrits complets*, vol. II (Paris: Macula, 2018), p. 1878.

One could go further in "naturalizing" this supernatural power of Hitchcock by recognizing in his narration the authority to which all directors — all auteurs — may implicitly lay claim. As Noa Steimatsky has recently described in detail, the superimposition solving *The Wrong Man* seems like a miracle largely because Henry Fonda prays to Christ just before the 'right man' is revealed.[21] Indeed, we see Fonda's shadow to the left of the icon of Christ who dominates the reverse shot of his gaze, as if Christ has taken over the narration. Then comes the famous superimposition... except that it is not a superimposition at all, but an extended lap-dissolve! This is hardly a special effect but rather a standard punctuation mark in fiction films since before Griffith, a way in which a director takes us to another place or another time.[22] Henry Fonda does not materialize his nemesis via special effect. If anything, Christ takes over the narration by initiating the dissolve to a simultaneous action elsewhere in the city that exonerates Fonda. Fonda has no vision; all he knows is that he prays and that later that night he is called to the police lineup where he comes face to face with his double. True, what was layered when he prayed becomes separated at the precinct; the superimposed faces that looked straight ahead (at us or at Christ) are pulled apart and shown in profile where they look at one another. But it is equally true that from their actual separation in the first instance (Fonda praying in his bedroom, the criminal walking the street to rob a store), they have been brought together in the precinct.

Morgan and Godard set Hitchcock up as an idol, a god whose mysterious power over his worlds and his audiences is held in awe. Bazin, as Morgan notes, finds such idolatry problematic; Hitchcock saves his soul through irony, black humor being a wedge that separates the director of the film from the director of the universe. The problem is that the irony in *The Wrong Man* is deadly serious, as Rose, Henry Fonda's wife, recognizes so clearly that she becomes catatonic.

Perhaps Bazin was drawn to this film because it was not a natural Hitchcock film; like a miracle, it rerouted Hitchcock's usual mechanism, his usual control. This is what William Rothman argues in *Must We Kill the Thing We*

21 Noa Steimatsky, *The Face on Film* (New York: Oxford University Press, 2016).

22 It must be mentioned that the first use of the term "*fondu*" in a French book on film technique prescribes it for ghosts: "*les brusques apparitions, les soudaines métamorphoses, obtenues par arrêt sont d'un effet moins saisissant que les substitutions graduelles et la progressive matérialisation de fantômes* [...] *On arrive à des résultats plus parfaits, en cinématographie, par la manoeuvre du diaphragme combiné avec une double impression. C'est ce qu'on appelle des vues* fondantes... *ou, pour abréger, le* fondu." Coustet, *Traité pratique de cinématographie,* (1) (1914), p. 76. Source found in Jean Giraud, *Le Lexique francais du cinéma des origines à 1930* (Paris: CNRS, 1958), p. 128.

Love: Emersonian Perfectionism and the Films of Alfred Hitchcock. Rothman latches onto the self-same moment while coming up with a distinct view.[23] He believes that in this film, and especially in this privileged moment, Hitchcock aims to stand back from his role as auteur so as to let God author the miracle that brings the right man to justice. I would put it this way: since all places and times are available to God's "eye," superimposition is his standard way of monitoring actions that humans see serially and from limited perspective. Rothman does not explore the theology at stake, but, unlike Morgan and Godard (whom he does not mention), he gives God a role distinct from and above Hitchcock's. Indeed, he suggests that Hitchcock so nominates God when, in the prologue he speaks to the audience to announce that this film, unlike all his others, comes directly from reality, that is, from the world as authored by God. Therefore, we should watch this film differently, without irony, and without applauding its director's skill at manipulating the world and the audience. The scene that produces "the right man" — the case in point — is then not a haunting superimposition as in a silent Swedish film (for Fonda does not feel the presence of his doppelganger), nor is it quite a standard narrative conjunction, transporting us to a parallel scene.

Whatever we may want to call the effect, whether superimposition or visual conjunction, it rises to a moral plane, because in the movement of a plot known only to God, the world rectifies what has been out of alignment; things come into phase. The optical work is precisely the phasing in of what had been blocked from view, the bringing into conjunction of two men with corresponding faces yet distinct destinies. This is one way of understanding the term "*rapprochement*" that Morgan rightly fastens on in Godard's vocabulary. *Rapprochement*, a key term for Bazin and Godard as well as for Baudelaire, Breton, and Pierre Reverdy, suggests the secular miracle of a metaphor, the flash of understanding when distant things are taken for — or as — each other. If Bresson, Dreyer, and Hitchcock (three directors with distinct religious backgrounds) all rigorously restricted themselves to laying out the intricate mechanism of the mundane, it was to "leave one part open to chance" when, in the grace of a moment, something comes into phase and is registered in the face on screen, or in the lighting, or in some stylistic trace like the vestiges of superimposition.

Another way to understand *rapprochement*, Morgan's explicit way, is to equate it with "juxtaposition." As a rhetorical effort of discourse, not

23 William Rothman, *Must We Kill the Thing We Love? Emersonian Perfectionism and the Films of Alfred Hitchcock* (New York: Columbia University Press, 2014), pp. 126–140.

an artistic achievement of perception.[24] A superimposition is, of course, a literal rapprochement, a forcible bringing together of distinct images, rather like a simultaneous version of the kind of sequential montage associated with Eisenstein's writings. Superimposition of voice atop image as well as of images atop one another constitute the method and fabric of Godard's *Histoire(s) du cinéma,* a truly major intervention in twentieth-century history, in the history of cinema, and in audio-visual language. By insisting on the simultaneous presentation of two or three (sometimes more) sources of information, Godard re-animates a moribund technique. Importantly, he employs it not in a film but in an extended (6-hour) videographic essay. Whereas the kind of superimpositions Bazin objected to in 1945 and the one that Hitchcock brought off so well in 1956 insist on a diegetic world that is, confoundedly, larger than the ordinary world of sense perception, and whereas 40 years later when "Godard introduces the superimposition of Hitchcock himself [...] he breaks the diegetic frame [...] Godard resurrects superimposition in his videographic practice not as cinema but for cinema."[25] Godard, then, gives superimposition an afterlife in a new mode of visual thinking made possible after cinema itself — and not just its techniques — has begun to expire in the digital age.

Let me recapitulate. Bazin's short essay on superimposition came out in *Écran Français* in two installments during August of 1945 just at the moment of "Ontology of the Photographic Image" and a year before "The Myth of Total Cinema." Organizing his collected works in 1958, he carefully positioned "Vie et mort de la surimpression" just behind these two famous essays in the first volume of *Qu'est-ce que le cinéma?,* a volume that carries the subtitle: *Ontologie et Langage.* Did he want the apparently incidental topic of superimposition to test the timeless axiom of cinema's realist ontology? Did he need to see how, or if, that axiom could function in films that deal with the supernatural and the oneiric, the kinds of films that would not normally be thought to support his theory? A half-century later, Godard

24 The French verb *rapprocher* can convey the sense of "bringing together to show concord and similarity," but it can be translated as "juxtaposition," which emphasizes contrast, even opposition. The word *juxtaposition* exists in French, but neither Godard nor Reverdy, whom he cites via Breton, use it. Reverdy's famous declaration is as follows: *"l'image est une création pure de l'esprit. Elle ne peut naître d'une comparaison mais du rapprochement de deux réalités plus ou moins éloignées. Plus les rapports de deux réalités rapprochées seront lointains et justes, plus l'image sera forte"* (1918). This formulation would seem to support the idea of a world that is continuous but so large that only poetic efforts can bring its various relations to light (as in Baudelaire's "Correspondences"). But it also supports the idea of a fragmented world whose elements can be forcefully brought together in a kind of collage...or montage.

25 Morgan, "The Afterlife of Superimposition," op. cit., pp. 138, 140.

de-ontologizes superimposition, making it a discursive technique for a videographic *language* that consists of images from various media as well as spoken and written words.[26] Of the two terms of his volume's title, "*ontologie*" and "*langage*," Bazin stresses the former, Godard the latter, while both men recognize the link and the fluctuating ratio between these concepts. Bazin monitored this ratio as it changed throughout the history of cinema, while Godard senses that history and the medium itself to have come to an end. Thus, "what cinema is" (its ontology and language) becomes evident within a plurality of histories, i.e., "Histoire(s) du cinéma."

Bazin thought of superimposition in relation to cinema's ontology since it brings two different states of being together on screen simultaneously. The director who can perform such an operation, such a miracle, controls the universe, claims Godard, and can direct the masses even "more than did Napoleon, Hitler, or Alexander the Great." But Godard thinks of superimposition not ontologically but in relation to language: a rhetorical device that animates new thoughts by bringing together whatever images and sounds one believes will achieve an effect, including the effect of truth. A master of rhetoric, Godard in his 1990s *Histoire(s)* controls not the universe but the discourse. Morgan suggests that while a godlike director such as Hitchcock hides behind the world he controls, where he cannot be interrogated, Godard, as historian and rhetor, is open to having his images questioned. He stands responsible for the aptness or truth of the line of thought his imagistic discourse projects.

Of all artforms, cinema has been the one in which ontology has been most at stake. This holds true even for those movies in which "the world as we know it" is clearly abrogated in favor of imagined places, for most fiction films present us with an image of a continuous world even when they are produced by a sequence of discontinuous shots. Certainly, the continuity effect almost always matters in classic cinema, as Bazin pointed out when he once confessed that the American term "continuity script" made more sense than its French equivalent, "*découpage*." Bazin believed in a continuous universe. To him, cutting (*découper*) allows filmmakers to organize distinct perceptions (always partial) that together aim to indicate a state of affairs. However, the world that filmmakers face before they begin their films and the one they aim to produce in the minds of viewers through cutting and pasting are continuous.

Godard, occasionally nostalgic but nevertheless certain that *cinema* has had its day, has looked for what might what come after it. And he has naturally turned to the electronic and then to the digital, which technologically have

26 Ibid., p. 139.

come after *film*. The continuity inherent in strips of celluloid film has now been replaced by the discontinuous bits assembled by computers to make video. Ones and zeros are sequenced to illuminate a monitor's pixels and constitute a window; multiple windows may be stacked atop one another, each of which may itself comprise layers of images, sounds, and written script. Video rather than cinema, such a process genuinely amounts to juxtaposition more than to correspondence or rapprochement. Thus, Godard's *Histoire(s)* prepare the digital by incorporating, even resurrecting, Hitchcock's *film world* within his own *video discourse*. Video makes an afterlife possible for superimposition — as well as for cinema in toto — but it does so on its own terms. Bazin would say that the technique of superimposition may have been resurrected, but that its function, hence its identity, is distinct.

If you assume that Bazin would reject a new function for superimposition or that he would reject new media, you would be hasty. True, he wrote while cinema was at its zenith, just as we write now anticipating or announcing its death. But in fact, Bazin had already imagined cinema's death, and quite directly, in an article saucily titled "Le cinéma est-il mortel?." Nonchalant about the threat of television in 1952, he mocks the belief that something as effete as Art (with a capital A) might be able to stand up both to the industrialists who wield power over entertainment and (far more important) to the popular imagination which may have something on its mind other than Art. Do not compare cinema to theater, he says, since theater will always be resuscitated, even when it goes through hard times, since it is nourished from "the games of children, from village festivals, from the irrepressible need of young men and women to 'play' in front of their assembled fellows."[27] In short, it is a natural human art, whereas cinema "is an *industrial Art*," and so it does not enjoy this immunity. "It was born not of man but of technology, and it depends completely on the latter and on its evolution."[28]

Bazin then dares to imagine that our belief in the existence of cinema has been an optical illusion, one that is "fleeting like the shadow traced by the sun." Sounding like Siegfried Zielinski,[29] he writes:

> Perhaps "the cinema" was in fact nothing but a stage in the vast evolution of the means of mechanical reproduction that had their origin in the 19th

27 André Bazin, "Le cinéma, est-il mortel?," in *France Observateur* (170) (13 August 1953); translated in *André Bazin's New Media* (Oakland, CA: University of California Press, 2014), p. 315.
28 Ibid.
29 Siegfried Zielinski, *Audiovisions: Cinema and Television as Entr'actes in History* (Amsterdam: Amsterdam University Press, 1999).

> century with photography and the phonograph and of which television is the most recent form. Perhaps it is only by way of one serendipitous cluster of technical, economic, and sociological convergences that the thing we call cinema has had the time to evolve toward indubitably aesthetic forms.[30]

Next, he projects that in the future a different set of circumstances might produce a different cluster of convergences, and thus quite different forms, putting the self-assurance of his own generation of critics in perspective and in question. Directly targeting his disciples at *Cahiers du Cinéma*, he writes this extraordinary passage:

> So perhaps in twenty years the "young critics" of some new form of spectacle that we cannot even imagine, and which can't be guaranteed to be 'an art,' will be reading our critical writings [of 1953] with a condescending smirk. Our views today could seem to them more naïve than the aesthetic sectarianism we find in our predecessors of the 1930s, who were properly outraged at the death throes of an art of the pure image which had finally reached maturity.[31]

Bazin wasn't far off in this prediction, which he bases on the 20 years separating the definitive death of a silent film aesthetic from the modern *cinéma d'auteur* that *Cahiers* championed in 1953 (i.e., *Viaggio in Italia* [*Journey to Italy*, Roberto Rossellini, 1954]). "Twenty years" later, in 1973, a very different editorial board at *Cahiers* would indeed repudiate their founder, André Bazin, and condescendingly dismiss as "naïve" the primacy of the auteur. But this occurred a bit in advance of the new forms of spectacle that Bazin imagined might develop from a convergence of revolutions in the economics of film (*Jaws* [Steven Spielberg, 1975], *Star Wars* [George Lucas, 1977], and global Hollywood), in the technical sector (VHS in 1977, digital processes by 1982) and in the sociology of entertainment (home theaters with viewers in control). So it was not until the 1990s that media scholars began to doubt the viability of "the thing we call cinema" and predict its mortality, this in the face of forms of spectacle, not necessarily aesthetic, produced by new media. Bazin does not seem to mind that a new paradigm

30 André Bazin, "Le cinéma est-il mortel"? op. cit., translated in *André Bazin's New Media*, op. cit., p. 315. I thank Blandine Joret for originally pointing out this passage to me; she deals with it in her own thesis, University of Amsterdam, 2013.
31 Ibid., p. 316.

would subtly arise, possibly eclipsing not only film art but film *tout court*. Still, you cannot live as if evolution has already made your era obsolete, and so he wisely concluded his essay by admonishing his colleagues: "In the meantime while waiting, let's just play dodgeball; I mean let's go to the cinema and treat it as an art."[32]

And this is just what he did in his daily criticism, where he examined and explicated the language and style of what he found on the screen. Scrutinizing thousands of species, registering changing techniques, he apprehended the evolution of genres, styles, and of film-art *in toto*. Unlike the pure cinephiles at *Cahiers*, Bazin frequently stepped back from films to examine cinema as a medium evolving from, and alongside, traditional artistic media (theater, painting, and especially the novel), and in relation to emerging technologies that amounted to the new media of the 1950s (television, 3D, Cinerama, 'Scope).[33]

The evolutionary science that Bazin subscribed to unconditionally, the result of his studies of geology and geography at the ENS Saint-Cloud, was by no means mechanistic. Not only did he believe the future to be open, he was attracted to phenomena that seemed to escape determining laws and that called genealogies into question. He was drawn to Jean Painlevé's surrealist scientific pieces for example, and to Claude Autant-Lara's adaptation of Jack London's classic story "To Build a Fire," shot on the Chrétien Hypogonar widescreen format in 1929 (*Construire un feu*). Such exceptions stand outside the main evolutionary lines. Like rare orchids, you might be able to analyze their DNA, but you can also treat them as miracles, single events that take place in some hole that opens up in the chain-link fence of historical development.

Exceptional works of art and literature have often been treated this way; genius rather than genus brings them unaccountably into existence, and they perdure as independent, self-regulating organisms. Pergolesi's "*Stabat*

32 Ibid.

33 The French term *technique* can refer to large-scale technological issues or small-scale issues of technique. We might think of "technological formats" (sound film, 3D, Scope, IMAX, etc.) in the way literary scholars do "modes," that is, as the material that determines the entire baseline of a whole range of artworks, regulating the way a public accesses them. Thus, iambic pentameter is a mode that Milton and others used for epic poems in English, while Henry Fielding addressed novels to audiences in the mode of prose with embedded dialogue. This analogy is not perfect, but it does set off modes like 3D from "special effects," which would be comparable to ostentatious techniques. The latter are similar to figures of speech that might come up in any mode or format. The history of special effects thus seems part of the evolution of the language of cinema (its technique) more than to the evolution of the medium *qua* medium.

Mater" or Le Douanier Rousseau's "Sleeping Gypsy" seem sui generis; they are aberrations if we insist on seeing them within the evolution of art. But why not treat them as miracles? So, while Bazin was an evolutionary thinker, he was not caught up in a single "*histoire*," the one he previewed in "Life and Death of the Superimposition" and then laid out in full dress in the hugely influential "Evolution of the Language of Cinema." For these essays deal with just the middle term of an interrelated constellation that constitutes his view of cinema: myth, language, and genius. Language evolves, but myths are timeless... and genius arises without consulting history.

And this is how Bazin treats *Ordet*, which concludes on a miracle and which is itself a kind of miracle. The faithful auteurist ("everything Dreyer made must be a revelation") accepts his uncommon or aberrant direction not just when but because it interrupts the standard current that carries cinema forward in its historical evolution. While there have always been films whose peculiarity wins them the label of "avant-garde" since they are ahead of the curve even while "they seem to conform to the idea we have of the evolution of cinematic art," *Ordet* does not point forward; indeed, it points at nothing but its own existence. It adopts what looks like

> an archaic style, the prolongation of silent expressionism into talking films. [However,] the notion of obsolescence becomes incongruous. *Ordet* does not show us an "out-of-date" aesthetic — no more than *Limelight* [Charlie Chaplin, 1952] does. Such works cannot be evaluated in relation to the evolution of film. They appear from time to time outside any historical frame of reference, like a pearl in an oyster — and the luster of this one is matchless![34]

On a par with certain inimitable works of poetry and painting, *Ordet* ignores conventions other than those it institutes itself. It ignores superimpositions because these speak to our senses; they are sensational in the etymological sense of that word. Whereas Dreyer's vision goes deeper. Within this universe, which is fully conscious of mystery, the supernatural does not loom up from outside. It is pure immanence, revealed in its extremity as the ambiguity of Nature and, above all, the ambiguity of death... The supernatural as it appeared in earlier films was an outgrowth of the profane eeriness that nourished quite a few German and Nordic films. Nothing in *Ordet* is linked to the marvelous. The religious meaning of the world bypasses what can be

34 André Bazin, "*Ordet*," in *Cinema of Cruelty*, op. cit., p. 27.

sensed [*sensibilité*]. *Ordet* is a sort of theological tragedy without the least concession to terror.[35]

The miracle of Inger's resurrection in the final scene of *Ordet* tests the faith of spectators in the movie theater in a way quite different from that of special effects. We interrogate the latter when we are incredulous before an image: "Is there really a phantom hovering near Sylvie?" or "Is that a genuine balloon following the little boy?" or "Is that lion actually in the cage with Charlie Chaplin?." Such questions crystallize, so to speak, the chemical constitution of film. Its molecular makeup binds realism to magic in varying ratios, balancing evidence and illusion, document and poetry. Critics may calculate the proportion of photographic realism to special effects in any given case, including the most extreme. If, as a mode, newsreel stands at one extreme, then animation stands at the other; in the first, the quotient of special effects approaches zero but remains latent, while in the second, the quotient of photographic realism does the same. Let us identify as cinema's DNA the elements that enable this combination of realism and magic.

Isolating DNA has become the standard technique by which evolutionary scientists distinguish species that appear to be near neighbors. Bazin played at being such a scientist in the cultural realm when he tried to demonstrate that cinema came into the world an infant among older cousins, two of which best display its ancestry: Emile Reynaud's hand-drawn glass plates projected in sequence and Etienne Jules Marey's chronophotography. These were close to cinema, in the way that Neanderthals were close to Homo Sapiens. A matter of some dispute among today's media archeologists, cinema's particular combination of photographic realism and imaginative illusion made it unquestionably a distinct entity to Bazin (and to most scholars in his day and ours).

From 1895–1897, that newborn invention quickly differentiated itself to become sufficiently stabilized so that it could be advertised and exported as a specific and spectacular cultural phenomenon. But this was just its first, precarious moment, when not even its putative parents (the Lumières) knew if it would ever walk or talk, or what it might resemble should it survive beyond its incubator. In an article on the stakes of 3D for cinema's future, Bazin claims that the medium's

> first [identity] crisis arose in 1897. It didn't take the public more than eighteen months to let it become just another spectacle demoted quickly from upscale café-concerts to fairground tents. The cinema as we know it

35 Ibid., p. 30. Translation corrected.

> resulted from a second birth, that of the cinema spectacle, with Méliès its principal artisan. The public's already sated curiosity required something of more durable interest to follow, and it got it in works composed and constructed for the imagination.[36]

Here Bazin, that fanatic for realism, recognized that it took someone like Méliès, a special effects man and a magician by trade, to animate the at-risk infant *Cinématographe*, so that it could develop, learn to walk, stride into the twentieth century, and grow into our most vigorous art. Méliès did more than inaugurate a beloved genre; he provided certain chemicals that cinema's DNA required so it could grow beyond infancy. Bazin pursued this analogy perhaps beyond its breaking point in declaring cinema to have coincidentally reached puberty in 1910 at just about the proper age for a human being, i.e., fifteen years after its 1895 birth.

> After 1910, cinema — 15 years old at the time, the age of puberty — began to pass from a primitive, pre-artistic stage into a continuous esthetic evolution that between 1925–1930 would lead to the great masterworks of silent cinema [...]. The cinema is still only at the age of a mature human being.[37]

In daring to compare the evolution of a technological medium with the growth of a human being, Bazin opens the door to a quasi-cybernetic account of the cinema's development. Audiences mature alongside the films made for them. He intimates that even the cinephilic public who greeted Epstein's *La chute de la maison Usher* (*The Fall of the House of Usher*) in 1927 were in some respects more gullible as well as more passionate, (i.e., more adolescent) than the mature post-WWII audiences for whom superimpositions and other "image effects" no longer worked their magic so readily.

To evoke "the beyond" after the coming of sound, and certainly after 1940, *mise en scène* should come before special effects. I wonder if Bazin thought of maturity of film style as analogous to maturity in religious belief. Puerile believers (not all of them children by any means) may be on the lookout for direct visions of the supernatural and are primed to respond to sensational (i.e., sensory) evocations of religious scenes. However, the more skeptical

36 André Bazin, "Aura-t-il une guerre en relief," in *France-Observateur* (153) (16 April 1953), translated in *André Bazin's New Media*, op. cit., pp. 245–246.

37 "Découpage," in *Vingt ans de cinéma à Rome* (Rome: Ateno, 1952), p. 359; reprinted in André Bazin, *Écrits complets,* vol. I, op., cit., p. 974.

Catholics and Protestants Bazin knew (including himself) had outgrown both early belief and subsequent rebellious disenchantment; if religion still spoke to them and commanded some measure of respect, it did so quietly, at most suggesting the supernatural or the unseen moral world. Those who believed in the sacramental power of a transcendent film — the critics at *Cahiers du cinema* watching, say, Rossellini's *Viaggio in Italia* — responded with nearly hyperbolic piety to the subtle suggestions of its *mise en scène*. I bring up *Viaggio in Italia* because it concludes on an apparent miracle, glimpsed in longshot, as a prelude to, and trigger for, the more important miracle of a couple's conversion from self-concern to at least momentary love for one other, the miracle of their remarriage. As Stanley Cavell has taught us, re-marriage, like the regaining of faith in maturity, trumps the obvious attractions of initial bliss or magic.

Bazin loved the sophistication of *Viaggio in Italia*, but he considered *Ordet* an even better case of a film that could only appeal to viewers who had developed a mature "doubting faith," the kind of faith that existentialist theologians of the day like Paul Tillich were thinking through for the skeptical modern age.[38] Of course, there would be apostates and unbelievers in cinema just as there are in religion...people who refuse miracles in any form, including purported aesthetic ones on screen.

Compared to the stark, unyielding, and protestant *Ordet*, *The Wrong Man*, made by the Jesuit-educated Hitchcock, flirts less with faith than with bad faith.[39] Bad Faith is not the same as the irony that we know to be the master of suspense's usual ethos; ironic authors present a story or perspective only to reveal its opposite. For instance, Hitchcock knowingly winks at us as when he thrillingly exposes evil in *Strangers on a Train* (1951) or *Shadow of a Doubt* (1943). But *The Wrong Man* is not a thriller, as Hitchcock tells us directly in his on-screen prologue; "In the picture you are going to see, every word is true."[40] He appeals to the viewer for a different kind of faith and indicates that this time he is a different kind of auteur, a sincere, not an ironic one.

Sincerity is visible in the low-key naturalism of the acting. We may not be surprised at how convincing Henry Fonda is, since he could always "act" like a natural man, but Vera Miles and Anthony Quayle are equal to him.

38 Paul Tillich, *The Courage To Be* (New Haven, CT: Yale University Press, 1954), p. 190.

39 In 1956, the year Hitchcock was shooting *The Wrong Man*, Eric Rohmer and Claude Chabrol published the first monograph on the master of suspense, emphasizing his Roman Catholic sensibility.

40 William Rothman astutely wonders why Hitchcock would say that a picture is composed of words, true or not. Rothman also slyly, even malevolently, suggests that the tiny figure speaking to us in silhouette may not be Hitchcock at all but a stand in.

They never pose or aim to stand out as do so many of Hitchcock's actors in other films of his. The location shots of the Stork Club, the subway, and the ordinary residential street in Queens seem borrowed from neorealism. And the camerawork is modest, except for the moment when Manny's nausea at being locked in a jail cell is expressed. Generally, Hitchcock went out of his way to depict an ordinary man in an ordinary way, sincerely and without embellishment. The real story was to unfold "as it really happened" in the normal course of events and without authorial intervention. That is its horror.

Yet sincerity, as Sartre had famously shown, readily edges into bad faith. Hitchcock may believe his prologue; nevertheless, he manipulates the drama just as he does in his ironic thrillers. He suggests (and may have pretended to himself) that Manny's prayer to Jesus miraculously brings about the exposure of the criminal. But no miracle takes place. While Manny prays, a man somewhere in his neighborhood is caught committing a robbery and so takes his place in jail. Through the lengthy lap dissolve, Hitchcock, the storyteller, says, "*Meanwhile*, elsewhere in the city, the Right Man...," but he structures the scene to suggest (in bad faith) that Manny's prayer causes, via superimposition, the Right Man to materialize out of the dark and into the light of justice.[41]

On the other hand, one miracle is said by Hitchcock to occur in the film; the last line of the film's dialogue points directly to it. "I guess I was hoping for a miracle," Manny mumbles dejectedly in leaving his deranged wife at the sanatorium. The nurse replies, "They happen, but it takes time." Hitchcock then delivers this miracle through cinema's oldest form of superimposition: atop the image of Manny walking down the dark corridor, the following words appears: "Two years later, Rose Balestrero walked out of the sanitarium — completely cured. Today she lives happily in Florida with Manny and the two boys... and what happened seems like a nightmare to them — but it did happen..."

Unlike Manny, on whose tortured supplicating face (and seemingly in whose mind) the Right Man had been superimposed, Rose had offered up nothing but a blank face, impenetrable to her husband, to Hitchcock, and to us. "She's in her own world," the nurse said. Yet somehow, she is instantaneously cured, living happily ever after. Hitchcock did not dare title his film, "The Wrong Man's Wife," but he concludes on the fate of this unknown woman whom Manny lives with but has never understood, Rose, whose life as wife and mother was inauthentic and in bad faith until she became

41 Unless one goes beyond Rothman's suggestion that this shot delivers the omniscient point of view of Jesus who sees not just Manny before him but everything else happening just then, including the robber walking toward the store.

paralyzed by the anguish brought on by Manny's ordeal. Abandoning her home, abandoning her identity, she hovered in ambiguity for two years until, in a miracle, she experienced what Sartre calls "self-recovery," a renewed love of life, on the hither side of bad faith. This miracle Hitchcock does not film.

Godard reveres Hitchcock, yet he pulls back the veil to expose him as a sort of "Wizard of Oz," the man operating the machinery that produces illusions for those who believe. What should we say, then, of Godard? We know what he would like us to say about him: that he is "authentic" to the limit, that he asserts his freedom and takes responsibility for it without illusions. But Sartre should make us leery of authenticity, "bad faith of the freedom that chooses to be always at a distance from itself, and that wills to hold itself in awe."[42]

Few texts hold themselves more in self-awe than *Histoire(s) du cinéma*. By revealing every image to be constructed, every image authored by Godard promises an honest image. Having undergone the dark night of loss of faith in cinema, then proclaiming that cinema, like God, is dead, Godard now brings it back in video. Hitchcock may be master of the *universe*, manipulating its continuity by cutting it into segments of space-time, but Godard is master of the *universe of images*. Surrounded in his Swiss cell by a library of films on tape, he controls this universe thanks to the technologies of discourse we hear at work: the whirring of the VCR and the staccato pounding of the typewriter.

I cannot help but conclude by noting the prominence of *The Wrong Man* and of *Ordet* in the *Histoire(s)*. Evidently, Godard continues to think alongside or against Bazin with whom he discussed these films 60 years ago. And alongside Cocteau as well, who appears in each of the eight sections of the *Histoire(s)*, sometimes via his photograph, sometimes via his films (especially *Le sang d'un poète, La belle et la bête,* and *Orphée*), and sometimes via his voice. Like Bazin, Godard recognized that Cocteau, the amateur, had gone straight to cinema's core. Indeed, in Godard's view, Cocteau reached an unsurpassed wisdom about this art in his final work, *Le testament d'Orphée* (*The Testament of Orpheus*, 1960), which premiered just a month before *À bout de souffle* (*Breathless*, Jean-Luc Godard, 1960) started shooting. In 1965, Godard designated *Testament* as the third best film made in the postwar era.

Despite its fireworks display of special effects, *Le testament d'Orphée* is more a work of discourse than of spectacle. As a discourse woven from fragments of spectacle, it could serve as a model for Godard's *Histoire(s)*. Like embossed designs embroidered onto the gown that his film throws

42 Sartre quoted by Ronald Santoni, *Bad Faith Good Faith* (Philadelphia: Temple University Press, 1995), p. 95 and note 44.

over reality, Cocteau's special effects transform reality into myth. Bazin, who died a year before he could see *Testament*, would have approved of the way its effects open our eyes not to a poetic world but to the world itself as seen by the poet. Godard on the other hand understood Cocteau's film to be a "Testament" to what cinema had been, a funereal discourse delivered in a film studio. Rather than a film, *Le testament d'Orphée* would seem to be an "histoire du cinema." Forty years later, Godard multiplied that "histoire," and he did so in video, on a monitor. Gone is the magic of cinema; what is left is just the video effect, which is hardly a special effect at all.

Godard's triumph is, then, both beautiful, powerful, and self-defeating. He has replaced the ontology of the photographic image with the language of new media. As for superimposition, no longer is an image from elsewhere super-*imposed* atop the illusions of a continuous world as happens in key moments in cinema; in the *Histoire(s)*, all images are *imposed*. In this mutation from superimposition to imposition, the spectator's act of faith or of doubt is replaced by an act of acceptance or refusal.

As for me, I accept Godard, even as I mourn his loss of faith.

Bibliography

Andrew, Dudley, ed. *André Bazin's New Media* (Berkeley: University of California Press, 2014).

Baxandall, Michael. *Patterns of Intention: On the Historical Explanation of Pictures* (New Haven, CT: Yale University Press, 1985).

Bazin, André. *Qu'est-ce que le cinéma? Ontologie et langage*, vol. 1 (Paris: Éditions du Cerfs, 1958).

———. *Écrits complets,* vols. 1 and 2, edited by Hervé Joubert-Laurencin (Paris: Macula, 2018).

Cocteau, Jean. *The Art of Cinema* (London: Marion Boyard, 1992).

Giraud, Jean. *Le lexique français du cinema des origins à 1930* (Paris: CNRS, 1958).

Morgan, Daniel. "The Afterlife of Superimposition." In *Opening Bazin,* edited by Dudley Andrew (New York: Oxford University Press, 2012): 127–141.

Rothman, William. *Must We Kill the Thing We Love? Emersonian Perfectionism and the Films of Alfred Hitchcock* (New York: Columbia University Press, 2014).

Santoni, Ronald. *Bad Faith Good Faith* (Philadelphia: Temple University Press, 1995).

Steimatsky, Noa. *The Face on Film* (New York: Oxford University Press, 2016).

Tillich, Paul. *The Courage to Be* (New Haven, CT: Yale University Press, 1954).

Zielinski, Siegfried. *Audiovisions: Cinema and Television as Entr'actes in History* (Amsterdam University Press, 1999).

About the Author

Dudley Andrew at Yale University is biographer of *André Bazin*, whose ideas he extends in his book *What Cinema Is!*, *Opening Bazin*, and in his editing and translating of themed collections of Bazin's writings. With two books on 1930s French cinema, Andrew was named Commandeur de l'Ordre des Arts et des Lettres. He is a member of the American Academy of Arts and Sciences and gained the Lifetime Achievement Award from the Society for Cinema and Media Studies. His current projects include issues in world cinema and in comparative arts.

Bibliography

Anon. "2 blade and 3 blade shutter difference." *Film-Tech Cinema System Forum*. Accessed 13 November 2021. http://www.film-tech.com/ubb/f1/t007944.html.

Anon. "300 Important Cameramen Becoming Recognized as Photographic Marvels," *Variety*, (4 January 1928): 6, 13.

Anon. "A Little Letter to the Inquisitive by the Inquiry Editor," *The Motion Picture Story Magazine* (April 1912): 162.

Anon. "Anglo-European Panorama," *The Daily News* (Perth), 19 December 1903.

Anon. "Attempts to Control Double Exposure Method," *American Cinematographer* 6 (September 1922): 4.

Anon. "Biggest Stage on Earth Devoted Entirely to Special Process Work," *American Cinematographer* (April 1929): 20–21, 35.

Anon. "Earle to Make Artistic Production of *The Rubaiyat of Omar Khayyam*," *Exhibitors Herald*, (2 July 1921): 106.

Anon. "Expansion with a Big E," *International Photographer* (February 1930): 34-35.

Anon. "Method of Taking Motion-Pictures." Google, 22 May 1916, no. 1,273,435. Accessed 17 November 2021. https://patents.google.com/patent/US1273435.

Anon. "Metro Uses Motion Painting," *Motion Picture News* (6 April 1918): 2065

Anon. "Movie Projector." *Wikipedia*. Accessed 13 November 2021. https://en.wikipedia.org/wiki/Movie_projector.

Anon. "Norse Saga in Pictures," *Los Angeles Times* (22 August 1920): III 23

Anon. "Obituary," *Los Angeles Times* (6 December 1932): 16

Anon. "Omar Khayyam Bows to Screen," *The Lewiston Daily Sun* (16 August 1922): 4

Anon. "Painted Sets Take Place of Costly Fabric," *Los Angeles Times* (23 December 1923): 25

Anon. "Pauline Periled in Auto Race," *Motography* 16 (17 October 1914): 525

Anon. "Producers Line-Up: Director's Suit Calls Up Strong Opposition," *Los Angeles Times* (17 August 1922): II 11.

Anon. "Rear Projection Big Advantage," *International Photographer* (April 1938): 30–33.

Anon. "Some Tricks of the Moving Picture Maker," *Scientific American* vol. 100, no 26, (26 June 1909): 476-477, 487.

Anon. "Split Reel Notes for Theater Men," *Motography* 5 (4 August 1917): 262.

Anon. "Who's Who and Where," *Film Fun* 335 (February 1917): n.p.

Anon. "World's Greatest Film is Screened," *Motography* 2 (11 July 1914): 38

Abbott, L.B. *Special Effects; Wire, Tape and Rubber Band Style* (Hollywood: ASC Press, 1984).

Abel, Richard. *The Red Rooster Scare: Making Cinema American, 1900–1910* (Berkeley, CA: University of California Press, 1999).

Adorno, Theodor W. *Aesthetic Theory*, edited by Gretel Adorno and Rolf Tiedermann, translated by Robert Hullot-Kentor (London: Athlone Press, 1997).
———. *Minima Moralia: Reflections from a Damaged Life* (London: Verso, 2005 [1951]).
Albera, François. "Eisenstein and the Theory of the Photogram." In *Eisenstein Rediscovered: Soviet Cinema of the 20's and 30's*, edited by Ian Christie and Richard Taylor (Oxford: Routledge, 1993): 194–203.
———. "Pour une épistémographie du montage: préalables," *Cinémas* 13, no. 2 (2002): 11–32.
———. "Le machinisme et la révolution dans *À nous la liberté!* de René Clair vus par trois périodiques de gauche." *1895. Revue d'histoire du cinema* 72 (Spring 2014): 108–27. https://doi.org/10.4000/1895.4808.
———. "First Discourses on Film and the Construction of a Cinematic Episteme." In *A Companion to Early Cinema*, edited by Nicolas Dulac, André Gaudreault, and Santiago Hidalgo (Oxford: Blackwell, 2021): 121–140.
Allard, Laurence. "*Express Yourself 3.0!* Le mobile comme technologie pour soi et quelques autres entre double agir communicationnel et continuum disjonctif soma-technologique." In *Téléphone mobile et création*, edited by Laurence Allard, Laurent Creton, and Roger Odin (Paris: Armand Colin, 2014), pp. 139–162.
Altman, Rick. "Toward a Theory of the History of Representational Technologies," *Iris* 2, no. 2 (1984): 111–125.
Amelunxen, Hubertus, Stefan Ilghaut, and Florian Rotzer, eds. *Photography After Photography: Memory and Representation in the Digital Age* (Amsterdam: G+B Arts, 1996).
Amyot, Jacques. "Vies Demetrius." Centre national des ressources textuelles et lexicales (see the entry "machine"). https://www.cnrtl.fr/definition/machine.
Anderson, Anthony. "*Omar Khayyam* A Picture Gallery," *Los Angeles Times* (20 December 1925): III 44.
Andrew, Dudley, ed. *André Bazin's New Media* (Berkeley: University of California Press, 2014).
Anon. "Les installations Eastman Kodak," *Phono-Cinéma-Revue* (April 1908): 28-29.
Anon. "Les trucs du Cinématographe," *Lectures pour tous* 9 (June 1908): 749-758.
Anon. "*Sunrise* Paves the Way," *Los Angeles Times* (8 December 1927): A9.
Aristotle. *Politics*, translated by Benjamin Jowett (New York: Modern Library, 1943).
———. *Analytics. Book II*, translated by Robin Smith (Indianapolis: Hackett, 1989).
———. *Metaphysics*. In *The Basic Works of Aristotle*, edited by Richard McKeon, translated by W.D. Ross (New York: The Modern Library, 2001): 681–986.
Arnheim, Rudolf. *Film Art* (Berkeley, CA: University of California Press, 1957).
Aumont, Jacques. *Limites de la fiction: considérations actuelles sur l'état du cinéma* (Montrouge: Bayard, 2014).

———. *Que reste-t-il du cinéma?* (Paris: Vrin, 2013).
Austin, J.L. *Sense and Sensibilia* (Oxford: Oxford University Press, 1962).
Avila, Eric. *Popular Culture in the Age of White Flight: Fear and Fantasy in Suburban Los Angeles*. (Berkeley, CA: University of California Press, 2006).
Bab, Julius. "Die Kinematographen-Frage," *Die Rheinlande* 22 (1912): 311-314.
Babin, Gustave. "Les coulisses du cinématographe," *L'Illustration* 3396 (28 March 1908); "Les coulisses du cinématographe," *L'Illustration* 3397 (4 April 1908); "Le théâtre cinématographique," *L'Illustration* 3427 (31 October 1908). Reproduced in *Les Grands Dossiers de l'Illustration: Le Cinéma* (Paris: SEFAG, 1987): 22–33.
Balio, Tino. *Hollywood in the New Millennium* (London: British Film Institute, 2013).
Bann, Stephen. *The True Vine: On Western Representation and the Western Tradition* (Cambridge: Cambridge University Press, 1989).
Barker, Jennifer M. *The Tactile Eye: Touch and Cinematic Experience* (Berkeley, CA: University of California Press, 2009).
Barthes, Roland. *Camera Lucida* (New York: Hill and Wang, 1981).
Baxandall, Michael. *Patterns of Intention: On the Historical Explanation of Pictures* (New Haven, CT: Yale University Press, 1985).
Bazin, André. *Qu'est-ce que le cinema? Ontologie et langage*, vol. 1 (Paris: Éditions du Cerfs, 1958).
———. *What is Cinema?*, vol. 1, translated by Hugh Gray (Los Angeles: University of California Press, 1967).
———. "The Life and Death of Superimposition." In *Bazin at Work: Major Essays and Reviews from the Forties and Fifties*, translated by Alain Piette and Bert Cardullo, edited by Bert Cardullo (New York: Routledge, 1997): 73–77.
———. *What is Cinema?* Translated by Timothy Barnard (Montréal: Caboose, 2009).
———. "Will CinemaScope Save the Cinema?" In *André Bazin's New Media*, edited and translated by Dudley Andrew (Berkeley, CA: University of California Press, 2014 [1953]): 267–287.
———. *Écrits complets*, vols. 1 and 2, edited by Hervé Joubert-Laurencin (Paris: Macula, 2018).
Bean, Jennifer. *The Fun Factory: The Keystone Film Company and the Emergence of Mass Culture* (Berkeley, CA: University of California Press, 2009).
Beaudelaire, Charles. "The Painter of Modern Life." In *The Painter of Modern Life and Other Essays*, translated and edited by Jonathan Mayne (London: Phaidon Press, 1965 [1863]).
Bechstein, Ludwig. *Die Darstellung der Tragödie Faust auf der Bühne* (Stuttgart: Brodhag'sche Buchhandlung, 1831).
Belisle, Brooke. "The Dimensional Image: Overlaps in Stereoscopic, Cinematic, and Digital Depth," *Film Criticism* 37-38, no. 3-1 (2013): 117–137.

Bellour, Raymond. *Le corps du cinéma. Hypnoses, émotions, animalités* (Paris: POL, 2009).

Belton, John. "If Film is Dead, What is Cinema?" *Screen* 55, no. 4 (Winter 2014): 460–470.

Belton, John. "Painting by the Numbers: The Digital Intermediate," *Film Quarterly* 61, no. 3 (Spring 2008): 58–65.

Belton, John. *Widescreen Cinema* (Cambridge, MA: Harvard University Press, 1992).

Benjamin, Walter. *Illuminations: Essays and Reflections*, translated by H. Zohn, edited by Hannah Arendt (New York: Schocken Books, 2007 [1950]).

Berardi, Franco "Bifo." *The Soul at Work: From Alienation to Autonomy* (Cambridge, MA and London: MIT Press, 2009).

Bergeret, Albert, and Felix Drouin. *Les Récréations photographiques* (Paris: Charles Mendel, 1893).

Bergson, Henri. *Creative Evolution*, translated by A. Mitchell (New York: Random House, 1944).

Bergstrom, Janet. "Invitation to Phantom." In *Phantom*, directed by Frederich W. Murnau. Flicker Alley (DVD) 2003, revised for BluRay 2019.

Berman, Marshall. *All That is Solid Melts Into Air* (New York: Penguin Books, 1982).

Berton, Mireille. "The 'Magism' of Cinema and Imaginary Spiritism in France at the Beginning of the Twentieth Century," *Early Popular Visual Culture* 13, no. 2 (2015): 113–133.

Bessy, Maurice. *Les truquages au cinéma* (Paris: Prisma, 1951).

Bigelow, Joe. "King Kong," *Variety* (6 March 1933). http://variety.com/1933/film/reviews/king-kong-2-1200410783.

Black, Joel. *The Reality Effect: Film Culture and the Graphic Imperative* (New York and London: Routledge, 2002).

Bloch, Ernst. *Héritage de notre temps*. Translated by Jean Lacoste (Paris: Pavot, 1978 [1935]).

———. *The Principle of Hope*, translated by Neville Plaice, Stephen Plaice, and Paul Knight (Cambridge, MA: MIT Press, 1986).

Bollman, Gladys, and Henry Bollman. *Motion Pictures for Community Needs: A Practical Manual of Information and Suggestion for Educational, Religious and Social Work* (New York: Holt, 1922).

Bolter, Jay David, and Richard Grusin. *Remediation: Understanding New Media* (Cambridge, MA: MIT Press, 2000).

Bordwell, David, Janet Staiger, and Kristin Thompson. *The Classical Hollywood Cinema. Film Style and Mode of Production to 1960* (New York: Columbia Press, 1985).

Braxton Soderman, Anton. "The Index and the Algorithm," *Differences* 18, no. 1 (2007): 153–186.

Brewster, David. *The Stereoscope: Its History, Theory, and Construction, With Its Application to the Fine and Useful Arts and to Education* (London: John Murry, 2012 [1856]).

Brown, Karl. *Adventures with D.W. Griffith* (New York: Farrar, Straus & Giroux, 1973).

Brown, William. "Man Without a Movie Camera – Movies Without Men: Towards a Posthumanist Cinema?" In *Film Theory and Contemporary Hollywood Movies*, edited by Warren Buckland (New York: Routledge, 2009): 66–85.

Brownlow, Kevin. *The Parade's Gone by...* (Berkeley, CA: University of California Press, 1968).

Bukatman, Scott. *Matters of Gravity: Special Effects and Supermen in the 20th Century* (Durham, NC: Duke University Press, 2002).

Cavell, Stanley. *The World Viewed* (Cambridge, MA: Harvard University Press, 1971).

Chambers, Gordon A. "Process Photography." In *Cinematographic Annual*, vol. 2, edited by Hal Hall (Hollywood: American Society of Cinematographers, 1931).

Chaplot, Charles. *La Photographie récréative et fantaisiste. Recueil de divertissements, trucs, passe-temps photographiques* (Paris: Charles Mendel, 1904).

Chateau, Dominique, and Martin Lefebvre. "Dance and Fetish: Phenomenology and Metz' Epistemological Shift," *October* 148 (2014): 101–130.

Chéroux, Clément. "'The Great Trade of Tricks': On Some Relations Between Conjuring Tricks, Photography, and Cinematography." In *Between Still and Moving Images*, edited by Laurent Guido and Olivier Lugon (New Barnet, UK: John Libbey Publishing, 2012): 81–96.

Chessex, Robert. "F.W. Murnau dreht Faust. Filmstadt Berlin um 1925," *Film Bulletin* 153 (April/May 1987): 53–63.

Cholodenko, Alan. "'First Principles' of Animation." In *Animating Film Theory*, edited by Karen Redrobe Beckman (Durham, NC and London: Duke University Press, 2014): 98–110.

Christout, Marie-Françoise. "Aspects de la féérie romantique: de *La Sylphide* (1832) à *La Biche au bois* (1845); chorégraphie, décors, trucs et machines," *Romantisme* 12, no. 38 (1982): 77–96. https://doi.org/10.3406/roman.1982.4577.

Clarke, Charles G. *Highlights and Shadows: The Memoirs of a Hollywood Cameraman* (Metuchen: Scarecrow Press, 1989).

Cocteau, Jean. *The Art of Cinema* (London: Marion Boyard, 1992).

Cohen-Séat, Gilbert. *Essai sur les principes d'une philosophie du cinéma I. Introduction générale: Notions fondamentales et vocabulaire de filmologie* (Paris: Presses Universitaire de France, 1946).

Cohl, Émile. "Dessins animés," *Larousse mensuel* 222 (August 1925): 861-864.

Coissac, G.-Michel. *Les coulisses du cinéma* (Paris: Paul Duval, 1929).

Collins, Jo, and John Jervis. "Introduction." In *Uncanny Modernity: Cultural Theories, Modern Anxieties*, edited by Jo Collins and John Jervis (London and New York: Palgrave MacMillan, 2006): 1–9.

Cook, Simon. "'Our Eyes, Spinning like Propellers': Wheel of Life, Curve of Velocities, and Dziga Vertov's 'Theory of the Interval,'" *October* 121 (Summer 2007): 79–91.

Copeland, Gene. "In a Persian Garden," *Picture Play Magazine* 4 (June 1920): 34–35, 92.

Corneille, Pierre. *Dessein de la tragédie d'Andromède, représentée sur le théâtre royal de Bourbon. Contenant l'ordre des scènes, la description des théâtres et des machines, et les paroles qui chantent en musique* (Rouen, 1650).

Cosandey, Roland. "Cinéma 1908: films à trucs et Film d'Art. Une campagne de *l'Illustration*." *Cinémateque* 3 (Spring-Summer 1993): 58–71.

Cotta Vaz, Mark, and Craig Barron. *The Invisible Art: The Legends of Movie Matte Painting* (San Francisco: Chronicle Books, 2002).

Coubertin, Pierre de. "Le sportif et le cinéma," *Bulletin du Bureau International de Pédagogie Sportive* 8 (1932): 7-9.

Crafton, Donald. *Before Mickey: The Animated Film 1898–1928* (Chicago: University of Chicago Press, 1993).

———. "The Veiled Genealogies of Animation and Cinema," *Animation: An International Journal* 6, no. 2 (July 2011): 93–110.

———. *Shadow of a Mouse: Performance, Belief, and World-Making in Animation* (Berkeley, CA: University of California Press, 2013).

Cram, Christopher. "Digital Cinema: The Role of the Visual Effects Supervision," *Film History* 24, no. 2 (January 2012): 169–186.

Crary, Jonathan. *Techniques of the Observer: On Vision and Modernity in the Nineteenth Century* (Cambridge, MA: MIT Press, 1990).

Creitzenbach, Wilheilm W. *Die Bühnengeschichte des Goethe'schen Faust* (Frankfurt am Main: Rütten & Loening, 1881).

Crump, Irving. "Interesting Picture Figures," *Motion Picture Magazine* (August 1914): 113-114.

Cubitt, Sean. "Phalke, Méliès and Special Effects Today," *Wide Angle* 21, no. 1 (January 2001): 114–130.

Currie, Gregory. *Image and Mind: Film, Philosophy, and Cognitive Science* (Cambridge: Cambridge University Press, 1995).

Dadoun, Roger. "*King Kong*: du monstre comme démonstration," *Littérature* 8 (Paris: Larousse, 1972).

Dahlquist, Marina. *The Invisible Seen in French Cinema Before 1917* (Stockholm: Aura förlag, 2001).

Danto, Arthur. "The End of Art." In *The Philosophical Disenfranchisement of Art* (New York: Columbia University Press, 2005): 81–115.

Darwin, Charles. *The Expression of Emotions in Man and Animal* (London: John Murray, 1872).

Dawn, Norman O. *Innovations in Cinematography* (Los Angeles: American Film Institute, 1977).

Day, Susan. "Walter Percy Day: Biography." Last modified December 2017. http://www.walterpercyday.org.

de Toth, André. *Fragments – Portraits from the Inside* (London: Faber and Faber, 1994).

Deleuze, Gilles. "Postscript on the Societies of Control." In *October: The Second Decade, 1986–1996*, edited by Rosalind Krauss et al. (Cambridge, MA: MIT Press, 1997): 443–447.

———. *Cinéma 2. The Image Time* (London: Athlone Press, 1989).

DeMille, Cecil B. "A Director Looks at 'Process-Shots'," *American Cinematographer* (November 1936): 458–459.

Derrida, Jacques. *Of Grammatology*, translated by Gayatri Charkravorty Spivak (Baltimore, MD: Johns Hopkins University Press, 1976).

———. "Plato's Pharmacy." In *Dissemination*, translated by Barbara Johnson (Chicago: University of Chicago Press, 1983): 61-171.

Descartes, René. *The Philosophical Writings of Descartes*, vol. 2, translated by John Cottingham et al. (Cambridge: Cambridge University Press, 1984).

———. *Méditations métaphysiques* (Paris: GF Flammarion, 1992).

*Direct Matin, no.*1841 (8 March 2016).

Doane, Mary Ann. "The Indexical and the Concept of Medium Specificity," *Difference: A Journal of Feminist Cultural Studies* 18, no. 1 (2007): 128–152.

Doherty, Tom. "*The Great Gatsby*," *Cineaste* 38, no. 4 (2013): 45–47.

Dubois, Philippe. *L'Acte photographique* (Paris/Brussels: Nathan/Labor, 1983).

Duchenne De Boulogne, Guillaume. *Mécanisme de la physionomie humaine ou analyse électro-physiologique de l'expression des passions* (Paris: J.-B. Baillière et fils, 1862).

Duckett, William, ed. *Dictionnaire de la conversation et de la lecture*, vol. 61, supplement to vol. 9 (Paris: Firmin-Didot, 1847).

Ducom, Jacques. *Le Cinématographe scientifique et industriel: Traité pratique de cinématographie.* (Paris: Geisler, 1911).

———. *Cinéma muet, sonore et parlant* (Paris: Albin Michel, 1931).

Dulac, Nicolas, ed. *Du media au postmédia: continuités, ruptures / From Media to Post-Media : Continuities and Ruptures.* (Lausanne: L'Âge d'Homme, Forthcoming.).

During, Elie. "Notes on the Bergsonian Cinematograph." In *Cine-dispositives*, edited by François Albera and Maria Tortajada (Amsterdam: Amsterdam University Press, 2015).

Earle, Ferdinand. "Screen Renaissance Through Motion Painting." In *The Blue Book of the Screen*, edited by Ruth Wing (Hollywood: Pacific Gravure Company, 1924): 345–348.

Ebert, Roger. "Why I Hate 3-D (And You Should Too)," *Newsweek*, (9 May 2010). https://www.newsweek.com/roger-ebert-why-i-hate-3d-movies-70247.

Edouart, Farciot. "The Transparency Projection Process," *American Cinematographer* (July 1932): 15, 39.

Egan, Jack. "Color Schemes," *Daily Variety*, (6 January 2005). Uploaded 15 June 2013. http://variety.com/2005/film/awards/color-schemes-1117915901/.

Eisenschitz, Bernard. "Le voyage à Hollywood," *Cinématographe* 75 (February 1982): 16.

Eisenstein, Sergei M. "On Stereocinema," translated by Sergey Levchin, *Public: Art/Culture/Ideas* 24, no. 47 (2013): 22.

———. *Film Form: Essays in Film Theory*, edited and translated by Jay Layda (New York: Houghton Mifflin Harcourt, 2014).

Eisner, Lotte H. *The Haunted Screen*. Translated by Goer Greaves (Berkeley, CA: University of California Press, 1969).

Elmaleh, Gad. "Jamie Bell: 'Tintin est très complexe'," Interview by La Rédaction, *Le Soir* (26 October, 2011). https://www.chicagomanualofstyle.org/tools_citationguide/citation-guide-1.html#cg-interview.

Enslin, Adolph. *Die ersten Theateraufführungen des Goethe'schen Faust* (Berlin: Verlag von Gebdrueder Paetel, 1880).

Eyman, Scott. *Five Cinematographers: Interviews with Karl Struss, Joseph Ruttenberg, James Wong Howe, Linwood Dunn and William H. Clothier* (Metuchen: Scarecrow Press, 1987).

Ezra, Elizabeth. *Georges Méliès: The Birth of the Auteur* (Manchester: Manchester University Press, 2000).

Fielding, Raymond. *The Technique of Special Effects Cinematography* (London, Boson: Focal Press, 1985).

Fihman, Guy. "Le cinéma date du jour où," In *Du cinéma selon Vincennes*, edited by Département d'études cinématographiques et audio-visuelles de l'Université Paris 8 (Paris: L'Herminier, 1979): 181–188.

Floury, Edmond. "La cuisine théâtrale," *La Revue théâtrale* 54 (1906): 1387–1388.

Fordham, Joe. "Last Man Standing," *Cinefex* 134 (July 2013): 94-116.

Fossati, Giovanna. *From Grain to Pixel: The Archival Life of Film in Transition* (Amsterdam: Amsterdam University Press, 2009).

Foucault, Michel. "Subjectivity and Truth." In *Ethics: Subjectivity and Truth*, edited by Paul Rabinow, translated by Robert Hurley et al. (New York: New Press, 1997).

Frazer, John. *Artificially Arranged Scenes: The Films of Georges Méliès* (Boston: G.K. Hall and co., 1979).

Frege, Gottlob. *The Foundations of Arithmetic*, translated by John L. Austin (New York: Harper, 1953).

Frenzel, Herbert A. *Geschichte des Theaters. Daten und Dokumente 1470-1890* (Munich: DTV, 1984).

Freud, Sigmund. "The Uncanny." In *The Penguin Freud Library: Art and Literature. Vol. 14*, edited by Albert Dickson, translated by James Strachey (London: Penguin Books, 1985 [1919]).

Freund, Karl. Interview by Frank Camino (1 June 1964). UCLA Special Collections, Collection 543, Box 2.

Frisby, David, and Mike Featherstone, eds. *Simmel on Culture: Selected Writings* (London: Sage, 1997).

Furmanek, Bob, and Greg Kintz. "An In-Depth Look at *House of Wax*." 3-D Film Archive. Accessed 16 November 2021. http://www.3dfilmarchive.com/House-of-Wax.

Galloway, Alexander R. *Protocol: How Control Exists After Decentralization* (Cambridge, MA: MIT Press, 2004).

Gaudreault, André. "Les détours du récit filmique (Sur la naissance du montage parallèle)," *Cahiers de la cinémathèque* 29 (1979): 88–107.

———. "Fragmentation et segmentation dans les 'vues animées': le corpus Lumière." In *Stop Motion, Fragmentation of Time*, edited by François Albera, Marta Braun, and André Gaudreault (Lausanne: Payot, 2002): 225–245.

———, with the assistance of Jean-Marc Lamotte. "Fragmentation and Segmentation in the Lumière 'Animated Views'," *The Moving Image* 3, no. 1 (Spring 2003): 110–131.

———. *Film and Attraction: From Kinetography to Cinema* (Urbana: University of Illinois, 2011).

Gaudreault, André, and Philippe Gauthier. "Les séries culturelles conférence-avec-projection et projection-avec-boniment: continuités et ruptures." In *Beyond the Screen: Institutions, Networks and Public of Early Cinema*, edited by Marta Braun et al. (Eastleigh: John Libbey, 2012): 233–238.

Gaudreault, André, and Philippe Marion. *The End of Cinema? A Medium in Crisis in the Digital Age*, translated by. Timothy Barnard. (New York: Columbia University Press, 2015).

Gautier, Théophile. *Histoire de l'art dramatique en France depuis vingt-cinq ans*, vol. 2 (Paris: Hetzel, 1853).

Gaynor, Janet. "The Reminiscences of Janet Gaynor." *Oral History Collection* (New York: Butler Library, 1958). Sound recording.

Gédéo, "Le Cinématographe chez soi – les souvenirs animés," *Phono-Cinéma-Revue* (April 1908) : 7-8.

Genette, Gérard. *Narrative Discourse: An Essay in Method*, translated by Jane E. Lewin (Ithaca: Cornell University Press, 1980).

Giddens, Anthony. *The Consequences of Modernity* (Palo Alto: Stanford University Press, 1990).

Giesen, Rolf. *Special Effects: Vom Spiegeleffect bis zur Computeranimation* (Ebersberg: 8½, 1994).

Ginisty, Paul. *La Féérie* (Paris: Hachette, 1910).

Giraud, Jean. *Le lexique français du cinéma: des origines à 1930* (Paris: Centre national de la recherche scientifique, 1958).

Godfrey, Chris. Vimeo. "The Great Gatsby VFX." Uploaded 15 June 2013. https://vimeo.com/68451324.

Goldner, Orville, and George Turner. *The Making of King Kong: The Story Behind a Film Classic* (New York: A.S. Barnes, 1975).

Gombrich, Ernst. *Art and Illusion: A Study in the Psychology of Pictorial Representation* (Oxford: Phaidon, 1960).

Greer Cohn, Robert, et al., eds. *Mallarmé in the Twentieth Century* (Madison: Farileigh Dickinson University Press, 1998).

Gregory, Carl Louis "Trick' Photography Methods Summarized," *American Cinematographer* 3 (June 1926): 9–10, 16–17, 20–22.

———. "Trick Photography," *Journal of the Society of Motion Picture Engineers* (SMPE) (September 1926): 99–107.

Grosoli, Marco. "Motion/Performance Capture and the Afterlife of the Index. A Reconsideration of André Bazin's 'Myth of Total Cinema'," *Recherches sémiotiques/Semiotic Inquiry* 31, nos. 1-2-3 (2011): 152–173.

Grundberg. Andy. "Ask It No Questions: The Camera Can Lie," *The New York Times*, (12 August 1990).

Guise, Chris. *Artbook, Les aventures de Tintin.* (Wellington and Brussels: Weta/Moulinsart, 2011).

Gunning, Tom. "The Cinema of Attractions: Early Film, its Spectator, and the Avant-Garde," *Wide Angle* 8, nos. 3-4 (1986): 63–70.

———. "An Aesthetic of Astonishment: Early Film and the (In)credulous Spectator," *Art and Text* 34 (Spring 1989): 31–45. Reproduced in *Viewing Positions: Ways of Seeing Film*, edited by Linda Williams (New Brunswick, NJ: Rutgers University Press, 1995): 114–133.

———. "ReNewing Old Technologies: Astonishment, Second Nature, and the Uncanny in Technology from the Previous Turn-of-the Century." In *Rethinking Media Change: The Aesthetics of Transition*, edited by David Thorburn and Henry Jenkins (Cambridge, MA: MIT Press, 2004): 39–59.

———. "What's the Point of an Index? Or, Faking Photographs," *Nordicom Review* 25, no. 1–2 (September 2004): 39–49.

———. "Uncanny Reflections, Modern Illusions: Sighting the Modern Optical Uncanny." In *Uncanny Modernity: Cultural Theories, Modern Anxieties*, edited by Jo Collins and John Jervis (London and New York: Palgrave MacMillan, 2006): 68–89.

———. “Moving Away from the Index: Cinema and the Impression of Reality,” *Differences* 18, no. 1 (2007): 29–52.

———. “The ‘Arrested’ Instant: Between Stillness and Motion.” In *Between Still and Moving Images,* edited by Laurent Guido and Olivier Lugon (New Barnet, UK: John Libbey Publishing, 2012): 21–31.

Gurevitch, Leon, and Miriam Ross. “Stereoscopic Media: Beyond Booms and Busts,” *Public* 24, no. 47 (2013): 83–92.

Hall, Mordaunt. “Murnau’s Drums and Fifes of Life,” *The New York Times* (2 October 1927): x 7.

Hammond, Paul. *Marvellous Méliès* (London: Gordon Fraser, 1974).

Hamus-Vallée, Réjane, and Caroline Renouard, *Les effets spéciaux au cinéma: 120 ans de créations en France et dans le monde* (Vanves: Armand Colin, 2018).

Hamus-Vallée, Réjane. *Les effets spéciaux* (Paris: Cahiers du cinéma/CNDP, 2004).

———. *Peindre comme un cinéaste. Pour une esthétique du Matte Painting.* Thèse d’HDR, Université Paris III Sorbonne Nouvelle, 2013.

Hansen, Miriam. “The Mass Production of the Senses: Classical Cinema as Vernacular Modernism,” *Modernism/Modernity* 6, no. 2 (1999): 59–77.

———. *Babel & Babylon: Spectatorship in American Silent Film* (Cambridge, MA: Harvard University Press, 1997).

Hardy-Carnac, Ted. “‘L’Écume des jours’ de Gondry: une romance écrasée sous un déluge d’effets visuels,” *Nouvel Obs* (29 April, 2013). https://leplus.nouvelobs.com/contribution/849216-l-ecume-des-jours-de-gondry-une-romance-ecrasee-sous-un-deluge-d-effets-visuels.html.

Harvith, Susan, and John Harvith, eds. *Karl Struss: Man with a Camera* (Cranbrook: Cranbrook Academy of Art/Museum, 1976).

Heffernan, Kevin. *Ghouls, Gimmicks, and Gold: Horror Movies and the American Movie Business, 1953–1968* (Durham, NC: Duke University Press, 2004).

Heinz Ewers, Hans. “Der Kientop,” *Der Kinematograph* 61 (6 February 1908).

Hennefeld, Maggie. *Specters of Slapstick and Silent Film Comediennes* (New York: Columbia University Press, 2018).

Herkt, Günther. “Dunning-Verfahren,”*Film-Kurier* 38 (19 September 1931): 1.

Herzog, Amy, John Richardson, and Carol Vernallis, eds. *The Oxford Handbook of Sound and Image in Digital Media* (Oxford: Oxford University Press, 2014).

Higgins, Scott. “3D in depth: *Coraline, Hugo* and a sustainable aesthetic,” *Film History* 16, no. 3 (2012): 196-209.

Holmes Jr., Oliver Wendell “Sun-Painting and Sun-Sculpture: With a Stereoscopic Trip Across the Atlantic.” In *Soundings from the Atlantic* (Boston: Ticknor and Sons, 1864): 166–227.

———. "The Stereoscope and the Stereograph." In *Soundings from the Atlantic* (Boston: Ticknor and Sons, 1864): 124–165.

Hood, Fred. "Interessante Tricks in der Kinematographie," *Erste Internationale Filmzeitung* 46 (11 November 1909), n.p.

Hunt, Thomas, ed. *Wellington College Register: 1859–July 1905* (Wellington College, Berkshire, 1906).

Hurst, Adrienne. Digital Media World. "Cutting Edge Transforms Gastby into Living Color." Accessed 9 November 2021. https://www.digitalmediaworld.tv/in-depth/423-the-great-gatsby-in-living-colour.

Hyman, John. *The Imitation of Nature* (Oxford: Basil Blackwell, 1989).

———. *The Objective Eye: Color, Form, and Reality in the Theory of Art* (Chicago and London: University of Chicago Press, 2006).

Isolani, Gertrud. "Gespräch mit Eugen Shufta," *Basler Nachrichten* (19 October 1965): 9.

Jackman, Fred W. "The Evolution of Special-Effects Cinematography from an Engineering Viewpoint," *Journal of the Society of Motion Picture Engineers* (SMPE) (September 1937): 293–302.

Jackson, Kenneth. *Crabgrass Frontier: The Suburbanization of the United States* (London: Oxford University Press, 1987).

Jameson, Frederic. *Archaeologies of the Future: The Desire Called Utopia and Other Science Fiction* (New York: Verso, 2005).

Jenny, Laurent. *La vie esthétique: Stases et flux* (Lagrasse: Verdier, 2013).

Jentsch, Ernst. "On the Psychology of the Uncanny." In *Uncanny Modernity: Cultural Theories, Modern Anxieties*, edited by Jo Collins and John Jervis (London and New York: Palgrave MacMillan, 2006): 216–227.

Jost, François. "The Promise of Genres," *Réseaux* 6, no. 1 (1998): 99–121. https://www.academia.edu/220307/The_promise_of_genre.

———. *La télévision du quotidien. Entre réalité et fiction* (Bruxelles and Paris: De Boeck Université/INA, 2001).

———. *Realtà/finzione. L'impero del falso* (Milan: Castoro editrice, 2003).

———. "Les Mondes de l'image: entre fiction et réalité," *Recherches sémiotiques/Semiotic Inquiry* 23, no. 1-2-3 (2003): 165–184.

———. "Loft Story: Big Brother and the Migration." In *Big Brother International*, edited by Ernest Mathis and Janet Jones (London: Wallflower Press, 2004): 105–1130.

Joubert-Laurencin, Hervé. *La lettre volante: quatre essais sur le cinéma d'animation* (Paris: Presses Sorbonne Nouvelle, 1997).

Joyce, James. *Ulysses* (Marmondsworth: Penguin, 1968 [1992]).

Jullier, Laurent. *L'écran postmoderne. Un cinéma de l'allusion et du feu d'artifice* (Paris: L'harmattan, 1997).

———. "L'ère numérique: vers l'évanescence du trucage." In *1895. Revue d'Histoire du Cinéma* 27 (September 1999): 113–120.

Kann. "*Sunrise* and Movietone," *Film Daily* 41, no. 72 (25 September 1927): 3-4,

Keil, Charlie, and Kristen Whissel, eds. *Editing and Special/Visual Effects* (New Brunswick, NJ: Rutgers University Press, 2016).

Kessler, Frank. "Actualités." In *Encyclopedia of Early Cinema*, edited by Richard Abel (London: Routledge, 2005): 6–7.

———. "The *Féérie* Between Stage and Screen." In *A Companion to Early Cinema*, edited by André Gaudreault, Nicolas Dulac, and Santiago Hidalgo (Oxford: Wiley-Blackwell, 2012): 64–79.

———. "La féerie: un spectacle paradoxal," *Lendemains* 152 (2013): 71–80.

Kilian, Eugen. *Goethes Faust auf der Bühne* (Munich and Leipzig: Georg Müller, 1907).

Kittler, Friedrich. *Optical Media: Berlin Lectures* (Cambridge, UK: Polity Press, 2010).

Klein, Norman M. *The Vatican to Vegas: A History of Special Effects* (New York: New Press, 2004).

Klinger, Barbara. "Beyond Cheap Thrills: 3D Cinema Today, the Parallax Debates and the 'Pop-Out'," *Public: Art/Culture/Ideas* 24, no. 47 (2013): 186–189.

Koszarki, Richard. *Karl Struss: New York to Hollywood* (Fort Worth, TX: Amon Carter Museum, 1995).

Kress, Eugène. "Trucs et illusions. Applications de l'optique et de la mécanique au cinématographe (c. 1912)," *1895. Revue d'histoire du cinéma* 27 (September 1999): 7–32.

———. *Conférences sur la cinématographie*, vol. 1 (Paris: Comptoir d'éditions de Cinéma-Revue, 1912).

La Fontaine, Jean de. *Nouvelles œuvres diverses* (Paris: Nepveu, 1820).

La politique des auteurs: les entretiens. (Paris: Cahiers du cinéma, 2001).

Laferrière, Dany. *L'art presque perdu de ne rien faire*. (Paris: Grasset, 2014).

Lafitte, Jacques. *Réflexions sur la science des machines* (Paris: Vrin, 1972).

Larousse mensuel 65 (1912)

Larsen, Egon. "Here is the Inside Story of the Magician of British Films," *Cavalcade* 21 (31 May 1949): 9.

Lee, Laura. *Japanese Cinema Between Frames* (London: Palgrave Macmillan, 2017).

Lefebvre, Martin. "The Art of Pointing. On Peirce, Indexicality, and Photographic Images." In *Photography Theory* (*The Art Seminar, II*), edited by James Elkins (New York : Routledge, 2007): 220–244.

Lefebvre, Martin, and Marc Furstenau. "Digital Editing and Montage: The Vanishing Celluloid and Beyond," *Cinémas* 13, no. 1-2 (2002): 69–107.

Lefebvre, Simon. "The Disappearance of the Surface." In *Screens*, edited by Dominique Chateau and José Moure (Amsterdam: Amsterdam University Press, 2016): 97–106.

Lefebvre, Thierry, "Avant-propos."In *1895. Revue d'Histoire du Cinéma* (September 1999): 5-6.

Leisen, James Mitchell. "Some Problems of the Art Director," *Transactions of the Society of Motion Picture Engineers* 12, no. 33 (April 9-14, 1928): 71-80.

Lenin, Vladimir I. "The Taylor System – Man's Enslavement by the Machine." In *Collected Works*, vol. 20 (London: Lawrence & Wishart, 1964): 152-154. Originally published in *Put Pravdv*, no. 35 (13 March 1914).

Lenk, Sabine. *Théâtre contre cinéma. Die Diskussion um Kino und Theater vor dem Ersten Weltkrieg in Frankreich* (Münster: MAKS Publikationen, 1989).

Leroi-Gouran, André. *Speech and Gesture*, translated by Anna Bostock Berger (Cambridge, MA: MIT Press, 1993).

Lescarboura, Austin C. *Behind the Motion Picture Screen* (New York: Scientific American Publishing Company/Munn & Company, 1919).

Leutrat, Jean-Louis. *L'alliance brisée. Le western des années 1920* (Lyon: Presses Universitaires de Lyon, 1985).

Limmer, Eric. "It's Crazy How Much of *The Wolf of Wall Street* is Actually CGI," *Gizmodo* (14 January 2014). Accessed 17 November 2021. https://gizmodo.com/its-crazy-how-much-of-the-wolf-of-wall-street-is-actua-1501402962.

Lippit, Akira Mizuta. *Atomic Light* (*Shadow Optics*.) (Minneapolis, MN: University of Minnesota Press, 2005).

Littré, Émile. *Dictionnaire de la langue française. Édition intégrale*, vol. 3 (Paris: Jean-Jacques Pauvert, 1956).

Löbel, Léopold. *La technique cinématographique: projection, fabrication des films* (Paris: Dunod et Pinat, 1912).

Loew, Katharina. "Tangible Spectres: 3D Cinema in the 1910s," *Film Criticism* 37–38, nos. 3-1 (2013): 87–115.

———. "The Spirit of Technology: Early German Thinking about Film," *New German Critique* 122 (Summer 2014): 129–148.

———. "Magic Mirrors: The Schüfftan Process." In *Special Effects: New Histories, Theories, Contexts*, edited by Dan North et al. (London: British Film Institute, 2015): 62–77.

———. *Special Effects and German Silent Film: Techno-Romantic Cinema* (Amsterdam: Amsterdam University Press, 2021).

Lopes, Dominic. *Understanding Pictures* (Oxford: Clarendon Press, 1996).

Lotman, Juri. *Culture and Explosion*, edited by Marina Grishakova, translated by Wilma Clark (New York: Mouton de Gruyter, 2009).

Lukács, Georg. "Thoughts on an Aesthetics of Cinema." In *German Essays on Film*, edited by Richard W. McCormick and Alison Guenther-Pal (New York and London: Continuum, 2004): 11–16.

Malthête, Jacques. "Quand Méliès n'en faisait qu'à sa tête," *1895. Revue d'histoire du cinéma* 27 (September 1999): 26.

———. "Kinematographic Views." In André Gaudreault, *Film and Attraction: From Kinematography to Cinema* (Urbana: University of Illinois, 2011): 132–152.

Mangolte, Bebette. "Analog Versus Digital, The Perennial Question of Shifting Technology and Its Implications for an Experimental Filmmaker's Odyssey." In *Camera Obscura, Camera Lucida: Essays in Honor of Annette Michelson*, edited. by Richard Allen and Malcolm Turvey, 261-274. (Amsterdam: Amsterdam University Press: 2003), pp. 261–274.

Mannoni, Octave. *Clefs pour l'imaginaire*. (Paris: Seuil, 1969).

Manovich, Lev. "What is Digital Cinema?" Accessed 7 November 2021. http://manovich.net/index.php/projects/what-is-digital-cinema.

———. *The Language of New Media* (Cambridge, MA: MIT Press, 2001).

Marcuse, Herbert. *One-Dimensional Man Studies in the Ideology of Advanced Industrial Society* (London: Routledge, 2002).

Margolis, Joseph. "Mechanical Reproduction and Cinematic Humanism." In *What, After All, is a Work of Art? Lectures in the Philosophy of Art* (University Park, PA: The Pennsylvania State University Press, 1999):101–127.

Marion, Philippe. "Spielberg au pays de la ligne claire: de la *graphiation* à la *performance capture*." In *La capture de mouvement ou le modelage de l'invisible*, edited by Marco Grosoli and Jean-Baptiste Massuet, (Rennes: Presses Universitaires de Rennes, 2013): 203–224.

———. "Les avatars du cinéma: De la caméra GoPro à la *Performance Capture*," *Communication & Langages* 182, no. 4 (December 2014): 61-76.

Marks, Laura U. *The Skin of the Film: Interculture Cinema, Embodiment, and the Senses* (Durham, NC: Duke University Press, 2000).

———. "How Electrons Remember." In *Touch: Sensuous Theory and Multisensory Media* (Minneapolis: University of Minnesota Press, 2002): 162–178.

Marx, Karl. *Capital: A Critical Analysis of Capitalist Production*, vol. 1, edited by Frederick Engels, translated by S. Moore and E. Aveling (New York: International Publishers, 1967).

Maynard, Patrick. "Drawings as Drawn: An Approach to Creativity in an Art." In *The Creation of Art: New Essays in Philosophical Aesthetics*, edited by Berys Gaut and Paisley Livingston (New York: Cambridge University Press, 2003): 53–88.

McClean, Shilo T. *Digital Storytelling. The Narrative Power of Visual Effects* (Cambridge, MA: MIT Press, 2007).

McLaren, Norman. Handwritten note reproduced in *Cinema* 57, no. 14 (January 1957): 12.

———. "Homage to George Méliès." In *Georges Méliès: First Wizard of Cinema (1896-1913)*, booklet accompanying a DVD, edited by David Shepard (Los Angeles: Flicker Alley, 2008).

Méliès, Georges. "Les vues cinématographiques." In *Annuaire général et international de la photographie* (Paris: Librairie Plon, 1907), pp. 362–392. Reprinted as "Kinematographic Views: A Discussion." In Gaudreault, André, *Film and Attraction: From Kinematography to Cinema* (Urbana: University of Illinois Press, 2011): 133–152.

———. "Les coulisses de la cinématographie. Doit-on le dire? M. Méliès, président du Syndicat des illustionistes de France combat ceux qui 'débinent' les trucs," *Photo-Ciné-Revue* (April 1908): 2–3.

Metz, Christian. "On the Impression of Reality in the Cinema." In *Film Language. A Semiotics of Cinema*, translated by Michael Taylor (Chicago: University of Chicago Press, 1974): 3–15.

———. "*Trucage* and the Film," translated by François Melzer, *Critical Inquiry* 3, no. 4 (Summer 1977): 657–675.

Mihailova, Mihaela, and John MacKay. "Frame Shot: Vertov's Ideologies of Animation." In *Animating Film Theory*, edited by Karen Redrobe Beckman (Durham, NC and London: Duke University Press, 2014): 145–166.

Minguet Batllori, Joan M. *Segundo de Chomón: The Cinema of Fascination* (Barcelona: Biblioteca de Catalunya, 2010).

Mitchell, William, J. *The Reconfigured Eye: Visual Truth in the Post-Photographic Era* (Cambridge, MA: MIT Press, 1994).

Mitry, Jean. *Histoire du cinéma: art et industrie* (Paris: Éditions Universitaires, 1967).

Morgan, Daniel. "Rethinking Bazin: Ontology and Realist Aesthetics," *Critical Inquiry* 32, no. 3 (2006): 443–481.

———. "The Afterlife of Superimposition." In *Opening Bazin*, edited by Dudley Andrew (New York: Oxford University Press, 2012): 127–141.

Morin, Edgar. *The Cinema, or the Imaginary Man*, translated by Lorraine Mortimer (Minneapolis, MN: University of Minnesota Press, 2005).

Moussaron, Jean-Pierre. "*King Kong* (1933) ou la migration des sens," *Modernités* 25 (Bordeaux: Presses Universitaire de Bordeaux, 2007): 251-264.

Moynet, Georges. *Trucs et décors: Explications raisonnées de tous les moyens employés pour produire des illusions théâtrales* (Paris: À la librairie illustrée, 1893).

Moussinac, Léon. "Diversité des films," *Ce soir* (3 April 1937). Reprinted in Valérie Vignaux and François Albera, eds. *Léon Moussinac; Critique et théoricien des arts; Anthologie critique* (Paris: Association Française de Recherche sur l'Histoire du Cinéma, 2014): 354-355.

Moynet, J. *L'envers du théâtre* (Paris: Librarie Hachette, [1873] 1888).

Mulvey, Laura. "A Clumsy Sublime," *Film Quarterly* vol. 60, no. 3 (2007): 3.

———. *Death 24x a Second: Stillness and the Moving Image* (London: Reaktion Books, 2006).

Musil, Robert. "Hier ist es schön." In *Prosa und Stück. Kleine Prosa, Aphorismen. Autobiographisches. Essays und Reden. Kritik* (Reinbek bei Hamburg : Rowohlt Verlag, 1978).

Musser, Charles. *Edison Motion Pictures, 1890–1900: An Annotated Filmography.* (Washington/Porderone: Smithsonian Institution Press / Le Giornate del Cinema muto, 1997).

Ndalianis, Angela. "Special Effects, Morphing Magic, and the 1990s Cinema of Attractions." In *Meta-Morphing: Visual Transformation and the Culture of Quick Change*, edited by Vivian Sobchak (Minneapolis, MN: University of Minnesota Press, 2000): 251–272.

Nieter, Hans. "The Schüfftan Process of Model Photography," *The Photographic Journal* (January 1930): 16–18.

North, Dan. *Performing Illusions: Cinema, Special Effects, and the Virtual Actor* (London: Wallflower, 2008).

North, Dan, Bob Rehak, and Michael S. Duffy, eds. *Special Effects: New Histories/ Theories/Contexts* (London: British Film Institute, 2015).

———. "Afterword: A Conversation with Lev Manovich." In *Special Effects: New Histories, Theories, Contexts*, edited by Dan North, Bob Rehab, and Michael S. Duffy (London: Palgrave, 2015): 267–275.

Odin, Roger. *De la fiction* (Bruxelles: De Boeck, 2000).

———. *Spaces of Communication. Elements of Semio-Pragmatics* (Amsterdam: Amsterdam University Press 2021 [2011]).

———. "Espaces de communication physiques, espaces de communication mentaux." In *D'un écran à l'autre: les mutations du spectateur*, edited by Jean Châteauvert and Gilles Delavaud (Paris: L'Harmattan, 2016): 331–342.

Oostrum, Kees van. "President's Desk: Inward Significance," *American Cinematographer.* (February 2018). https://ascmag.com/blog/presidents-desk/presidents-desk-inward-significance

Païni, Dominique. "The Wandering Gaze: Hitchcock's Use of Transparencies." In *Hitchcock and Art: Fatal Coincidences*, edited by Dominique Païni and Guy Cogeval, translated by Guy Conolly et al. (Montréal: The Montreal Museum of Fine Arts, 2000): 51–78.

Pape, Helmut. "Charles S. Peirce on Objects of Thought and Representation," *Noûs* 24, no. 3 (1990): 375–395.

Paquet, Alfons. "Vom Werte und Unwerte des Kinos." In *Kino-Debatte*, edited Anton Kaes (Munich: DTV, 1978): 62–66.

Paul, William. "The Aesthetics of Emergence," *Film History* 5, no. 3 (1993): 321–355.

Peirce, Charles S. *The Collected Papers of Charles Sanders Peirce*, vols. 1-8, edited by Charles Hartshorne and Paul Weiss (vols. 1-6); and Arthur Burks (vols. 7-8) (Cambridge, MA: Harvard University Press, 1931-1958).

———. *The Essential Peirce*, vol. 2, edited by Peirce Edition Project (Bloomington: Indiana University Press, 1998).

Percheron, Daniel, and Marc Vernet. "Entretien avec Christian Metz," *ÇA cinéma* 7-8 (May 1975): 18-51.

Percy Day, Walter. "The Origin and Development of the Matte Shot," *The Photographic Journal* (October 1948): 209–211.

Petley, Julian. *Capital and Culture: German Cinema 1933-45* (London: British Film Institute, 1979).

Pierson, Michelle. *Special Effects: Still in Search of Wonder* (New York: Columbia University Press, 2002).

Pliny. *Natural History*, Vol 10. Translated by H. Rackham. Loeb Classical Library (Cambridge, MA: Harvard University Press, 1961).

Pougin, Arthur. *Dictionnaire historique du théâtre et des arts qui s'y rattachent. Poétique, musique, danse, pantomime, décor, costume, machinerie, acrobatisme; jeux antiques, spectacles forains, divertissements scéniques, fêtes publiques, réjouissances populaires, carrousels, courses, tournois, etc., etc., etc.* (Paris: Firmin-Didot, 1883).

Prince, Stephen. *Digital Visual Effects in Cinema: The Seduction of Reality* (New Brunswick, NJ: Rutgers University Press, 2012).

Purse, Lisa. *Digital Imagining in Popular Cinema* (Edinburgh: Edinburgh University Press, 2019).

Récanati, François. *La transparence et l'énonciation* (Paris: Seuil, 1979).

Redman, Jeanne. "Art Effects in Cinema," *Los Angeles Times* (11 August 1920): III 4.

Rhodes, John David. *Spectacle of Property: The House in American Film* (Minneapolis, MN: University of Minnesota Press, 2017).

Richter, Hans. "Film as an Original Art Form." In *Introduction to the Art of the Movies*, edited by Lewis Jacobs (New York: Octagon, 1970): 282–288.

Ricoeur, Paul. *Time and Narrative*, 3 vols. Translated by Kathleen McLaughlin and David Pellauer. (Chicago: University of Chicago, 1984-88).

Roberts, Alan. "The Film Look: It's not Just Jerky Motion..." *BBC R&D White Paper* (December 2002). http://downloads.bbc.co.uk/rd/pubs/whp/whp-pdf-files/WHP053.pdf.

Rodowick, David Norman. *The Virtual Life of Film* (Cambridge, MA: Harvard University Press, 2007).

Rogers, Ariel. *Cinematic Appeals: The Experience of New Movie Technologies* (New York: Columbia University Press, 2013).

———. *On the Screen: Displaying the Moving Image, 1926–1942* (New York: Columbia University Press, 2019).

Romanello, Linda, and Marc Loftus. "The Evolution of the Digital Intermediate," *Post Magazine* (2 April 2015). http://lightiron.com/about/news/evolution-digital-intermediate/.

Rosen, Philip. *Change Mummified Cinema: Historicity, Theory* (Minneapolis: University of Minnesota Press, 2001).

Ross, Miriam "The 3D Aesthetic: *Avatar* and Hyper-Haptic Visuality," *Screen* 53, no. 4 (2012): 381–397.

———. *3D Cinema: Optical Illusions and Tactile Experiences* (New York: Palgrave MacMillan, 2015).

Rossell, Deac. "Projectors." In *Encyclopedia of Early Cinema*, edited by Richard Abel (Abingdon: Taylor & Francis, 2005): 539.

Rothman, William. *Must We Kill the Thing We Love? Emersonian Perfectionism and the Films of Alfred Hitchcock* (New York: Columbia University Press, 2014).

Sabbattini, Nicola. *Pratica di fabricar scene e macchine ne' teatri.* Ravenne, 1638. Translated by Maria and René Canavaggia and Louis Jouvet. *Pratique pour fabriquer scènes et machines de théâtre* (Neufchâtel: Ides et Calendes, 1977).

Sadoul, Georges. *Histoire générale du cinéma. Vol. 1: L'invention du cinéma* (Paris: Denoël, 1946).

———. *Georges Méliès* (Paris: Seghers, 1970).

Salt, Barry. *Film Style and Technology: History and Analysis* (London: Starword, 1983).

———. "The Shape of 1999: The Stylistics of American Movies at the End of the Century," *New Review of Film and Television Studies* 2, no. 1 (May 2004): 61–86.

Sandberg, Mark. *Living Pictures, Missing Persons: Mannequins, Museums, and Modernity* (Princeton, NJ: Princeton University Press, 2003).

Sandifer, Philip. "Out of the Screen and Into the Theater: 3-D Film as Demo," *Cinema Journal* 50 no. 3 (2011): 62–78.

Santoni, Ronald. *Bad Faith Good Faith* (Philadelphia: Temple University Press, 1995).

Schallert, Edwin. "*Sunrise* Rare Art Feature. Murnau Production Ultra in Treatment. Camera Effects Exercise Dazzling Influence. Brilliant Audience Applauds Principals," *Los Angeles Times* (1 December 1927): A9.

Schatz, Dennis. *World of Inventors: Thomas Edison* (San Diego: Silver Dolphin Books, 2009).

Schier, Flint. *Deeper Into Pictures* (Cambridge: Cambridge University Press, 1986).

Schroter, Jens. *3D: History, Theory and Aesthetics of the Transplane Image*, translated by B. Pichon and D. Rudvtsky (London: Bloomsbury, 2014).

Schumann, Max. "Die Zauberposse auf dem modernen Theater," *Masken. Wochenschrift des Düsseldorfer Schauspielhauses* 6, no. 40 (1911): 632–637.

Schwartz, Vanessa. *Spectacular Realities: Early Mass Culture in Fin-de-Siècle Paris* (Berkeley, CA: University of California Press, 1998).

Sconce, Jeffrey. *Haunted Media: Electronic Presence from Telegraphy to Television* (Durham, NC: Duke University Press, 2000).

Seeber, Guido *Der praktische Kameramann. Vol. 2: Der Trickfilm in seinen grundsätzlichen Möglichkeiten.* (Frankfurt am Main: Deutsches Filmmuseum, 1979).

Seeber, Guido. "Gefilmte Projektion," *Filmtechnik* 4 (5 August 1925): 76-78.

Selfiecity. "Introduction: Investing the style of self-portraits (selfies) in five cities across the world." 2014. http://selfiecity.net/.

Shields, David S. *Still: American Silent Motion Picture Photography* (Chicago: University of Chicago Press, 2013).

Sierrek, Karl. "C'est beau ici: se regarder voir dans le film de famille." In *Le film de famille, usage privé, usage public*, edited by Roger Odin (Paris: Méridiens Klincksieck, 1999): 63–77.

Slotkin, Richard. *Gun Fighter Nation: The Myth of the Frontier in Twentieth-Century America* (New York: Atheneum, 1992).

Smith, Jacob. *The Thrill Makers: Celebrity, Masculinity, and Stunt Performance* (Berkeley, CA: University of California Press, 2012).

Sobchack, Vivan. *The Address of the Eye: A Phenomenology of Film Experience* (Princeton, NJ: Princeton University Press, 1992).

Sontag, Susan. *Regarding the Pain of Others* (New York: Farrar, Straus & Giroux, 2003).

Souriau, Étienne. "La structure de l'univers filmique," *Revue internationale de filmologie* 7-8 (2e trimestre, 1951): 231-240.

Souriau, Étienne. *L'Univers filmique* (Paris: Flammarion, 1953).

Spadoni, Robert. *Uncanny Bodies: The Coming of Sound Film and the Origins of the Horror Genre* (Berkeley, CA: University of California Press, 2007).

Spigel, Lynn. *Make Room for Television: Television and the Family Ideal in Postwar American* (Chicago: University of Chicago Press, 1992).

Stamp, Shelley. *Movie-Struck Girls: Women and Motion Picture Culture after the Nickelodeon* (Princeton, NJ: Princeton University Press, 2000).

Stansky, Peter. *The First Day of the Blitz: September 7, 1940* (New Haven, CT: Yale University Press, 2007).

Steimatsky, Noa. *The Face on Film* (New York: Oxford University Press, 2016).

Struss, Karl. "Dramatic Cinematography," *Transactions of the the Society of Motion Picture Engineers* 12, no. 34 (April 9-14, 1928): 317-319.

Struss, Karl. "Recollections of Karl Struss: An Oral History of the Motion Picture in America." Interview by John Dorr. UCLA Special Collections 1969 (9, 24, 30 July, 13–27 August 1968).

Stull, William. "Process Cinematography." In *The Complete Photographer*, vol. 8, edited by Willard D. Morgan (New York: New York Educational Alliance, 1943): 2994–3005.

———. "Producers Pool Composite Process Patents," *American Cinematographer* 11 (November 1936): 2994–3005.

Tabet, Frédéric. "*Circulation technique entre l'art magique et le cinématographe avant 1906*," Doctoral dissertation (Université Paris-Est, 2011).

Talbot, Frederick A. *Moving Pictures: How They Are Made and Worked* (Philadelphia: J.B. Lippincott Company, 1912).

Telotte, Jay P. "Animating Space: Disney, Science, and Empowerment," *Science Fiction Studies* 35, no. 104 (February 2008): 48-59.

Theisen, Earl. "In the Realm of Tricks and Illusions," *The International Photographer* (June 1934): 8–11.

———. "The Evolution of the Motion Picture Story, Parts I and II" *The International Photographer* (April and May 1936): 16–17; 22–23; 12–13, 27.

Tillich, Paul. *The Courage to Be* (New Haven, CT: Yale University Press, 1954).

Tortajada, Maria. "'The Cinematographic Snapshot: Rereading Etienne-Jules Marey." In *Cinema Beyond Film: Media Epistemology in the Modern Era*, edited by François Albera and Maria Tortajada (Amsterdam: Amsterdam University Press, 2010): 79–96.

———. "Photography/Cinema: Complementary Paradigms in the Early Twentieth Century." In *Between Still and Moving Images*, edited by Laurent Guido and Olivier Lugon (New Barnet: John Libbey, 2012): 81–96.

Trotter, David. "Stereoscopy: Modernism and the 'Haptic'," *Critical Quarterly* 46, no. 4 (2004): 39–42.

Turner, Georges E. *The Making of* King Kong*: The Story Behind a Film Classic* (New York: Ballantine Books, 1976).

———., ed. *The ASC Treasury of Visual Effects* (Hollywood: American Society of Cinematographers, 1983).

Turnock, Julie. "The Screen on the Set: The Problem of Classical-Studio Rear Projection," *Cinema Journal* 51, no. 2 (Winter 2012): 157–162.

———. "Patient Research on the Slapstick Lots: From Trick Men to Special Effects Artist in Silent Hollywood," *Early Popular Visual Culture* 13, no. 2 (May 2015): 152–173.

———. *Plastic Reality: Special Effects, Technology, and the Emergence of 1970s Blockbuster Aesthetics* (New York: Columbia University Press, 2015).

Turquety, Benoît. "On Viewfinders, Tape Slicers and Video Assist Systems: Questioning the History of Techniques and Technology in Cinema." In *Technology and Film Scholarship: Experience, Study, History*, edited by André Gaudreault and Santiago Hidalgo (Amsterdam: Amsterdam University Press, 2015): 239–259.

Uzanne, Octave. "Un des 'premiers amis du cinéma' Monsieur Octave Uzanne," *Phono-Cinéma-Revue* (April 1908): 9-10

Valleiry, François. "Doit-on le dire? Réponse à M. Méliès," *Phono-Ciné-Gazette*, no. 78 (15 June 1908).

Van Lier, Henri. *Philosophie de la photographie* (Paris: Les Cahiers de la photographie, 1983).

Vertov, Dziga. *The Writings of Dziga Vertov*, edited by Annette Michelson (Berkeley, CA: University of California Press, 1984).

Veyne, Paul. *Les Grecs ont-ils cru à leurs mythes? Essai sur l'imagination constituante* (Paris: Seuil, 1983).

Vignaux, Valérie, and François Albera, eds., *Léon Moussinac; Critique et théoricien des arts; Anthologie critique* (Paris: Association Française de Recherche sur l'Histoire du Cinéma, 2014).

Vilder, Anthony. *The Architectural Uncanny. Essays in the Modern Unhomely* (Cambridge, MA: MIT Press, 1992).

Virno, Paolo. *Déjà Vu and the End of History*, translated by David Broder (London: Verso, 2015).

Visentin, Hélène. "Des tableaux vivants à la machine d'architecture dans les entrées royales lyonnaises (XVIe – XVIIe siècles)," *Dix-septième siècle* 212 (March 2001): 419–428.

Warm, Hermann. *Murnau* (Berkeley, CA: University of California Press, 1973).

Warner, Marina. *Phantasmagoria: Spirit Visions, Metaphor, and Media into the Twenty-First Century* (Oxford: Oxford University Press, 2006).

Weiberg, Birk. "Roy J. Pomeroy, Dunning Process Co., Inc., and Paramount Publix Corporation vs. Warner Bros Pictures, Inc., Vitaphone Corporation, and Frederick Jackman: How the Movie Industry Turned to Rear Projection." Unpublished conference paper (Seattle, WA: Society for Cinema and Media Studies Annual Conference, 2014).

Whissel, Kristen. *Spectacular Digital Effects. CGI and Contemporary Cinema* (Durham, NC: Duke University Press, 2014).

———. "Digital 3D, Parallax Effects, and the Construction of Film Space in *Tangled* 3D and *Cave of Forgotten Dreams* 3D." In *Screen Space Reconfigured*, edited by Susanne Ø. Saether and Synne Tollerud Bull (Amsterdam: Amsterdam University Press, 2020): 77–103.

———. "*Dial M for Murder*: The Detective Thriller, the Postwar Uncanny, and 3D Cinema," *New Review of Film and Television Studies* 20, no. 1 (Winter 2022): 49-62.

Whitman, Philip H. "Pans and Tilts," *American Cinematographer* 2, no. 20 (1 November 1921): 8.

Wiley, G. Harrison. "Does the Camera Lie?," *Photoplay Magazine* 4 (September 1923): 30–33, 115–119.

———. "In Lieu of Genii: Modern Cameraman Surpasses Magician of Ancient Days," *Photodramatist* 7 (December 1922): 11–12, 42.

Williams, Frank. "Trick Photography," *Transactions of the Society of Motion Picture Engineers* 12, no. 34 (April 9-14, 1928): 537-540.

Williams, Raymond. *Keywords: A Vocabulary of Culture and Society* (London: Fontana, 1983).

Wolf, Mark. *Abstracting Reality: Art, Communication and Cognition in the Digital Age* (Lanham, MD: University Press of America, 2000).

Wollheim, Richard. "Seeing-as, Seeing-in, and Pictorial Representation." In *Art and its Objects: An Introduction to Aesthetics* (New York: Harper and Row, 1968): 137–151.

———. "On Drawing an Object." In *On Art and the Mind: Essays and Lectures* (London: Allen Lane, 1973): 3–30.

York, Cal. "How They Do It!" *Photoplay* 29, no. 5 (April 1926): 28-31.

Zielinski, Siegfried. *Audiovisions: Cinema and Television as Entr'actes in History* (Amsterdam University Press, 1999).

Žižek, Slavoj. *In Defence of Lost Causes* (London: Verso, 2008).

Index

For Product Safety Concerns and Information please contact our EU representative GPSR@taylorandfrancis.com Taylor & Francis Verlag GmbH, Kaufingerstraße 24, 80331 München, Germany

Batch number: 10722100

Printed by Printforce, the Netherlands